Fundamentals of the Finite Element Method

HARTLEY GRANDIN, Jr.
Worcester Polytechnic Institute

Fundamentals of the Finite Element Method

Macmillan Publishing Company
New York

Collier Macmillan Publishers
London

Copyright © 1986, Macmillan Publishing Company, a division of Macmillan, Inc.

PRINTED IN THE UNITED STATES OF AMERICA

All rights reserved. No part of this book may be reproduced or transmitted in any form or by any means, electronic or mechanical, including photocopying, recording, or any information storage and retrieval system, without permission in writing from the publisher.

Macmillan Publishing Company
866 Third Avenue, New York, New York 10022

Collier Macmillan Canada, Inc.

Library of Congress Cataloging in Publication Data

Grandin, Hartley.
    Fundamentals of the finite element method.

    Bibliography: p.
    Includes index.
    1. Finite element method. I. Title.
TA347.F5G73   1986   620'.0015153'53   85-5084
ISBN 0-02-345480-6

Printing: 1 2 3 4 5 6 7 8   Year: 6 7 8 9 0 1 2 3 4 5

ISBN 0-02-345480-6

Preface

The finite element method has become widely accepted as a valuable technique for solving complex problems. Originating in the field of structural engineering, the method is now recognized as having a general basis with wide applicability to many additional areas of engineering and mathematical physics such as steady-state and time-dependent heat transfer, fluid flow, and electric potential problems.

The increasing adoption of computers and the availability of finite element computer code has put this method of analysis into the hands of engineers at all levels of industry. The undergraduate engineering student should be provided with an opportunity for an introduction to the method.

This book has been written to introduce the basic concepts of a limited but very fundamental portion of the finite element method. The format of the presentation is designed to:

1. Bring the reader into the subject with a reliance on engineering and mathematical skills at the undergraduate level.
2. Illustrate applications of important concepts and procedures with simple examples, most of which can be done by hand calculation.

3. Show how the technique is ultimately put to use by providing listings and descriptions of simple, yet functional computer programs.

The text material is intended for a one-semester course for students at the senior-year level.

The inclusion of the topic on matrix analysis of structures, especially as Chapter 1 in a text on the finite element method, may be somewhat controversial. There are some who believe that this is a technique used only by the civil engineer; this is unfortunate. Engineers designing vehicle chassis frames or aircraft structures, and so on, appreciate the value of this analytical tool. Unfortunately, this topic is taught primarily in civil engineering departments and its value is not seen by students in other disciplines. All engineers having anything to do with stress analysis should be familiar with the technique, especially since computer code is so readily available. There is an equally important additional reason for wanting to include the technique. A fair amount of miscomprehension and confusion exists regarding the distinction between the two methods of analysis. The matrix analysis of structures and the finite element method are fundamentally different in their development and only through a study of both topics can the differences be appreciated. On the other hand, there are many similarities in the two techniques; the matrix equations, programming methods, and the equation-solving routines can be identical. Since an understanding of matrix structural methods depends only on the typical sophomore-level strength-of-materials course plus some elementary matrix-handling operations, students make progress quickly. Even the students with very limited computer programming experience become very enthusiastic about their new capabilities and begin to appreciate the computer. At the same time, much of the groundwork gets accomplished for the eventual computer implementation of the finite element method.

The introduction to the finite element method really begins with Chapter 2 with a study involving the elasticity problem. Although this is not the simplest problem type to begin with, it is my experience that it is the area best understood by the largest number of students. The chapter includes a review of basic relationships and discussions on work, strain, and potential energy, and approximation methods using the minimum potential energy principle. The level of sophistication is raised with an introduction to the variational formulation. We proceed to justify the potential energy principle and then provide the basis for the finite element discretization of the region and show the applicability of the method to all problem types having the same form of governing differential equation.

Chapter 3 is an application of the theory to the one-dimensional problem. This is a very important chapter because it discusses the rationale for approximation functions; shows, through simple examples, how the system equations are derived; and considers the accuracy of the approximations. The generality of the method is emphasized by showing the application to both the elasticity and the heat flow problem.

Chapters 4 through 8 use the two-dimensional elasticity problem as the vehicle for the development of triangle and quadrilateral elements. Considerable detail is included in these chapters. In this as in all other chapters, the example problems are simple enough to allow for hand computation but complete enough to illustrate the principles. I believe that going through each step by hand is necessary for a thorough understanding. A computer program in Appendix 3 involves the three-node triangle and is a real asset in illustrating the behavior of a discretized system and for checking results of hand calculations. I generally require a project assignment in my course. This has involved altering the computer program to include a different element type.

The alternative Galerkin approximation method is not introduced until Chapter 9 because of the importance of repetition of the similar processes in development of the triangle and quadrilateral elements in Chapters 4 through 8. Chapter 10 details the application of the method using the two-dimensional heat transfer problem; a computer program is also included.

The book ends with a preview of some three-dimensional elements. By this time the student has a good understanding of the steps involved in the development. Because of the complexity of the expressions for these elements, there is no longer any such thing as a simple example suitable for hand calculation, so that illustrative examples are not included. However, although the presentations appear to be brief, they do show the student the added complexity brought about by the third dimension, and they include sufficient detail to allow the student a good start on developing computer code for the elements described.

H. G., Jr.

Contents

Preface v
Principal Notation xv

INTRODUCTION 1

    I.1 Historical Note 1
    I.2 What is the Finite Element Method? 3
    I.3 Current Applicability of the Finite Element Method 6

1 BRIEF EXPOSURE TO MATRIX ANALYSIS OF STRUCTURES 9

    1.1 Introduction 9
    1.2 Two-Force Member Element: Uniaxial Displacement 10
    1.3 Two-Force Member Element: General Plane Displacement 19
    1.4 Coordinate Transformation 21
    1.5 Beam Element in a Plane 30
    1.6 Axially Loaded Beam Element 34

1.7	An Equation Solver: Gauss Elimination	43
1.8	Stiffness Matrix in Banded Matrix Format	45

2 PRELIMINARY TOPICS 55

2.1	Introduction	55
2.2	Coordinate Reference Systems	56
2.3	Definition of Strain	60
2.4	Stress–Strain Relations	61
2.5	Work and the Force Energy Potential Function	63
2.6	Strain Energy	70
2.7	Principle of Minimum Potential Energy	73
2.8	Concepts Involved in Application of the Principle of Minimum Potential Energy	74
2.9	Discussion of the Principle of Minimum Potential Energy	75
2.10	Introduction to the Calculus of Variations and the Variational Formulation	77
2.11	Equilibrium Elasticity Variational Formulation: Strain and Force Potential Energies	83
2.12	Approximate Solutions: The Rayleigh–Ritz Method	84
2.13	The Finite Element Model Concept	89
2.14	Determining the Functional: Some Examples	91
2.15	Alternatives to the Variational Calculus Approach	102

3 AN ELEMENT FOR THE ONE-DIMENSIONAL PROBLEM 107

3.1	Introduction	107
3.2	The Uniaxial Stress Element Model	107
3.3	Step 1. Assumed Displacement Function	108
3.4	Step 2. Derivation of Element Strain and Strain Energy	111
3.5	Step 3. Derivation of Applied Load Potential Energy Functions	112
3.6	Step 4. Summation of Energy Terms	116
3.7	Step 5. Application of the Principle of Minimum Potential Energy	116
3.8	Modeling a Continuous One-Dimensional System with Finite Elements	117
3.9	Assembling the Elements	119
3.10	One-Dimensional Elasticity Example	122
3.11	Interpretation of Results and Some General Conclusions	125
3.12	Finite Element Model for One-Dimensional Heat Flow	127
3.13	One-Dimensional Heat Flow with Convection	133

CONTENTS   xi

    3.14 Programming Nonzero Dirichlet (Essential) Boundary Conditions   139
    3.15 Conclusion   140

4 DEVELOPMENT OF THE CONSTANT-STRAIN TRIANGULAR ELEMENT   145

    4.1 Introduction   145
    4.2 The Triangle Element: Global Coordinates   146
    4.3 The Triangle Element: Local Coordinates   160
    4.4 Local-to-Global Coordinate Transformation   165
    4.5 Assembling the Elements   167

5 INTERPOLATION FORMULAS AND NUMERICAL INTEGRATION   181

    5.1 Introduction   181
    5.2 Interpolation Formula: One Independent Variable   182
    5.3 Interpolation Formula: Two Independent Variables   184
    5.4 Interpolation Applied to a Three-Node Triangle Element   187
    5.5 Integration by Gauss Quadrature   189
    5.6 Change of Integration Variable: The Jacobian Determinant   195

6 ISOPARAMETRIC THREE-NODE TRIANGULAR ELEMENT   201

    6.1 Introduction   201
    6.2 Development of the Stiffness Matrix   202
    6.3 Determining the Equivalent Nodal Forces   205
    6.4 Integration Formulas for Triangle Natural Coordinates   209
    6.5 Concluding Remarks on the Element Development   214
    6.6 The Computer Program and Application Examples   215
    6.7 Data Reduction for Best Results   230
    6.8 Element and Node Numbering   231
    6.9 Taking Advantage of Loading and Region Symmetry   232

7 ISOPARAMETRIC FOUR-NODE QUADRILATERAL ELEMENT   239

    7.1 Introduction   239
    7.2 The Element Natural Coordinates   240

7.3	The Interpolation Formula	241
7.4	Development of the Element Strain–Displacement Matrix	242
7.5	The Element Stiffness Matrix	250
7.6	Distributed Surface Tractions: Equivalent Nodal Forces	256
7.7	Distributed Body Forces: Equivalent Nodal Forces	262
7.8	Assembling the Elements and Solving for Stresses	264
7.9	Comparing the Four-Node Quadrilateral and the Three-Node Triangle	272
7.10	Triangle Subassembly and Condensation	276

8 HIGHER-ORDER ELEMENTS AND ADDITIONAL CAPABILITIES 281

8.1	Isoparametric Six-Node Triangle Stiffness Matrix	281
8.2	Equivalent Nodal Forces: Surface Traction	285
8.3	Isoparametric Eight-Node Quadrilateral Stiffness Matrix	291
8.4	Equivalent Nodal Forces	295
8.5	Using Higher-Order Elements	299
8.6	Convergence Comparison of Four Elements: An Example	300
8.7	Skew Boundary Conditions	302
8.8	Temperature-Induced Strain	304

9 THE GALERKIN AND OTHER WEIGHTED RESIDUAL METHODS 315

9.1	Introduction	315
9.2	Approximation Techniques for Differential Equations: Weighted Residual Methods	316
9.3	Compatibility Requirements	322
9.4	Reduction of Order	323
9.5	Finite Element Solution of the Differential Equation	324
9.6	Jump Conditions in System Properties	331
9.7	Laplace's Equation: Two Dimensions	333
9.8	Poisson's Equation: Two Dimensions	334

10 HEAT TRANSFER AND FLUID FLOW IN TWO DIMENSIONS 337

10.1	Governing Equations for Steady-State Heat Flow	337
10.2	The Element Equation	339
10.3	Selecting the Element	341
10.4	Determination of Element Matrices	342

	10.5	Assembling the Element Equations	348
	10.6	Comment on Boundary Conditions	351
	10.7	Solving the Equations	351
	10.8	Two-Dimensional Steady-State Irrotational Incompressible Flow	360
	10.9	Heat and Fluid Flow with Applied Source: Poisson's Equation	365
	10.10	Conclusion	366

11 AXISYMMETRIC STRESS ANALYSIS 369

11.1	Introduction	369
11.2	The Basic Equations for the Element	370
11.3	Axisymmetric Elasticity Equations	372
11.4	The Element Displacement Functions	373
11.5	Evaluation of the Element Stiffness Matrix	375
11.6	The Nodal Load Vectors	376
11.7	Assembly, Boundary Conditions, Unsymmetrical Loads, and Anisotropy	378

12 THREE-DIMENSIONAL STRESS ELEMENTS 381

12.1	Introduction	381
12.2	The Elasticity Equations	382
12.3	The Four-Node Tetrahedron	383
12.4	The Isoparametric Eight-Node Solid	387

APPENDIXES

A1 MATRIX DEFINITIONS AND OPERATIONS 395

A2 COMPUTER PROGRAMS IN APPLESOFT BASIC AND TURBO PASCAL FOR PLANE FRAME ANALYSIS 399

A2.1	The Element	399
A2.2	Major Characteristics of the Program	399
A2.3	Sample Program Run	401
A2.4	Program Listings in Applesoft BASIC	405
	FRAME-2D	405
	FRAME-2D FILE EDITOR	414
	FRAME-2D DISTORTION PLOT	417

	A2.5 Program Listings in TURBO Pascal	419

A3 COMPUTER PROGRAMS IN APPLESOFT BASIC AND TURBO PASCAL FOR PLANE STRESS ANALYSIS — 435

- A3.1 The Element — 435
- A3.2 Application Suggestions — 435
- A3.3 Sample Program Run — 436
- A3.4 Program Listings in Applesoft BASIC — 440
  - STRESS-3NODE — 440
  - TRIANGLE FILE EDITOR — 449
  - TRIANGLE DISTORTION PLOT — 453
- A3.5 Program Listings in TURBO Pascal — 454

A4 INPUT FILE AND RESULTS FOR ICES STRUDL-II PROGRAM — 469

A5 DERIVATIVES OF THE SHAPE FUNCTIONS — 473

- A5.1 The Six-Node Isoparametric Triangle — 473
- A5.2 The Eight-Node Isoparametric Element — 474

A6 NODAL BODY FORCES FOR THE FOUR-NODE QUADRILATERAL — 475

A7 COMPUTER PROGRAMS IN APPLESOFT BASIC AND TURBO PASCAL FOR TWO-DIMENSIONAL HEAT TRANSFER — 477

- A7.1 The Element — 477
- A7.2 Program Capability — 477
- A7.3 Sample Program Run — 478
- A7.4 Program Listing of HEAT-3NODE in Applesoft BASIC — 482
- A7.5 Program Listing in TURBO Pascal — 490

A8 PARTIAL LIST OF COMMERCIAL FINITE ELEMENT PROGRAMS — 504

SELECTED BIBLIOGRAPHY — 507

ANSWERS TO SELECTED EXERCISES — 511

INDEX — 525

Principal Notation

A	area of two-dimensional region
$[A]$	rotation matrix
$[A_2]$	rotation matrix $= \begin{bmatrix} [A] & [0] \\ [0] & [A] \end{bmatrix}$
	$[A_3] = \begin{bmatrix} [A] & [0] & [0] \\ [0] & [A] & [0] \\ [0] & [0] & [A] \end{bmatrix}$
$[B]$	strain–displacement matrix, $\{\epsilon\} = [B]\{q\}$
B_x, B_y	body force components (force per unit volume)
$[C]$	constitutive matrix, $\{\sigma\} = [C]\{\epsilon\}$
E	modulus of elasticity
$|J|$	Jacobian determinant
k	scalar value of stiffness or thermal conductivity
k_{ij}	row i, column j entry in element stiffness matrix

$[k]$	element stiffness matrix
$[k_{cd_x}], [k_{cd_y}]$	x- and y-direction heat conduction matrix
$[k_{cv}]$	boundary convection matrix
$[K]$	global assemblage stiffness matrix
K_{ij}	row i, column j entry in assemblage stiffness matrix
L, l	lengths
L_1, L_2, L_3	triangle area (natural) coordinates
$[N]$	shape function (interpolation function) matrix
q_{cd}	conduction heat flow
q_{cv}	convection heat flow
$\{q\}$	element nodal displacement vector
$\{q_t\}$	element nodal temperature vector
$\{Q_b\}$	applied boundary heat flux vector
$\{Q\}_{BF}$	equivalent body force element nodal load vector
$\{Q_{cv}\}$	element convective heat flow vector
$\{Q\}_{NF}$	applied force element nodal load vector
$\{Q\}_T$	equivalent stress traction element nodal load vector
$\{Q\}_{temp}$	equivalent thermal element nodal load vector
$\{r\}$	assemblage global nodal displacement vector
$\{R\}$	assemblage global nodal load vector
T_i	temperature of node i
T_x, T_y	stress traction components on element boundary (force per area)
u_i, v_i, w_i	displacements of node i in x, y, and z directions
U	strain energy
V	potential energy
δ	variational operator
$\{\epsilon\}$	strain vector
ν	Poisson's ratio
Π	functional of the problem
$\{\sigma\}$	stress vector
Ω	three-dimensional region of integration

Introduction

I.1 HISTORICAL NOTE

The finite element method has evolved in the past 25 years from a specialized technique for aircraft frame analysis to a general numerical solution technique applicable to a broad range of physical problems. The relatively brief history of the finite element method begins in the early 1940s. Aircraft structural engineers were struggling for better approaches to the analysis of the increasingly complex airframes. Many engineers made significant contributions in the early stages of the finite element method. The sampling of published works (listed in the References at the end of this introduction) of a few researchers in the period from 1941 to 1965 provides a good view of the state of the art and the progress during those years.

With the development of high-performance aircraft, the prediction of the response of these new structures to static and dynamic loading was becoming more difficult. A major difficulty centered on the prediction of the stiffness of the shell-type structures made up of beam-type elements called spars and ribs, plus stiffeners (tension/compression bars) and the metal skin. Beam and bar stiffness was easily calculated; the metal skin stiffness, however, was not. In 1941, A. Hrennikoff proposed a technique to simulate a section of the metal skin with a framework of bars of known stiffness. The problem centered on determining the proper framework pattern to use for correct modeling of the stiffness properties of the skin. The success of the technique was quite good provided that the region being modeled was rectangular; equivalence was very difficult to achieve with irregular boundaries. In 1956, Turner, Clough, Martin, and Topp presented a method of estimating the in-plane stiffness of triangular- and quadrilateral-shaped pieces (elements) of the metal skin. The technique was based on an assumed strain field in the element, a radical departure from the Hrennikoff model and a foundation procedure for much future development. At about this same time, S. Levy in the United States (1953) and J. Argyris and S. Kelsey in Great Britain (1960) were refining the matrix structural analysis methods and implementing the digital computer to solve the many simultaneous equations generated by the technique. The emerging capability of the digital computer was paramount to the implementation of these new techniques. Without the computer, these newly discovered methods would have been of little value.

Returning to the year 1943, the eminent German mathematician Richard Courant presented a discussion of the numerical solution of a variational problem using, as one approach, the Rayleigh–Ritz method. This paper formulated the basis for the very much later expansion of the finite element method to problems other than structures. More than two decades elapsed before the mathematical concepts presented by Courant were formulated into practical numerical solution techniques that, with the all-important computer, could solve practical, nonstructural-type problems. In 1965, O. C. Zienkiewicz and Y. K. Cheung argued that with the solution of elasticity problems by the finite element method interpreted as a minimizing of the total applied load and elastic potential energy functional, other problems possessing a corresponding functional must be solvable by the same technique. In their paper they discuss the solution of the Poisson equation with applications to the torsion of triangular, rectangular, and a hollow bimetallic shaft and also heat transfer in the wall of an axisymmetric pressure vessel.

Subsequent development of the method has proceeded at a rapid pace, and the present applicability, dependability, and acceptance of the method is becoming far reaching. Computers, from large mainframes to single-user micros, with commercially available programs able to solve steady-state or transient structural, elasticity, heat transfer, and fluid flow problems, are becoming easier to use and cost-effective not only for the large corporation, but also for the individual consultant as well. Today, this has to be considered a standard analysis technique.

I.2 WHAT IS THE FINITE ELEMENT METHOD?

The finite element method is a numerical solution technique applicable to a broad range of physical problems, the variables of which are related by means of algebraic, differential, or integral equations. The student in continuum mechanics soon discovers that finding a solution that satisfies a differential equation throughout a region, and also yields the boundary conditions, is a very difficult and often an impossible task for all but the most elementary problems. Even the search for an approximate solution, assuming a polynomial function for example, is generally not practical for the entire region. The finite element method addresses this difficulty by dividing a region into small subregions so that the solution within each subregion can be represented by a function very much simpler than that required for the entire region. The subregions are joined together mathematically by enforcing conditions that make each element boundary compatible with each of its neighbors while satisfying the region boundary requirements.

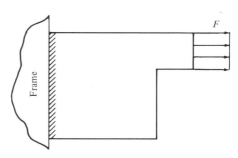

(a) Thin plate with uniformly distributed force, F.

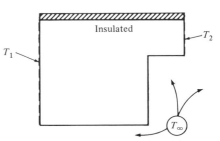

FIGURE I.1 Examples illustrating two problem types readily solved with the finite element method.

(b) Heat-conducting region with boundary conditions of known heat flow, temperature, and convection.

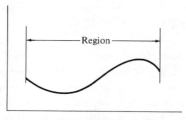

(a) Continuous polynomial function.

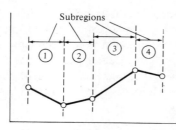

(b) Approximation of function (a) with linear functions in four subregions.

FIGURE I.2 Subdivision of a continuous polynomial of high degree into simpler linear functions.

Consider the following two problems. In Figure I.1(a), a thin plate welded to a "rigid" frame is subjected to the uniformly distributed force F. *Problem*: Determine the stress field in the plate.

In Figure I.1(b), the boundaries of the heat-conducting region have temperature specified on two edges, no heat flow (insulation) on one edge, and convection to the surroundings on the remaining edges. *Problem*: Determine the temperature field in the plate.

To determine the solution functions that yield the boundary conditions and satisfy the governing differential equations everywhere in the region for each problem is a very difficult, perhaps impossible task. The solution functions for each case must represent considerable gradient variations, especially in the vicinities of the boundary corners.

The basic concept behind the finite element method is subdivision of a region into sufficiently small regions so that the solution in each small region (element) can be represented by a simple function. For example, the continuous polynomial function in the region shown in Figure I.2(a) might be approximated with linear functions in each of the four subregions shown in Figure I.2(b). In this example, only the continuity of the function value is enforced at the junction (node) of the subregions; the derivatives of the subregion functions are not continuous. With a sufficiently large number of subregions (elements), the assembly of the

I.2 WHAT IS THE FINITE ELEMENT METHOD?

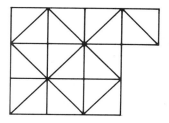

(a) Thin plate with triangular subregions (elements).

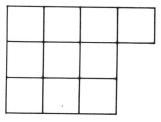

(b) Quadrilateral elements.

FIGURE I.3 "Simple function" subregions assembled to approximate the solution throughout the problem region.

individual element solutions can represent a very high order solution over the complete region.

Returning to the problems posed in Figure I.1, the finite element analysis of the elasticity or heat transfer problem would involve the subdivision (or discretization) of the region into small regions called elements. In Figure I.3(a), the region has been divided into triangles; in Figure I.3(b), quadrilaterals have been used.

In Figure I.4, the region has been divided into a nonuniform mesh of triangular elements. You may anticipate that the reason for this subdivision is based on the need for smaller elements in regions of a rapidly changing gradient (slope) of the solution; regions of small change do not require fine subdivision. In terms of economy of computer memory and calculation time, you want to limit the number of elements in your model.

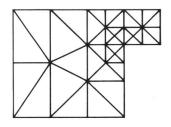

FIGURE I.4 Element refinement in location of high gradient.

The triangle and quadrilateral are commonly used elements in plane analysis. This book includes a thorough development of each of these elements, with applications to elasticity, heat transfer, and fluid flow.

I.3 CURRENT APPLICABILITY OF THE FINITE ELEMENT METHOD

Well-tested commercially available computer code exists to solve structural problems involving two- and three-dimensional solids, plates, and shells. Material properties can be linear elastic isotropic, linear elastic anisotropic, nonlinear elastic, viscoelastic, or plastic; the deformations can be small strain or large strain. Programs are also available for rock and soil mechanics, fluid flow, heat transfer, acoustics, and electromagnetism. Some solve problems that have a close or sequential coupling of the problem types. Problems can be static or transient. Detailed description of available programs can be obtained from handbooks such as the *Finite Element Systems* [edited by C. A. Brebbia (1982)] from computer manufacturers, or directly from the code distributor. The distributors of a few well-known large general-purpose finite element programs are listed in Appendix 8.

To a large degree, the ease and successful application of finite element analysis code depends on the supporting pre- and postprocessing programs. The preprocessing programs assist the user in generating the element mesh in the region of the problem. In Figure I.5, a mechanical spring, formed from flat stock, is shown modeled with quadrilateral and triangular elements of thickness equal to the part. An analysis of the spring was to be made to determine its spring rate. Because the finite element analysis code requires that every element and every element corner be assigned a number and the coordinates of every element corner be supplied, determining and typing in these input data can be difficult

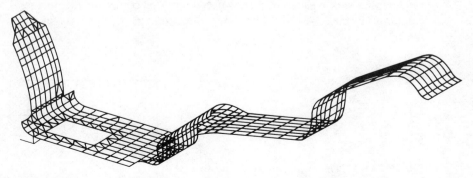

FIGURE I.5 Spring modeled with triangle and quadrilateral elements.

and tedious. The preprocessing program reduces this effort by assisting the user in drawing the outline of the spring, and then with simple commands it automatically generates the mesh, numbers the elements and nodes, defines coordinates, and places all this information in a format that the analysis program can read directly.

Following the analysis of a fairly large problem, examining column after column, row after row of hard-copy output can be a dizzying experience. The postprocessing programs assist in the interpretation of the results, very often by incorporating graphical displays.

REFERENCES

ARGYRIS, J., and S. Kelsey. *Energy Theorems and Structural Analysis*. London: Butterworth & Company (Publishers) Ltd., 1960. Reprinted by Plenum Publishing Corporation, 1968.

BREBBIA, C. A., ed. *Finite Element Systems: A Handbook*. New York: Springer-Verlag, New York, Inc., 1982.

COURANT, R. "Variational Methods for Solution of Problems of Equilibrium and Vibration." *Bull. Am. Math. Soc.* 49 (1943), 1–43.

HRENNIKOFF, A. "A Solution of Problems in Elasticity by the Framework Method." *J. Appl. Mech.* 8 (1941), 169–175.

LEVY, S. "Structural Analysis and Influence Coefficients for Delta Wings." *J. Aeronaut. Sci.* 20, No. 7 (1953), 449.

TURNER, M., R. CLOUGH, H. MARTIN, and L. TOPP. "Stiffness and Deflection Analysis of Complex Structures." *J. Aeronaut. Sci.* 23, No. 9 (1956), 805–823.

ZIENKIEWICZ, O. C., and Y. K. CHEUNG. "Finite Elements in the Solution of Field Problems." *The Engineer* 220 (1965), 507–510.

Brief Exposure to Matrix Analysis of Structures

1.1 INTRODUCTION

The development of the finite element method to solve practical problems originated in the early 1940s in the field of structural analysis. Since then, a sound mathematical basis has been developed for application of the method to the governing equations corresponding to a broad class of problems, including elasticity, heat transfer, and fluid flow. This book will parallel the historical development by first introducing a brief summary of matrix structural analysis. In so doing, we develop an understanding of this very important and commonly used technique while learning many of the procedures common to the implementation of structural analysis as well as the general finite element method. In addition, we gain the knowledge necessary to appreciate the difference between the matrix structural analysis formulation and the more general differential-equation-based finite element method.

1.2 TWO-FORCE MEMBER ELEMENT: UNIAXIAL DISPLACEMENT

Consider a two-force prismatic bar of length L, cross-sectional area A, and elastic modulus E. The position of the bar is restricted to the fixed X axis. The ends of the bar are identified as *nodes*. The nodes are the points of attachment to other elements, and the points in an assemblage of elements for which displacements are sought. In this initial discussion of the two-force member element, forces will be applied to the nodes only, and the displacements of all nodes will be confined to the X axis. Refer to Figure 1.1, where the displacement of nodes 1 and 2 is represented by the symbols u_1 and u_2, respectively.

We will establish the following conventions:

1. Forces and displacements are defined to be positive when they are acting in the positive coordinate direction. Figure 1.1 shows all forces and displacements in the positive direction. [Note that for equilibrium of a two-force member, this sign convention would require the force on one end (node) to be positive and the force on the other end to be negative.]
2. The position of a node in the undeformed structure is taken to be the reference position for that node. A node displacement, therefore, is the change of position of the node as the structure deforms during load application.

Upon referring again to Figure 1.1 of the two-force member (tensile bar) having length L, cross-sectional area A, and material elastic modulus E, if this member were to have equal and opposite forces of magnitude F applied to the end nodes, we know from elementary mechanics of materials that the member would undergo a change in

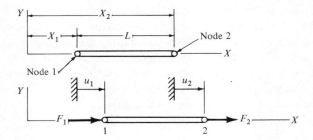

FIGURE 1.1 Sign convention for forces and displacements of two-force member (bar) element.

1.2 TWO-FORCE MEMBER ELEMENT: UNIAXIAL DISPLACEMENT

length according to the relationship

$$u_2 - u_1 = \delta = \frac{FL}{AE} \quad (1.1)$$

We will define the *stiffness*, k, of the member as the ratio of the force acting on the member to the change in length of the member:

$$k = \frac{F}{\delta}$$

or from (1.1),

$$k = \frac{AE}{L} \quad (1.2)$$

Applying this definition to the two-force member shown in Figure 1.1, we can express the force at each node of the member in terms of the nodal displacements, u_1 and u_2, and the member stiffness, k.

$$F_1 = k(u_1 - u_2)$$
$$F_2 = k(u_2 - u_1)$$

These two equations can be put into matrix form as follows:

$$\begin{Bmatrix} F_1 \\ F_2 \end{Bmatrix} = k \begin{bmatrix} 1 & -1 \\ -1 & 1 \end{bmatrix} \begin{Bmatrix} u_1 \\ u_2 \end{Bmatrix} \quad (1.3a)$$

or

$$\{Q\} = [k]\{q\} \quad (1.3b)$$

where

$\{Q\} = [F_1 \quad F_2]^T$, element nodal load vector

$[k]$ = element stiffness matrix [be careful to distinguish between the matrix $[k]$ and the scalar stiffness, k],

$\{q\} = [u_1 \quad u_2]^T$, element nodal displacement vector

Note: The transpose of the vectors is shown simply to save space.

Consider an assemblage of two of these two-force member elements. In Figure 1.2 the element nodes are numbered 1, 2, and 3. Recall that the individual elements have two nodes each and that equation (1.3) is applicable to each of the elements. When looking at each of the elements individually, the two nodes are always labeled 1 and 2. This is called the *local node numbering* of the generic element.

The node numbering of the assemblage of elements is identified as *global node numbering*. A mapping array that identifies the global numbers of the local nodes 1 and 2 of each individual element can be constructed as shown in Table 1.1 The necessity of tagging the global node numbers to each node of each element may not be clear to the reader at this point. This will be explained later when element equations are combined.

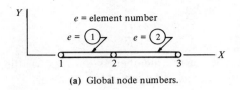

(a) Global node numbers.

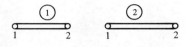

(b) Local node numbers.

FIGURE 1.2 Node numbering for the two-force element.

TABLE 1.1 Global–Local Node Mapping

Element Number, e	Global Node Number of:	
	Local Node 1	**Local Node 2**
1	1	2
2	2	3

Observe that we identify only two displacements in each element: the displacement of node 1 in the X direction, u_1, and the displacement of node 2, u_2, also in the X direction. We do not concern ourselves in the matrix method of structural analysis with displacements of points in an element other than the node points. This is not the case in the finite element method, which we investigate later.

Definition. *A degree of freedom in a structure is an independent displacement of a node.*

For this two-force member element, we are involving node displacements only, and since each node can move only along the X axis, each node has one degree of freedom. Therefore, each element possesses two degrees of freedom.

Equation (1.3) is the general force–displacement relation for the two-force member element, and it can be applied to all members in an assemblage of elements. For clarity, we will apply an element number identifier, e, to the equation:

$$\begin{Bmatrix} F_1 \\ F_2 \end{Bmatrix}_e = k_e \begin{bmatrix} 1 & -1 \\ -1 & 1 \end{bmatrix} \begin{Bmatrix} u_1 \\ u_2 \end{Bmatrix}_e \quad (1.3c)$$

where e denotes the element number.

1.2 TWO-FORCE MEMBER ELEMENT: UNIAXIAL DISPLACEMENT

Consider now the application of equation (1.3c) to the assemblage of two elements shown in Figure 1.2. For element 1:

$$\begin{Bmatrix} F_1 \\ F_2 \end{Bmatrix}_1 = k_1 \begin{bmatrix} 1 & -1 \\ -1 & 1 \end{bmatrix} \begin{Bmatrix} u_1 \\ u_2 \end{Bmatrix}_1$$

For element 2:

$$\begin{Bmatrix} F_1 \\ F_2 \end{Bmatrix}_2 = k_2 \begin{bmatrix} 1 & -1 \\ -1 & 1 \end{bmatrix} \begin{Bmatrix} u_1 \\ u_2 \end{Bmatrix}_2$$

Note that the form of each equation is identical (same local node numbers), but a subscript identifies the particular element. The next step is to replace the local node numbers in these equations with the correct global node numbers. Refer to Table 1.1. Note that local node 2 of element 1 has a global number 2, and that local node 1 of element 2 also has global number 2. Local node 2 of element 2 has a global number 3. One other change in notation is to be made, we will add the element number, e, to the node force subscript.

$$\begin{Bmatrix} F_1 \\ F_2 \end{Bmatrix}_e \Rightarrow \begin{Bmatrix} F_{1e} \\ F_{2e} \end{Bmatrix}$$

If we now replace the element local node number in each equation with the correct global node number, we will obtain the following:

$$\begin{Bmatrix} F_{11} \\ F_{21} \end{Bmatrix} = k_1 \begin{bmatrix} 1 & -1 \\ -1 & 1 \end{bmatrix} \begin{Bmatrix} u_1 \\ u_2 \end{Bmatrix}$$

$$\begin{Bmatrix} F_{22} \\ F_{32} \end{Bmatrix} = k_2 \begin{bmatrix} 1 & -1 \\ -1 & 1 \end{bmatrix} \begin{Bmatrix} u_2 \\ u_3 \end{Bmatrix}$$

(1.4)

where, remember, F_{ij} is defined as the force acting on global node i of element j. The correctly labeled forces are shown on each of the elements in Figure 1.3.

The assemblage of two elements shown in Figure 1.4 has three nodes (and three degrees of freedom). Define the resultant force acting on global node i as R_i. Note that a free-body diagram of the region surrounding node 2 as shown in Figure 1.4 would involve the externally applied load R_2 and the forces exerted by the adjacent elements.

Equilibrium of the region yields

$$R_2 = F_{21} + F_{22}$$

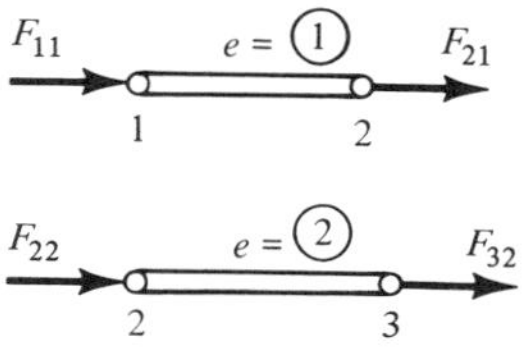

FIGURE 1.3 Description of nodal forces.

14 BRIEF EXPOSURE TO MATRIX ANALYSIS OF STRUCTURES

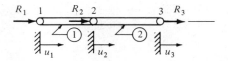

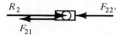

FIGURE 1.4 Free-body diagram of region surrounding node 2.

Also, for nodes 1 and 3 we obtain

$$R_1 = F_{11}$$
$$R_3 = F_{32}$$

The objective is to combine the two equations (1.4) to obtain one equation that relates the nodal forces R_i to the global displacements u_i of the assemblage. One approach that is good for gaining insight into the process, but not practical for real problem solutions, consists of the following procedure:

1. "Enlarge" equations (1.4) to include all global displacements. This yields

$$\begin{Bmatrix} F_{11} \\ F_{21} \\ 0 \end{Bmatrix} = k_1 \begin{bmatrix} 1 & -1 & 0 \\ -1 & 1 & 0 \\ 0 & 0 & 0 \end{bmatrix} \begin{Bmatrix} u_1 \\ u_2 \\ u_3 \end{Bmatrix}$$

$$\begin{Bmatrix} 0 \\ F_{22} \\ F_{32} \end{Bmatrix} = k_2 \begin{bmatrix} 0 & 0 & 0 \\ 0 & 1 & -1 \\ 0 & -1 & 1 \end{bmatrix} \begin{Bmatrix} u_1 \\ u_2 \\ u_3 \end{Bmatrix}$$

2. Sum the two equations in step 1:

$$\begin{Bmatrix} F_{11} \\ F_{21} \\ 0 \end{Bmatrix} + \begin{Bmatrix} 0 \\ F_{22} \\ F_{32} \end{Bmatrix} = k_1 \begin{bmatrix} 1 & -1 & 0 \\ -1 & 1 & 0 \\ 0 & 0 & 0 \end{bmatrix} \begin{Bmatrix} u_1 \\ u_2 \\ u_3 \end{Bmatrix}$$

$$+ k_2 \begin{bmatrix} 0 & 0 & 0 \\ 0 & 1 & -1 \\ 0 & -1 & 1 \end{bmatrix} \begin{Bmatrix} u_1 \\ u_2 \\ u_3 \end{Bmatrix}$$

3. Factor out the displacement vector:

$$\begin{Bmatrix} R_1 \\ R_2 \\ R_3 \end{Bmatrix} = \begin{Bmatrix} F_{11} \\ F_{21} + F_{22} \\ F_{32} \end{Bmatrix} = \begin{bmatrix} k_1 & -k_1 & 0 \\ -k_1 & k_1 + k_2 & -k_2 \\ 0 & -k_2 & k_2 \end{bmatrix} \begin{Bmatrix} u_1 \\ u_2 \\ u_3 \end{Bmatrix}$$

or

$$\{R\} = [K]\{r\} \tag{1.5}$$

1.2 TWO-FORCE MEMBER ELEMENT: UNIAXIAL DISPLACEMENT

Consider now the application of equation (1.3c) to the assemblage of two elements shown in Figure 1.2. For element 1:

$$\left\{ \begin{array}{c} F_1 \\ F_2 \end{array} \right\}_1 = k_1 \begin{bmatrix} 1 & -1 \\ -1 & 1 \end{bmatrix} \left\{ \begin{array}{c} u_1 \\ u_2 \end{array} \right\}_1$$

For element 2:

$$\left\{ \begin{array}{c} F_1 \\ F_2 \end{array} \right\}_2 = k_2 \begin{bmatrix} 1 & -1 \\ -1 & 1 \end{bmatrix} \left\{ \begin{array}{c} u_1 \\ u_2 \end{array} \right\}_2$$

Note that the form of each equation is identical (same local node numbers), but a subscript identifies the particular element. The next step is to replace the local node numbers in these equations with the correct global node numbers. Refer to Table 1.1. Note that local node 2 of element 1 has a global number 2, and that local node 1 of element 2 also has global number 2. Local node 2 of element 2 has a global number 3. One other change in notation is to be made, we will add the element number, e, to the node force subscript.

$$\left\{ \begin{array}{c} F_1 \\ F_2 \end{array} \right\}_e \Rightarrow \left\{ \begin{array}{c} F_{1e} \\ F_{2e} \end{array} \right\}$$

If we now replace the element local node number in each equation with the correct global node number, we will obtain the following:

$$\left\{ \begin{array}{c} F_{11} \\ F_{21} \end{array} \right\} = k_1 \begin{bmatrix} 1 & -1 \\ -1 & 1 \end{bmatrix} \left\{ \begin{array}{c} u_1 \\ u_2 \end{array} \right\}$$

$$\left\{ \begin{array}{c} F_{22} \\ F_{32} \end{array} \right\} = k_2 \begin{bmatrix} 1 & -1 \\ -1 & 1 \end{bmatrix} \left\{ \begin{array}{c} u_2 \\ u_3 \end{array} \right\}$$

(1.4)

where, remember, F_{ij} is defined as the force acting on global node i of element j. The correctly labeled forces are shown on each of the elements in Figure 1.3.

The assemblage of two elements shown in Figure 1.4 has three nodes (and three degrees of freedom). Define the resultant force acting on global node i as R_i. Note that a free-body diagram of the region surrounding node 2 as shown in Figure 1.4 would involve the externally applied load R_2 and the forces exerted by the adjacent elements.

Equilibrium of the region yields

$$R_2 = F_{21} + F_{22}$$

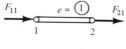

FIGURE 1.3 Description of nodal forces.

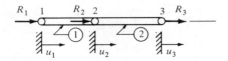

FIGURE 1.4 Free-body diagram of region surrounding node 2.

Also, for nodes 1 and 3 we obtain

$$R_1 = F_{11}$$
$$R_3 = F_{32}$$

The objective is to combine the two equations (1.4) to obtain one equation that relates the nodal forces R_i to the global displacements u_i of the assemblage. One approach that is good for gaining insight into the process, but not practical for real problem solutions, consists of the following procedure:

1. "Enlarge" equations (1.4) to include all global displacements. This yields

$$\begin{Bmatrix} F_{11} \\ F_{21} \\ 0 \end{Bmatrix} = k_1 \begin{bmatrix} 1 & -1 & 0 \\ -1 & 1 & 0 \\ 0 & 0 & 0 \end{bmatrix} \begin{Bmatrix} u_1 \\ u_2 \\ u_3 \end{Bmatrix}$$

$$\begin{Bmatrix} 0 \\ F_{22} \\ F_{32} \end{Bmatrix} = k_2 \begin{bmatrix} 0 & 0 & 0 \\ 0 & 1 & -1 \\ 0 & -1 & 1 \end{bmatrix} \begin{Bmatrix} u_1 \\ u_2 \\ u_3 \end{Bmatrix}$$

2. Sum the two equations in step 1:

$$\begin{Bmatrix} F_{11} \\ F_{21} \\ 0 \end{Bmatrix} + \begin{Bmatrix} 0 \\ F_{22} \\ F_{32} \end{Bmatrix} = k_1 \begin{bmatrix} 1 & -1 & 0 \\ -1 & 1 & 0 \\ 0 & 0 & 0 \end{bmatrix} \begin{Bmatrix} u_1 \\ u_2 \\ u_3 \end{Bmatrix}$$

$$+ k_2 \begin{bmatrix} 0 & 0 & 0 \\ 0 & 1 & -1 \\ 0 & -1 & 1 \end{bmatrix} \begin{Bmatrix} u_1 \\ u_2 \\ u_3 \end{Bmatrix}$$

3. Factor out the displacement vector:

$$\begin{Bmatrix} R_1 \\ R_2 \\ R_3 \end{Bmatrix} = \begin{Bmatrix} F_{11} \\ F_{21} + F_{22} \\ F_{32} \end{Bmatrix} = \begin{bmatrix} k_1 & -k_1 & 0 \\ -k_1 & k_1 + k_2 & -k_2 \\ 0 & -k_2 & k_2 \end{bmatrix} \begin{Bmatrix} u_1 \\ u_2 \\ u_3 \end{Bmatrix}$$

or

$$\{R\} = [K]\{r\} \tag{1.5}$$

1.2 TWO-FORCE MEMBER ELEMENT: UNIAXIAL DISPLACEMENT

where

$\{R\}$ = vector of assemblage global nodal forces

$\{r\}$ = vector of assemblage global nodal displacements

$[K]$ = assembled system global stiffness matrix

Application of this method should enable you to develop matrix equations for systems containing more than two elements. Although this technique works well for hand calculation, it is not practical for computer implementation. A preferred technique will be illustrated in Example 1.2.

Once the matrix equations are developed for the structure, the next step is to substitute boundary conditions and solve for the unknown displacements. At every node in the structure, you will be given the externally applied load or the specified nodal displacement. You will know one or the other, not both at each node. As an example, consider two elements having the same cross-sectional area, length, and elastic modulus. The nodal loads and displacement constraints are defined in Figure 1.5.

Substituting known values of load and displacement into equation (1.5) yields

$$\begin{Bmatrix} R_1 \\ 10 \\ -15 \end{Bmatrix} = \frac{AE}{L} \begin{bmatrix} 1 & -1 & 0 \\ -1 & 2 & -1 \\ 0 & -1 & 1 \end{bmatrix} \begin{Bmatrix} 0 \\ u_2 \\ u_3 \end{Bmatrix}$$

The dimension of the matrix equation can be reduced by removing the unknown force R_1 and the row of the stiffness matrix associated with R_1 (row 1). In addition, the zero displacement allows the removal of the column of the stiffness matrix associated with u_1 (column 1). The reduced matrix equation thus involves known nodal forces and unknown nodal displacements.

$$\begin{Bmatrix} 10 \\ -15 \end{Bmatrix} = \frac{AE}{L} \begin{bmatrix} 2 & -1 \\ -1 & 1 \end{bmatrix} \begin{Bmatrix} u_2 \\ u_3 \end{Bmatrix}$$

The inverse of

$$\begin{bmatrix} 2 & -1 \\ -1 & 1 \end{bmatrix} \quad \text{is} \quad \begin{bmatrix} 1 & 1 \\ 1 & 2 \end{bmatrix}$$

FIGURE 1.5 Two-element loaded structure.

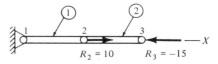

Thus we solve for the displacements:

$$\left\{\begin{matrix} u_2 \\ u_3 \end{matrix}\right\} = \frac{L}{AE}\begin{bmatrix} 1 & 1 \\ 1 & 2 \end{bmatrix}\left\{\begin{matrix} 10 \\ -15 \end{matrix}\right\} = \frac{L}{AE}\left\{\begin{matrix} -5 \\ -20 \end{matrix}\right\}$$

$$u_2 = -\frac{5L}{AE}$$

$$u_3 = -\frac{20L}{AE}$$

Expanding the original matrix equation for R_1 in terms of the nodal displacements, we obtain

$$R_1 = -\frac{AE}{L}u_2$$

Substituting the nodal displacement, we arrive at the value of the support reaction:

$$R_1 = 5$$

Example 1.1 Solve for the node 2 displacement and the support reactions for the structure attached to the rigid walls at nodes 1 and 3 in Figure 1.6.

Solution. For element 1, the aluminum cylinder:

$$k_1 = \left(\frac{AE}{L}\right)_1$$

$$= \frac{(2)(10)(10^6)}{12}$$

$$= 1.667(10^6) \text{ lb/in.}$$

For element 2, the steel cylinder:

$$k_2 = \left(\frac{AE}{L}\right)_2$$

$$= \frac{(1)(30)(10^6)}{16}$$

$$= 1.875(10^6) \text{ lb/in.}$$

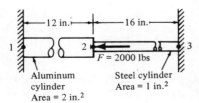

FIGURE 1.6 Structure modeled with two-force elements.

1.2 TWO-FORCE MEMBER ELEMENT: UNIAXIAL DISPLACEMENT

The boundary conditions are

$$u_1 = 0 = u_3$$

Substitute into equation (1.5):

$$\begin{Bmatrix} R_1 \\ -2000 \\ R_3 \end{Bmatrix} = 10^6 \begin{bmatrix} 1.667 & -1.667 & 0 \\ -1.667 & 3.542 & -1.875 \\ 0 & -1.875 & 1.875 \end{bmatrix} \begin{Bmatrix} 0 \\ u_2 \\ 0 \end{Bmatrix} \quad (a)$$

Reduce the stiffness matrix by removing rows 1 and 3 and columns 1 and 3. This leaves

$$-2000 = 10^6 (3.542 u_2)$$

or

$$u_2 = -564(10^{-6}) \text{ in.}$$

Substitute u_2 into equation (a) and solve for R_1 and R_3:

$$R_1 = (10^6)(-1.667) u_2 = 942 \text{ lb}$$
$$R_3 = (10^6)(-1.875) u_2 = 1058 \text{ lb} \quad \square$$

The next example illustrates a more efficient technique for assembling the global stiffness matrix from the individual element stiffness matrices.

Example 1.2 Derive the global assemblage stiffness matrix for the three element structure shown in Figure 1.7.

Solution. First, we construct a table that identifies the global numbers associated with the local nodes of each element (Table 1.2).

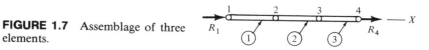

FIGURE 1.7 Assemblage of three elements.

TABLE 1.2 Node Number Map

Element Number	Global Node Number of:	
	Local Node 1	Local Node 2
1	1	2
2	2	3
3	3	4

Next, to appreciate the general concept, write the equation $\{Q\} = [k]\{q\}$ for each element, using local and then global numbers.

$$\begin{Bmatrix} F_{11} \\ F_{21} \end{Bmatrix} = \begin{bmatrix} k_1 & -k_1 \\ -k_1 & k_1 \end{bmatrix} \begin{Bmatrix} u_1 \\ u_2 \end{Bmatrix}_1 \qquad \begin{Bmatrix} F_{11} \\ F_{21} \end{Bmatrix} = \begin{bmatrix} k_1 & -k_1 \\ -k_1 & k_1 \end{bmatrix} \begin{Bmatrix} u_1 \\ u_2 \end{Bmatrix}$$

$$\begin{Bmatrix} F_{12} \\ F_{22} \end{Bmatrix} = \begin{bmatrix} k_2 & -k_2 \\ -k_2 & k_2 \end{bmatrix} \begin{Bmatrix} u_1 \\ u_2 \end{Bmatrix}_2 \qquad \begin{Bmatrix} F_{22} \\ F_{32} \end{Bmatrix} = \begin{bmatrix} k_2 & -k_2 \\ -k_2 & k_2 \end{bmatrix} \begin{Bmatrix} u_2 \\ u_3 \end{Bmatrix}$$

$$\begin{Bmatrix} F_{13} \\ F_{23} \end{Bmatrix} = \begin{bmatrix} k_3 & -k_3 \\ -k_3 & k_3 \end{bmatrix} \begin{Bmatrix} u_1 \\ u_2 \end{Bmatrix}_3 \qquad \begin{Bmatrix} F_{33} \\ F_{43} \end{Bmatrix} = \begin{bmatrix} k_3 & -k_3 \\ -k_3 & k_3 \end{bmatrix} \begin{Bmatrix} u_3 \\ u_4 \end{Bmatrix}$$

— element number   — element number
— local node number   — global node number

The assemblage matrix equation has the form [recall equation (1.5)]

$$\begin{Bmatrix} R_1 \\ R_2 \\ R_3 \\ R_4 \end{Bmatrix} = \begin{Bmatrix} F_{11} \\ F_{21} + F_{22} \\ F_{32} + F_{33} \\ F_{43} \end{Bmatrix} = \begin{bmatrix} k_{11} & k_{12} & k_{13} & k_{14} \\ k_{21} & k_{22} & k_{23} & k_{24} \\ k_{31} & k_{32} & k_{33} & k_{34} \\ k_{41} & k_{42} & k_{43} & k_{44} \end{bmatrix} \begin{Bmatrix} u_1 \\ u_2 \\ u_3 \\ u_4 \end{Bmatrix}$$

Note that the address of entries in the global stiffness matrix is

row number $= i$ of the dual subscript on F_{ij}

$\qquad\qquad = $ global node number i supporting external loading R_i

column number $=$ subscript i on displacement u_i

$\qquad\qquad\qquad = $ global node i

The global stiffness matrix is formed by inserting the element stiffness matrix entries in their proper location. The location of each element stiffness matrix entry in the global stiffness matrix can be identified by tagging the element matrix rows and columns with the global numbers associated with local nodes 1 and 2 of each element. For this example we would have the following:

Element 1	Element 2	Element 3
① ② ← global number 1 2 ← local number $\begin{bmatrix} k_1 & -k_1 \\ -k_1 & k_1 \end{bmatrix} \begin{matrix} 1\ ① \\ 2\ ② \end{matrix}$	② ③ 1 2 $\begin{bmatrix} k_2 & -k_2 \\ -k_2 & k_2 \end{bmatrix} \begin{matrix} 1\ ② \\ 2\ ③ \end{matrix}$	③ ④ 1 2 $\begin{bmatrix} k_3 & -k_3 \\ -k_3 & k_3 \end{bmatrix} \begin{matrix} 1\ ③ \\ 2\ ④ \end{matrix}$

The global numbers now serve as the global address of each element entry in the assemblage stiffness matrix.

For example, global stiffness element k_{22} is the sum of all the element stiffness values that have the global address 2, 2. Thus

$$k_{22} = k_1 + k_2$$
$$k_{23} = -k_2 \quad \text{etc.}$$

Doing this for all k_{ij} yields

$$[K] = \begin{bmatrix} k_1 & -k_1 & 0 & 0 \\ -k_1 & k_1 + k_2 & -k_2 & 0 \\ 0 & -k_2 & k_2 + k_3 & -k_3 \\ 0 & 0 & -k_3 & k_3 \end{bmatrix} \quad \square$$

1.3 TWO-FORCE MEMBER ELEMENT: GENERAL PLANE DISPLACEMENT

Consider the two-force member of Section 1.2 positioned arbitrarily in the global XY plane instead of being confined to the X axis. Two coordinate systems are established:

1. The global coordinate XY axes, chosen for convenient representation of the entire structure.
2. The local coordinate axes xy, selected for convenient representation of the properties of the element. For a two-force member, it is common to have the x axis placed along the length of the element.

Refer to Figure 1.8. Let F_{1X} be the component of force $\bar{F}_1$ on node 1 in the global X direction and let F_{1x} be the component in the local x direction. F_{1Y} is the component of the force on node 1 in the global Y direction, and so on.

The nodal displacements are represented by

u_{iX} = displacement of node i in the X direction (global)

u_{ix} = displacement of node i in the x direction (local)

v_{iY} = displacement of node i in the Y direction (global)

v_{iy} = displacement of node i in the y direction (local)

(note that $F_{1y} = 0 = F_{2y}$ for this element)

The sign conventions are the same as discussed earlier: Positive forces and displacements are in the positive coordinate directions. From our work with this element in Section 1.2, we see that equation (1.3) can be used to express the force–displacement relation in terms of components in the local xy coordinate system.

$$\begin{Bmatrix} F_{1x} \\ F_{2x} \end{Bmatrix} = k \begin{bmatrix} 1 & -1 \\ -1 & 1 \end{bmatrix} \begin{Bmatrix} u_{1x} \\ u_{2x} \end{Bmatrix} \quad (1.6)$$

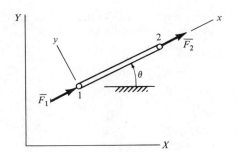

FIGURE 1.8 General position of element in the XY plane.

We want to expand this two-degree-of-freedom equation into a four-degree-of-freedom equation that incorporates the two force components at each node and the two orthogonal displacement components of each node of this two-node element. This can be done by including two additional equations:

$$F_{1y} = 0 \quad \text{and} \quad F_{2y} = 0$$

in the matrix equation (1.6). This results in the following matrix equation:

$$\begin{Bmatrix} F_{1x} \\ F_{1y} \\ F_{2x} \\ F_{2y} \end{Bmatrix} = k \begin{bmatrix} 1 & 0 & -1 & 0 \\ 0 & 0 & 0 & 0 \\ -1 & 0 & 1 & 0 \\ 0 & 0 & 0 & 0 \end{bmatrix} \begin{Bmatrix} u_{1x} \\ v_{1y} \\ u_{2x} \\ v_{2y} \end{Bmatrix} \quad (1.7)$$

or

$$\{Q\}_{xy} = [k]_{xy} \{q\}_{xy} \quad (1.8)$$

where

$\{Q\}_{xy}$ = column vector of element node forces acting in the local x and y directions

$[k]_{xy}$ = element stiffness matrix for the local xy coordinate system

$\{q\}_{xy}$ = column vector of the local coordinate direction displacement components

In regard to the stiffness matrix:

1. It is always square.
2. It is always symmetric for linear systems.
3. The diagonal elements are always positive or zero.

In a general structure, many elements are involved, and they would be oriented with different values of angle θ (see Figure 1.8). To develop a system equation such as equation (1.5), the equations of all the elements must be referred to a common global coordinate system. This requires a coordinate transformation relationship that can be applied to equation

(1.8) for each element in the structure. This transformation relationship is developed in the next section.

1.4 COORDINATE TRANSFORMATION

A vector can be expressed in terms of any coordinate system set of unit basis vectors ($\hat{i}$, $\hat{j}$ for xy, etc.; see Figure 1.9).

$$\bar{V} = [\hat{I} \quad \hat{J}]\begin{Bmatrix} V_X \\ V_Y \end{Bmatrix} = [\hat{i} \quad \hat{j}]\begin{Bmatrix} V_x \\ V_y \end{Bmatrix}$$

where

$$[\hat{i} \quad \hat{j}] = [\hat{I} \quad \hat{J}]\begin{bmatrix} \cos\theta & -\sin\theta \\ \sin\theta & \cos\theta \end{bmatrix} = [\hat{I} \quad \hat{J}][A]$$

with θ measured positive counterclockwise from the global X axis. Substitution yields

$$\bar{V} = [\hat{I} \quad \hat{J}]\begin{Bmatrix} V_X \\ V_Y \end{Bmatrix} = [\hat{I} \quad \hat{J}][A]\begin{Bmatrix} V_x \\ V_y \end{Bmatrix}$$

Equating the coefficients of the same basis vectors gives us

$$\begin{Bmatrix} V_X \\ V_Y \end{Bmatrix} = [A]\begin{Bmatrix} V_x \\ V_y \end{Bmatrix} \quad (1.9)$$

This is the transformation relationship for any vector. Note that the elements of the square matrix $[A]$ are the direction cosines of the angles between the unit vectors of the two coordinate systems.

This makeup of the matrix is quite universal, but be careful; some authors might reverse the interpretation by writing

$$[\hat{I} \quad \hat{J}] = [\hat{i} \quad \hat{j}][\underline{A}]$$

The $[\underline{A}]$ matrix in this equation is the transpose of the $[A]$ in equation (1.9).

Note that $[A]^T = [A]^{-1}$ for the special case of a rotation of an orthogonal coordinate system.

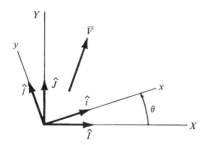

FIGURE 1.9 Definition of unit basis vectors.

Consider now transforming the force vector of equation (1.7) from the local xy coordinate system to the global XY coordinate system.

$$\begin{Bmatrix} F_{1X} \\ F_{1Y} \end{Bmatrix} = [A] \begin{Bmatrix} F_{1x} \\ F_{1y} \end{Bmatrix} \quad \text{(node 1 force)}$$

$$\begin{Bmatrix} F_{2X} \\ F_{2Y} \end{Bmatrix} = [A] \begin{Bmatrix} F_{2x} \\ F_{2y} \end{Bmatrix} \quad \text{(node 2 force)}$$

"Stack" these two relationships to obtain the complete nodal element force vector:

$$\begin{Bmatrix} F_{1X} \\ F_{1Y} \\ F_{2X} \\ F_{2Y} \end{Bmatrix} = \begin{bmatrix} [A] & [0] \\ [0] & [A] \end{bmatrix} \begin{Bmatrix} F_{1x} \\ F_{1y} \\ F_{2x} \\ F_{2y} \end{Bmatrix}$$

or

$$\{Q\}_{XY} = [A_2]\{Q\}_{xy} \quad (1.10)$$

where

$$[A_2] = \begin{bmatrix} \cos\theta & -\sin\theta & 0 & 0 \\ \sin\theta & \cos\theta & 0 & 0 \\ 0 & 0 & \cos\theta & -\sin\theta \\ 0 & 0 & \sin\theta & \cos\theta \end{bmatrix}$$

Recall equation (1.8):

$$\{Q\}_{xy} = [k]_{xy}\{q\}_{xy}$$

Substitute this equation into (1.10):

$$\{Q\}_{XY} = [A_2][k]_{xy}\{q\}_{xy} \quad (1.11)$$

The transformation matrix $[A_2]$ can also be applied to the displacement vector.

$$\begin{Bmatrix} u_{1X} \\ v_{1Y} \\ u_{2X} \\ v_{2Y} \end{Bmatrix} = [A_2] \begin{Bmatrix} u_{1x} \\ v_{1y} \\ u_{2x} \\ v_{2y} \end{Bmatrix}$$

or

$$\{q\}_{XY} = [A_2]\{q\}_{xy}$$

Solve for $\{q\}_{xy}$:

$$\{q\}_{xy} = [A_2]^{-1}\{q\}_{XY}$$
$$= [A_2]^T\{q\}_{XY}$$

1.4 COORDINATE TRANSFORMATION

Substitute in equation (1.11):

$$\{Q\}_{XY} = [A_2][k]_{xy}[A_2]^T \{q\}_{XY} \quad (1.12)$$

Equation (1.8) is a vector equation and thus has the same form for any coordinate system reference. We can therefore write it relative to the XY coordinate system as

$$\{Q\}_{XY} = [k]_{XY} \{q\}_{XY}$$

Equate this to equation (1.12):

$$[k]_{XY}\{q\}_{XY} = [A_2][k]_{xy}[A_2]^T \{q\}_{XY}$$

or

$$[k]_{XY} = [A_2][k]_{xy}[A_2]^T \quad (1.13)$$

This relationship transforms the local element stiffness matrix (written originally relative to the local coordinate system) to the global coordinate system. You may recognize this as the form of the transformation of second-order tensors.

The following examples illustrate this coordinate transformation of the stiffness matrix.

Example 1.3 Consider the element shown in Figure 1.10, where the element is parallel to the global Y axis and node 1 is labeled. Construct the stiffness matrix relative to the global XY system.

Solution. The local xy system is applied with the origin at node 1. The stiffness matrix of this element relative to the local system is known to be [from equation (1.7)]

$$[k]_{xy} = \frac{AE}{L} \begin{bmatrix} 1 & 0 & -1 & 0 \\ 0 & 0 & 0 & 0 \\ -1 & 0 & 1 & 0 \\ 0 & 0 & 0 & 0 \end{bmatrix}$$

The stiffness matrix relative to the global XY system can be obtained

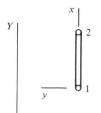

FIGURE 1.10 Two-force element rotated 90° to X coordinate axis.

by using equation (1.13):

$$[k]_{XY} = [A_2][k]_{xy}[A_2]^T$$

where $[A_2]$ for $\theta = 90°$ is

$$[A_2] = \begin{bmatrix} 0 & -1 & 0 & 0 \\ 1 & 0 & 0 & 0 \\ 0 & 0 & 0 & -1 \\ 0 & 0 & 1 & 0 \end{bmatrix}$$

$$[k]_{XY} = \begin{bmatrix} 0 & -1 & 0 & 0 \\ 1 & 0 & 0 & 0 \\ 0 & 0 & 0 & -1 \\ 0 & 0 & 1 & 0 \end{bmatrix} \begin{bmatrix} 1 & 0 & -1 & 0 \\ 0 & 0 & 0 & 0 \\ -1 & 0 & 1 & 0 \\ 0 & 0 & 0 & 0 \end{bmatrix}$$

$$\times \begin{bmatrix} 0 & 1 & 0 & 0 \\ -1 & 0 & 0 & 0 \\ 0 & 0 & 0 & 1 \\ 0 & 0 & -1 & 0 \end{bmatrix} \frac{AE}{L}$$

$$= \begin{bmatrix} 0 & 0 & 0 & 0 \\ 0 & 1 & 0 & -1 \\ 0 & 0 & 0 & 0 \\ 0 & -1 & 0 & 1 \end{bmatrix} \frac{AE}{L} \qquad \square$$

Example 1.4 For the structure shown in Figure 1.11, develop the stiffness matrix of each element relative to the global XY coordinate system.

Solution. Consider element 1 shown in Figure 1.12. Define (arbitrarily) global node 3 as local node 1 of this element.

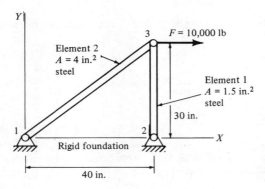

FIGURE 1.11 Simple structure of two-force elements.

1.4 COORDINATE TRANSFORMATION

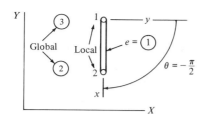

FIGURE 1.12 Local node numbering of element 1.

The stiffness matrix relative to the local xy system is defined by equation (1.7).

$$[k]_{xy} = k_1 \begin{bmatrix} 1 & 0 & -1 & 0 \\ 0 & 0 & 0 & 0 \\ -1 & 0 & 1 & 0 \\ 0 & 0 & 0 & 0 \end{bmatrix}$$

where $k_1 = (AE/L)_1$. The rotation matrix $[A_2]$ of equation (1.10) is

$$[A_2] = \begin{bmatrix} 0 & 1 & 0 & 0 \\ -1 & 0 & 0 & 0 \\ 0 & 0 & 0 & 1 \\ 0 & 0 & -1 & 0 \end{bmatrix}$$

Applying equation (1.13) gives us

$$[k]_{XY} = [A_2][k]_{xy}[A_2]^T$$

$$= \begin{bmatrix} 0 & 1 & 0 & 0 \\ -1 & 0 & 0 & 0 \\ 0 & 0 & 0 & 1 \\ 0 & 0 & -1 & 0 \end{bmatrix} \begin{bmatrix} 1 & 0 & -1 & 0 \\ 0 & 0 & 0 & 0 \\ -1 & 0 & 1 & 0 \\ 0 & 0 & 0 & 0 \end{bmatrix}$$

$$\times \begin{bmatrix} 0 & -1 & 0 & 0 \\ 1 & 0 & 0 & 0 \\ 0 & 0 & 0 & -1 \\ 0 & 0 & 1 & 0 \end{bmatrix} k_1$$

Thus the stiffness matrix relative to the global XY system for element 1 is

$$[k]_{XY} = \begin{bmatrix} 0 & 0 & 0 & 0 \\ 0 & 1 & 0 & -1 \\ 0 & 0 & 0 & 0 \\ 0 & -1 & 0 & 1 \end{bmatrix} k_1$$

Now consider element 2 (Figure 1.13). Define the global node 1 as local node 1. Following the same procedures as those used for ele-

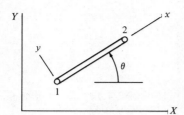

FIGURE 1.13 Local node numbering of element 2.

ment 1, we first define the local stiffness matrix.

$$[k]_{xy} = k_2 \begin{bmatrix} 1 & 0 & -1 & 0 \\ 0 & 0 & 0 & 0 \\ -1 & 0 & 1 & 0 \\ 0 & 0 & 0 & 0 \end{bmatrix}$$

The rotation matrix $[A_2]$ is

$$[A_2] = \begin{bmatrix} 0.8 & -0.6 & 0 & 0 \\ 0.6 & 0.8 & 0 & 0 \\ 0 & 0 & 0.8 & -0.6 \\ 0 & 0 & 0.6 & 0.8 \end{bmatrix}$$

Using equation (1.13), we have

$$[k]_{XY} = [A_2][k]_{xy}[A_2]^T$$

$$= [A_2] \begin{bmatrix} 1 & 0 & -1 & 0 \\ 0 & 0 & 0 & 0 \\ -1 & 0 & 1 & 0 \\ 0 & 0 & 0 & 0 \end{bmatrix}$$

$$\times \begin{bmatrix} 0.8 & 0.6 & 0 & 0 \\ -0.6 & 0.8 & 0 & 0 \\ 0 & 0 & 0.8 & 0.6 \\ 0 & 0 & -0.6 & 0.8 \end{bmatrix} k_2$$

Thus the element 2 global system stiffness matrix is

$$[k]_{XY} = \begin{bmatrix} 0.64 & 0.48 & -0.64 & -0.48 \\ 0.48 & 0.36 & -0.48 & -0.36 \\ -0.64 & -0.48 & 0.64 & 0.48 \\ -0.48 & -0.36 & 0.48 & 0.36 \end{bmatrix} k_2$$

□

Example 1.5 Assemble the element stiffness matrices of the structure in Example 1.4 to obtain the global system stiffness matrix.

1.4 COORDINATE TRANSFORMATION

Solution. It may prove beneficial to go back and reread Section 1.2 and to review the discussion of the assembly of the equations of two elements. The general concepts involved are the same for this problem, but we will develop a more sensible method of constructing the global stiffness matrix.

The first step in the matrix assembly process is to assign numbers to each displacement variable. On a local reference basis, the four displacements (two displacements for each node) will be labeled consecutively 1 through 4.

Local node number:	1	1	2	2
Local displacement:	u_1	v_1	u_2	v_2
Local displacement number:	1	2	3	4

Next, assign numbers to each global displacement. Since there are three nodes in this problem, there are six displacements (before applying constraint boundary conditions).

Global node number:	1	1	2	2	3	3
Global displacement:	u_1	v_1	u_2	v_2	u_3	v_3
Global displacement number:	1	2	3	4	5	6

Observe that the global displacement number is related to the global node number as follows:

$$\begin{aligned}\text{global displacement number in } X \text{ direction} &= 2(\text{global node number}) - 1 \\ \text{global displacement number in } Y \text{ direction} &= 2(\text{global node number})\end{aligned} \quad (1.14)$$

The next step is to establish the node numbering "map" along the lines of Table 1.1 in Section 1.2. A map that identifies the global number of each local node of every element can be established with the array $GN(I, J)$. The row (first subscript) of the array will define the element number, the column (second subscript) will define local node 1 or 2. For this two-element problem we would have

$$GN(I, J) \Rightarrow \begin{array}{c|cc} I \backslash J & 1 & 2 \\ \hline 1 & 3 & 2 \\ 2 & 1 & 3 \end{array} \begin{array}{l} \leftarrow \text{element 1} \\ \leftarrow \text{element 2} \end{array}$$

$$\text{local node 1} \nearrow \quad \nwarrow \text{local node 2}$$

Next, select the local stiffness matrix of one of the elements and assign the address of each entry in the element matrix. The local address is simply made using the local displacement numbers 1 through 4. The global address is made using the global displacement numbers derived from equation (1.14). This global address designation is the address of the local element entry value in the global stiffness matrix. For example, the global and local addresses for the matrix of element 1 are labeled as follows:

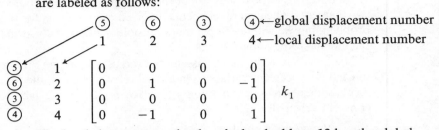

The local element entry that has the local address 12 has the global address 56 and thus is added to the global stiffness matrix in row 5, column 6. The stiffness matrix for element 2 is

$$\begin{array}{c} \\ \\ ①\\ ②\\ ⑤\\ ⑥ \end{array} \begin{array}{c} ① \quad\quad ② \quad\quad ⑤ \quad\quad ⑥ \\ 1 \quad\quad 2 \quad\quad 3 \quad\quad 4 \\ 1 \\ 2 \\ 3 \\ 4 \end{array} \left[\begin{array}{cccc} 0.64 & 0.48 & -0.64 & -0.48 \\ 0.48 & 0.36 & -0.48 & -0.36 \\ -0.64 & -0.48 & 0.64 & 0.48 \\ -0.48 & -0.36 & 0.48 & 0.36 \end{array} \right] k_2$$

The procedure now consists of adding the elements from each local stiffness matrix that have a common global address. This creates the global stiffness matrix. The resulting global matrix is

$$[K] = \left[\begin{array}{cccccc} 0.64k_2 & 0.48k_2 & 0 & 0 & -0.64k_2 & -0.48k_2 \\ 0.48k_2 & 0.36k_2 & 0 & 0 & -0.48k_2 & -0.36k_2 \\ 0 & 0 & 0 & 0 & 0 & 0 \\ 0 & 0 & 0 & k_1 & 0 & -k_1 \\ -0.64k_2 & -0.48k_2 & 0 & 0 & 0.64k_2+0 & 0.48k_2+0 \\ -0.48k_2 & -0.36k_2 & 0 & -k_1 & 0.48k_2+0 & 0.36k_2+k_1 \end{array} \right]$$

□

Example 1.6 Continuing with the structure shown in Example 1.4, construct the load and displacement vectors. Then, using the global stiffness matrix developed in Example 1.5, solve for the displacement of node 3.

Solution. At this point it may prove beneficial to return to Section 1.2 and reread the discussion of the construction of the global equations for a simpler structure. We are reminded that the global force vector is simply an array of the resultant externally applied loads at each node, and the global displacement vector is simply an array of the coordinate direction displacements of each node.

Returning to our current problem, we find that the resultant externally applied loads acting on the structure in Figure 1.11 are

1.4 COORDINATE TRANSFORMATION

placed into the *load vector* $\{R\}$ as follows:

$$\{R\} = \begin{Bmatrix} R_{1X} \\ R_{1Y} \\ R_{2X} \\ R_{2Y} \\ R_{3X} \\ R_{3Y} \end{Bmatrix} = \begin{Bmatrix} R_{1X} \\ R_{1Y} \\ R_{2X} \\ R_{2Y} \\ 10{,}000 \\ 0 \end{Bmatrix}$$

Note that the foundation reaction forces R_{1X}, R_{1Y}, R_{2X}, and R_{2Y} are unknown. Next, the *displacement vector* $\{r\}$ is

$$\{r\} = \begin{Bmatrix} u_1 \\ v_1 \\ u_2 \\ v_2 \\ u_3 \\ v_3 \end{Bmatrix}_{XY} = \begin{Bmatrix} 0 \\ 0 \\ 0 \\ 0 \\ u_3 \\ v_3 \end{Bmatrix}_{XY}$$

Substituting these vectors into the load–displacement relationship given as

$$\{R\} = [K]\{r\} \tag{1.5}$$

we have

$$\begin{Bmatrix} R_{1X} \\ R_{1Y} \\ R_{2X} \\ R_{2Y} \\ 10{,}000 \\ 0 \end{Bmatrix} = \begin{bmatrix} 0.64k_2 & 0.48k_2 & 0 & 0 & -0.64k_2 & -0.48k_2 \\ 0.48k_2 & 0.36k_2 & 0 & 0 & -0.48k_2 & -0.36k_2 \\ 0 & 0 & 0 & 0 & 0 & 0 \\ 0 & 0 & 0 & k_1 & 0 & -k_1 \\ -0.64k_2 & -0.48k_2 & 0 & 0 & 0.64k_2 & 0.48k_2 \\ -0.48k_2 & -0.36k_2 & 0 & -k_1 & 0.48k_2 & 0.36k_2 + k_1 \end{bmatrix} \times \begin{Bmatrix} 0 \\ 0 \\ 0 \\ 0 \\ u_{3X} \\ v_{3Y} \end{Bmatrix} \tag{1.15}$$

The solution of this equation follows the procedure employed in Section 1.2. First, reduce the stiffness matrix by removing rows associated with the unknown forces and the columns corresponding to the zero displacement values. This leaves the following equation:

$$\begin{Bmatrix} 10,000 \\ 0 \end{Bmatrix} = \begin{bmatrix} 0.64k_2 & 0.48k_2 \\ 0.48k_2 & 0.36k_2 + k_1 \end{bmatrix} \begin{Bmatrix} u_{3X} \\ v_{3Y} \end{Bmatrix}$$

Calculating the bar stiffness constants yields

$$k_1 = \left(\frac{AE}{L}\right)_1 = \frac{(1.5)(30)(10^6)}{30} = 1.5(10)^6 \text{ lb/in.}$$

$$k_2 = \left(\frac{AE}{L}\right)_2 = 2.4(10^6) \text{ lb/in.}$$

By substituting these values into the stiffness matrix, the unknown displacements can be determined by inverting the 2 × 2 matrix. These values are

$$u_{3X} = 0.0103 \text{ in.}$$
$$v_{3Y} = -0.0050 \text{ in.}$$

The foundation reactions can be determined by substituting the displacements just determined back into equation (1.15). These values are (Figure 1.14)

$$\begin{Bmatrix} R_{1X} \\ R_{1Y} \\ R_{2X} \\ R_{2Y} \end{Bmatrix} = \begin{Bmatrix} -10,000 \\ -7,500 \\ 0 \\ 7,500 \end{Bmatrix} \text{ lb}$$

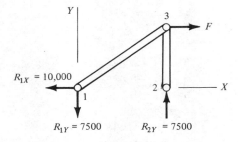

FIGURE 1.14 Foundation reactions

□

1.5 BEAM ELEMENT IN A PLANE

The two-force element does have practical application in structural analysis, but it is a very limiting element, since it cannot provide flexural rigidity. A second element to study is the beam element. We will consider initially an element free of axial loading and then, in Section 1.6,

1.5 BEAM ELEMENT IN A PLANE

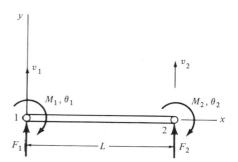

FIGURE 1.15 Forces and displacements for the beam element.

consider the general frame element that will support axial shear and bending forces.

The beam element is shown in Figure 1.15. In our discussion, forces F_i and the couples M_i will act in one plane and only at the end nodes of the beam. The node force F_i will remain perpendicular to the element; we will not include (or allow) axial loading. Also, the local xy coordinate system will remain parallel to the global XY system. This means that all the elements in a structure will be connected in a straight line.

The displacements at each node include a linear displacement v_i in the y (or Y) direction and the neutral axis rotation θ_i. The beam is assumed to have a uniform flexural stiffness EI over L. The xy coordinate system shown is the local system for this element. The origin is always at local node 1.

The load–stiffness–displacement relationship for this element is the same form symbolically as for the element studied previously:

$$\{Q\} = [k]\{q\} \tag{1.3b}$$

or

$$\begin{Bmatrix} F_1 \\ M_1 \\ F_2 \\ M_2 \end{Bmatrix} = \begin{bmatrix} k_{11} & k_{12} & k_{13} & k_{14} \\ k_{21} & k_{22} & k_{23} & k_{24} \\ k_{31} & k_{32} & k_{33} & k_{34} \\ k_{41} & k_{42} & k_{43} & k_{44} \end{bmatrix} \begin{Bmatrix} v_1 \\ \theta_1 \\ v_2 \\ \theta_2 \end{Bmatrix}$$

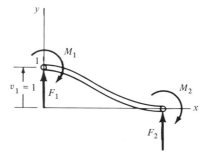

FIGURE 1.16 Unit displacement of node 1 in the y direction.

The problem now is the determination of the values k_{ij} of the element local stiffness matrix. This was rather simple for the two-force member. The task becomes increasingly difficult with increasing complexity of the elements. Here is one approach.

Let us force a displacement configuration on this element. Require that $v_1 = 1$ unit and all other displacements = 0. The element would be deformed as shown in Figure 1.16, and substituting into the load–displacement relation, we have

$$\begin{Bmatrix} F_1 \\ M_1 \\ F_2 \\ M_2 \end{Bmatrix} = \begin{bmatrix} k_{11} & k_{12} & k_{13} & k_{14} \\ k_{21} & k_{22} & - & - \\ k_{31} & - & - & - \\ k_{41} & - & - & k_{44} \end{bmatrix} \begin{Bmatrix} 1 \\ 0 \\ 0 \\ 0 \end{Bmatrix}$$

The expansion yields

$$F_1 = k_{11}$$
$$M_1 = k_{21}$$
$$F_2 = k_{31}$$
$$M_2 = k_{41}$$

In other words, the stiffness matrix entries equal the forces (couples) required for the imposed displacements.

For linear systems we can use the superposition technique to determine the "forces" (see Figure 1.17). Using beam deflection equations from any strength-of-materials text, we derive the following:

$$v_1 = 1 = \delta_1 + \delta_2$$
$$1 = \frac{F_1 L^3}{3EI} + \frac{M_1 L^2}{2EI}$$
$$\theta_1 = 0 = \gamma_1 + \gamma_2$$
$$0 = \frac{F_1 L^2}{2EI} + \frac{M_1 L}{EI}$$

Solving these two equations yields

$$F_1 = \frac{12EI}{L^3} = k_{11}$$
$$M_1 = \frac{-6EI}{L^2} = k_{21}$$

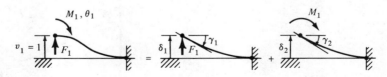

FIGURE 1.17 Superposition of deflections.

1.5 BEAM ELEMENT IN A PLANE

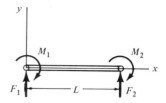

FIGURE 1.18 Free-body diagram of beam element.

Solving for the other forces using equilibrium equations, we have (see Figure 1.18)

$$\sum F_y = 0$$

$$F_2 = -F_1 = \frac{-12EI}{L^3} = k_{31}$$

$$\sum M = 0$$

$$M_2 = -(F_1 L + M_1)$$

$$M_2 = \frac{-6EI}{L^2} = k_{41}$$

Note that we have now calculated the elements of column 1 of the stiffness matrix. Since the matrix is symmetric, we thus have the first row as well. (However, a second independent calculation would be desirable for a check.)

To obtain the second column, we impose the displacement $\theta = 1$ unit, all others equal to zero. The deformed element is shown in Figure 1.19. From the stiffness equation, we obtain the following:

$$F_1 = k_{12}$$
$$M_1 = k_{22}$$
$$F_2 = k_{32}$$
$$M_2 = k_{42}$$

Solving this statically indeterminate problem (using superposition again

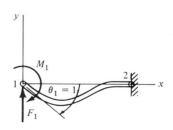

FIGURE 1.19 Unit rotation of node.

if you wish), we find the following expressions for the matrix elements:

$$F_1 = \frac{-6EI}{L^2} = k_{12}$$

$$M_1 = \frac{4EI}{L} = k_{22}$$

$$F_2 = \frac{6EI}{L^2} = k_{32}$$

$$M_2 = \frac{2EI}{L} = k_{42}$$

Continuing this process of imposing one independent unit displacement at a time and solving for the required forces, we can determine all the elements of the stiffness matrix. The resulting matrix is as follows:

$$\{Q\} = [k]\{q\}$$

$$\begin{Bmatrix} F_1 \\ M_1 \\ F_2 \\ M_2 \end{Bmatrix} = \frac{EI}{L^3} \begin{bmatrix} 12 & -6L & -12 & -6L \\ -6L & 4L^2 & 6L & 2L^2 \\ -12 & 6L & 12 & 6L \\ -6L & 2L^2 & 6L & 4L^2 \end{bmatrix} \begin{Bmatrix} v_1 \\ \theta_1 \\ v_2 \\ \theta_2 \end{Bmatrix} \quad (1.16)$$

This completes the development of the element stiffness matrix. The solution of a problem modeled with these elements follows the procedure discussed earlier for the two-force member element: The continuous beam is discretized into the elements, the element stiffness matrices are calculated, the global stiffness matrix is assembled, boundary conditions (loads and displacements) are substituted, the matrix equation is reduced, and the displacements are determined by "inverting" the reduced stiffness matrix.

1.6 AXIALLY LOADED BEAM ELEMENT

The final structural element we will consider is the general beam element capable of supporting axial loading in addition to the shear force and bending couple. As before, our development will be confined to the XY plane.

The development of this element is now very simple because for small deflection theory the force–displacement relationships are linear and the principle of superposition applies. Thus this general element local stiffness matrix is obtained by superposition of the simple beam stiffness matrix, Section 1.5, and the two-force element stiffness matrix, Section 1.3.

Figure 1.20 defines the axially loaded beam element (often referred to as the beam–column element or bar element) with the conventions for positive loads and displacements. The element local coordinate system is

1.6 AXIALLY LOADED BEAM ELEMENT

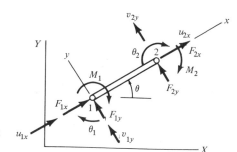

FIGURE 1.20 Axially loaded beam element.

shown at an arbitrary angle to the global axes because for this element each end node can displace in two coordinate directions. Therefore, adding the linear displacement degrees of freedom to the angular rotation of each end, we have an element with six degrees of freedom.

The element stiffness matrix is constructed by superposition of equations (1.7) and (1.16). We thus develop the following local coordinate force–displacement relationship for the axially loaded beam element:

$$\begin{Bmatrix} F_{1x} \\ F_{1y} \\ M_1 \\ F_{2x} \\ F_{2y} \\ M_2 \end{Bmatrix} = \frac{AE}{L} \begin{bmatrix} 1 & 0 & 0 & -1 & 0 & 0 \\ 0 & \frac{12I}{AL^2} & \frac{-6I}{AL} & 0 & \frac{-12I}{AL^2} & \frac{-6I}{AL} \\ 0 & \frac{-6I}{AL} & \frac{4I}{A} & 0 & \frac{6I}{AL} & \frac{2I}{A} \\ -1 & 0 & 0 & 1 & 0 & 0 \\ 0 & \frac{-12I}{AL^2} & \frac{6I}{AL} & 0 & \frac{12I}{AL^2} & \frac{6I}{AL} \\ 0 & \frac{-6I}{AL} & \frac{2I}{A} & 0 & \frac{6I}{AL} & \frac{4I}{A} \end{bmatrix} \begin{Bmatrix} u_{1x} \\ v_{1y} \\ \theta_1 \\ u_{2x} \\ v_{2y} \\ \theta_2 \end{Bmatrix}$$

(1.17)

Solution procedures for problems modeled with this element are identical in principle to the procedures followed with the two-force element. However, the calculation details are considerably more complex because of the additional degrees of freedom and the increased complexity of the local stiffness matrix [compare (1.7) and (1.17)].

To review these procedures briefly, consider the problem of solving for the support reactions for the continuous rod, bent, built into the vertical wall, pinned at the floor, and loaded with a couple at the bent section (Figure 1.21). Two elements will be used.

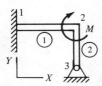

FIGURE 1.21 Statically indeterminate structure.

Summarizing the steps involved, we must do the following:
1. Calculate the 6 × 6 stiffness matrix of element 1. Since the local xy coordinate axes are parallel to the global XY axes, the local and global stiffness matrices are identical.
2. Calculate the stiffness matrix of element 2. The local stiffness matrix must be rotated to the global axes. If node 3 is taken as the local coordinate origin, ϕ is 90° and the rotation matrix $[A]$ for the three degrees of freedom at each node is

$$[A] = \begin{bmatrix} \cos\phi & -\sin\phi & 0 \\ \sin\phi & \cos\phi & 0 \\ 0 & 0 & 1 \end{bmatrix} = \begin{bmatrix} 0 & -1 & 0 \\ 1 & 0 & 0 \\ 0 & 0 & 1 \end{bmatrix}$$

and the "stacked" rotation matrix is

$$[A_2] = \begin{bmatrix} [A] & [0] \\ [0] & [A] \end{bmatrix}$$

Using equation (1.13) yields

$$[k]_{XY} = [A_2][k]_{xy}[A_2]^T$$

3. Construct the assemblage global stiffness matrix and load vector. Each entry in the element stiffness matrix is placed in the proper location in the global assemblage stiffness matrix using the global displacement numbers (the global stiffness matrix address) of the three nodal displacements at each node of the beam–column element. These numbers are calculated using the following formulas:

$$\text{global displacement number in the } X \text{ direction} = 3(\text{global node number}) - 2$$

$$\text{global displacement number in the } Y \text{ direction} = 3(\text{global node number}) - 1 \qquad (1.18)$$

$$\text{global rotational displacement number} = 3(\text{global node number})$$

4. Apply the displacement constraints, reduce the equations, and solve for the displacements.
5. Multiply the element stiffness matrices and the element nodal displacements to obtain the forces acting on each element.

1.6 AXIALLY LOADED BEAM ELEMENT

A longhand solution of even this simple two-element problem is a very long, laborious process. The practical necessity of using a computer should be very clear. In Appendix 2, a program written for the two-force member and beam–column element is described and an example involving the structure in Figure 1.21 is presented to illustrate the data input procedures and interpretation of results. The following example is included to introduce the reader to the potential of the program and to show the importance to the designer of this analysis tool.

Example 1.7 Using the computer program in Appendix 2, analyze the two simple models of bicycle frame design shown in Figure 1.22. Compare the stiffness of each frame and the loads on the ends of each member in the frames.

Solution. The first step in the application of the computer program is the labeling of each element and node in the structure. This is shown in Figure 1.23.

All coordinates of the nodes and all material and section properties of the frame members are fed into the program and the results presented on the following pages were calculated by an Apple computer in approximately 2 minutes for each design. Review the output for each problem. Note the node map and boundary condition information.

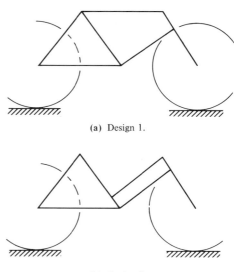

(a) Design 1.

(b) Design 2.

FIGURE 1.22 Two bicycle frame designs.

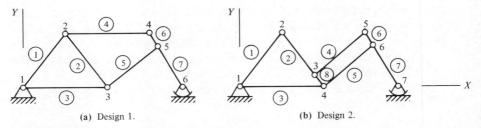

FIGURE 1.23 Element and node numbering of the two designs.

Computer Printout for Design 1

PLANE FRAME ANALYSIS: INPUT DATA

***** ELEMENT GLOBAL NODE NUMBERS ****

ELEMENT NUMBER	GLOBAL NODE NUMBERS OF		ELEMENT TYPE
	LOCAL 1	LOCAL 2	
1	1	2	2
2	2	3	2
3	1	3	2
4	2	4	2
5	3	5	2
6	4	5	2
7	5	6	2

**** NODE GLOBAL COORDINATES AND BOUNDARY DISPLACEMENTS ****

NODE	X	Y	BOUNDARY DISPLACEMENTS		
			U	V	THETA
1	0	0	0	0	
2	10	15			
3	20	0			
4	30	15			
5	32	11.25			
6	38	0		0	

****NODAL FORCES AND COUPLES ****

NODE	X FORCE	Y FORCE	COUPLE
2	0	-125	0
4	0	-25	0

**** MATERIAL AND SECTION PROPERTIES ****

ELEMENT	ELASTIC MODULUS	AREA	AREA MOMENT OF INERTIA
1	30000000	.1	.01
2	30000000	.15	.02
3	30000000	.15	.02
4	30000000	.15	.02
5	30000000	.15	.02
6	30000000	.3	.1
7	30000000	.3	.1

1.6 AXIALLY LOADED BEAM ELEMENT

PLANE FRAME ANALYSIS: RESULTS

```
*******************************************
**             DISPLACEMENTS             **
*******************************************
```

NODE	XDISPL	YDISPL	THETA, RAD.
1	0	0	9.2E-05
2	1.919E-03	-2.133E-03	2.29E-04
3	2.91E-04	-3.046E-03	2.97E-04
4	1.703E-03	-6.677E-03	-3.24E-04
5	3.21E-03	-5.852E-03	-5.19E-04
6	.014041	0	-1.19E-03

```
*******************************************
**           ELEMENT NODE FORCES         **
*******************************************
```

--NOTE--NODE NUMBERS ARE GLOBAL, LISTED
IN ASCENDING LOCAL NUMBER ORDER

EL	NODE	AXIAL LOAD	SHEAR LOAD	COUPLE
1	1	118	-1	-2
1	2	-119	0	2
2	2	35	-4	29
2	3	-36	3	34
3	1	-66	-1	1
3	3	65	0	13
4	2	48	4	-33
4	4	-49	-5	-66
5	3	-58	9	-49
5	5	57	-10	-108
6	4	40	33	65
6	5	-41	-34	-209
7	5	46	-25	315
7	6	-47	24	0

Note: All units are consistent with the input units: in this case, pounds and inches.

There should be no question about the interpretation of the displacement results. The Distortion Plot program in Appendix 2 was used to generate the undeformed and deformed configuration of the frame (Figure 1.24). This plot can be very helpful in quickly identifying regions of relative flexibility.

The interpretation of the element end loads requires care and an understanding of the sign convention and the orientation convention for the local coordinate system. In Figure 1.25 the forces and couples

40 BRIEF EXPOSURE TO MATRIX ANALYSIS OF STRUCTURES

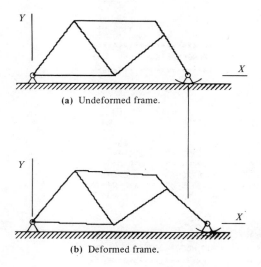

FIGURE 1.24 Distortion of frame from BASIC program.

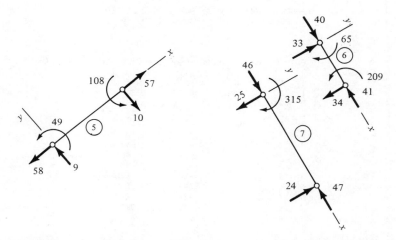

FIGURE 1.25 Element nodal loads on selected members in the frame of design 1.

have been applied to selected members in the frame. Note that each member should have an equilibrium force system applied, and that the assembly of elements should result in a force summation equal to the resultant externally applied load.

Note: Round-off error in the program *output* caused the element nodal forces to vary by 1 unit.

1.6 AXIALLY LOADED BEAM ELEMENT

Computer Printout for Design 2

PLANE FRAME ANALYSIS: INPUT DATA

***** ELEMENT GLOBAL NODE NUMBERS ****

ELEMENT NUMBER	GLOBAL NODE NUMBERS OF		ELEMENT TYPE
	LOCAL 1	LOCAL 2	
1	1	2	2
2	2	3	2
3	1	4	2
4	3	5	2
5	4	6	2
6	5	6	2
7	6	7	2
8	3	4	2

**** NODE GLOBAL COORDINATES AND BOUNDARY DISPLACEMENTS ****

NODE	X	Y	BOUNDARY DISPLACEMENTS		
			U	V	THETA
1	0	0	0	0	
2	10	15			
3	17.5	3.75			
4	20	0			
5	30	15			
6	32	11.25			
7	38	0		0	

****NODAL FORCES AND COUPLES ****

NODE	X FORCE	Y FORCE	COUPLE
2	0	-125	0
5	0	-25	0

**** MATERIAL AND SECTION PROPERTIES ****

ELEMENT	ELASTIC MODULUS	AREA	AREA MOMENT OF INERTIA
1	30000000	.1	.01
2	30000000	.15	.02
3	30000000	.15	.02
4	30000000	.15	.02
5	30000000	.15	.02
6	30000000	.3	.1
7	30000000	.3	.1
8	30000000	.15	.02

PLANE FRAME ANALYSIS: RESULTS

```
*****************************************
**              DISPLACEMENTS          **
*****************************************
```

NODE	XDISPL	YDISPL	THETA, RAD.
1	0	0	1.606E-03
2	.017765	-.012613	2.405E-03
3	-2.71E-03	-.026006	-4.31E-04
4	2.95E-04	-.023912	-7.34E-04

```
5    -.016185    -.011668    -9.64E-04
6    -.012412    -9.652E-03  -1.152E-03
7    5.545E-03   0           -1.823E-03
```

```
*******************************************
**           ELEMENT NODE FORCES          **
*******************************************
```

```
--NOTE--NODE NUMBERS ARE GLOBAL,LISTED
IN ASCENDING LOCAL NUMBER ORDER


EL   NODE   AXIAL LOAD   SHEAR LOAD   COUPLE

1     1        106          -9          66
1     2       -107           8          92
2     2        71           32         -93
2     3       -72          -33        -345
3     1       -67           13         -67
3     4        66          -14        -207
4     3       113          -13         120
4     5      -114           12          82
5     4      -133           -6          63
5     6       132            5          32
6     5         6          101         -83
6     6        -7         -102        -349
7     6        46          -25         315
7     7       -47           24           0
8     3        75          -82         224
8     4       -76           81         143
```

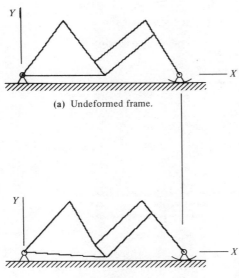

FIGURE 1.26 Distortion of design 2.

(a) Undeformed frame.

(b) Deformed frame.

1.7 AN EQUATION SOLVER: GAUSS ELIMINATION

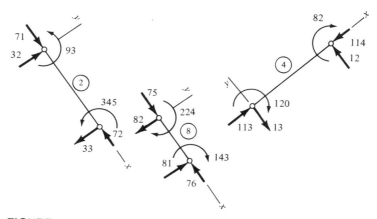

FIGURE 1.27 Node forces on selected members in the design 2 frame.

Reviewing the displacement results and comparing them to the results for design 1, we note that design 2 is considerably more flexible than design 1, and the points of maximum deflection in each design are not in the same location. The computer plots of the undeformed and deformed frame clearly reveal the "weak" section in design 2 in the region of node 3 (see Figure 1.26).

The element loading results show that design 2 frame members must support greater shear, axial, and couple loadings. In Figure 1.27 the loads on each end of selected members have been labeled. The reader should completely label the forces for the remaining elements. □

1.7 AN EQUATION SOLVER: GAUSS ELIMINATION

When the system of equations is completely assembled in matrix form,
$$[K]\{r\} = \{R\}$$
we have to "invert" the stiffness matrix to solve for the displacement vector. In practice, the actual inverse matrix is not determined because the process is very inefficient with regard to time, computer storage, and accuracy. Instead, alternative procedures are employed that operate on the equations to put the stiffness matrix in upper triangular form. We discuss the popular Gauss elimination method by way of the following simple example.

Given the following matrix equation, we wish to solve for the $\{r\}$ vector.

$$\begin{bmatrix} 4 & 3 & 1 \\ 3 & 2 & -1 \\ 1 & -1 & 4 \end{bmatrix} \begin{Bmatrix} r_1 \\ r_2 \\ r_3 \end{Bmatrix} = \begin{Bmatrix} 4 \\ 12 \\ 8 \end{Bmatrix}$$

STEP 1. Divide row 1 by its main diagonal term:
$$\begin{bmatrix} 1 & \frac{3}{4} & \frac{1}{4} \\ 3 & 2 & -1 \\ 1 & -1 & 4 \end{bmatrix}\{r\} = \begin{Bmatrix} 1 \\ 12 \\ 8 \end{Bmatrix}$$

STEP 2. Multiply row 1 by -3 and add to row 2:
$$\begin{bmatrix} 1 & \frac{3}{4} & \frac{1}{4} \\ 0 & -\frac{1}{4} & -\frac{7}{4} \\ 1 & -1 & 4 \end{bmatrix}\{r\} = \begin{Bmatrix} 1 \\ 9 \\ 8 \end{Bmatrix}$$

STEP 3. Multiply row 1 by -1 and add to row 3:
$$\begin{bmatrix} 1 & \frac{3}{4} & \frac{1}{4} \\ 0 & -\frac{1}{4} & -\frac{7}{4} \\ 0 & -\frac{7}{4} & \frac{15}{4} \end{bmatrix}\{r\} = \begin{Bmatrix} 1 \\ 9 \\ 7 \end{Bmatrix}$$

STEP 4. Divide row 2 by its diagonal term:
$$\begin{bmatrix} 1 & \frac{3}{4} & \frac{1}{4} \\ 0 & 1 & 7 \\ 0 & -\frac{7}{4} & \frac{15}{4} \end{bmatrix}\{r\} = \begin{Bmatrix} 1 \\ -36 \\ 7 \end{Bmatrix}$$

STEP 5. Multiply row 2 by $\frac{7}{4}$ and add row 2 to row 3:
$$\begin{bmatrix} 1 & \frac{3}{4} & \frac{1}{4} \\ 0 & 1 & 7 \\ 0 & 0 & \frac{64}{4} \end{bmatrix}\{r\} = \begin{Bmatrix} 1 \\ -36 \\ -\frac{224}{4} \end{Bmatrix}$$

STEP 6. Solve for r_3:
$$\frac{64}{4}r_3 = \frac{-224}{4} \qquad r_3 = -3.5$$

STEP 7. Back-substitute r_3 into row 2 and solve for r_2:
$$r_2 + 7r_3 = -36$$
$$r_2 + 7(-3.5) = -36$$
$$r_2 = -11.5$$

STEP 8. Back-substitute into row 1 and solve for r_1:
$$r_1 + \frac{3}{4}r_2 + \frac{1}{4}r_3 = 1$$
$$r_1 + \frac{3}{4}(-11.5) + \frac{-3.5}{4} = 1$$
$$r_1 = 10.5$$

It is always a good idea to substitute the values determined to check on the accuracy of the result.

$$\begin{bmatrix} 4 & 3 & 1 \\ 3 & 2 & 1 \\ 1 & -1 & 4 \end{bmatrix} \begin{Bmatrix} 10.5 \\ -11.5 \\ -3.5 \end{Bmatrix} = \begin{Bmatrix} 4.0 \\ 12.0 \\ 8.0 \end{Bmatrix}$$

The example shown above has used a symmetric matrix, which is representative of the structural stiffness matrix. The Gauss elimination method will work on an unsymmetric matrix as well.

1.8 STIFFNESS MATRIX IN BANDED MATRIX FORMAT

The structures program in Appendix 2 was written for a microcomputer for two reasons. First, personal ownership of these computers is increasing, and a program such as this should be a valuable addition to the engineer-user's library. The second and equally important reason is that the relatively slow speed and small memory of the microcomputer give the student a better appreciation of the importance of efficient programming techniques.

In developing a computer program for the microcomputer to solve structures problems, we soon realize that memory is a real problem, and the assembled global stiffness matrix is recognized as a major user of memory. Consider, for example, that a structure modeled with N nodes produces a stiffness matrix with $(3N)^2$ matrix entries. (An Apple II + with 48k will run out of memory with approximately 15 nodes.)

Upon a reexamination of the stiffness matrix, we become aware that storing the complete stiffness matrix is not necessary. For one thing, the matrix is symmetric. Using this characteristic alone, we could reduce the storage requirements appreciably if we elect not to store the identical entries on the opposite side of the diagonal. In addition, we observe that the assembled stiffness matrix frequently has all the nonzero entries in a region close to the diagonal. Refer to the assembled 4×4 stiffness matrix of Example 1.2. Note the band of three diagonal rows of nonzero entries. Note also that because of the symmetry, this matrix of 16 entries can be completely specified by storing only the seven unique nonzero entries, which results in a significant reduction in memory requirement.

The banding characteristic of the stiffness matrix depends on the global node numbering scheme of the structure model. Recall that the global displacement numbers are the assembled matrix entry addresses and that they are calculated using formulas (1.14) for the two-force member or the beam element and (1.18) for the axially loaded beam element. To achieve a narrow band of matrix entries, we must maintain a minimum difference between the two global node numbers on each element. For example, if an axially loaded beam element is numbered according to Figure 1.28(a), the 6×6 element stiffness matrix would

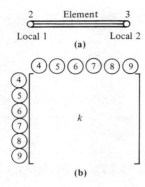

FIGURE 1.28 Global displacement numbers of beam–column element with consecutive node numbers.

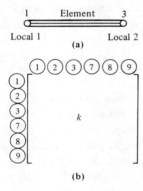

FIGURE 1.29 Global displacement numbers of beam–column element.

have the global matrix addresses defined in Figure 1.28(b). If the same element is assigned the global node numbers shown in Figure 1.29(a), the element global matrix addresses would be as shown in Figure 1.29(b). You must appreciate that as the difference between the node numbers for an element increases, the element entries are placed farther apart in the global matrix. This increases the width of the band of the global matrix and increases the memory storage requirement.

To take advantage of the symmetry and bandedness of the global stiffness matrix, we store the global matrix in banded form as shown in Figure 1.30. For a structure modeled with the axially loaded beam element (beam–column element) with N nodes, the matrix would have $3N$ rows, and the number of columns, BW, would be determined using the formula

$$\text{BW} = 3(\text{maximum node number difference}) + 3 \qquad (1.19)$$

The structures program in Appendix 2 incorporates this storage scheme and uses a modified Gauss routine to solve for the displacements when the stiffness matrix is in band matrix form.

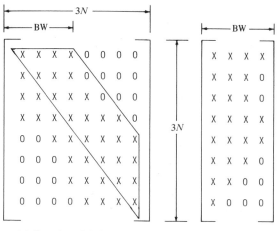

(a) Complete global matrix. (b) Banded matrix format.

FIGURE 1.30 Comparison of the complete assemblage stiffness matrix $[K]$ and the corresponding band matrix.

PROBLEMS

1.1 Determine the matrix product.

$$\begin{bmatrix} 1 & 2 \\ 3 & 4 \end{bmatrix} \begin{Bmatrix} 5 \\ 6 \end{Bmatrix}$$

1.2 Determine the inverse of the matrix.

$$\begin{bmatrix} 4 & 2 \\ 2 & 3 \end{bmatrix}^{-1}$$

1.3 Refer to Figure P1.3. Assemble the two local stiffness matrices into the global stiffness matrix, assuming the following two node mappings:
(a)

Element Number	Global Node Number of:	
	Local Node 1	Local Node 2
1	1	3
2	3	2

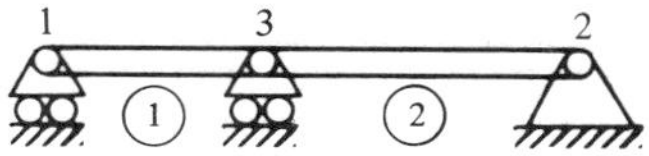

FIGURE P1.3

$[k]_1 = \begin{bmatrix} 1 & 2 \\ 2 & 3 \end{bmatrix}$

$[k]_2 = \begin{bmatrix} 4 & 5 \\ 5 & 6 \end{bmatrix}$

48 BRIEF EXPOSURE TO MATRIX ANALYSIS OF STRUCTURES

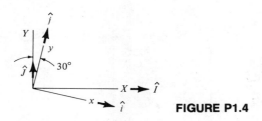

FIGURE P1.4

(b)

Element Number	Global Node Number of:	
	Local Node 1	Local Node 2
1	3	1
2	3	2

1.4 Obtain the rotation matrix $[A]$ of equation (1.9) for the axes orientation shown in Figure P1.4.

1.5 Using the rotation matrix of Problem 1.4, transform the vector

$$\overline{F} = 10\hat{I} + 5\hat{J}$$

to the basis vectors $\hat{i}$ and $\hat{j}$ of the xy coordinate system.

1.6 The structure shown in Figure P1.6 is modeled with two two-force elements. The material and geometric properties of each element are as follows:

Element Number	Elastic Modulus, E (psi)	Area (in.^{2})
1	$30(10)^6$	2
2	$30(10)^6$	3

(a) Construct the individual element stiffness matrices.
(b) Assemble the global stiffness matrix.
(c) Construct the global displacement vector $\{r\}$ and the global load vector $\{R\}$.
(d) Write the assembled global displacement equation in the form of equation (1.5).

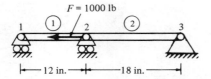

FIGURE P1.6

PROBLEMS

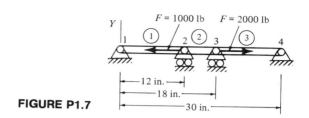

FIGURE P1.7

(e) Substitute known boundary displacements and known loads.
(f) Reduce the matrix equation and solve for the displacements of nodes 1 and 2.

1.7 The structure in Figure P1.7 is modeled with three elements. The material and geometric properties of each element are defined as follows:

Element Number	Elastic Modulus, E (psi)	Area (in.²)
1	$10(10)^6$	4
2	$30(10)^6$	3
3	$10(10)^6$	6

Assemble the matrix force–displacement equation, substitute boundary conditions, and solve for the displacements of nodes 2 and 3.

1.8 Figure P1.8 defines the location of a two-force member (element). The member is a steel tube with a 3-in. outside diameter and a $\frac{1}{8}$-in. wall thickness. Element local nodes 1 and 2 and the local xy coordinate reference are indicated. Determine the following:
(a) The local stiffness matrix for the element.
(b) The transformation (rotation) matrix $[A_2]$ for this local and global axis combination.
(c) The element stiffness matrix relating the global XY forces to the global XY displacements.

1.9 Figure P1.9 shows the member of Problem P1.8 in the same location. However, the local node numbering has been reversed. Do the following:
(a) Sketch the local xy reference axes.
(b) Determine the rotation matrix $[A_2]$.

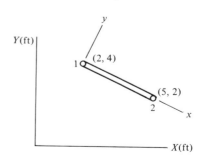

FIGURE P1.8

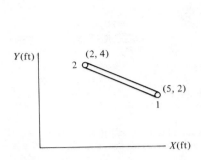

FIGURE P1.9

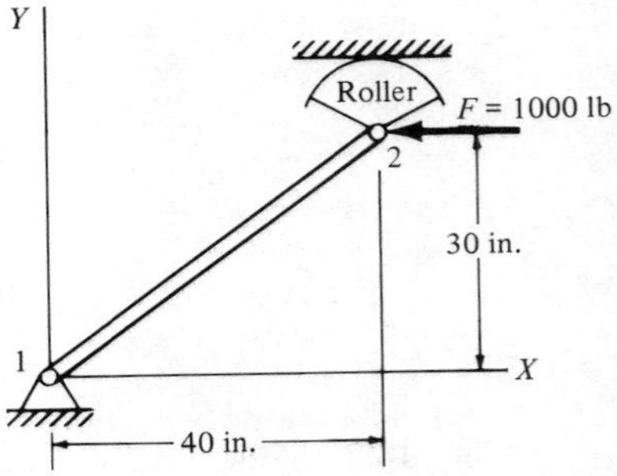

FIGURE P1.10

    (c) Determine the local reference stiffness matrix.
    (d) Determine the global reference stiffness matrix.

1.10 Figure P1.10 shows an 8-in.-diameter aluminum tube with a $\frac{1}{4}$-in. wall thickness. End 1 is attached to the "ground" with a smooth pin, end 2 is attached to a rigid roller with a smooth pin.
    (a) Determine the XY reference force–displacement matrix equation (obtain the stiffness matrix).
    (b) Using the matrix equation, solve for the displacement of end 2 resulting from the application of the 1000-lb force.
    (c) Solve for the axial and shear force at each end of the rod.
    (d) Check your answer using a direct application of equilibrium equations.

1.11 The structure shown in Figure P1.11 is modeled with two two-force elements. The material and geometric properties are defined as follows:

Element Number	Elastic Modulus, E (psi)	Area (in.^{2})
1	$30(10)^6$	1.5
2	$30(10)^6$	2.0

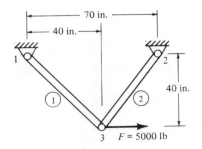

FIGURE P1.11

PROBLEMS

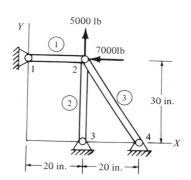

FIGURE P1.12

Using the following map of element node numbering:

Element Number	Global Node Number of:	
	Local Node 1	Local Node 2
1	1	3
2	3	2

solve for the following:
(a) Each element local stiffness matrix.
(b) Each element global stiffness matrix (rotate to XY reference).
(c) The assembled global stiffness matrix.
(d) The reduced global force–displacement equation.
(e) The displacement components (XY) of node 3.
(f) The axial load in each element (shear load = 0).

1.12 Figure P1.12 shows three two-force members connected together and to the "ground" with smooth pins. The material, cross-sectional area, and node numbering map for each element are defined in the table.

Forces are applied to node 2 as indicated. Solve for the displacement of node 2 and the axial and shear force exerted on the ends of each member.

Element Number	Material	Area (in.^{2})	Global Node Number of:	
			Local Node 1	Local Node 2
1	Aluminum	1.5	1	2
2	Steel	1.0	3	2
3	Aluminum	2.0	2	4

1.13 A uniform steel shaft has a diameter of 1 in. Figure P1.13 shows four different loadings and support conditions for this shaft. Using two beam elements, solve for the linear and angular displacements of the nodes and the reactions at the supports for each of the four conditions. In each case, clearly derive the individual element stiffness matrices, the assembled global stiffness matrix, the reduced equation, and the solution.

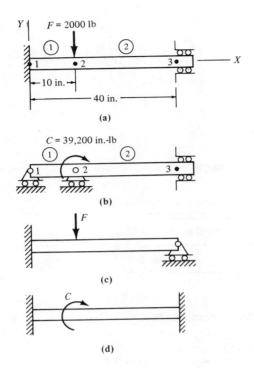

FIGURE P1.13

1.14 The structure shown in Figure P1.14 is made by bending a 1-in.-diameter steel rod. For the support and loading condition shown, solve for the linear and angular displacement of a point on the 90° bend (global node 2) and the support reactions. The structure is shown modeled with two axial beam elements.

1.15 Consider the structure shown in Figure P1.15. An aluminum beam is rigidly attached to the wall at the left end and pinned (smooth pin) to the ground support at the right end. In addition, a $\frac{1}{2}$-in.-diameter steel rod is attached to the beam and joined to the wall with smooth pins. Determine the bending moment distribution in the beam and the axial load in the rod resulting from the application of the 5000-lb force. (A computer solution is recommended.)

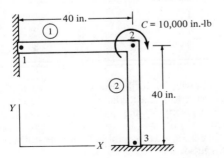

FIGURE P1.14

PROBLEMS

53

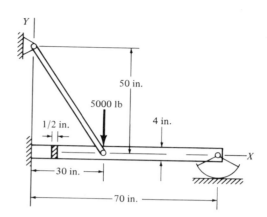

FIGURE P1.15

1.16 Label the force magnitudes on the ends of the bicycle frame members in Example 1.7, design 1.

1.17 Label the force magnitudes on the ends of the bicycle frame members in Example 1.7, design 2.

1.18 *Project.* Use the computer program to redesign the bicycle frame design 2 of Example 1.7 so that the frame member loading is less and the flexibility is reduced.

CHAPTER 2

Preliminary Topics

2.1 INTRODUCTION

The discussion of the matrix analysis of structures in the first chapter introduced the reader to many of the terms and techniques common to the finite element method: matrix manipulation, coordinate transformation, generation of stiffness matrices, force–displacement equations, and methods of solving the many simultaneous equations. A knowledge of these topics will make the presentation of the development of the finite element method more efficient.

In this book, the finite element method is presented in a manner that hopefully will show the reader the broad applicability of the method to physical problems. A major emphasis, however, will be on the application to elasticity problems. In this chapter we review some of the important elasticity relationships that will be used in the subsequent development. After a discussion of work, force energy potential, strain energy, and the application of the principle of minimum potential energy, the concept of obtaining an approximate solution is introduced.

Beginning with Section 2.10, the mathematical level is raised with introduction of the calculus of variations. The discussion presents the basis of the variational formulation corresponding to the differential equation and boundary conditions of a physical problem. The Rayleigh–Ritz approximation method is introduced, and the motivation for the mathematical discretization of the region is presented. These sections are vital to the understanding of the generality of the finite element method as an approximate solution technique for a broad range of problem types.

2.2 COORDINATE REFERENCE SYSTEMS

There are three different classifications of coordinate systems used in the finite element method of analysis; two of them, the global coordinate system and the local coordinate system, were used in Chapter 1 and are easily understood. The third coordinate system, the natural system, is a little more difficult to comprehend. This system is used extensively and must be appreciated. We introduce this system here together with a summary of the characteristics of the other two systems.

Global Coordinate System. This coordinate system is the frame of reference for the entire continuum (see Figure 2.1); all points are located in space using the global coordinate system. There is only one global system for a particular analysis. The coordinate axes have dimensions of length.

Local Coordinate System. This system is attached to the element. Its orientation relative to the element does not change regardless of the orientation of the element relative to the global system. It is established for convenience in developing element relationships. This system was introduced in our earlier discussion of the axially loaded element and, as we saw then, there is a local coordinate system for every element in the continuum (or structure). The dimensions (and units) of the local system are the same as for the global system.

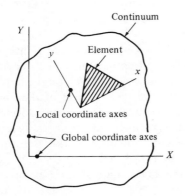

FIGURE 2.1 Coordinate systems.

2.2 COORDINATE REFERENCE SYSTEMS

Natural Coordinate System. This system consists of dimensionless coordinates that identify positions in elements without regard to the element size or shape. Two examples will help to illustrate the meaning of this coordinate system.

Example 2.1 Assume that we wish to identify the location of a point A in a rod of length l (Figure 2.2). One way would be to say that this point is so many inches to the right of end 1 or to the left of end 2. We could make this location specification dimensionless by saying that point A is a certain percentage of the rod length from end 1. A statement of this nature identifies the location independent of units and of the physical dimension of the rod. We call this location variable the *natural coordinate*. Note that this particular example, the rod, has two possible natural coordinates, a fraction of rod length measured from end 1 and also one from end 2. These coordinates are not independent; however, their sum is unity.

From Figure 2.2:

$$X_A = X_1 + S = X_1 + \frac{S}{l}l = X_1 + \frac{S}{l}(X_2 - X_1)$$

$$= X_1\left(1 - \frac{S}{l}\right) + \frac{S}{l}X_2 \tag{2.1}$$

Define two natural coordinates, L_1 and L_2:

$$L_1 = 1 - \frac{S}{l}$$

$$L_2 = \frac{S}{l}$$

Thus we can write equation (2.1) as

$$X_A = L_1 X_1 + L_2 X_2 \tag{2.2}$$

Observe the following:

1. L_1 and L_2 range in value from 0 to 1.
2. L_1 and L_2 are not independent:

$$L_1 + L_2 = 1$$

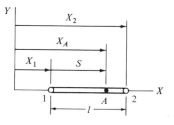

FIGURE 2.2 Coordinates of point A.

3. If point A is at end 1, $L_1 = 1$, $L_2 = 0$. If point A is at end 2, $L_1 = 0$, $L_2 = 1$. ☐

Example 2.2 Determine the natural coordinates of the point P in the rod (Figure 2.3).

Solution. The natural coordinate L_1 will be chosen so that its value varies from zero at end 2 to unity at end 1 (Figure 2.4). This yields the following:

$$L_2 = \tfrac{1}{4} \qquad L_1 = \tfrac{3}{4}$$

Note that L_1 is measured from node 2 and L_2 from node 1.

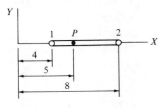

FIGURE 2.3 Global coordinates of P.

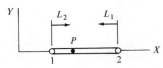

FIGURE 2.4 Natural coordinates for element. ☐

Example 2.3 As another example, consider the triangle shown in Figure 2.5. The total area of the triangle is A. Lines connecting the triangle vertices and the point P form three triangles in the interior.

The natural coordinates can be defined as the ratio of the area of each of these interior triangles to the total area A.

$$L_1 = \frac{A_1}{A}$$
$$L_2 = \frac{A_2}{A}$$
$$L_3 = \frac{A_3}{A}$$

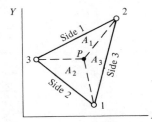

FIGURE 2.5 Area coordinates of P.

2.2 COORDINATE REFERENCE SYSTEMS

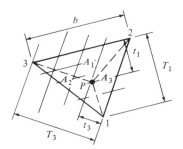

FIGURE 2.6 Triangle dimensions.

Note that the sum of these coordinates equals 1; thus only two of the coordinates are independent. □

It may be more appealing physically to view the triangle natural coordinates as ratios of triangle altitudes rather than areas. Referring to Figure 2.6, we can derive the following from similar triangles:

$$L_1 = \frac{A_1}{A} = \frac{\frac{1}{2}bt_1}{\frac{1}{2}bT_1}$$

$$L_1 = \frac{A_1}{A} = \frac{t_1}{T_1}$$

$$L_2 = \frac{A_2}{A} = \frac{t_2}{T_2}$$

$$L_3 = \frac{A_3}{A} = \frac{t_3}{T_3}$$

Thus we can think of the triangle natural coordinates as a percentage of the distance from a side to the opposite vertex. Remember, only two of the coordinates are independent.

Example 2.4 Determine the natural coordinate values of the vertices and midside points (nodes) of the triangle shown in Figure 2.7.

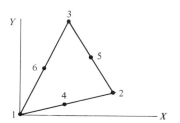

FIGURE 2.7 Points to be located using natural coordinates.

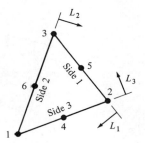

FIGURE 2.8 Identification of sides and natural coordinates of triangle.

Solution. The first step is to identify the sides of the triangle (Figure 2.8). The vertices and midside nodes have been assigned numbers according to a standard convention (this convention will be explained later). The numbering of each side corresponds to the number of the opposite vertex. For example, side 1 is the side opposite vertex 1.

The next step is the establishment of the natural coordinates. L_1 is to be measured from side 1; it is zero for all points on side 1 and unity for vertex (node) 1. Similarly, L_2 is measured from side 2 and L_3 is measured from side 3. Recalling the definition of the natural coordinates (refer to Figure 2.6), we find that the solution is

Point	L_1	L_2	L_3
1	1	0	0
2	0	1	0
3	0	0	1
4	$\frac{1}{2}$	$\frac{1}{2}$	0
5	0	$\frac{1}{2}$	$\frac{1}{2}$
6	$\frac{1}{2}$	0	$\frac{1}{2}$

□

2.3 DEFINITION OF STRAIN

The intention of this section is to set forth the definitions of the strain components corresponding to small displacement theory. Anyone wishing to review the derivations will find the details in any complete elasticity or strength-of-materials text (e.g., Little, 1973; Wang, 1953).

We will consider only small deformations in the continuum. The symbols u, v, and w represent the X, Y, and Z coordinate components of the displacement of a point in the continuum. The longitudinal strain

2.4 STRESS–STRAIN RELATIONS

components in the X, Y, and Z directions are

$$\epsilon_X = \frac{\partial u}{\partial X}$$
$$\epsilon_Y = \frac{\partial v}{\partial Y} \qquad (2.3)$$
$$\epsilon_Z = \frac{\partial w}{\partial Z}$$

The shearing strains are

$$\gamma_{XY} = \frac{\partial u}{\partial Y} + \frac{\partial v}{\partial X}$$
$$\gamma_{YZ} = \frac{\partial v}{\partial Z} + \frac{\partial w}{\partial Y} \qquad (2.4)$$
$$\gamma_{ZX} = \frac{\partial w}{\partial X} + \frac{\partial u}{\partial Z}$$

These equations can be written in matrix form as

$$\begin{Bmatrix} \epsilon_X \\ \epsilon_Y \\ \epsilon_Z \\ \gamma_{XY} \\ \gamma_{YZ} \\ \gamma_{ZX} \end{Bmatrix} = \begin{bmatrix} \frac{\partial}{\partial X} & 0 & 0 \\ 0 & \frac{\partial}{\partial Y} & 0 \\ 0 & 0 & \frac{\partial}{\partial Z} \\ \frac{\partial}{\partial Y} & \frac{\partial}{\partial X} & 0 \\ 0 & \frac{\partial}{\partial Z} & \frac{\partial}{\partial Y} \\ \frac{\partial}{\partial Z} & 0 & \frac{\partial}{\partial X} \end{bmatrix} \begin{Bmatrix} u \\ v \\ w \end{Bmatrix} \qquad (2.5)$$

2.4 STRESS–STRAIN RELATIONS

The equations that relate stress to strain (often referred to as the *constitutive equations*) are also developed in the literature. Again, we will summarize equations for particular conditions of continuum stress and strain for linearly elastic, homogeneous, and isotropic material.

Uniaxial Stress. The simplest stress–strain relationship is for the case of uniaxial stress (Figure 2.9):

$$\sigma_X = E\epsilon_X \qquad (2.6)$$

where E is the modulus of elasticity. Note that in this case the strain in the Y and Z directions is not zero, but rather

$$\epsilon_Y = \epsilon_Z = -\nu\epsilon_X$$

where ν is Poisson's ratio.

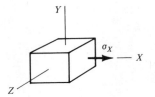

FIGURE 2.9 Tensile stress on X face in X direction.

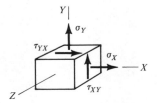

FIGURE 2.10 Normal and shear stresses on the X and Y faces of a differential solid.

Plane Stress. Refer to Figure 2.10. For this case the normal and shear stress components act in two coordinate directions only. Note that in general the longitudinal strain is nonzero in all coordinate directions and

$$\sigma_Z = 0 = \tau_{ZX} = \tau_{ZY}$$

The relations between stress and strain are

$$\sigma_X = \frac{E}{1-\nu^2}(\epsilon_X + \nu\epsilon_Y)$$

$$\sigma_Y = \frac{E}{1-\nu^2}(\epsilon_Y + \nu\epsilon_X)$$

$$\tau_{XY} = G\gamma_{XY}$$

These constitutive equations written in matrix form are

$$\begin{Bmatrix} \sigma_X \\ \sigma_Y \\ \tau_{XY} \end{Bmatrix} = \frac{E}{1-\nu^2} \begin{bmatrix} 1 & \nu & 0 \\ \nu & 1 & 0 \\ 0 & 0 & \frac{1-\nu}{2} \end{bmatrix} \begin{Bmatrix} \epsilon_X \\ \epsilon_Y \\ \gamma_{XY} \end{Bmatrix} \quad (2.7a)$$

or

$$\{\sigma\} = [C]_\sigma \{\epsilon\} \quad (2.7b)$$

where $\{\sigma\}$ and $\{\epsilon\}$ are the stress and strain vectors and $[c]_\sigma$ is the constitutive matrix for plane stress.

Plane Strain. Referring to Figure 2.11, we note that the stress in the Z direction is not zero but is the value necessary to establish the linear strain in the Z direction to be zero. Also,

$$\gamma_{ZX} = \gamma_{ZY} = 0 = \epsilon_Z$$

2.5 WORK AND THE FORCE ENERGY POTENTIAL FUNCTION

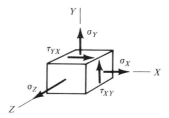

FIGURE 2.11 General stress at a point for plane strain condition.

The stress–strain equations are

$$\sigma_X = \frac{E}{(1+\nu)(1-2\nu)}\left[(1-\nu)\epsilon_X + \nu\epsilon_Y\right]$$

$$\sigma_Y = \frac{E}{(1+\nu)(1-2\nu)}\left[\nu\epsilon_X + (1-\nu)\epsilon_Y\right]$$

$$\sigma_Z = \frac{\nu E}{(1+\nu)(1-2\nu)}(\epsilon_X + \epsilon_Y) = \nu(\sigma_X + \sigma_Y)$$

$$\tau_{XY} = G\gamma_{xy}$$

Writing the XY-plane relationships in matrix form, we have

$$\begin{Bmatrix} \sigma_X \\ \sigma_Y \\ \tau_{XY} \end{Bmatrix} = \frac{E}{(1+\nu)(1-2\nu)} \begin{bmatrix} 1-\nu & \nu & 0 \\ \nu & 1-\nu & 0 \\ 0 & 0 & \frac{1-2\nu}{2} \end{bmatrix} \begin{Bmatrix} \epsilon_X \\ \epsilon_Y \\ \gamma_{XY} \end{Bmatrix}$$

(2.8a)

or

$$\{\sigma\} = [C]_\epsilon \{\epsilon\} \qquad (2.8b)$$

where $[C]_\epsilon$ is the constitutive matrix for plane strain.

We postpone consideration of the general three-dimensional constitutive relationships until later chapters. For the present it will be sufficient to concentrate on plane analysis. This will allow an examination of the finite element method without introducing unnecessary manipulative complexity.

2.5 WORK AND THE FORCE ENERGY POTENTIAL FUNCTION

Work and potential energy are introduced and discussed in detail in elementary mechanics texts. We review here the definitions and discuss some concepts that generally pose difficulty.

The work done by a force, $\overline{F}$, undergoing a change in position, $\overline{R}$ (Figure 2.12), is defined as

$$W = \int_{\overline{R}_1}^{\overline{R}_2} \overline{F} \cdot d\overline{R}$$

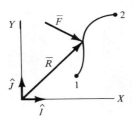

FIGURE 2.12 Position and path of a moving force, $\bar{F}$.

This scalar value depends in general on the path of the integration as well as the initial and final positions. Expanding $\bar{F}$ and $d\bar{R}$ into their rectangular coordinate components and looking at the expression for the differential work yields

$$\bar{F} = F_X\hat{I} + F_Y\hat{J} \qquad d\bar{R} = dX\hat{I} + dY\hat{J}$$
$$dW = F_X\,dX + F_Y\,dY$$

It is important to emphasize that dW cannot be treated as an exact differential. The integration is performed along a particular path, and the result depends in general on the path as well as the endpoints of the integration. However, it is also important to recognize that there are some forces which create an integral that depends on the endpoints but is independent of the path. These forces are called *conservative forces*. In these cases the differential work is an exact differential, and the work of the force can be evaluated as the difference of a scalar function of position between the path endpoints.

Let us create a function ϕ, which we shall call a *work function*. This function will exist only if we are dealing with a conservative force. This function will depend on the spatial coordinates, and the scalar difference of the function evaluated at two different coordinate positions will be equal to the work done by the conservative force. Thus the work W_c, done by a conservative force, $\bar{F}_c$, can be evaluated in two ways: as a line integration and/or as the simple subtraction of two scalar magnitudes, ϕ_1 and ϕ_2.

$$W_c = \int_{R_1}^{R_2} \bar{F}_c \cdot d\bar{R} = \phi_2 - \phi_1 \qquad (2.9)$$

Evaluating work by subtracting the values of ϕ_2 and ϕ_1 is very appealing, certainly much easier than evaluating the integral. It is for this reason that the function is derived. Since ϕ is a function of position only, it is an exact differential.

$$d\phi = \frac{\partial \phi}{\partial X}\,dX + \frac{\partial \phi}{\partial Y}\,dY = \bar{\nabla}\phi \cdot d\bar{R} \qquad (2.10)$$

where

$$\bar{\nabla} = \hat{I}\frac{\partial}{\partial X} + \hat{J}\frac{\partial}{\partial Y}$$

is the vector differential operator, del.

2.5 WORK AND THE FORCE ENERGY POTENTIAL FUNCTION

Equating the differential change $d\phi$ to the differential work, we have

$$d\phi = dW_c$$

or

$$\frac{\partial \phi}{\partial X} dX + \frac{\partial \phi}{\partial Y} dY = F_{cX} dX + F_{cY} dY$$

Because the differential displacements dX and dY are arbitrary, the force components must be related to the work function ϕ in the following manner:

$$\frac{\partial \phi}{\partial X} = F_{cX} \qquad \frac{\partial \phi}{\partial Y} = F_{cY} \qquad (2.11)$$

Thus $\overline{F}_c$ is related to ϕ as follows:

$$\overline{F}_c = \overline{\nabla} \phi \qquad (2.12)$$

Given the force components, equations (2.11) are used to determine the work function.

Example 2.5 Given the constant force $\overline{F}_c = 10\hat{I}$, solve for the scalar work function ϕ.

Solution. From equations (2.11),

$$F_{cX} = \frac{\partial \phi}{\partial X} = 10 \quad \text{and} \quad F_{cY} = \frac{\partial \phi}{\partial Y} = 0$$

Integrating, we obtain $\phi = 10X + g(Y)$, from F_{cX}. From F_{cY}, $g(Y)$ must be a constant. Thus

$$\phi = 10X + C$$

If ϕ is arbitrarily set equal to zero at the initial position, X_i, then $C = -10X_i$ and

$$\phi = 10(X - X_i)$$

or

$$\phi = 10u$$

where u is the displacement of $\overline{F}_c$ in the X direction. □

Example 2.6 Solve for the work function, ϕ, of the force $\overline{F}_c = 10X\hat{I}$.

Solution. Follow the procedure used in Example 2.5:

$$F_{cX} = \frac{\partial \phi}{\partial X} = 10X \quad \text{and} \quad F_{cY} = \frac{\partial \phi}{\partial Y} = 0$$

$\phi = 5X^2 + g(Y)$, from F_{cX}. From F_{cY},

$$g(Y) = C$$

Thus

$$\phi = 5X^2 + C \qquad \square$$

For the examples given above, we can demonstrate that the integral evaluation of the work for each force is equal to the difference of the derived scalar function ϕ at the endpoints of the path of the force.

For the constant force of Example 2.5, evaluate the work integral if the force moves in the X direction from the position $X = 5$ to the position $X = 20$.

$$W = \int_{\bar{R}_1}^{\bar{R}_2} \bar{F}_c \cdot d\bar{R} = \int_{X_1}^{X_2} F_{cX} \, dX = \int_5^{20} 10 \, dX \qquad W = 150$$

For the same force, evaluate the corresponding change in the work function ϕ.

$$\phi_2 - \phi_1 = 10 X_2 + C - (10 X_1 + C) = 10(X_2 - X_1)$$
$$= 10(20 - 5)$$
$$\phi_2 - \phi_1 = 150$$

which is the same as the calculated W.

We do not find the work function ϕ used in problem solving. Rather, the use of energy principles requires the establishment of a different scalar function, V, which defines the potential (capacity) of the conservative force to do work. This new function is called the *potential energy function*, and we shall show that the potential energy function, V, is the negative of the work function, ϕ.

Note the following definition:

Definition. *The work done by a conservative force plus its change in potential for doing work equals zero.*

Thus we can write

$$W + (V_2 - V_1) = 0$$

or

$$(\phi_2 - \phi_1) + (V_2 - V_1) = 0$$

Considering differential changes yields

$$d\phi + dV = 0$$
$$d(\phi + V) = 0$$
$$\phi + V = C$$

Letting $C = 0$ gives us

$$\phi = -V$$

Thus we have the relation between the work function and the force energy potential:

$$V = -\phi$$

Recalling the relationship between the conservative force and the work

2.5 WORK AND THE FORCE ENERGY POTENTIAL FUNCTION

function, we have

$$\bar{F}_c = \bar{\nabla}\phi$$

We can now express the relationship between the conservative force and its potential energy function:

$$\bar{F}_c = -\bar{\nabla}V$$

You must remember that for the potential energy function to exist, the work of the force must be path independent. From the vector identity

$$\bar{\nabla} \times \bar{\nabla}B(X,Y) = 0$$

where $B(X,Y)$ is any scalar function of position, a necessary but not sufficient condition for the existence of the potential energy function V is

$$\bar{\nabla} \times \bar{F}_c = 0$$

Example 2.7 Solve for the potential energy function of the constant force $\bar{F} = 3\hat{i} + 7\hat{j}$.

Solution. We note first that $\bar{\nabla} \times \bar{F} = 0$. Second, we use the relation $\bar{F} = -\bar{\nabla}V$ to obtain

$$\frac{\partial V}{\partial X} = -3 \quad \text{and} \quad \frac{\partial V}{\partial Y} = -7$$

Integrating the first expression gives us

$$V = -3X + g(Y)$$

Substitution into the second expression yields

$$\frac{\partial g(Y)}{\partial Y} = -7$$

from which

$$g(Y) = -7Y + C$$

Thus the potential function becomes

$$V = -3X + -7Y + C$$

where C is an arbitrary constant. □

The function V, the force energy potential, defines the capacity of a force to do work. It is expressed in terms of the X and Y coordinates of the force. The arbitrary constant, C, is determined by establishing the magnitude of V at some particular coordinates X_i, Y_i of the force. For our particular situation we will encounter the force $\bar{F}$ applied to a deformable continuum. In this case, we will establish the force energy potential to be zero when X_i and Y_i are the force position coordinates of the undeformed continuum. For example,

$$V = 0 = -3X_i - 7Y_i + C \quad C = 3X_i + 7Y_i$$

and
$$V = -3(X - X_i) - 7(Y - Y_i)$$
where the quantities $(X - X_i)$ and $(Y - Y_i)$ represent the displacement of the force in the X and Y directions, respectively, as the continuum deforms. If we let
$$X - X_i = u \quad \text{and} \quad Y - Y_i = v$$
we obtain
$$V = -3u - 7v$$
as the potential energy of the force in terms of its displacement.

You should appreciate at this point that *the potential energy expression of any force that has a magnitude independent of the displacement of the force is simply the negative of the scalar dot product of the force and the displacement vector.*
$$V = -F_X u - F_Y v \tag{2.13}$$

Example 2.8 A uniformly distributed load is applied to a deformable bar on the X axis (Figure 2.13). Solve for the potential energy function of the distributed load.

Solution. We assume that the bar deformation could be such that points in the bar move with differing displacements. We concentrate our attention first on a differential force $d\overline{F}$ acting at an arbitrary position along the axis.

The potential energy of the force $d\overline{F}$ changes as the force moves from an initial position X_i to a general position X as shown in the figure. The first task is to develop an expression for the energy potential of the constant differential force $d\overline{F}$.

The potential function of a differential force would have a differential magnitude, dV. Thus, relating $d\overline{F}$ and dV, we have
$$d\overline{F} = -\overline{\nabla}(dV)$$

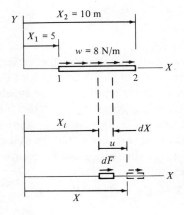

FIGURE 2.13 Distributed loading in X direction on continuum deformable in the X direction.

2.5 WORK AND THE FORCE ENERGY POTENTIAL FUNCTION

For our one-dimensional situation,

$$dF = -\frac{d(dV)}{dX}$$

Finding the indefinite integral gives us

$$dV = -dF(X) + C$$

Letting the differential energy potential equal zero when the force is in the undeformed position X_i, we have

$$dV = -dF(X - X_i)$$

where $X - X_i$ is the displacement, u, of the force along the X axis. Thus we have

$$dV = -dF\,u$$

We now proceed to imagine writing the expression for dV for all differential forces along the bar. Note that each force will move with its own displacement, and we assumed initially that the displacements of points in the bar were dependent on their location in the bar. Let us assume, as an example, that the bar displacement relationship is linear. We then have

$$u = C_0 + C_1 X$$

The differential potential energy expression then becomes

$$dV = -dF(C_0 + C_1 X)$$

We now want to integrate this expression over the length of the rod. This is accomplished by expressing the force dF at the arbitrary position on the rod as

$$dF = w\,dX = 8\,dX$$

(Note that if the distributed loading were not constant, we would simply substitute the X dependence in this expression.) Thus we have

$$dV = -8(C_0 + C_1 X)\,dX$$

$$V = -8\left(C_0 X + C_1 \frac{X^2}{2}\right)\Big|_5^{10}$$

$$= -8(5C_0 + 37.5C_1)$$

$$= -40C_0 - 300C_1 \quad \text{joules}$$

The end result depends on the values of the constants C_0 and C_1. For example, if $C_1 = 0$ and $C_0 = u$, then all points in the rod have the same displacement and

$$V = -40u \quad \text{joules}$$

Note that this is the same as the value that we would obtain using the resultant of the distributed load acting at a point that undergoes a displacement u. □

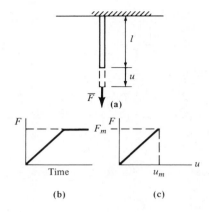

FIGURE 2.14 Force–deflection relationship for tensile bar.

2.6 STRAIN ENERGY

Consider the tensile bar shown in Figure 2.14. Force $\bar{F}$ is applied to the free end, gradually increasing in magnitude in the manner defined by the force–time diagram. As the force slowly increases, the bar elongates and finally the end attains a displacement u_m when the force is F_m. If the deflection of the end of the bar varies linearly with the applied force magnitude as indicated in Figure 2.14(c), the work done by the force is the area under the curve:

$$W = \int \bar{F} \cdot d\bar{R} = \int_0^{u_m} F_m \frac{u}{u_m} \, du = \frac{1}{2} F_m u_m$$

Application of the conservation of mechanical energy to this bar system indicates that the internal energy of the system (bar) must increase an amount equal to the work of the force. A gradual reduction to zero of the force on the rod would result in a return of a perfectly elastic bar to its original undeformed state; the work of the force for this assumed reversible process is

$$W = \int_{u_m}^{0} F_m \frac{u}{u_m} \, du = -\frac{1}{2} F_m u_m$$

and the energy of the system (bar) returns to the value (usually taken to be zero) at the original undeformed state. This returnable energy that is stored in a deformed elastic body is called *strain energy*.

For a general development of the expression for the strain energy in a continuum, the reader is referred once again to textbooks on elasticity or strength of materials (e.g., Little, 1973; Wang, 1953). We include here only a summary of the strain energy equations.

2.6 STRAIN ENERGY

The strain energy stored in a differential element of volume $d\Omega$ of a continuum subject to a uniaxial stress state is

$$dU = \tfrac{1}{2}\sigma_X \epsilon_X \, d\Omega$$

For a biaxial stress state,

$$dU = \tfrac{1}{2}(\sigma_X \epsilon_X + \sigma_Y \epsilon_Y) \, d\Omega$$

For the case of pure shear,

$$dU = \tfrac{1}{2}\tau_{XY}\gamma_{XY} \, d\Omega$$

Note: γ_{XY} is the shear strain. It is twice the magnitude of the off-diagonal elements of the strain tensor.

The strain energy stored in the element $d\Omega$ under a general three-dimensional stress state is

$$dU = \tfrac{1}{2}(\sigma_X \epsilon_X + \sigma_Y \epsilon_Y + \sigma_Z \epsilon_Z + \tau_{XY}\gamma_{XY} + \tau_{YZ}\gamma_{YZ} + \tau_{ZX}\gamma_{ZX}) \, d\Omega$$

This equation can be put into matrix form:

$$dU = \tfrac{1}{2}\{\epsilon\}^T\{\sigma\} \, d\Omega \tag{2.14}$$

where

$$\{\epsilon\} = \begin{Bmatrix} \epsilon_X \\ \epsilon_Y \\ \epsilon_Z \\ \gamma_{XY} \\ \gamma_{YZ} \\ \gamma_{ZX} \end{Bmatrix} \qquad \{\sigma\} = \begin{Bmatrix} \sigma_X \\ \sigma_Y \\ \sigma_Z \\ \tau_{XY} \\ \tau_{YZ} \\ \tau_{ZX} \end{Bmatrix}$$

For our initial study of the equilibrium problem in elasticity, we will be concerned with plane stress and/or plane strain. For these cases we involve only three components of the stress and strain vectors. However, the symbolic expression for the strain energy remains the same:

$$U = \int_\Omega \tfrac{1}{2}\{\epsilon\}^T\{\sigma\} \, d\Omega \tag{2.15}$$

where Ω is the region of the integration and

$$\{\epsilon\} = \begin{Bmatrix} \epsilon_X \\ \epsilon_Y \\ \gamma_{XY} \end{Bmatrix} \qquad \{\sigma\} = \begin{Bmatrix} \sigma_X \\ \sigma_Y \\ \tau_{XY} \end{Bmatrix}$$

Equation (2.15) involves both the stress and strain components. This is not a convenient form for our purposes. We will want to express the strain energy in terms of strain alone. To accomplish this, simply substitute into equation (2.15) the appropriate constitutive equation from Section 2.4. These constitutive equations can be written in symbolic form.

For equation (2.7b),
$$\{\sigma\} = [C]_\sigma \{\epsilon\}$$
For equation (2.8b),
$$\{\sigma\} = [C]_\epsilon \{\epsilon\}$$
Substitution of either of these equations into equation (2.15) gives us the energy formula in terms of strain alone.

For plane stress:
$$U = \int_\Omega \tfrac{1}{2}\{\epsilon\}^T [C]_\sigma \{\epsilon\}\, d\Omega \qquad (2.16)$$

For the case of uniaxial stress in the X direction, substitution of the strain relationship $\epsilon_Y = -\nu \epsilon_X$ in expression (2.16) results in the following:
$$U = \int_\Omega \frac{E}{2} \epsilon_X^2 \, d\Omega \qquad (2.17a)$$

For the simplest case of uniform cross section and strain dependent on the X coordinate alone, we obtain the strain energy expression:
$$U = \frac{EA}{2} \int_X \epsilon_X^2 \, dX \qquad (2.17b)$$

For plane strain:
$$U = \int_\Omega \tfrac{1}{2}\{\epsilon\}^T [C]_\epsilon \{\epsilon\}\, d\Omega \qquad (2.18)$$

Example 2.9 The deflection of the neutral axis of a beam (Figure 2.15) is assumed to be
$$v = \frac{v_e}{l^2} X^2$$
Solve for the strain energy in the beam, assuming plane uniaxial stress and small beam deflection theory.

Solution. Strain energy,
$$U = \int_\Omega \tfrac{1}{2}\{\epsilon\}^T [C]_\sigma \{\epsilon\}\, d\Omega$$

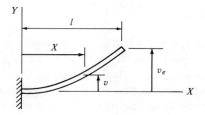

FIGURE 2.15 Deflection (exaggerated) of cantilever beam of length l.

From beam deflection theory, the bending moment at any section of the beam is obtained from the second derivative of the neutral-axis displacement:

$$M = EI\frac{d^2v}{dX^2} = EI\frac{d^2}{dX^2}\left(\frac{v_e}{l^2}X^2\right)$$

$$= 2EI\frac{v_e}{l^2}$$

The stress in the beam is obtained from the bending moment as

$$\sigma_X = \frac{My}{I}$$

where I is the area moment of inertia about the neutral axis and y is the distance from the neutral axis to any point in the cross section. The strains corresponding to the uniaxial flexure stress are obtained as

$$\epsilon_X = \frac{\sigma_X}{E} = \frac{My}{EI} = 2EI\frac{v_e}{l^2}\frac{y}{EI} \qquad \epsilon_X = \frac{2v_e}{l^2}y$$

$$\epsilon_Y = -\nu\epsilon_X = -\nu\frac{2v_e}{l^2}y$$

If the shear strain is considered to be negligible, the strain energy for this uniaxial flexure stress state can be calculated using expression (2.17a) or the general expression (2.16). We will use (2.16) to illustrate its use.

$$U = \int_\Omega \frac{1}{2}\{\epsilon\}^T[C]_\sigma\{\epsilon\}\,d\Omega$$

$$= \int_\Omega \frac{1}{2}\left(\frac{2v_e}{l^2}y\right)^2 [1\ -\nu\ 0]\begin{bmatrix}1 & \nu & 0\\ \nu & 1 & 0\\ 0 & 0 & \frac{1-\nu}{2}\end{bmatrix}\begin{Bmatrix}1\\ -\nu\\ 0\end{Bmatrix}\frac{E}{1-\nu^2}\,d\Omega$$

$$= \iint_{l\,A}\frac{2Ev_e^2}{l^4}y^2\,dA\,dX = \frac{2Ev_e^2}{l^3}\int_A y^2\,dA \qquad \text{where } \int_A y^2\,dA = I$$

$$= 2EI\frac{v_e^2}{l^3} \qquad \square$$

2.7 PRINCIPLE OF MINIMUM POTENTIAL ENERGY

Recall that in Section 2.5 we showed how to derive a potential energy function for a conservative force system. In Section 2.6 we presented equations for determining the strain energy magnitude in a deformed linearly elastic, homogeneous isotropic continuum. (This strain energy is often referred to as potential energy. It is preferable, however, to distinguish this energy from the force potential energy.) We now set forth the principle of minimum potential energy, which serves as the basis of the development of the finite element method of stress analysis.

Definition. *The* principle of minimum potential energy *states that of all the geometrically possible configurations that a body can assume, the true one, corresponding to the satisfaction of stable equilibrium, is identified by a minimum value for the total potential energy (strain energy plus force energy potential).*

Assigning the symbol Π to the total potential energy, we have

$$\Pi = U + V \qquad (2.19)$$

The true configuration of the deformed elastic continuum yields a minimum value of Π. This is the basis of the development of approximation techniques for the equilibrium elasticity problem.

2.8 CONCEPTS INVOLVED IN APPLICATION OF THE PRINCIPLE OF MINIMUM POTENTIAL ENERGY

Given a continuum physically constrained against rigid body motion and subjected to known loads, our problem becomes one of determining the displacements throughout the continuum. The steps in the solution involve the following:

1. The manner of displacement throughout the continuum is assumed by creating a continuous displacement function containing n arbitrary constants, the only requirement being that the function satisfy the geometric boundary conditions. Selection of this displacement function is a crucial step, because the resulting displacements can be correct only if the displacement function is the true one. Often, a solution is accepted based on a "reasonable," but not perfect choice of displacement function. A simple polynomial is a commonly assumed trial function.
2. The assumed displacement function is used in equations (2.3) and (2.4) to develop relationships for strain, and the resulting expressions are substituted into the appropriate strain energy equation (2.16) or (2.18).
3. Using the assumed displacement function and equation (2.13), potential energy expressions are derived for all the (constant) applied loads. These expressions will contain the n arbitrary constants in the assumed displacement function. As mentioned previously, the force

energy potential is established to be zero when the continuum is undeformed; this permits the potential function to be in terms of the displacement instead of the position coordinates.

4. The energy terms are summed. This results in an equation involving the n unknown arbitrary constants of the assumed displacement function.
5. Apply the principle of minimum potential energy. Differentiate the energy equations with respect to each of the n unknown arbitrary constants, and set each of the resulting equations to zero to establish the minimum energy state. This yields n simultaneous linear equations, which can be solved for the n unknown constants.

2.9 DISCUSSION OF THE PRINCIPLE OF MINIMUM POTENTIAL ENERGY

An axially loaded tensile member will be used as an example in discussing the application of the minimum potential energy principle. Referring to Figure 2.16, when force F is applied to the bar, it increases in magnitude from zero to F_m while the end of the bar displaces from the undeformed position an amount u_E. In this example we will assume that the force F_m is known and that the displacement throughout the bar is to be determined.

Following the procedure outlined in Section 2.8, the first step is the assumption of a continuous displacement function that satisfies the boundary conditions of the problem. For this very simple problem we can feel comfortable with assuming a linear trial function for the displacement:

$$u(X) = C_0 + C_1 X$$

To satisfy the zero displacement requirement at $X = 0$, the constant C_0 must be zero; thus we have

$$u(X) = C_1 X$$

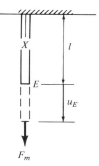

FIGURE 2.16 Elongation of tensile bar.

We now have a family of continuous functions, and the problem then centers on finding the correct value of the constant C_1.

It is important to understand the philosophy behind the search for the correct value of the arbitrary constant in the trial function. The process in effect amounts to causing a *variation of the trial function* while maintaining a constant value of the applied load. The correct value of the constant is that value which causes the total energy to be a minimum.

Proceeding to the second step of our procedure, we obtain the X-direction longitudinal strain in the region (bar) by differentiating the displacement function:

$$\epsilon_X = \frac{du}{dX} = C_1$$

and the strain energy for this uniaxial stress case is found using expression (2.17):

$$U = \frac{EA}{2}\int_X \epsilon_X^2 \, dX = \frac{EA}{2}\int_X C_1^2 \, dX = \frac{EA}{2}C_1^2 \int_0^l dX$$

$$= \frac{EAC_1^2}{2} l$$

Step 3 of our procedure requires the establishment of the expressions for the force energy potential. From (2.13) we obtain

$$V = -F_m u(l) = -F_m C_1 l$$

In step 4 we sum the strain energy and the force potential energy. The total energy becomes

$$\Pi = \frac{EA}{2} C_1^2 l - F_m C_1 l$$

It is instructive to look at the variation of this energy total as the trial function constant C_1 is varied. A plot of the total energy, Π, versus the trial function constant, C_1, would appear as shown in Figure 2.17. Note that the energy begins at zero, decreases, reaches a minimum, and then increases. The principle of minimum potential energy states that the stable equilibrium state coincides with the minimum value of the total energy; this defines the correct value of the constant.

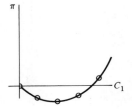

FIGURE 2.17 Plot of total energy, Π, versus trial function constant, C.

The final step involves the differentiation of the energy with respect to, in this case, the single unknown constant and setting the equation equal to zero:

$$\frac{d\Pi}{dC_1} = 0 = EAC_1 - F_m$$

Solving for C_1, we have

$$C_1 = \frac{F_m}{EA}$$

Substituting this value of C_1 into the trial displacement function, we have

$$u(X) = \frac{F_m X}{EA}$$

which we recognize as the correct solution for the tensile bar.

This simple example incorporates the basic principles with a very minimum of complexity. You must realize that this problem and the beam problem of Example 2.9 are single-degree-of-freedom problems; only one constant is required to establish the trial function for the displacement throughout the continuum. The solution of real problems is not as easy. Complex geometry and boundary conditions make the task of creating one good displacement function for the entire region quite impossible.

In the next sections we look at the justification of the minimum potential energy principle and put forth the rationale for discretizing the body and approximating the solution throughout the region with not one, but rather, many piecewise continuous functions.

2.10 INTRODUCTION TO THE CALCULUS OF VARIATIONS AND THE VARIATIONAL FORMULATION

In Section 2.7 the principle of minimum potential energy was stated without providing any mathematical justification for the validity of the minimization of the sum of strain energy and force potential energy. In this section we hope to show, after a brief discussion of variational calculus, how a problem description involving the equilibrium equations and boundary conditions can be cast into the alternative "energy" form, the minimization of which reproduces the equilibrium equations. The discussion will be limited to the one-dimensional region.

Given a definite integral that has an integrand involving a single independent variable, x, a function $u(x)$, and possibly a derivative of

$u(x)$ with respect to x:

$$\Pi = \int_{x_1}^{x_2} f(x, u(x), u'(x))\, dx \qquad (2.20)$$

where $u'(x)$ is du/dx, the basic problem in variational calculus is to determine the function $u(x)$ that causes the functional, Π, to be either a maximum or a minimum (a stationary value). The $u(x)$ function will be required to satisfy certain boundary conditions. We will assume initially that $u(x_1)$ and $u(x_2)$ are prescribed.

To visualize how Π might vary with variations of the function $u(x)$, we will imagine replacing $u(x)$ and $u'(x)$ with the functions $w(x)$ and $w'(x)$, which are defined as follows:

$$\begin{aligned} w(x) &= u(x) + \epsilon\eta(x) \\ w'(x) &= u'(x) + \epsilon\eta'(x) \end{aligned} \qquad (2.21)$$

where $\eta(x)$ may be any continuously differentiable function of x which, for defined end points on $u(x)$, is equal to zero at the limits of the integral. The "constant," ϵ, is a parameter that controls the magnitude of the difference between the functions $u(x)$ and $w(x)$ and it can be varied in a continuous manner. The relationship between the functions $u(x)$ and $w(x)$ can be visualized as sketched in Figure 2.18.

Important: We define $u(x)$ to be the function that minimizes the functional Π.

We now wish to consider the variation of the functional Π with changes in the parameter ϵ.

$$\Pi(\epsilon) = \int_{x_1}^{x_2} F(x, w(\epsilon), w'(\epsilon))\, dx \qquad (2.22)$$

If we imagine to vary the parameter ϵ and evaluate $\Pi(\epsilon)$, a curve of

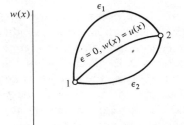

FIGURE 2.18 Dependence of function $w(x)$ on parameter.

2.10 THE CALCULUS OF VARIATIONS AND THE VARIATIONAL FORMULATION

$\Pi(\epsilon)$ versus ϵ would, by definition of $u(x)$, attain an extreme value when $w(x) = u(x)$, or in other words, when $\epsilon = 0$.

$$\frac{d\Pi}{d\epsilon}(\epsilon) = \frac{d}{d\epsilon}\int_{x_1}^{x_2} F(x, w, w')\, dx = \int_{x_1}^{x_2} \frac{d}{d\epsilon} F(x, w, w')\, dx \quad (2.23)$$

Performing the required differentiation, we obtain the following:

$$\frac{dF}{d\epsilon} = \frac{\partial F}{\partial w}\left(\frac{\partial w}{\partial \epsilon}\right) + \frac{\partial F}{\partial w'}\left(\frac{\partial w'}{\partial \epsilon}\right) = \frac{\partial F}{\partial w}\eta + \frac{\partial F}{\partial w'}\eta'$$

At $\epsilon = 0$, we have

$$\left.\frac{\partial F}{\partial w}\right|_{\epsilon=0} = \frac{\partial f}{\partial u} \quad \text{etc.}$$

so that

$$\left.\frac{dF}{d\epsilon}\right|_{\epsilon=0} = \frac{\partial f}{\partial u}\eta + \frac{\partial f}{\partial u'}\eta' \quad (2.24)$$

Substituting (2.24) back into (2.23), we have the stationary value of the functional defined as follows:

$$\left.\frac{d\Pi}{d\epsilon}\right|_{\epsilon=0} = 0 = \int_{x_1}^{x_2} \left(\frac{\partial f}{\partial u}\eta + \frac{\partial f}{\partial u'}\eta'\right) dx$$

Integrating the second term by parts yields

$$\delta 0 = \int_{x_1}^{x_2}\left[\frac{\partial f}{\partial u} - \frac{d}{dx}\left(\frac{\partial f}{\partial u'}\right)\right]\eta\, dx + \left.\frac{\partial f}{\partial u'}\eta\right|_{x_1}^{x_2} \quad (2.25)$$

For the stated requirement that $\eta(x_1) = 0 = \eta(x_2)$, we have

$$\int_{x_1}^{x_2}\left[\frac{\partial f}{\partial u} - \frac{d}{dx}\left(\frac{\partial f}{\partial u'}\right)\right]\eta\, dx = 0 \quad (2.26)$$

For arbitrary $\eta(x)$, the definite integral of the functional Π attains an extreme value only if the integrand of the definite integral satisfies the following differential equation:

$$\frac{\partial f}{\partial u} - \frac{d}{dx}\left(\frac{\partial f}{\partial u'}\right) = 0 \quad (2.27)$$

Equation (2.27) is called the *Euler equation* of the variational formulation defined by (2.20).

An extension of this problem can be made if we return to equation (2.25) and let the $u(x)$ remain undefined at end 2. This in turn requires $\eta(x)$ to be zero only at the specified "forced" boundary value at end 1.

Again, the boundary term in equation (2.25) must be zero for arbitrary $\eta(x)$, and this can only happen if

$$\left.\frac{\partial f}{\partial u'}\right|_{x=x_2} = 0 \tag{2.28}$$

Thus we see that in addition to satisfying the Euler equation, the integrand function, f, must also satisfy the constraint defined by (2.28).

Example 2.10 The length of line between two points A and B (Figure 2.19) is defined by

$$S = \int_s \sqrt{dx^2 + dy^2}$$

$$= \int \sqrt{1 + \left(\frac{dy}{dx}\right)^2}\, dx$$

$$= \int_{x_A}^{x_B} \sqrt{1 + (y')^2}\, dx$$

Solve for the function $y(x)$, which minimizes the length of the line between the two points A and B.

Solution. The integrand $f(x, y, y')$ of the functional S is $\sqrt{1 + (y')^2}$. For the definite integral to attain a minimum value, this integrand function must satisfy the Euler equation (2.27). Performing the differentiation, we obtain

$$\frac{\partial f}{\partial y} = 0, \qquad \frac{\partial f}{\partial y'} = \frac{y'}{\sqrt{1 + (y')^2}}$$

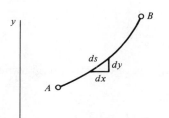

FIGURE 2.19 Arbitrary line between two points.

2.10 THE CALCULUS OF VARIATIONS AND THE VARIATIONAL FORMULATION

Substitution into (2.27) yields

$$\frac{-d}{dx}\left(\frac{y'}{\sqrt{1+(y')^2}}\right) = 0$$

Solving this equation gives us

$$\frac{y'}{\sqrt{1+(y')^2}} = \text{constant}$$

or

$$y = c_1 x + c_2$$

which is the equation of a straight line. □

Return to the equation (2.20) and let us define a variation of the definite integral by the symbol δ:

$$\delta \Pi = \delta \int_{x_1}^{x_2} f(x, u(x), u'(x)) \, dx \qquad (2.29)$$

The variation of the functional Π occurs as a result of a variation of the function $u(x)$, as discussed previously. It is important to distinguish between the differential of $u(x)$ as the change of $u(x)$ corresponding to a differential change dx along the curve defined by $u(x)$ and the variation of $u(x)$ as the change in this function as we move from one curve to another at a constant x. The variation of $u(x)$ is to be denoted by δu and to be equal to the difference between the two curves $w(x)$ and $u(x)$. From equation (2.21), this variation is

$$\delta u = \epsilon \eta(x) \qquad (2.30)$$

In a manner analogous to differentiation, variation and integration are commutative, and the variation of sums, products, and so on, obeys the laws of differentiation. Also, it may be shown that the variation of the differential is equal to the differential of the variation (Arpici, 1966):

$$\delta(du) = d(\delta u) \qquad (2.31)$$

Returning to (2.29), we can write

$$\delta \Pi = \int_{x_1}^{x_2} \delta f(x, u(x), u'(x)) \, dx$$

The variation of the integrand can be expanded in a Taylor's series with x assumed to be constant. Dropping the higher-order terms, we have

$$\delta f = \frac{\partial f}{\partial u} \delta u + \frac{\partial f}{\partial u'} \delta u' \qquad (2.32)$$

Thus we have

$$\delta\Pi = \int_{x_1}^{x_2} \left(\frac{\partial f}{\partial u} \delta u + \frac{\partial f}{\partial u'} \delta u' \right) dx = \int_{x_1}^{x_2} \left[\frac{\partial f}{\partial u} \delta u + \frac{\partial f}{\partial u'} \frac{d}{dx}(\delta u) \right] dx$$

Integrating the second term by parts yields

$$\delta\Pi = \int_{x_1}^{x_2} \left[\frac{\partial f}{\partial u} - \frac{d}{dx}\left(\frac{\partial f}{\partial u'} \right) \right] \delta u\, dx + \left. \frac{\partial f}{\partial u'} \delta u \right|_{x_1}^{x_2} \qquad (2.33)$$

Comparison of (2.33) with (2.25) shows that they are proportional, with the arbitrary variation $\delta u(x)$ related to $\eta(x)$. In addition, we can state that the vanishing of the variation of the functional,

$$\delta\Pi = 0 \qquad (2.34)$$

is a necessary condition for the existence of the extreme value of the functional.

In summary, we can say the following. Given the functional, or the variational formulation of a problem, the corresponding differential equation formulation of the problem may be obtained directly through application of the Euler equation, or it may be obtained by the indirect process of variation of the functional.

For the most part, we are generally interested in the reverse process. For physical problems it is easy to derive the differential equation. The corresponding variational formulation, or functional, for most physical problems is not readily apparent or easily determined. For some problems, the attempt to determine the functional has not been successful. Very simply, the basic idea behind the determination of the functional is the following.

Recalling equations (2.33) and (2.34), you should appreciate that the Euler equation in the variational form of (2.33) yields the differential equation of the problem. If we have the differential equation of the problem, why not substitute that into (2.33) in the place of the Euler equation and operate on the integrand to bring the variational operator outside the integral sign? For the case of fixed end conditions and a differential equation represented by

$$Lu(x) = p(x)$$

where L is a differential operator, the starting point for the determination of the functional would be

$$\delta\Pi = 0 = \int_{x_1}^{x_2} [Lu(x) - p(x)] \delta u\, dx \qquad (2.35)$$

The functional, Π, for this problem will be determined when the variational operator is brought from its location inside the integral in (2.35) to the outside. In most cases, this is not easily done.

In Section 2.14 we look at some examples of this formal determination of the functional. In the next section, however, we will show that the functional for equilibrium elasticity problems is the sum of the strain energy and the force potential energy.

2.11 EQUILIBRIUM ELASTICITY VARIATIONAL FORMULATION: STRAIN AND FORCE POTENTIAL ENERGIES

After having discussed a formal procedure for the development of the variational form corresponding to the differential equation of any given physical problem, we now throw somewhat of a curve by describing how the functional for equilibrium elasticity problems has been derived by an alternative procedure.

The *principle of virtual work* is used as the basis for the construction of the functional, Π, for equilibrium elasticity problems. The details of the development of the functional will not be included here. The interested reader is referred to Wang (1953) or Gallagher (1975).

Virtual work is defined as the work done by a force undergoing a virtual displacement. A *virtual displacement* can be thought of as the variation of the displacement function. It is an artificial displacement constrained only by the boundary conditions of the problem. It is assumed that the variations can be small enough so that changes in the force resulting from the virtual displacements create higher-order terms that can be neglected.

The application of the principle of virtual work proceeds along the lines of evaluating the change in the strain energy in a body in equilibrium resulting from the virtual work of the applied loads undergoing the virtual displacements. The result is that the change of strain energy equals the virtual work:

$$\delta U = \delta W \tag{2.36}$$

If the potential energy, V, of the applied loads is zero at the undeformed condition of the body, we can let $W = -V$ and (2.36) can be written as

$$\delta(U + V) = 0 \tag{2.37}$$

Comparing (2.37) to (2.34), we see that the functional for the elasticity problem is the sum of the strain energy and the applied load potential energies. Given this functional, it can be shown (as a test of its correctness) that the variation of the functional yields the elasticity equilibrium equations. (Alternatively, the equilibrium equations may be obtained by direct substitution of the integrand of the functional into the Euler equation.)

2.12 APPROXIMATE SOLUTIONS: THE RAYLEIGH–RITZ METHOD

If the variational form of a problem exists (i.e., we have the functional, Π, for the problem), an approximate solution for the dependent variable can be made by creating a solution function that minimizes the functional. The "creation" of the solution function involves selecting a function that:

1. Satisfies the boundary conditions.
2. Contains arbitrary constants that can be "adjusted."

For example, if we assume a trial solution function

$$\tilde{u}(x) = \sum_{i=1}^{n} c_i \phi_i(x)$$

and substitute this into the functional of (2.20), we will have

$$\Pi_a = \int_{x_1}^{x_2} f\left(x, \sum_{i=1}^{n} c_i \phi_i(x), \sum_{i=1}^{n} c_i \phi_i'(x)\right) dx$$

where Π_a is the approximated value of the functional Π. The functional now involves the n arbitrary constants, c_i. These constants are determined by finding the extreme of the functional with respect to each constant according to

$$\frac{\partial \Pi}{\partial c_1} = 0 \quad \frac{\partial \Pi}{\partial c_2} = 0 \quad \text{etc.} \quad (2.38)$$

You must appreciate that the resulting solution $\tilde{u}(x)$ is, by definition, an approximate solution. Whereas the exact minimization of the functional yields a solution that satisfies the differential equation of the problem throughout the region, this approximate solution will only approximately satisfy the equations.

Example 2.11 Given a rod of uniform cross-sectional area A, length L, and elastic modulus E with forces of equal magnitude F at the midpoint and end (Figure 2.20), solve for the displacement function, $u(x)$, using the Rayleigh–Ritz approximation.

Solution. For uniaxial strain, the strain energy in the rod is

$$U = \frac{EA}{2} \int_X \epsilon_x^2 \, dx = \frac{EA}{2} \int_0^L \left(\frac{du}{dx}\right)^2 dx$$

2.12 APPROXIMATE SOLUTIONS: THE RAYLEIGH–RITZ METHOD

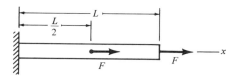

FIGURE 2.20 Axially loaded rod of uniform section.

The potential energy of the applied loads is

$$V = -Fu\left(\frac{L}{2}\right) - Fu(L)$$

The functional for this problem is $U + V$:

$$\Pi = U + V = \frac{EA}{2}\int_0^L \left(\frac{du}{dx}\right)^2 dx - Fu\left(\frac{L}{2}\right) - Fu(L)$$

We now have to assume a trial function for $u(x)$. As a simple example, let us assume the function

$$\tilde{u}(x) = c_0 + c_1 x + c_2 x^2$$

For those who understand the physics of this problem, it is apparent that there is a discontinuity in the strain at $L/2$. This cannot be modeled well with this function, but we will proceed and see what happens.

The boundary condition has to be satisfied: $u(0) = 0$ yields $c_0 = 0$. Thus we have the approximate function as

$$\tilde{u}(x) = c_1 x + c_2 x^2$$

Also, the derivative becomes

$$\tilde{u}'(x) = c_1 + 2c_2 x$$

Substituting into the functional and solving for the constants using (2.38), we obtain

$$c_1 = \frac{9F}{4EA} \qquad c_2 = \frac{-3F}{4EA}$$

and the solution function is

$$\tilde{u}(x) = \frac{9F}{4EA} x - \frac{3F}{4EAL} x^2$$

The corresponding function for strain is obtained by differentiating:

$$\frac{d\tilde{u}}{dx} = \frac{9F}{4EA} - \frac{6F}{4LEA} x$$

Figure 2.21 compares the approximate solution with the exact solution. We observe a remarkable agreement between the exact and approximate solutions for the displacement. However, as we expected, the strain does not agree well with this theoretical problem. To obtain a better model of the strain throughout the body with one function will require many more terms in the assumed polynomial trial function. This is one of the difficulties with attempting an approximate solution function for the entire region, and it is the prime motivating factor for the concept of discretization of the region.

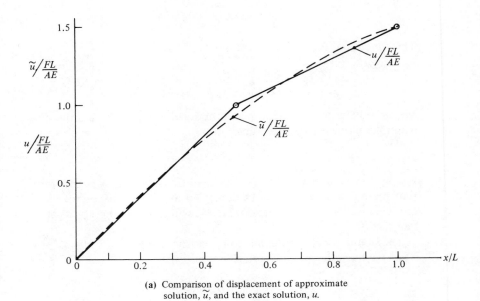

(a) Comparison of displacement of approximate solution, $\tilde{u}$, and the exact solution, u.

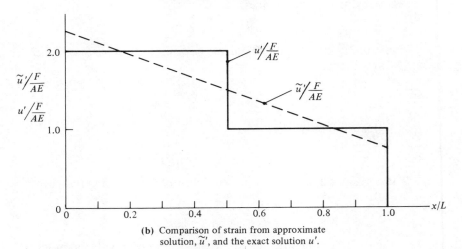

(b) Comparison of strain from approximate solution, $\tilde{u}'$, and the exact solution u'.

FIGURE 2.21

2.12 APPROXIMATE SOLUTIONS: THE RAYLEIGH–RITZ METHOD

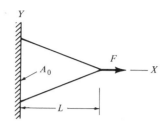

FIGURE 2.22 Tapered bar with end load.

Example 2.12 A tapered tensile (Figure 2.22) bar has a cross-sectional area defined by

$$A = A_0 \frac{L - X}{L}$$

The bar is loaded in tension with a force at its tip ($X = L$). Assuming that the displacement in the X direction can be approximated by the function (which satisfies fixed boundary constraints)

$$u = C_1 X + C_2 X^2$$

determine an expression for the deflection of the end of the bar in terms of the applied force, material constants, and geometry.

Solution

1. Determine the strain energy in the rod. Differentiate the displacement expression to obtain ϵ_x and, in this case of uniaxial stress, use (2.17) for the strain energy.

$$U = \frac{E}{2} \int_{\text{VOL}} \epsilon_x^2 \, d\Omega = \frac{E}{2} \int_0^L \epsilon_x^2 A \, dX$$

$$= \frac{EA_0}{2} \int_0^L \frac{L - X}{L} (C_1 + 2C_2 x)^2 \, dX$$

$$= \frac{EA_0}{12L} \left(3C_1^2 L^2 + 4C_1 C_2 L^3 + 2C_2^2 L^4 \right)$$

2. Determine the energy potential of the applied force. From (2.13),

$$V = -Fu(L) = -F\left(C_1 L + C_2 L^2\right)$$

3. Sum the energies to obtain the total energy:

$$\Pi = \frac{EA_0}{12L} \left(3C_1^2 L^2 + 4C_1 C_2 L^3 + 2C_2^2 L^4 \right) - FC_1 L - FC_2 L^2$$

4. Minimize energy with respect to the arbitrary constants C_1 and C_2:

$$\frac{\partial \Pi}{\partial C_1} = 0 = \frac{EA_0}{12L}(6C_1L^2 + 4C_2L^3) - FL$$

$$\frac{\partial \Pi}{\partial C_2} = 0 = \frac{EA_0}{12L}(4C_1L^3 + 4C_2L^4) - FL^2$$

5. Solving the two equations for C_1 and C_2 yields

$$C_1 = 0 \qquad C_2 = \frac{3F}{EA_0L}$$

and the solution becomes

$$u_a = \frac{3FX^2}{EA_0L}$$

where the subscript a denotes it as an approximate solution.

The theoretical solution to this problem can be obtained from the differential equation

$$\frac{d}{dX}\left(AE\frac{du}{dX}\right) = 0$$

with the boundary conditions

$$u(0) = 0 \quad \text{and} \quad \left[AE\frac{du}{dX}\right]_{x=0} = F$$

yielding

$$u = \frac{FL}{A_0E}\ln\frac{L}{L-X}$$

The theoretical and approximate solutions are compared in Figure 2.23. Considering the simplicity of the assumed solution function, we have a remarkably good estimate of the displacement except near the region of the physically impossible condition of a force applied to a tip of zero area.

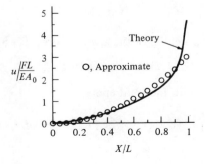

FIGURE 2.23 Comparison of theoretical and approximate solutions for tapered bar. □

2.13 THE FINITE ELEMENT MODEL CONCEPT

With the simple example in Section 2.12, we were able to show that application of the Rayleigh–Ritz procedure in the minimization of a functional is conceptually rather simple, but for solutions of real problems with odd-shaped regions and all kinds of loadings, the chance of finding a solution function that would satisfy the differential equations throughout the entire region is rather slim.

As an alternative to modeling the entire region with one very complex trial function, the idea is presented of discretizing the region into "finite elements," using a simpler trial function for each element. For example, if we were to reconsider the problem in Example 2.11 we could perhaps divide the region into two (or more) elements. If we then assumed a linear trial function for each element, we feel confident that we would be able to model the known linearity of the displacement throughout each section and also the known constancy of the strain. However, we would not, at this point, predict the accuracy of the result.

In the next and subsequent chapters, we will look at the development of some standard commonly used elements and show how they are applied to obtain the approximate solution. The first element will be the one-dimensional spatial element (also called the *bar element* or *uniaxial element*).

The following example illustrates the technique of assuming multiple trial functions for a region.

Example 2.13 Consider the structure shown in Figure 2.24. The stepped shaft is placed between two rigid supports and a force is applied at the juncture of the two sections. Assume that the displacement in the X direction in section 1 can be approximated by the linear function

$$u_1 = C_1 + C_2 X$$

and the displacement in section 2 can be approximated by a second

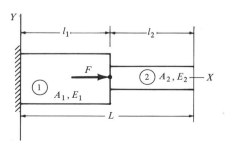

FIGURE 2.24 Stepped shaft with axial force and rigid end constraints.

linear function:
$$u_2 = C_3 + C_4 X$$
Solve for the four constants.

Solution. 1. First, we must satisfy the boundary conditions. They are

(a) $u_1(0) = 0 = C_1$

(b) $u_1(l_1) = u_2(l_1)$
$$C_2 l_1 = C_3 + C_4 l_1$$

(c) $u_2(L) = 0 = C_3 + C_4 L$

Including these boundary conditions in the equations, we have two linear equations for the region, each equation involving the same unknown constant.

$$u_1 = C_2 X$$
$$u_2 = C_2 l_1 \frac{L - X}{L - l_1}$$

From these displacement relations, we differentiate to obtain the strain ϵ_X in each section.

$$\epsilon_{X_1} = C_2$$
$$\epsilon_{X_2} = -C_2 \frac{l_1}{l_2}$$

2. The strain energy in the shaft is the sum of the strain energies in each section.

$$U = U_1 + U_2 = \int \frac{EA_1}{2} \epsilon_{X_1}^2 \, dX + \int \frac{EA_2}{2} \epsilon_{X_2}^2 \, dX$$

$$= \frac{E_1 A_1}{2} \int_0^{l_1} C_2^2 \, dX + \frac{E_2 A_2}{2} \int_{l_1}^{l_2} C_2^2 \left(\frac{l_1}{l_2}\right)^2 dX$$

$$= \frac{E_1 A_1}{2} l_1 C_2^2 + \frac{E_2 A_2}{2} \frac{l_1^2}{l_2} C_2^2$$

3. The force energy potential function
$$V = -Fu(l_1) = -FC_2 l_1$$

4. The total energy
$$\Pi = \frac{E_1 A_1}{2} l_1 C_2^2 + \frac{E_2 A_2}{2} \frac{l_1^2}{l_2} C_2^2 - FC_2 l_1$$

5. Minimize the energy with respect to the unknown constant C_2.

$$\frac{d\Pi}{dC_2} = 0 = E_1 A_1 l_1 + E_2 A_2 \frac{l_1^2}{l_2} C_2 - F l_1$$

$$C_2 = \frac{F l_2}{E_1 A_1 l_2 + E_2 A_2 l_1}$$

To test this result, substitute the data from Example 1.1.

$$C_2 = \frac{-2000(16)(10)^{-6}}{10(2)(16) - 30(1)(12)}$$

$$= -47.1(10)^{-6}$$

Substituting into u_1 and solving at $X = l_1$ yields

$$u_1(l_1) = -47.1(10^{-6})(12) = -564(10^{-6}) \text{ in.}$$

which checks. □

2.14 DETERMINING THE FUNCTIONAL: SOME EXAMPLES

In Section 2.10 we suggested the manner in which we might proceed to determine the functional of a problem by starting with the differential equation and inverting the process of the variation of the functional. This, it was pointed out, can be a difficult process and may prove to be impossible for some cases.

To illustrate the procedure, we examine four different problem types. Beginning with the differential equation of each problem, we proceed to develop the corresponding functional form of each problem.

Uniaxial Stress. Consider a bar of variable cross-sectional area $A(x)$ in equilibrium under the action of the forces at each end and the distributed load $T(x)$ (force per unit length; see Figure 2.25). The differential equation arising from the equilibrium of the differential element is

$$\frac{d}{dx} A\sigma_x + T(x) = 0$$

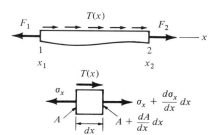

FIGURE 2.25 Loading on bar and stresses on differential element.

For the uniaxial stress condition, the stress σ_x is related to the strain as follows:

$$\sigma_x = E\epsilon_x = E\frac{du}{dx}$$

where E is the modulus of elasticity of the bar material. Substitution yields a differential equation in terms of the unknown displacement function $u(x)$.

$$\frac{d}{dx}AE\frac{du}{dx} + T(x) = 0 \qquad (2.39)$$

If we now substitute (2.39) into (2.35), we will have the starting point of the derivation of the functional Π corresponding to the differential equation of this problem.

$$\delta\Pi = 0 = \int_{x_1}^{x_2}\left[\frac{d}{dx}AE\frac{du}{dx} + T(x)\right]\delta u\,dx$$

$$= \int_{x_1}^{x_2}\left(\frac{d}{dx}AE\frac{du}{dx}\right)\delta u\,dx + \int_{x_1}^{x_2}T(x)\delta u\,dx$$

If we integrate the first term by parts, we obtain

$$\delta\Pi = AE\frac{du}{dx}\delta u\bigg|_{x_1}^{x_2} - \int_{x_1}^{x_2}AE\frac{du}{dx}\frac{d}{dx}\delta u\,dx + \int_{x_1}^{x_2}T(x)\delta u\,dx$$

Interchanging the order of differentiation and variation in the second term yields

$$\delta\Pi = AE\frac{du}{dx}\delta u\bigg|_{x_1}^{x_2} - \int_{x_1}^{x_2}AE\frac{du}{dx}\delta\left(\frac{du}{dx}\right)dx$$

$$+ \int_{x_1}^{x_2}T(x)\delta u\,dx \qquad (2.40)$$

Now let us examine each term. The first term is the boundary term arriving out of the integration by parts of the region integral. This term contains the variation of the displacement at the boundary, $\delta u(x_i)$, together with the first derivative of the displacement function. From the discussion in Section 2.10 we recall that in the process of determining a stationary value for a functional, variations of the unknown function were required to be zero at the boundary when the function values were specified at the boundary. If, for this problem, displacements are specified at the boundaries x_1 and x_2, the variation δu, and thus the boundary term, will be zero. If the displacements are not specified at the boundary, the boundary terms may remain in the functional.

Before further consideration of the quantities in the boundary term of this problem, we should discuss the general classification of these boundary value quantities. In the general case, after integration by parts, the resulting boundary term contains the variation of the dependent function and may also contain variations of derivatives of the dependent function: for example,

$$\delta u, \quad \delta \frac{du}{dx}, \quad \delta\left(\frac{d^2 u}{dx^2}\right), \quad \ldots$$

In addition to these quantities, which contain the variation operator, the boundary term also contains derivatives of the dependent function without the variation operator. All the quantities with the variational operator are referred to as *primary, essential,* or *Dirichlet boundary conditions*. Perhaps "essential" is the most descriptive title because a unique solution requires specification of these boundary quantities on some portion of the boundary. All other functions of the dependent variable, those that do not involve the variation operator, are called *secondary, natural,* or *Neumann boundary conditions*. Perhaps "natural" is a good title because the quantities arise as a natural consequence of the integration process.

Returning to the boundary term for this problem, we find that the displacement function contains the variation operator and thus the displacement is the essential boundary condition for this problem. For a unique solution the displacement must be specified at one of the two boundaries as a minimum. The coefficient of the δu_i quantity is

$$\left. AE \frac{du}{dx} \right|_{x_i}$$

The derivative du/dx is classified as the natural boundary condition since it does not contain the variation operator. The derivative du/dx is recognized as the strain, and the product of strain and AE yields a force. Therefore, this boundary quantity can be interpreted as a force acting on the end of the region (bar). Note that a positive du/dx will correspond to a force at end 1 in the negative coordinate direction and a force at end 2 in the positive coordinate direction. If we establish a convention of positive forces acting in the positive coordinate direction, the boundary term of (2.40) can be written as

$$F\delta u \big|_{x_1}^{x_2} = F_2 \delta u_2 + F_1 \delta u_1$$

where F_i is the force at the boundary and is defined by

$$F_1 = -AE \frac{du}{dx}\bigg|_{x=x_1} \qquad F_2 = AE \frac{du}{dx}\bigg|_{x=x_2} \qquad (2.41)$$

The value of this boundary term depends, of course, on the boundary values. As stated previously, if the boundary displacements are specified, δu, and thus the boundary term, are zero. Note that the natural boundary value, the support force, is not known when the essential boundary value is specified. Alternatively, if the boundary is free, that is, the boundary force is zero, then, again, the corresponding boundary term is zero. Note in this case that if the force, the natural boundary condition, is specified, the boundary displacement, the essential boundary condition, is unknown.

One final comment on this boundary term. The variation is on the primary variable, the displacement in this case. Since the boundary force is constant, the variation operator may be placed on the product of the force and displacement:

$$F_2 \delta u_2 + F_1 \delta u_1 = \delta [F_2 u_2 + F_1 u_1]$$

Returning to (2.40), we note the following. First, in the second term,

$$\frac{du}{dx} \delta \left(\frac{du}{dx} \right) = \frac{1}{2} \delta \left(\frac{du}{dx} \right)^2$$

and for fixed limits of integration, the variational operator can be taken outside the integral sign. Second, for the third term, the distributed loading, $T(x)$, is constant with a variation of the displacement function, so that the operator can be taken outside this integral as well. After multiplying each term by -1 (so that each term can be interpreted as energy), we can write the equation (2.40) as

$$\delta \Pi = \delta \left[-F_2 u_2 - F_1 u_1 + \int_{x_1}^{x_2} \frac{AE}{2} \left(\frac{du}{dx} \right)^2 dx - \int_{x_1}^{x_2} T(x) u(x) \, dx \right]$$

and the functional of this problem is

$$\Pi = -F_2 u_2 - F_1 u_1 + \int_{x_1}^{x_2} \frac{AE}{2} \left(\frac{du}{dx} \right)^2 dx - \int_{x_1}^{x_2} T(x) u(x) \, dx \quad (2.42)$$

The functional Π is thus defined in (2.42). We recognize the first two terms as the potential energies of the applied end loads, the third term as the strain energy, and the fourth term as the potential energy of the distributed loading.

It might prove to be a beneficial exercise to show that (2.42) is the variational form of the differential equation (2.39) by formally obtaining the variation of (2.42). We proceed in the manner described in the introduction to variational calculus in Section 2.10.

The variation of the displacement function is represented by $\epsilon \eta(x)$, where ϵ is an arbitrary constant small enough to allow us to discard the higher-order terms in a Taylor's series, and $\eta(x)$ is a continuous function with possible constraints at the region boundaries. Substitution of this displacement variation into (2.42) yields

$$\Pi + \delta \Pi = -F_2(u_2 + \epsilon \eta_2) - F_1(u_1 + \epsilon \eta_1) + \int_{x_1}^{x_2} \frac{AE}{2} \left(\frac{du}{dx} + \epsilon \frac{d\eta}{dx} \right)^2 dx$$
$$- \int_{x_1}^{x_2} T(x) [u(x) + \epsilon \eta(x)] \, dx$$

2.14 DETERMINING THE FUNCTIONAL: SOME EXAMPLES

Expanding and keeping only the terms linear in ϵ gives us

$$\delta\Pi = -F_2\epsilon\eta_2 - F_1\epsilon\eta_1 + \int_{x_1}^{x_2} AE\epsilon \frac{du}{dx}\left(\frac{d\eta}{dx}\right) dx$$
$$- \int_{x_1}^{x_2} T(x)\epsilon\eta(x)\, dx$$

Integrating the third term by parts yields

$$\delta\Pi = -F_2\epsilon\eta_2 - F_1\epsilon\eta_1 + AE\frac{du}{dx}\epsilon\eta\bigg|_{x_1}^{x_2} - \epsilon\int_{x_1}^{x_2} \frac{d}{dx} AE\frac{du}{dx}\eta(x)\, dx$$
$$-\epsilon\int_{x_1}^{x_2} T(x)\eta(x)\, dx$$

For any nonzero function $\eta(x)$ with arbitrary values η_1 and η_2 at the boundaries, the following relationships must hold for $\delta\Pi$ to be zero. First, we have the relations that define the magnitude of the first derivative of the displacement function at the boundaries in terms of the applied forces at the boundary

$$-F_2 + AE\frac{du}{dx}\bigg|_{x_2} = 0$$

$$-F_1 - AE\frac{du}{dx}\bigg|_{x_1} = 0$$

Second, we have the equation that we recognize as the differential equation of the problem:

$$\frac{d}{dx} AE\frac{du}{dx}\, dx + T(x)\, dx = 0$$

We have come full circle in this discussion. We started with the differential equation, derived the corresponding functional, from which we retrieved the differential equation and the required boundary relationships. In Chapter 3 we describe the use of the functional form to establish the finite element method solution procedures.

Steady-State Uniaxial Heat Flow. Consider a region of variable cross-sectional area $A(x)$ with heat flow Q (energy/time) at the ends and a source heat flux, $H(x)$ (energy/time–length), distributed along the x direction (Figure 2.26).

Application of the energy balance to the differential element with the longitudinal heat flux $q(x)$, (energy/time–cross-sectional area), and the applied heat flux, $H(x)$, yields the following equation:

$$\frac{d}{dx} Aq - H(x) = 0$$

From Fourier's law we are able to express the constitutive equation

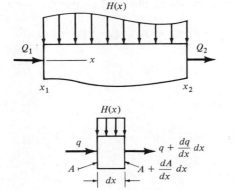

FIGURE 2.26 Distributed heat source, $H(x)$; total heat flow, Q_1 and Q_2, at boundaries; heat flow, q, per unit area at boundaries of the differential element.

relating the longitudinal heat flux to the temperature gradient:

$$q = -k\frac{dT}{dx}$$

where k is the thermal conductivity (energy/time–length–temperature) and T is the temperature. Substitution of Fourier's law into the differential equation yields

$$\frac{d}{dx}Ak\frac{dT}{dx} + H(x) = 0 \qquad (2.43)$$

Comparing (2.43) with (2.39), we observe that they are identical in form.

To obtain the equivalent variational form of this differential equation, we proceed as before by substituting the differential equation into (2.35).

$$\delta \Pi = 0 = \int_{x_1}^{x_2} \left[\frac{d}{dx} AK \frac{dT}{dx} + H(x) \right] \delta T \, dx$$

$$= \int_{x_1}^{x_2} \left(\frac{d}{dx} Ak \frac{dT}{dx} \right) \delta T \, dx + \int_{x_1}^{x_2} H(x) \delta T \, dx$$

Integrating the first term by parts (and multiplying all terms by -1), we obtain

$$\delta \Pi = -\left. Ak\frac{dT}{dx}\delta T \right|_{x_1}^{x_2} + \int_{x_1}^{x_2} Ak \frac{dT}{dx} \frac{d}{dx}\delta T \, dx - \int_{x_1}^{x_2} H(x) \delta T \, dx$$

(2.44a)

The first term is the boundary term. From our previous discussion we should recognize that the temperature, T, is the essential boundary condition and that δT must be zero at the boundary where T is specified. We should note also that the temperature gradient, dT/dx, is the natural boundary value. From Fourier's law, the temperature gradient times the thermal conductivity and the cross-sectional area yields heat flow. There-

2.14 DETERMINING THE FUNCTIONAL: SOME EXAMPLES

fore, this boundary quantity can be interpreted as the heat flowing into or out of the ends of the region. Insulated ends, for example, would make the boundary terms zero. For the case when the boundaries (ends) are not insulated and the temperatures are not specified, we include the boundary terms. Equation (2.44a) can now be written as follows:

$$\delta \Pi = \delta(Q_2 T_2 - Q_1 T_1) + \int_{x_1}^{x_2} Ak \frac{dT}{dx} \frac{d}{dx} \delta T \, dx - \int_{x_1}^{x_2} H(x) \delta T \, dx \tag{2.44b}$$

where the variation of Q_i, the heat flow at end i, is zero. The sign convention for positive heat flow is positive in the positive coordinate direction. The heat flow at the ends of the region are thus defined in terms of the temperature gradient as follows:

$$Q_1 = -Ak \frac{dT}{dx}\bigg|_{x=x_1} \quad \text{and} \quad Q_2 = -Ak \frac{dT}{dx}\bigg|_{x=x_2} \tag{2.45}$$

It is possible to cast the boundary terms in a form that will enable one to model convection at the ends of the region. From Newton's law of cooling, we can express the convective heat flow at boundaries 1 and 2 as

$$Q_1 = Q_{1c} = hA(T_\infty - T_1) \quad \text{and} \quad Q_2 = Q_{2c} = hA(T_2 - T_\infty) \tag{2.46}$$

where the constant h is the convection heat transfer coefficient (energy/time–area–temperature) and T_∞ is the ambient temperature of the medium contacting the boundary of the region. The use of either or both equations (2.46) for the boundary terms will model convective heat flow at the end(s) of the region.

Perhaps we should include the possibility that the distributed heat flux, $H(x)$, could also be attributed to convection and modeled with Newton's law of cooling. The energy added, by convection, to the region of length dx would be

$$H(x) \, dx = h(P \, dx)(T_\infty - T(x))$$
$$H(x) = hP(T_\infty - T(x)) \tag{2.47}$$

where for a given value of x, P is the perimeter of the cross section, T_∞ is the ambient temperature, and $T(x)$ is the temperature in the region. It must be emphasized that we are modeling a one-dimensional heat flow. The temperature is dependent on the x coordinate only; it cannot vary in directions normal to the x axis at any particular cross section.

Returning to (2.44b), we can proceed to bring the variational operators outside the integrals (procedure identical to stress problem), and finally write the functional for the one-dimensional heat transfer problem as

$$\Pi = Q_2 T_2 - Q_1 T_1 + \int_{x_1}^{x_2} \frac{Ak}{2} \left(\frac{dT}{dx}\right)^2 dx - \int_{x_1}^{x_2} H(x) T(x) \, dx \tag{2.48}$$

where (2.47) can be substituted for $H(x)$ and (2.46) can be used in either of the two boundary terms. Application examples in Chapter 3 will illustrate solutions with a variety of combinations of boundary conditions.

Steady-State Flow Through Permeable Media. A permeable substance of variable cross-sectional area $A(x)$ has fluid flow V (volume/time) across the ends of the region and a fluid flux $F(x)$ (volume/time–length) distributed along the length of the region (Figure 2.27). Applying the conservation of mass to the differential element of the medium, we obtain the following equation:

$$\frac{d}{dx}Av - F(x) = 0$$

where v is the velocity and $F(x)$ is the distributed flux. If we assume that the fluid in the material has a velocity proportional to the pressure gradient, we can write the constitutive relationship

$$v = -g\frac{dp}{dx}$$

where g is the coefficient of permeability. Substituting this relationship into the differential equation, we obtain

$$\frac{d}{dx}Ag\frac{dp}{dx} + F(x) = 0 \qquad (2.49)$$

Compare (2.49) with (2.39) and (2.43); they are identical in form. The equivalent variational form of the differential equation is obtained in exactly the manner employed for the previous two cases. The resulting expressions would follow:

$$\delta\Pi = -\left.Ag\frac{dp}{dx}\delta p\right|_{x_1}^{x_2} + \int_{x_1}^{x_2} Ag\frac{dp}{dx}\delta\left(\frac{dp}{dx}\right)dx - \int_{x_1}^{x_2} F(x)\delta p\, dx$$

where we recognize the pressure, p, to be the essential boundary condition and the pressure gradient, dp/dx, the natural boundary condition.

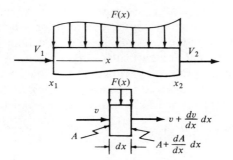

FIGURE 2.27 Distributed flow source, $F(x)$; total flow rate, V_i, at the boundaries; flow velocity, v, at the differential element boundaries.

2.14 DETERMINING THE FUNCTIONAL: SOME EXAMPLES

The functional becomes

$$\Pi = V_2 p_2 - V_1 p_1 + \int_{x_1}^{x_2} \frac{Ag}{2} \left(\frac{dp}{dx}\right)^2 dx - \int_{x_1}^{x_2} F(x) p(x) \, dx \qquad (2.50)$$

where the boundary terms have been rewritten in terms of the volume flow rate and pressure at the boundaries. Employing the sign convention of positive flow acting in the positive coordinate direction, we can express the boundary flow rate values as follows:

$$V_1 = -\left. Ag \frac{dp}{dx} \right|_{x=x_1} \quad \text{and} \quad V_2 = -\left. Ag \frac{dp}{dx} \right|_{x=x_2} \qquad (2.51)$$

We observe once again that the boundary terms will be zero if the boundary flow rates are zero or if the boundary pressure is specified.

Bending of a Beam. We will consider a beam undergoing small transverse displacements as a result of the distributed loading, $w(x)$ (force/length; see Figure 2.28). From the equilibrium of the differential beam element and the relationship between the (small) curvature and bending couple, we have the following three equations:

$$\begin{aligned} w \, dx + dV &= 0 \\ V \, dx + dM &= 0 \\ EI \frac{d^2 v}{dx^2} &= M \end{aligned} \qquad (2.52)$$

Combining these three equations, we have the governing fourth-order differential equation:

$$\frac{d^2}{dx^2} EI \frac{d^2 v}{dx^2} - w(x) = 0 \qquad (2.53)$$

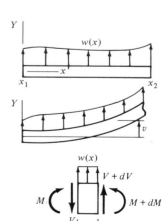

FIGURE 2.28 Beam deflection with distributed loading $w(x)$.

Once again we venture in search of the variational form by substituting this differential equation into (2.35).

$$\delta\Pi = 0 = \int_{x_1}^{x_2}\left[\frac{d^2}{dx^2}EI\frac{d^2v}{dx^2} - w\right]\delta v\,dx$$

$$= \int_{x_1}^{x_2}\left(\frac{d^2}{dx^2}EI\frac{d^2v}{dx^2}\right)\delta v\,dx - \int_{x_1}^{x_2}w\,\delta v\,dx$$

Integrating the first term by parts gives us

$$\delta\Pi = \frac{d}{dx}EI\frac{d^2v}{dx^2}\delta v\bigg|_{x_1}^{x_2} - \int_{x_1}^{x_2}\left(\frac{d}{dx}EI\frac{d^2v}{dx^2}\right)\delta\left(\frac{dv}{dx}\right)dx - \int_{x_1}^{x_2}w\,\delta v\,dx$$

Integrating the second term by parts gives us

$$\delta\Pi = \frac{d}{dx}EI\frac{d^2v}{dx^2}\delta v\bigg|_{x_1}^{x_2} - EI\frac{d^2v}{dx^2}\delta\left(\frac{dv}{dx}\right)\bigg|_{x_1}^{x_2}$$

$$+ \int_{x_1}^{x_2}EI\frac{d^2v}{dx^2}\delta\left(\frac{d^2v}{dx^2}\right)dx - \int_{x_1}^{x_2}w\,\delta v\,dx$$

or

$$\delta\Pi = -\left[\frac{d}{dx}EI\frac{dv}{dx}\delta v + EI\frac{dv}{dx}\delta\left(\frac{dv}{dx}\right)\right]_{x_1}^{x_2}$$

$$+ \delta\int_{x_1}^{x_2}\frac{EI}{2}\left(\frac{d^2v}{dx^2}\right)^2 dx - \delta\int_{x_1}^{x_2}wv\,dx \qquad (2.54)$$

Examining the boundary terms, we note that the variational operator acting on the displacement and derivative of displacement (the slope of the deflection curve) defines these two functions as the essential boundary conditions. The coefficients of the essential boundary conditions are the natural boundary conditions and they bear the following physical interpretations:

$$\frac{d}{dx}EI\frac{d^2v}{dx^2}\bigg|_{x_i} = \text{boundary shear force } V_i$$

$$EI\frac{d^2v}{dx^2}\bigg|_{x_i} = \text{boundary bending couple } M_i$$

If we substitute the boundary force symbols into (2.54), the boundary terms can be interpreted as the potential energies of the boundary shear forces and bending couples. The third term in (2.54) is the strain energy

2.14 DETERMINING THE FUNCTIONAL: SOME EXAMPLES 101

in the beam, the fourth term is the energy potential of the distributed beam loading. The functional can thus be written as follows:

$$\Pi = -\left(V_2 v_2 + V_1 v_1 + M\frac{dv}{dx}\bigg|_{x_2} + M\frac{dv}{dx}\bigg|_{x_1}\right)$$

$$+ \int_{x_1}^{x_2} \frac{EI}{2}\left(\frac{d^2v}{dx^2}\right)^2 dx - \int_{x_1}^{x_2} wv\, dx \quad (2.55)$$

The Two-Dimensional Laplace Equation. Many physical problems are governed by the familiar Laplace equation

$$\nabla^2 \phi = 0 \quad (2.56)$$

where the two-dimensional Laplace operator is

$$\nabla^2 = \frac{\partial^2}{\partial x^2} + \frac{\partial^2}{\partial y^2}$$

Substituting into (2.35), we have

$$\delta\Pi = 0 = \int_A \nabla^2\phi\, \delta\phi\, dA \quad (2.57)$$

Employing the vector identity, we have

$$\nabla^2\phi\, \delta\phi = \overline{\nabla} \cdot \delta\phi\, \overline{\nabla}\phi - \overline{\nabla}\delta\phi \cdot \overline{\nabla}\phi$$

We can substitute this into (2.52) and obtain the following:

$$\delta\Pi = 0 = \int_A \overline{\nabla} \cdot \delta\phi\, \overline{\nabla}\phi\, dA - \int_A \overline{\nabla}\delta\phi \cdot \overline{\nabla}\phi\, dA$$

Applying Gauss's theorem to the first term yields

$$\delta\Pi = 0 = \int_S \delta\phi\frac{\partial\phi}{\partial n}\, ds - \delta\int_A \frac{\overline{\nabla}\phi \cdot \overline{\nabla}\phi}{2}\, dA$$

where $\partial\phi/\partial n$ is the derivative normal to the boundary, S, and is also recognized to be the natural boundary condition of this problem. The essential boundary condition is the value of the dependent variable, ϕ, on the boundary.

We should recall, once again, that if only essential boundary conditions are specified, whether homogeneous or not, the functional of the problem becomes

$$\Pi = \int_A \frac{\overline{\nabla}\phi \cdot \overline{\nabla}\phi}{2}\, dA = \int_A \left[\left(\frac{\partial\phi}{\partial x}\right)^2 + \left(\frac{\partial\phi}{\partial y}\right)^2\right] dA \quad (2.58)$$

2.15 ALTERNATIVES TO THE VARIATIONAL CALCULUS APPROACH

For those of you who are in the position of having this reading be your first exposure to variational calculus, you are probably a bit troubled about upcoming topics in this book. Do not give up yet. The next few chapters deal with element formulation for the familiar elasticity problem using the energy minimization principle. In Chapter 9 we take a look at alternative methods of formulating the basic element equations. As an alternative to the variational calculus, we will employ techniques called weighted residual methods. These methods are very logical and practical and provide a procedure for developing the basic equations even when a variational functional cannot be determined.

PROBLEMS

2.1 Given the following expressions defining the displacement of a point in a continuum subjected to plane stress, solve for the expressions defining the strain in the continuum.

$$u = C_0 + C_1 X + C_2 Y$$
$$v = C_3 + C_4 X + C_5 Y$$

2.2 Solve for the expressions defining the stress in the continuum of Problem 2.1.

2.3 Solve for the global, local, and natural coordinate values of the point located at the midpoint of side 2–3 of the triangle shown in Figure P2.3.

2.4 The points on the line 1–2 in Figure P2.4 have a displacement in the X direction according to

$$u = C_0 + C_1 X$$

Express the displacement function in terms of the displacements of end nodes 1 and 2.

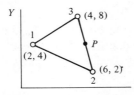

FIGURE P2.3

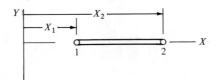

FIGURE P2.4

PROBLEMS

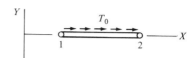

FIGURE P2.5

Hint:

$$u_1 = C_0 + C_1 X_1$$
$$u_2 = C_0 + C_1 X_2$$

Solve for C_0 and C_1 in terms of u_1, u_2, X_1, and X_2 and substitute back into the original displacement relationship:

$$u = C_0 + C_1 X.$$

2.5 The force on the line shown in Figure P2.5 is distributed uniformly along the length. The magnitude per unit length is T_0. Solve for the potential energy function for the distributed load in terms of the displacements of the end nodes using the linear displacement function of Problem 2.4.

2.6 The force on the line of Problem 2.5 is distributed according to

$$T(X) = C_0 + C_1 X \quad \text{with } T(X_1) = 0, \quad T(X_2) = T_0$$

Solve for the potential energy function of this loading in terms of the end node displacements for the linear displacement function of Problem 2.4.

2.7 The displacement of any point in the deformable continuum is given by

$$u = XY(10^{-5}) \quad \text{in.}$$

Solve for the energy potential of the uniformly distributed load acting normal to side 1–2 (Figure P2.7). Assume zero energy potential at the undeformed position.

2.8 Using the principle of minimum potential energy, obtain an estimate of the deflection of the end of the cantilever beam shown in Figure P2.8, assuming the following (estimates) of the deflection curve of the neutral axis.

(a) $v = C_0 + C_1 X + C_2 X^2$

(b) $v = C_0\left(1 - \cos\dfrac{\pi X}{2l}\right)$

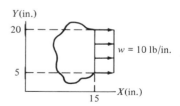

FIGURE P2.7

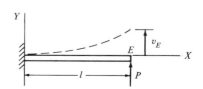

FIGURE P2.8

FIGURE P2.9

2.9 A rod of uniform cross section A and weight density ρ is suspended vertically from one end as shown in Figure P2.9.
  (a) Knowing that the strain is not constant in this continuum, suggest some reasonable displacement functions for the rod.
  (b) Select the best displacement function (your judgment) from part (a) and derive the strain energy expression for this rod in terms of the end displacement.
  (c) Using the same displacement function as in part (b), solve for the potential energy function in terms of the end displacement.
  (d) Use the principle of minimum potential energy to obtain the deflection (approximation) for the end of the rod.

2.10 A 1000-lb force is applied to the square plate shown in Figure P2.10 at the point $X = 10$, $Y = 0$. Assume that the plate material has an elastic modulus and Poisson's ratio of

$$E = 10^6 \text{ psi}$$
$$\nu = 0$$

We wish to solve for an expression for the displacement throughout the continuum constrained as shown in the figure.

As a guess, try the following functions for displacement and evaluate the unknown constants:

$$u = C_1 + C_2 X + C_3 Y + C_4 X^2 + C_5 Y^2$$
$$v = 0$$

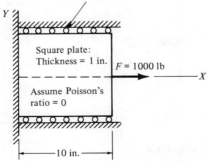

FIGURE P2.10

PROBLEMS

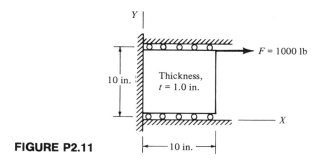

FIGURE P2.11

Having found the displacement function, solve for the expressions for the strain and stress throughout the plate.

Procedure:

1. Solve for the constants C_1, C_2, and C_5 to satisfy the displacement boundary conditions and symmetry.
2. Define strain ϵ_x, ϵ_y, and γ_{xy} using (2.5).
3. Derive the strain energy using (2.16).
4. Define force F potential energy ($X = 10$ in., $Y = 0$).
5. Sum energies, minimize Π with respect to C_3 and C_4 and solve for the constants.

2.11 A 1000-lb force is applied to the corner of the steel plate shown in Figure P2.11.

$$E = 30(10)^6 \text{ psi}$$
$$\nu = 0.3$$

Assume the following displacement functions for the continuum and solve for the constants using the procedure outlined in Problem 2.10.

$$u = C_1 + C_2 X + C_3 Y + C_4 XY$$
$$v = 0$$

Solve for the strain and stress in the plate for the assumed function. Compare with an alternative stress calculation.

CHAPTER 3

An Element for the One-Dimensional Problem

3.1 INTRODUCTION

In Section 2.13 we presented the thought that rather than try to use a single approximation function, such as a high-degree polynomial, for the solution of a problem, it may be possible to discretize the solution region and work with simpler functions for each smaller "space." In this chapter we show that this concept is valid and practical; the technique is termed the *finite element method*.

We first illustrate the finite element method solution technique for problems having the simplest type of region, one that has only one independent spatial variable. We also focus initially on the equilibrium elasticity problem, although as was pointed out in Section 2.14, this will lead to the solution procedure for the heat flow problem or any other problem that has a differential equation of the same form.

3.2 THE UNIAXIAL STRESS ELEMENT MODEL

Figure 3.1 shows the element of our discussion. Forces are applied parallel to the X axis, with positive values in the positive coordinate direction. The displacements of all points in the element are confined to the X axis.

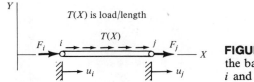

FIGURE 3.1 Distributed loading on the bar element: global node numbers i and j.

This element appears to be identical to the element of Section 1.2: same shape, loading and displacement freedom, and same sign conventions. There is one obvious difference, however; this element supports a spatially dependent distributed load. There is one other very important difference. Recall that the load deformation relation (stiffness) of the structural element in Section 1.2 was defined (known) to be AE/L. In the finite element continuum modeling, we do not begin with the stiffness of the element. Instead, we apply the Rayleigh–Ritz approximation method to the functional of the problem. Compare this element to the region of the uniaxial stress problem in Section 2.14; they are identical. Thus the functional for this element is defined in expression (2.42), and it requires determination of the element node (boundary) force energy potential, the distributed load energy potential, and the strain energy in the element. In this chapter we discuss the steps involved in:

1. The determination of each of these energy terms in the functional, Π (sum of the energies).
2. Application of the Rayleigh–Ritz procedure to define the force–displacement equations for this element.
3. Establishment of, using the Rayleigh–Ritz procedure over the entire region, the basis of assembly of the individual element equations.

To determine the force–displacement equation for the element, we follow the five steps outlined in Section 2.8. In summary, they are

1. Assume a displacement function for the region defined by the size of the element.
2. Derive the strain and strain energy.
3. Derive the applied load energy potentials.
4. Sum the energy terms.
5. Minimize the total energy with the constants in the assumed displacement function.

3.3 STEP 1. ASSUMED DISPLACEMENT FUNCTION

The motivation for discretization of the region into "small" elements is to be able to model the solution function for the entire domain with multiple elementary functions rather than with a single complex function.

3.3 STEP 1. ASSUMED DISPLACEMENT FUNCTION

Noting that the strain energy term in the functional of the elasticity problem involves the first derivative of the displacement function, and realizing that the solution over the domain should yield displacement continuity (no gaps), we should select a displacement function which, at a minimum, has piecewise continuous first derivatives over the entire domain.

The simplest suitable function is the linear function. The question now arises as to the method of expressing this function. Referring to Section 2.2, we are reminded that three different coordinate systems can be used. We will examine the form of the linear function using the global and local coordinate systems. The use of the natural system will be postponed.

Using the Global Coordinate System. For any point in the element located at the spatial X coordinate (Figure 3.2) we will assume the linear displacement function with the two arbitrary coefficients α_1 and α_2:

$$u(X) = \alpha_1 + \alpha_2 X \tag{3.1}$$

Next, we convert the α_i coefficients into quantities that have physical meaning. Substituting the nodal displacement, u_i, and the coordinate, X_i, into (3.1), we have

$$u_i = \alpha_1 + \alpha_2 X_i \tag{3.2}$$

Doing the same for the node j yields

$$u_j = \alpha_1 + \alpha_2 X_j$$

Now solve for the α_i coefficients:

$$\alpha_1 = \frac{u_i X_j - u_j X_i}{X_j - X_i}$$
$$\alpha_2 = \frac{u_j - u_i}{X_j - X_i} \tag{3.3}$$

Substituting (3.3) into (3.1) gives us

$$u = \frac{X_j - X}{X_j - X_i} u_i + \frac{X - X_i}{X_j - X_i} u_j \tag{3.4a}$$

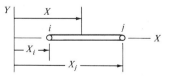

FIGURE 3.2 Spatial coordinates for points in the element.

This can be written in matrix form as

$$u = \begin{bmatrix} \dfrac{X_j - X}{X_j - X_i} & \dfrac{X - X_i}{X_j - X_i} \end{bmatrix} \begin{Bmatrix} u_i \\ u_j \end{Bmatrix} \quad (3.4b)$$

or

$$u = \begin{bmatrix} N_1 & N_2 \end{bmatrix} \{q\} = [N]\{q\} \quad (3.4c)$$

The coefficients N_1 and N_2 of the nodal displacements u_1 and u_2, respectively, are called *interpolation, shape,* or *basis functions*. The matrix $[N]$ is called the *interpolation, shape,* or *basis matrix*, and $\{q\}$ is called the *nodal displacement vector*. Observe what we have here: a function defining the displacement of any point in the element in terms of the nodal displacements. This particular function is stated in terms of the global X coordinate and the global node numbers i and j.

Using the Local Coordinate System. The linear displacement function written in terms of the local coordinates would be

$$u(x) = \alpha_1 + \alpha_2 x \quad (3.5)$$

Substituting nodal values yields

$$\begin{aligned} u_1 &= \alpha_1 \\ u_2 &= \alpha_1 + \alpha_2 x_2 \end{aligned} \quad (3.6)$$

where node numbers 1 and 2 are the local node numbers (see Figure 3.3). Substituting (3.6) into (3.5) gives us

$$u = \begin{bmatrix} \dfrac{x_2 - x}{x_2} & \dfrac{x}{x_2} \end{bmatrix} \begin{Bmatrix} u_1 \\ u_2 \end{Bmatrix} \quad (3.7)$$

Compare (3.7) with (3.4). The form of (3.7) is obviously simpler than (3.4). Note that x_2 defines the length of the element and is equal to $X_2 - X_1$. Note also that there is an obvious relationship between the local and global coordinates.

$$X = X_1 + x \quad (3.8)$$

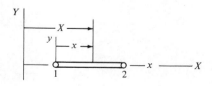

FIGURE 3.3 Local system on the element and local nodes 1 and 2.

3.4 STEP 2. DERIVATION OF ELEMENT STRAIN AND STRAIN ENERGY

Refer back to Section 2.3, which contains the definition of strain in a continuum. For our very simple element, we have only one strain component because we have defined an element having displacements in one direction only. The strain for this case is [from equations (2.3)]

$$\epsilon_X = \frac{du}{dX} \tag{3.9}$$

Note that the differentiation is with respect to the global spatial coordinate. Observe that differentiation with respect to the local coordinate yields the correct value of strain also, because from equation (3.8),

$$dX = 0 + dx$$

Substituting equation (3.4), the displacement in terms of the global X, into (3.9), we have

$$\epsilon_X = \frac{du}{dX} = \frac{u_j - u_i}{X_j - X_i} = \frac{1}{X_j - X_i}[-1 \quad 1]\begin{Bmatrix} u_i \\ u_j \end{Bmatrix}$$

Note that $X_j - X_i$ = the length of the element, L. Thus

$$\epsilon_X = \frac{1}{L}[-1 \quad 1]\begin{Bmatrix} u_i \\ u_j \end{Bmatrix} \tag{3.10a}$$

or

$$\epsilon_X = [B]\{q\} \tag{3.10b}$$

where the matrix $[B]$ is called the *strain–displacement matrix*.

Similarly, recognizing that $x_2 = L$, substituting equation (3.7), the displacement in terms of the local x, into (3.9) yields

$$\epsilon_X = \frac{du}{dX} = \frac{du}{dx} = \text{equation (3.10)}$$

If you have a physical appreciation of the meaning of strain, you should be satisfied that this is a correct expression for strain for this assumed displacement field. Note that the strain is constant throughout the length of this element.

Having derived the expression for the strain in the element, we can substitute into the strain energy relationships given in Section 2.6. For this simple uniaxial element, the strain energy is defined by (2.17) [which is equivalent to the strain energy expression in the functional defined in the expression (2.42)].

$$U = \int_X \frac{EA}{2} \epsilon_X^2 \, dX \tag{2.17}$$

Now substitute the expression (3.10) into (2.17). If you need assistance in doing this, note the following:

$$\epsilon_X^2 = \epsilon_X \epsilon_X = \epsilon_X^T \epsilon_X = \frac{1}{L}[u_i \quad u_j]\begin{Bmatrix} -1 \\ 1 \end{Bmatrix}\left(\frac{1}{L}\right)[-1 \quad 1]\begin{Bmatrix} u_i \\ u_j \end{Bmatrix}$$

Performing the matrix multiplication, we obtain the expression for the strain energy

$$U = \left(\frac{EA}{2}\right)\left(\frac{1}{L}\right)^2 [u_i \quad u_j]\begin{bmatrix} 1 & -1 \\ -1 & 1 \end{bmatrix}\begin{Bmatrix} u_i \\ u_j \end{Bmatrix}\int_{X_i}^{X_j} dX$$

Integrating yields

$$U = \frac{EA}{2L}[u_i \quad u_j]\begin{bmatrix} 1 & -1 \\ -1 & 1 \end{bmatrix}\begin{Bmatrix} u_i \\ u_j \end{Bmatrix} \quad (3.11a)$$

or

$$U = \tfrac{1}{2}\{q\}^T[k]\{q\} \quad (3.11b)$$

where

$$[k] = \frac{EA}{L}\begin{bmatrix} 1 & -1 \\ -1 & 1 \end{bmatrix}$$

Thus (3.11) is the expression for the strain energy in the element. Note that the expression is the same regardless of the coordinate system used. The advantage of one coordinate system over the other will not become apparent until we study a more complex element.

3.5 STEP 3. DERIVATION OF APPLIED LOAD POTENTIAL ENERGY FUNCTIONS

On this particular element, forces are applied to each end node, and a load is distributed arbitrarily along its length. This next step involves the determination of the energy potential function for each of these force systems. A review of Sections 2.5 and 2.12 might be helpful as we develop these expressions. We will look at the node forces and the distributed load separately.

The Node Force Energy Potential Function. From our discussion in Section 2.5, recall that the potential energy V_i for the constant magnitude force F_i at node i is equal to the negative of the product of the force and its displacement if the energy is taken to be zero at the undeformed position. Thus we have [refer to (2.13)]

$$V_i = -F_i u_i$$

3.5 STEP 3. DERIVATION OF APPLIED LOAD POTENTIAL ENERGY FUNCTIONS

Similarly, the potential energy of the force at node j is

$$V_j = -F_j u_j$$

The potential energy of the element node forces can be summed and written in matrix form as

$$V_{NF} = -\begin{bmatrix} F_i & F_j \end{bmatrix} \begin{Bmatrix} u_i \\ u_j \end{Bmatrix} \quad (3.12)$$

The Distributed Load Energy Potential. We assume first that the distributed force per unit length, $T(X)$, is a function of the global X coordinate. The differential force acting on a length dX is

$$d\bar{F} = T(X)\,dX\,\hat{\imath}$$

The differential energy potential becomes (review Example 2.8)

$$dV_T = -T(X)\,dX\,u(X)$$

To obtain the energy potential, substitute the global description of the distributed load, $T(X)$, and the displacement function for the element written in terms of the global variable [refer to expression (3.4)]. Integration of the resulting expression yields the distributed load energy potential function, V_T:

$$V_T = -\int_{X_i}^{X_j} T(X) \begin{bmatrix} \dfrac{X_j - X}{X_j - X_i} & \dfrac{X - X_i}{X_j - X_i} \end{bmatrix} \begin{Bmatrix} u_i \\ u_j \end{Bmatrix} dX \quad (3.13a)$$

or simplifying, using (3.4c), we have

$$V_T = -\int_{X_i}^{X_j} T(X)[N]\,dX\,\{q\} \quad (3.13b)$$

Example 3.1 The element shown in Figure 3.4 is subjected to a uniformly distributed loading,

$$T(X) = w \quad \text{a constant}$$

Using the linear displacement relationship equation (3.4), derive the energy potential function of this distributed loading.

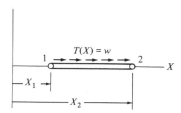

FIGURE 3.4 Uniform load distribution.

Solution. Equation (3.13) defines the energy potential function for the displacement field specified. Substituting $T(X) = w$, we have

$$V_T = -\int_{X_1}^{X_2} w \left[\frac{X_2 - X}{X_2 - X_1} \quad \frac{X - X_1}{X_2 - X_1} \right] \begin{Bmatrix} u_1 \\ u_2 \end{Bmatrix} dX$$

$$= -\frac{w}{X_2 - X_1} \int_{X_1}^{X_2} [X_2 - X \quad X - X_1] \, dX \begin{Bmatrix} u_1 \\ u_2 \end{Bmatrix}$$

$$= -\frac{w}{X_2 - X_1} \left[\frac{(X_2 - X_1)^2}{2} \quad \frac{(X_2 - X_1)^2}{2} \right] \begin{Bmatrix} u_1 \\ u_2 \end{Bmatrix}$$

$$= -w(X_2 - X_1) \left[\tfrac{1}{2} \quad \tfrac{1}{2} \right] \begin{Bmatrix} u_1 \\ u_2 \end{Bmatrix}$$

$$= -\frac{w(X_2 - X_1)}{2} [1 \quad 1] \begin{Bmatrix} u_1 \\ u_2 \end{Bmatrix}$$

Note that the resultant of the distributed loading is $w(X_2 - X_1)$. Notice also that if we compare the answer with equation (3.12), we have, for this particular distributed loading, an energy expression which is equal to that obtained for nodal forces each of which are equal to one-half of the total load. □

It may be that a situation will exist where the distributed load is given in terms of the local x coordinate. In that case the differential potential function dV can be written in terms of the local coordinate displacement function, $u(x)$, as follows:

$$dV_T = -T(x) \, dx \, u(x)$$

where $u(x)$ is defined by equation (3.7). Substitution yields

$$V_T = \int_0^{x_2} -T(x) \left[\frac{x_2 - x}{x_2} \quad \frac{x}{x_2} \right] \begin{Bmatrix} u_1 \\ u_2 \end{Bmatrix} dx \qquad (3.14)$$

Examine equations (3.13) and (3.14). They are very important. What they give you is the loads that must be applied to each node to obtain a loading that is work-equivalent to the distributed load along the element. Often, you are required to calculate these equivalent nodal loads when using a particular commercial finite element program.

Example 3.2 The element shown in Figure 3.5 is subjected to a distributed loading which varies from zero at node 1 to a maximum at node 2 according to

$$T(x) = Cx$$

where x is the local coordinate.

Solve for the nodal forces which, for this element, would be the energy equivalent of the distributed loading.

3.5 STEP 3. DERIVATION OF APPLIED LOAD POTENTIAL ENERGY FUNCTIONS

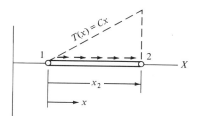

FIGURE 3.5 Linear load distribution.

Solution. Beginning with equation (3.14) and substituting the distributed loading function, we have

$$V_T = -\int_0^{x_2} Cx \left[\frac{x_2 - x}{x_2} \quad \frac{x}{x_2} \right] \begin{Bmatrix} u_1 \\ u_2 \end{Bmatrix} dx$$

$$= -\frac{C}{x_2} \int_0^{x_2} [x_2 x - x^2 \quad x^2] \, dx \begin{Bmatrix} u_1 \\ u_2 \end{Bmatrix}$$

$$= -\frac{C}{x_2} \left[\frac{x_2^3}{6} \quad \frac{x_2^3}{3} \right] \begin{Bmatrix} u_1 \\ u_2 \end{Bmatrix}$$

$$= -\frac{Cx_2^2}{2} \left[\tfrac{1}{3} \quad \tfrac{2}{3} \right] \begin{Bmatrix} u_1 \\ u_2 \end{Bmatrix}$$

Compare this answer with equation (3.12). Note that integration of this loading over the length of the element yields a resultant load of $Cx_2^2/2$. Therefore, the equivalent nodal loads for this loading are

equivalent $F_1 = \tfrac{1}{3}$ total load

equivalent $F_2 = \tfrac{2}{3}$ total load

for this particular distributed loading. ☐

We can derive an expression that will be convenient to use for the determination of equivalent nodal forces for any linear load distribution, $T(x) = C_0 + C_1 x$ (Figure 3.6). Note that if we express the value of the distributed loading at nodes 1 and 2 and solve for the constants C_0 and C_1 in terms of those values, we can write the load distribution in terms of the shape matrix, $[N]$, and node values of $T(x)$:

$$T(x) = [N] \begin{Bmatrix} T(1) \\ T(2) \end{Bmatrix}$$

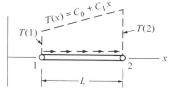

FIGURE 3.6 Arbitrary linearly distributed loading.

Substitution into (3.14) will lead to the following expression for the energy potential:

$$V_T = -\frac{L}{6}[T(1) \quad T(2)]\begin{bmatrix} 2 & 1 \\ 1 & 2 \end{bmatrix}\begin{Bmatrix} u_1 \\ u_2 \end{Bmatrix}$$

Thus the nodal forces that should be applied to local nodes 1 and 2 are defined by the following:

$$\{Q\}_T = \begin{Bmatrix} F_1 \\ F_2 \end{Bmatrix}_T = \frac{L}{6}\begin{bmatrix} 2 & 1 \\ 1 & 2 \end{bmatrix}\begin{Bmatrix} T(1) \\ T(2) \end{Bmatrix} \qquad (3.15)$$

3.6 STEP 4. SUMMATION OF ENERGY TERMS

Gather the expressions for the strain energy and the load potential energies. The sum represents the total energy in our single-element system.

$$\boxed{\text{total energy, } \Pi} = \boxed{\begin{array}{c}\text{strain energy, } U\\ (3.11)\end{array}} + \boxed{\begin{array}{c}\text{nodal force}\\ \text{potential energy}\\ (3.12)\end{array}}$$

$$+ \boxed{\begin{array}{c}\text{distributed force potential}\\ \text{energy}\\ (3.13), (3.14)\end{array}}$$

$$\Pi = U + V_{\text{NF}} + V_T$$

$$= \tfrac{1}{2}[u_i \quad u_j][k]\begin{Bmatrix} u_i \\ u_j \end{Bmatrix} - [F_i \quad F_j]\begin{Bmatrix} u_i \\ u_j \end{Bmatrix}$$

$$- \int_{x_i}^{x_j} T(X)[N_1 \quad N_2]\, dX \begin{Bmatrix} u_i \\ u_j \end{Bmatrix} \qquad (3.16)$$

3.7 STEP 5. APPLICATION OF THE PRINCIPLE OF MINIMUM POTENTIAL ENERGY

Reread Sections 2.9 and 2.12. Our task now is to determine the values of the nodal displacements that make Π of equation (3.16) a minimum. In the example in Section 2.9 there was one unknown displacement. Our current element energy equation contains two unknown nodal displacements. The correct values of these displacements minimize the energy stored in the force-element system.

Two equations for these two unknowns can be obtained by differentiating the energy equation with respect to each node displacement and setting the resulting equations equal to zero. For example, if we first differentiate (3.16) with respect to u_i, we will obtain the following:

$$\frac{\partial \Pi}{\partial u_i} = \frac{1}{2}\begin{bmatrix}1 & 0\end{bmatrix}[k]\begin{Bmatrix}u_i\\u_j\end{Bmatrix}$$
$$+ \frac{1}{2}[u_i \quad u_j][k]\begin{Bmatrix}1\\0\end{Bmatrix}$$
$$- [F_i \quad F_j]\begin{Bmatrix}1\\0\end{Bmatrix} - \int_{x_i}^{x_j} T(X)[N_1 \quad N_2]\,dX\begin{Bmatrix}1\\0\end{Bmatrix}$$
$$= 0 \quad\quad (3.17)$$

Note that the first two terms are equal scalars. Differentiating equation (3.16) again with respect to u_j and combining with equation (3.17) yields the following matrix equation:

$$[k]\begin{Bmatrix}u_i\\u_j\end{Bmatrix} = \begin{Bmatrix}F_i\\F_j\end{Bmatrix} + \int_{x_i}^{x_j} T(X)\begin{bmatrix}N_1\\N_2\end{bmatrix}dX \quad\quad (3.18a)$$

or

$$[k]\{q\} = \{Q\}_{\text{NF}} + \{Q\}_T \quad\quad (3.18b)$$

where

$[k]$ = element stiffness matrix = $\dfrac{EA}{L}\begin{bmatrix}1 & -1\\-1 & 1\end{bmatrix}$

$\{q\}$ = element nodal displacement vector

$\{Q\}_{\text{NF}}$ = vector of forces applied directly to element nodes

$\{Q\}_T$ = vector of nodal forces that are derived from and equivalent to the distributed loading on the element

$$\{Q\}_T = \int_{x_i}^{x_j} T(x)[N]^T dx \quad\quad (3.18c)$$

Comparing the stiffness matrix with the stiffness matrix for the two-force-member structural element, equation (1.2), we see that they are identical.

3.8 MODELING A CONTINUOUS ONE-DIMENSIONAL SYSTEM WITH FINITE ELEMENTS

Consider the uniaxial continuum in Figure 3.7 with applied loads on the interior of the region and having boundaries A and B with displacements and/or forces temporarily undefined. We wish to discuss how we might discretize the region and solve for the displacements using our simple linear element.

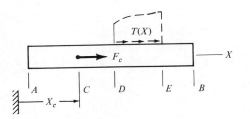

FIGURE 3.7 One-dimensional continuum.

Guidance for the proper method of modeling this problem with our linear finite element will arise out of the application of the functional of this problem to the entire region. We recall the functional from expression (2.42):

$$\Pi = -F_2 u_2 - F_1 u_1 + \int_{x_1}^{x_2} \frac{EA}{2} \left(\frac{du}{dX} \right)^2 dX - \int_{x_1}^{x_2} T(X) u(X) \, dX$$

Remember that the first two terms define the potential energy of the forces only at the boundaries, the third term is the total strain energy in the body, and the last term is the energy potential of all the forces applied to the interior of the region. The determination of the potential energy of distributed loadings and concentrated forces has been discussed in Section 2.5. We see in Figure 3.7 that we have a distributed load $T(X)$ and a concentrated force F_c at location C. Referring to the energy functional, it may at first appear that there is no term accounting for the energy of concentrated forces within the region. This, however, is not the case. The concentrated force is actually a distributed loading of extremely high magnitude over a very small distance. A mathematical expression for the distributed form of a concentrated loading is

$$T(X) = F_c \Delta(X - X_c)$$

where $\Delta(X - X_c)$ is the Dirac delta function with the following defined properties:

$$\Delta(X - X_c) = \begin{cases} 0 & \text{for } X \neq X_c \\ \infty & \text{for } X = X_c \end{cases}$$

$$\int_{-\infty}^{\infty} \Delta(X - X_c) \, dX = 1$$

(3.19)

Thus, for this "distributed" load at C, the energy potential is found using the last term of the functional:

$$V_{F_c} = -\int_{x_1}^{x_2} F_c \Delta(X - X_c) u(X) \, dX$$

$$= -F_c u(X_c)$$

Therefore, the energy potentials of all concentrated loads in the region are derived from the last term in the functional and have the same form as that of any other constant force.

We now should have some guidelines to follow in discretizing the region. We must consider the following:

1. The element displacement function is linear and the strain and stress are therefore constant throughout the element for given values of its end node displacements [refer to expression (3.10)]. Thus one element cannot model a strain or stress gradient.
2. Since the unknowns of this problem are the displacements, and because wherever a concentrated load is applied to the region we will have an energy expression involving the displacement of that point, we should make that point an element node point.

Referring to Figure 3.7, we should realize that the force in the region is constant between points A and C. Thus the strain and stress are constant and a linear function can represent the correct solution for that region. One element between points A and C will be sufficient. The same can be said for the sections between C and D and also between E and B. Between D and E, however, the stress and strain will depend on the X coordinate. Obviously, this section cannot be modeled with one linear element. The model in general will improve as the number of elements used increases. How many should be used? That decision has to be made after studying solution results. An example in Section 3.10 will show the effect that the number of elements (mesh size) has on the results.

3.9 ASSEMBLING THE ELEMENTS

In Chapter 1 the technique of assembling element stiffness equations to obtain a stiffness equation for the entire structure was based on the equilibrium of nodes in the structure (Section 1.2). It is the intent now to show that the assembly of the stiffness equations for a continuum is developed on the basis of minimization of the total energy of the entire continuum (all the elements modeling the continuum) and the applied forces.

Refer to Figure 3.8. The continuum example shown in Figure 3.7 is used again. Based on the discussion in Section 3.8, the region has been discretized into the fewest number of elements that we would select for

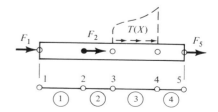

FIGURE 3.8 Continuum modeled with elements (node and element numbers).

an initial (very approximate) solution. We now evaluate the energy functional for the entire region of this simple model.

$$\Pi = \boxed{\begin{array}{c}\text{boundary force}\\\text{potential}\\\text{energy}\end{array}} + \boxed{\begin{array}{c}\text{total}\\\text{strain}\\\text{energy}\end{array}} + \boxed{\begin{array}{c}\text{interior loading}\\\text{potential}\\\text{energy}\end{array}}$$

We take the functional from (2.42) and then apply it to a discretized region as follows:

$$\Pi = \boxed{-F_2 u_2 - F_1 u_1} + \sum_{e=1}^{4} \tfrac{1}{2}[u_i \quad u_j][k]\begin{Bmatrix}u_i\\u_j\end{Bmatrix}$$

$$+ \boxed{\sum_{n=1}^{2} \int_{x_i}^{x_j} T(X)[N]\,dX \begin{Bmatrix}u_i\\u_j\end{Bmatrix}} \qquad (3.20)$$

Substituting the values associated with the system of Figure 3.8, we have

$$\Pi = \boxed{-F_5 u_5 - F_1 u_1} + \begin{array}{c} \tfrac{1}{2}[u_1 \quad u_2][k_1]\begin{Bmatrix}u_1\\u_2\end{Bmatrix}\\ +\\ \tfrac{1}{2}[u_2 \quad u_3][k_2]\begin{Bmatrix}u_2\\u_3\end{Bmatrix}\\ +\\ \tfrac{1}{2}[u_3 \quad u_4][k_3]\begin{Bmatrix}u_3\\u_4\end{Bmatrix}\\ +\\ \tfrac{1}{2}[u_4 \quad u_5][k_4]\begin{Bmatrix}u_4\\u_5\end{Bmatrix}\end{array}$$

$$+ \boxed{\begin{array}{c}-F_2 u_2\\+\\ -\int_{x_3}^{x_4} T(X)[N]\,dX \begin{Bmatrix}u_3\\u_4\end{Bmatrix}\end{array}} \qquad (3.21)$$

We now define the global system displacement vector $\{r\}$:

$$\{r\} = [u_1 \quad u_2 \quad u_3 \quad u_4 \quad u_5]^T$$

3.9 ASSEMBLING THE ELEMENTS

Expression (3.21) can now be written as

$$\Pi = -[F_1 \quad 0 \quad 0 \quad 0 \quad F_5]\{r\}$$

$$+ \tfrac{1}{2}\{r\}^T \begin{bmatrix} [k]_1 & & & & 0 \\ & [k]_2 & & & \\ & & [k]_3 & & \\ & & & [k]_4 & \\ 0 & & & & \end{bmatrix} \{r\}$$

$$-[0 \quad F_2 \quad 0 \quad 0 \quad 0]\{r\}$$

$$-\left[0 \quad 0 \quad \int_{x_3}^{x_4} T(X)N_1\,dX \quad \int_{x_3}^{x_4} T(X)N_2\,dX \quad 0\right]\{r\}$$

(3.22)

Next, we apply the Rayleigh–Ritz minimization scheme and minimize the functional with respect to the five nodal displacements. In a manner similar to the process for the single element in Section 3.7, we combine the five equations and write them in matrix form as

$$[K]\{r\} = \begin{Bmatrix} F_1 \\ F_2 \\ 0 \\ 0 \\ F_5 \end{Bmatrix} + \begin{Bmatrix} 0 \\ 0 \\ \int T(X)N_1\,dX \\ \int T(X)N_2\,dX \\ 0 \end{Bmatrix} \qquad (3.23)$$

We can write this as a general matrix equation that is valid for all one-dimensional elasticity problems:

$$[K]\{r\} = \{R\}_{\text{NF}} + \{R\}_T \qquad (3.24)$$

For a system modeled with $n - 1$ elements and n nodes:

1. $[K]$ (note the uppercase K) is the assembled $n \times n$ stiffness matrix.
2. $\{r\}$ is the $n \times 1$ displacement vector including all nodes.
3. $\{R\}_{\text{NF}}$ is the $n \times 1$ force vector of all nodal forces (including zero).
4. $\{R\}_T$ is the $n \times 1$ vector from (finite-valued) load distributions (force per unit length).

You should recognize that the process of assembly (addition) of the global stiffness matrix is identical to the method employed for the structural problem in Chapter 1. The solution procedures are the same also; substitute the boundary conditions, reduce the equation, and solve for the displacements.

3.10 ONE-DIMENSIONAL ELASTICITY EXAMPLE

A bar is fixed at the left end and subjected to a linearly distributed loading as shown in Figure 3.9. The boundary conditions are defined in Figure 3.10.

Using equation (3.24), solve for the displacement, stress, and strain in the rod. Begin with a single-element model, and then "refine the mesh" and determine the results using additional numbers of elements. The basic equation is

$$[K]\{r\} = \{R\}_{\text{NF}} + \{R\}_T$$

Single-Element Model. The element is defined in Figure 3.11.

$$[K] = \frac{EA}{L}\begin{bmatrix} 1 & -1 \\ -1 & 1 \end{bmatrix} = (10^6)\begin{bmatrix} 1 & -1 \\ -1 & 1 \end{bmatrix}$$

$$\{r\} = [u_1 \quad u_2]^T = [0 \quad u_2]^T$$

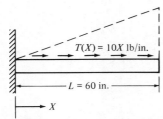

$T(X) = 10X$ lb/in.

$L = 60$ in.

X

Cross-sectional area, $A = 2$ in.^{2}
Elastic modulus, $E = 30(10^6)$ psi

FIGURE 3.9 Tensile rod with distributed load.

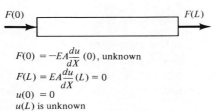

$F(0) = -EA\dfrac{du}{dX}(0)$, unknown
$F(L) = EA\dfrac{du}{dX}(L) = 0$
$u(0) = 0$
$u(L)$ is unknown

FIGURE 3.10 Boundary conditions.

3.10 ONE-DIMENSIONAL ELASTICITY EXAMPLE

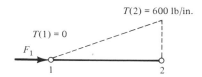

FIGURE 3.11 Single-element model.

$$\{R\}_{NF} = \begin{Bmatrix} F_1 \\ 0 \end{Bmatrix}$$

$$\{R\}_T = \begin{Bmatrix} F_1 \\ F_2 \end{Bmatrix}_T = \frac{L}{6}\begin{bmatrix} 2 & 1 \\ 1 & 2 \end{bmatrix}\begin{Bmatrix} T(1) \\ T(2) \end{Bmatrix} \quad \text{from (3.15)}$$

$$\{R\}_T = \frac{60}{6}\begin{bmatrix} 2 & 1 \\ 1 & 2 \end{bmatrix}\begin{Bmatrix} 0 \\ 600 \end{Bmatrix} = \begin{Bmatrix} 6{,}000 \\ 12{,}000 \end{Bmatrix}$$

Substitution into (3.24) yields

$$(10^6)\begin{bmatrix} 1 & -1 \\ -1 & 1 \end{bmatrix}\begin{Bmatrix} 0 \\ u_2 \end{Bmatrix} = \begin{Bmatrix} F_1 \\ 0 \end{Bmatrix} + \begin{Bmatrix} 6{,}000 \\ 12{,}000 \end{Bmatrix} = \begin{Bmatrix} F_1 + 6{,}000 \\ 12{,}000 \end{Bmatrix}$$

Reducing and solving, we obtain $u_2 = 0.012$ in.
The strain is determined from (3.10):

$$\epsilon_X = \frac{du}{dX} = [B]\{q\} = \frac{1}{L}[-1 \quad 1]\begin{Bmatrix} u_1 \\ u_2 \end{Bmatrix} = \frac{1}{60}[-1 \quad 1]\begin{Bmatrix} 0 \\ 0.012 \end{Bmatrix}$$

$$= 2(10^{-4}) \text{ in./in.}$$

The stress is calculated from (2.6):

$$\sigma_X = E\epsilon_X = 30(10^6)2(10^{-4}) = 6{,}000 \text{ psi}$$

Two-Element Model. The two elements selected are defined in Figure 3.12.

The element stiffness matrices are

$$[k_1] = \frac{EA}{L/2}\begin{bmatrix} 1 & -1 \\ -1 & 1 \end{bmatrix} = 2(10^6)\begin{bmatrix} 1 & -1 \\ -1 & 1 \end{bmatrix} = [k_2]$$

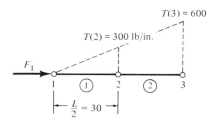

FIGURE 3.12 Two-element model.

The assembled global stiffness matrix is

$$[K] = 2(10^6) \begin{bmatrix} 1 & -1 & 0 \\ -1 & 2 & -1 \\ 0 & -1 & 1 \end{bmatrix}$$

The system displacement vector is

$$\{r\} = [u_1 \quad u_2 \quad u_3]^T$$

The nodal force vector is

$$\{R\}_{NF} = [F_1 \quad 0 \quad 0]^T$$

$\{R\}_T$ is the vector of nodal forces equivalent to the distributed loading. To determine this vector, apply formula (3.15) to each element distribution and sum the forces at the common nodes. For element 1:

$$\begin{Bmatrix} F_1 \\ F_2 \end{Bmatrix} = \frac{L/2}{6} \begin{bmatrix} 2 & 1 \\ 1 & 2 \end{bmatrix} \begin{Bmatrix} T(1) \\ T(2) \end{Bmatrix} = \frac{30}{6} \begin{bmatrix} 2 & 1 \\ 1 & 2 \end{bmatrix} \begin{Bmatrix} 0 \\ 300 \end{Bmatrix} = \begin{Bmatrix} 1500 \\ 3000 \end{Bmatrix} \text{ lb}$$

For element 2:

$$\begin{Bmatrix} F_2 \\ F_3 \end{Bmatrix} = \frac{30}{6} \begin{bmatrix} 2 & 1 \\ 1 & 2 \end{bmatrix} \begin{Bmatrix} T(2) \\ T(3) \end{Bmatrix} = \frac{30}{6} \begin{bmatrix} 2 & 1 \\ 1 & 2 \end{bmatrix} \begin{Bmatrix} 300 \\ 600 \end{Bmatrix}$$

$$= \begin{Bmatrix} 6000 \\ 7500 \end{Bmatrix} \text{ lb}$$

Combining the forces at each element yields

$$\{R\}_T = \{R_1\}_T + \{R_2\}_T$$

$$= \begin{Bmatrix} 1500 \\ 3000 \\ 0 \end{Bmatrix} + \begin{Bmatrix} 0 \\ 6000 \\ 7500 \end{Bmatrix} = \begin{Bmatrix} 1500 \\ 9000 \\ 7500 \end{Bmatrix}$$

Substituting into (3.24) gives us

$$2(10^6) \begin{bmatrix} 1 & -1 & 0 \\ -1 & 2 & -1 \\ 0 & -1 & 1 \end{bmatrix} \begin{Bmatrix} u_1 \\ u_2 \\ u_3 \end{Bmatrix} = \begin{Bmatrix} F_1 + 1500 \\ 9000 \\ 7500 \end{Bmatrix}$$

Substituting the boundary condition, $u_1 = 0$, and reducing and solving, we obtain

$$\begin{Bmatrix} u_2 \\ u_3 \end{Bmatrix} = \begin{Bmatrix} 0.00825 \\ 0.012 \end{Bmatrix} \text{ in.}$$

Solving for the element stresses yields

$$\sigma_{X1} = E[B]\{q\}_1 = \frac{30}{30}(10^6)[-1 \quad 1] \begin{Bmatrix} 0 \\ 8.25(10^{-3}) \end{Bmatrix}$$

$$= 8250 \text{ psi}$$

$$\sigma_{X2} = E[B]\{q\}_2 = \frac{30}{30}(10^6)[-1 \quad 1] \begin{Bmatrix} 8.25(10^{-3}) \\ 12.0(10^{-3}) \end{Bmatrix}$$

$$= 3750 \text{ psi}$$

An additional computation has been made using a four (equal-length)-element model. The results from this model as well as those from the single- and two-element model are discussed in the next section.

3.11 INTERPRETATION OF RESULTS AND SOME GENERAL CONCLUSIONS

We should spend some time examining the results of the example in Section 3.10. Although the example was very simple, proper appreciation of the characteristics of the solution will shed light on the understanding of more complex elements and modeling situations.

The displacement values have been plotted in Figure 3.13. The fact that the one- and two-element solutions match the exact value for the element node points might come as a surprise. The reason these nodal values are correct rests in the fact that the element nodal forces were calculated to be the energy equivalent to the distributed load based on the assumed linear displacement field of the element. The Figure 3.13 points out two very important characteristics that must be remembered:

1. Although the node values of displacement are good, the values at locations between the nodes are poor.
2. Secondary quantities, such as strain and stress, are derived from the slope of the displacement curve. A good solution requires a sufficient number of nodes to model the first derivative of the displacement function.

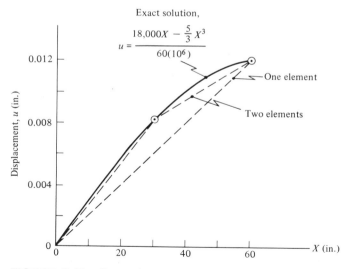

FIGURE 3.13 Comparison of region displacement predictions of one- and two-element models with exact values.

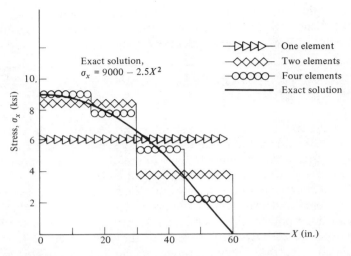

FIGURE 3.14 Comparison of region stress, σ_x, predictions of one-, two-, and four-element models with the exact solution.

Figure 3.14 compares the derived stress quantity with the exact value for this problem. We note the following:

1. The derived stresses are constant throughout each element. This is understandable because the assumed displacement function is linear.
2. The best approximation of the stress occurs near the midpoint of the element, not at the nodes. This is a direct consequence of the displacement derivative being better between the nodes than at the nodes.
3. The stress is not continuous across element boundaries. As a result, equilibrium is not satisfied across element boundaries. In addition, a free-body diagram of a single element, such as that shown in Figure 3.15, reveals that equilibrium is not satisfied for each element. The equilibrium equations that we started with in the variational procedure can be satisfied only in an approximate sense. As the number of elements used increases, and the discontinuity in the stress lessens across element boundaries, the approximation of equilibrium improves.

One final comment. The question was raised earlier as to when one knows when the approximate solution is good enough. This comes primarily from trying solutions and comparing results with an eye toward minimizing interelement discontinuity. In Figure 3.16 we have a plot of the stress at the wall location versus the number of nodes in the model. The decision to accept a solution is based on the appearance of convergence of several trials.

3.12 FINITE ELEMENT MODEL FOR ONE-DIMENSIONAL HEAT FLOW

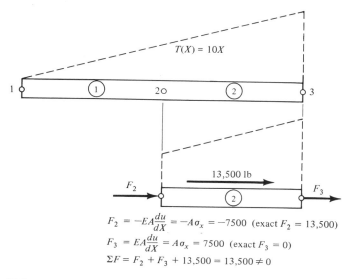

FIGURE 3.15 Free-body diagram of element 2 of the two-element model showing that equilibrium is not satisfied.

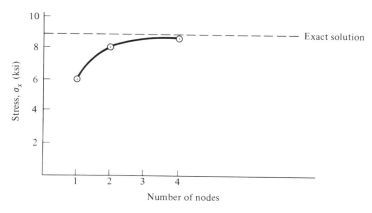

FIGURE 3.16 Convergence of the stress, σ_X, at location $X = 0$.

3.12 FINITE ELEMENT MODEL FOR ONE-DIMENSIONAL HEAT FLOW

As we discuss the modeling of the one-dimensional heat flow problem, you should observe the many similarities with the uniaxial stress problem in regard to the fundamental basis, procedure, and ultimately the final

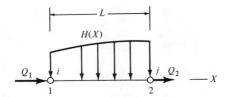

FIGURE 3.17 Two-node heat flow element with total boundary heat flows Q_1 and Q_2 (energy/time) and distributed heat flux (energy/length-time).

form of the equations. This discussion will assume familiarity with the uniaxial stress element development. Familiar concepts will be used with comments kept to a minimum.

The starting point for the finite element model is the problem functional as defined by (2.48):

$$\Pi = Q_2 T_2 - Q_1 T_1 + \int_{x_1}^{x_2} \frac{Ak}{2} \left(\frac{dT}{dx}\right)^2 dx - \int_{x_1}^{x_2} H(x) T(x)\, dx$$

(2.48)

Repeating the procedure followed for the stress problem, we derive the finite element equations for the single element and then assemble the element equations for the entire region.

The element is shown in Figure 3.17. Node numbers i and j are the global node numbers of local nodes 1 and 2. We will assume constant cross-sectional area and thermal conductivity throughout the element.

The first step in the procedure is to assume a function that defines the dependent variable throughout the element. Similar to the stress problem, we assume a linear variation of the dependent variable. In this case, this means a linear variation of the temperature in the element.

$$T = [N]\{q_t\} \tag{3.25a}$$

where the interpolation coefficients in the shape matrix $[N]$ have been defined in (3.4b) in terms of the global coordinates or (3.7) in terms of the local coordinates. The vector $\{q_t\}$ is defined as the nodal temperature vector.

$$\{q_t\} = \begin{bmatrix} T_i & T_j \end{bmatrix}^T \tag{3.25b}$$

Note that the third term of (2.48) requires the derivative of temperature. A review of Section 3.4 reminds us that the derivative of the shape matrix $[N]$ yields the following:

$$\frac{dT}{dx} = \frac{1}{L}\begin{bmatrix} -1 & 1 \end{bmatrix}\{q_t\} \tag{3.26a}$$

3.12 FINITE ELEMENT MODEL FOR ONE-DIMENSIONAL HEAT FLOW

or

$$\frac{dT}{dx} = [B]\{q_t\} \qquad (3.26b)$$

Substitution of (3.25) and (3.26) into (2.48) yields the following:

$$\Pi = -[Q_i \quad -Q_j]\{q_t\} + \frac{Ak}{2}\int_{x_i}^{x_j}\{q_t\}^T[B]^T[B]\{q_t\}\,dx$$
$$- \int_{x_i}^{x_j} H(x)[N]\{q_t\}\,dx$$

or

$$\Pi = -[Q_i \quad -Q_j]\{q_t\} + \frac{Ak}{2L}\{q_t\}^T\begin{bmatrix}1 & -1\\-1 & 1\end{bmatrix}\{q_t\}$$
$$- \int_{x_i}^{x_j} H(x)[N]\{q_t\}\,dx$$

We now apply the familiar Ritz procedure of differentiation of this scalar equation with respect to each nodal temperature and setting each of the two equations equal to zero. Finally, assembly in matrix form establishes the governing equation for the single element.

$$\frac{Ak}{L}\begin{bmatrix}1 & -1\\-1 & 1\end{bmatrix}\{q_t\} = \begin{Bmatrix}Q_i\\-Q_j\end{Bmatrix} + \int_{x_1}^{x_2}H(x)[N]^T\,dx \qquad (3.27a)$$

or

$$[k_{cd}]\{q_t\} = \{Q_t\}_N + \{Q_t\}_H \qquad (3.27b)$$

where

$[k_{cd}]$ = element conduction matrix

$\{q_t\}$ = nodal temperature vector

$\{Q_t\}_N$ = nodal heat flow vector

$\{Q_t\}_H$ = nodal heat flow vector equivalent to the distributed flux

Note that the stress stiffness matrix and the heat conduction matrix differ only by the ratio of the elastic modulus and conduction coefficient.

The equation for an assemblage of elements can now be derived, as explained in Section 3.9, by applying the Rayleigh–Ritz procedure to the functional for the entire region. The assembled equation would have the following form:

$$[K_{cd}]\{r_t\} = \{R_t\}_N + \{R_t\}_H \qquad (3.28)$$

where

$[K_{cd}]$ = assembled conduction matrix
$\{r_t\}$ = assembled nodal temperature vector
$\{R_t\}_N$ = nodal heat flow at boundaries and node sources
$\{R_t\}_H$ = distributed heat flux vector

The distributed heat flux vector is the heat flow problem equivalent of the distributed load nodal force vector. This vector may be determined by integration, but why not apply the formula (3.15) derived in Section 3.5 for calculating equivalent nodal force values for linearly distributed loads? For the distributed flux, the formula for the equivalent nodal heat flow vector would be

$$\{Q_t\}_H = \begin{Bmatrix} Q_1 \\ Q_2 \end{Bmatrix}_H = \frac{L}{6}\begin{bmatrix} 2 & 1 \\ 1 & 2 \end{bmatrix}\begin{Bmatrix} H(1) \\ H(2) \end{Bmatrix} \qquad (3.29)$$

The assembly of the conduction matrices and the nodal force vectors follows exactly the procedures discussed in the stress problem. The following example illustrates the procedure.

Example 3.3 A continuum with uniform cross-sectional area of 2 in.^{2} and thermal conductivity of 3 Btu/hr-in.-°F has heat flow in the x direction only (Figure 3.18). The right end is insulated. The left end is maintained at 100°F, and the system has the linearly distributed heat flux shown.

Use a two-element system and estimate the temperature at the node points and the heat flow at the left boundary.

Solution. The basic equation:

$$[K_{cd}]\{r_t\} = \{R_t\}_N + \{R_t\}_H \qquad (3.28)$$

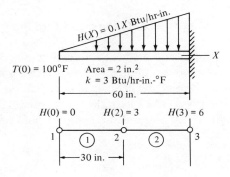

FIGURE 3.18 One-dimensional heat flow problem and a two-element model.

3.12 FINITE ELEMENT MODEL FOR ONE-DIMENSIONAL HEAT FLOW

For elements of equal length, the assembled stiffness matrix is

$$[K_{cd}] = 0.2 \begin{bmatrix} 1 & -1 & 0 \\ -1 & 2 & -1 \\ 0 & -1 & 1 \end{bmatrix}$$

The system temperature vector is

$$\{r_t\} = [T_1 \quad T_2 \quad T_3]^T$$

The nodal heat source vector is

$$\{R_t\}_N = \{Q_t\}_{N①} + \{Q_t\}_{N②} = \begin{Bmatrix} Q_1 \\ -Q_2 \end{Bmatrix}_{N①} + \begin{Bmatrix} Q_2 \\ -Q_3 \end{Bmatrix}_{N②} = \begin{Bmatrix} Q_1 \\ 0 \\ 0 \end{Bmatrix}$$

The distributed heat flux vector is, using formula (3.29),

$$\{R_t\}_H = \{Q_t\}_{H①} + \{Q_t\}_{H②} = \begin{Bmatrix} Q_1 \\ Q_2 \end{Bmatrix}_{H①} + \begin{Bmatrix} Q_2 \\ Q_3 \end{Bmatrix}_{H②}$$

$$= \begin{Bmatrix} 15 \\ 30 \\ 0 \end{Bmatrix} + \begin{Bmatrix} 0 \\ 60 \\ 75 \end{Bmatrix} = \begin{Bmatrix} 15 \\ 90 \\ 75 \end{Bmatrix}$$

Substituting into (3.28) yields

$$0.2 \begin{bmatrix} 1 & -1 & 0 \\ -1 & 2 & -1 \\ 0 & -1 & 1 \end{bmatrix} \begin{Bmatrix} T_1 \\ T_2 \\ T_3 \end{Bmatrix} = \begin{Bmatrix} Q_1 + 15.0 \\ 90.0 \\ 75.0 \end{Bmatrix}$$

To solve this equation, we substitute the Dirichlet (essential) boundary condition, $T_1 = 100$ °F. To solve the resulting equation using an equation solver such as Gauss elimination, we force the boundary condition, $T_1 = 100$ °F, by placing unity on the diagonal of the first row of the conduction matrix and substituting zeros for all other first row entries. In addition, the first row entry on the vector on the right side is replaced with 100.

$$\begin{bmatrix} 1 & 0 & 0 \\ -1 & 2 & -1 \\ 0 & -1 & 1 \end{bmatrix} \begin{Bmatrix} T_1 \\ T_2 \\ T_3 \end{Bmatrix} = \begin{Bmatrix} 100 \\ 450 \\ 375 \end{Bmatrix}$$

Solving this equation, we obtain the nodal temperatures:

$$\begin{Bmatrix} T_1 \\ T_2 \\ T_3 \end{Bmatrix} = \begin{Bmatrix} 100 \\ 925 \\ 1300 \end{Bmatrix}$$

The heat flow Q_1 at the left boundary can be found by substituting the node temperatures back into the system equation:

$$0.2(T_1 - T_2) = Q_1 + 15$$

or

$$Q_1 = 150 \text{ Btu/hr}$$

Note that the assumption of a linear temperature gradient in the element limits the modeling capability because the calculated heat flow will be constant throughout the element. Using Fourier's law, the heat flux (heat flow per unit area) in element 1 can be determined.

$$q_{①} = -k\frac{dT}{dx} = -k[B]\{q_t\} = \frac{-3}{30}[-1 \quad 1]\begin{Bmatrix} 100 \\ 925 \end{Bmatrix}$$

$$= -82.5 \text{ Btu/hr-in.}^2$$

or a total heat flow of 165 Btu/hr. For element 2 we would obtain $q_{②} = -37.5$ Btu/hr-in.^{2} or a total heat flow of 75 Btu/hr.

A comparison of the estimated results with the exact solution is made in Figure 3.19, and we observe the same characteristics of the solution as were discussed in Section 3.10.

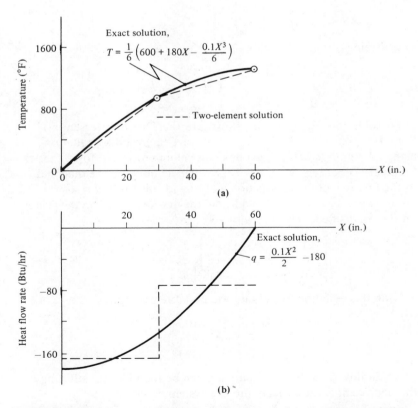

FIGURE 3.19 Comparison of exact solution and the two-element solution.

3.13 ONE-DIMENSIONAL HEAT FLOW WITH CONVECTION

In the derivation of the functional for the heat flow problem in Section 2.14, it was mentioned that convection at the end boundaries could be represented by substitution of equation (2.46) in the boundary heat flow vector; convection along the length of the region could be modeled using (2.47) in the distributed flux term.

Consider an element (Figure 3.20) which has the left and right ends exposed to an environment with ambient temperatures T_{∞_L} and T_{∞_R} and convection coefficients h_L and h_R, respectively. Also consider the length of the element exposed to environment of temperature T_{∞_H} and coefficient h_H. Recall equation (3.27):

$$[k_{cd}]\{q_t\} = \{Q_t\}_N + \{Q_t\}_H \tag{3.27}$$

If one end of the element is exposed to convection heat gain or loss, this nodal heat flow should be evaluated with the convection heat transfer relationship (2.46). For example, assume that convection occurs only at local node 1.

$$\{Q_t\}_N = \begin{Bmatrix} hA(T_{\infty_L} - T_1) \\ -Q_2 \end{Bmatrix} = \begin{Bmatrix} hAT_{\infty_L} \\ -Q_2 \end{Bmatrix} - \begin{Bmatrix} hAT_1 \\ 0 \end{Bmatrix}$$

$$= \begin{Bmatrix} hAT_{\infty_L} \\ -Q_2 \end{Bmatrix} - hA\begin{bmatrix} 1 & 0 \\ 0 & 0 \end{bmatrix}\begin{Bmatrix} T_1 \\ T_2 \end{Bmatrix} \tag{3.30a}$$

or

$$\{Q_t\}_N = \{Q_{cv}\}_L - [k_{cv}]_L\{q_t\} \tag{3.30b}$$

If only the right end has convection, then by substituting $Q_2 = hA(T_2 - T_{\infty_R})$, we have

$$\{Q_t\}_N = \begin{Bmatrix} Q_1 \\ hAT_{\infty_R} \end{Bmatrix} - hA\begin{bmatrix} 0 & 0 \\ 0 & 1 \end{bmatrix}\begin{Bmatrix} T_1 \\ T_2 \end{Bmatrix} \tag{3.31a}$$

or

$$\{Q_t\}_N = \{Q_{cv}\}_R - [k_{cv}]_R\{q_t\} \tag{3.31b}$$

where in (3.30) and (3.31), $\{Q_{cv}\}$ and $[k_{cv}]$ are the convection heat flow vectors and matrices, respectively.

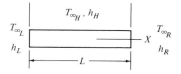

FIGURE 3.20 Region with all sides exposed to convection with ambient temperatures and convection coefficients.

For convection along the length of the element, the distributed heat flux can be replaced with the convection heat transfer relationship as follows:

$$\{Q_t\}_H = \int_0^L H(x)[N]^T dx = \int_0^L hP(T_{\infty_H} - T(x))[N]^T dx$$

where P is the perimeter of the exposed section and assumed constant over the length of the element:

$$\{Q_t\}_H = hPT_{\infty_H}\int_0^L [N]^T dx - hP\int_0^L [N]^T[N] dx \{q_t\}$$

Recalling the shape function matrix in terms of local coordinates, we have

$$[N] = \begin{bmatrix} 1 - \dfrac{x}{L} & \dfrac{x}{L} \end{bmatrix}$$

we can determine the following integrals:

$$\int_0^L [N]^T dx = L\begin{bmatrix} \tfrac{1}{2} & \tfrac{1}{2} \end{bmatrix}^T$$

$$\int_0^L [N]^T[N] dx = \dfrac{L}{6}\begin{bmatrix} 2 & 1 \\ 1 & 2 \end{bmatrix}$$

We thus have the following relationship for the distributed flux:

$$\{Q_t\}_H = \dfrac{hPLT_\infty}{2}_H \begin{Bmatrix} 1 \\ 1 \end{Bmatrix} - \dfrac{hPL}{6}\begin{bmatrix} 2 & 1 \\ 1 & 2 \end{bmatrix}\begin{Bmatrix} T_1 \\ T_2 \end{Bmatrix} \quad (3.32a)$$

or

$$\{Q_t\}_H = \{Q_{cv}\}_H - [k_{cv}]_H\{q_t\} \quad (3.32b)$$

where again we have a convection heat flow vector and matrix.

Now let us put these relationships together for a particular situation. Assume a single element with convection on the left boundary and along the length and heat flow Q_2 at the right boundary. From (3.27)

$$[k_{cd}]\{q_t\} = \{Q_t\}_N + \{Q_t\}_H$$

$$= \{Q_{cv}\}_L - [k_{cv}]_L\{q_t\} + \{Q_{cv}\}_H - [k_{cv}]_H\{q_t\}$$

$$[k_{cd}]\begin{Bmatrix} T_1 \\ T_2 \end{Bmatrix} = \begin{Bmatrix} hAT_{\infty_L} \\ -Q_2 \end{Bmatrix} - hA\begin{bmatrix} 1 & 0 \\ 0 & 0 \end{bmatrix}\begin{Bmatrix} T_1 \\ T_2 \end{Bmatrix}$$

$$+ \dfrac{hPLT_{\infty_H}}{2}\begin{Bmatrix} 1 \\ 1 \end{Bmatrix} - \dfrac{hPL}{6}\begin{bmatrix} 2 & 1 \\ 1 & 2 \end{bmatrix}\begin{Bmatrix} T_1 \\ T_2 \end{Bmatrix}$$

The square matrices can be added to form a thermal stiffness matrix resulting from conduction and convection. This yields

$$[[k_{cd}] + [k_{cv}]_L + [k_{cv}]_H]\{q_t\} = \{Q_{cv}\}_L + \{Q_{cv}\}_H$$

3.13 ONE-DIMENSIONAL HEAT FLOW WITH CONVECTION

where the convection vectors and matrices are defined in (3.30) and (3.32). Alternatively, if the element has convection on the right face,

$$[[k_{cd}] + [k_{cv}]_R + [k_{cv}]_H]\{q_t\} = \{Q_{cv}\}_R + \{Q_{cv}\}_H$$

where the convection matrices and vectors are defined in (3.31) and (3.32).

The following example will illustrate the application and assembly of the element equations for the simple model of a fin.

Example 3.4 The heat flow in the rectangular fin shown in Figure 3.21 is to be modeled as a one-dimensional problem. The left end of the fin is maintained at a temperature of 200°C and all other surfaces are exposed to an ambient temperature of 50°C. The convection coefficient for all exposed surfaces is 0.02 W/cm² · °C. The thermal conductivity of the material is 4 W/cm · °C.

Using first a single-element model and then a two-element model, estimate the temperature of the exposed end of the fin and the heat dissipated.

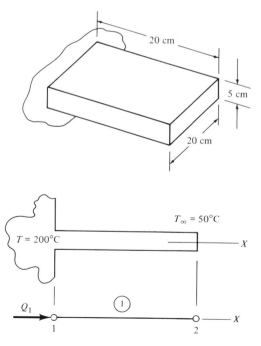

FIGURE 3.21 Rectangular fin geometry and the single-element model.

Solution
Single-Element Model Calculate the conduction matrix:

$$[k_{cd}] = \frac{Ak}{L}\begin{bmatrix} 1 & -1 \\ -1 & 1 \end{bmatrix} = \frac{100(4)}{20}\begin{bmatrix} 1 & -1 \\ -1 & 1 \end{bmatrix}$$

$$= \begin{bmatrix} 20 & -20 \\ -20 & 20 \end{bmatrix}$$

The element has convection at the right face only. The convection matrix for heat flow from the right face is

$$[k_{cv}]_R = hA\begin{bmatrix} 0 & 0 \\ 0 & 1 \end{bmatrix} = 0.02(100)\begin{bmatrix} 0 & 0 \\ 0 & 1 \end{bmatrix}$$

$$= \begin{bmatrix} 0 & 0 \\ 0 & 2 \end{bmatrix}$$

Next, calculate the convection matrix for heat flow from all other sides.

$$[k_{cv}]_H = \frac{hPL}{6}\begin{bmatrix} 2 & 1 \\ 1 & 2 \end{bmatrix} = \frac{0.02(50)(20)}{6}\begin{bmatrix} 2 & 1 \\ 1 & 2 \end{bmatrix}$$

$$= \begin{bmatrix} 6.7 & 3.3 \\ 3.3 & 6.7 \end{bmatrix}$$

Determine the convection vector for right-face convection.

$$\{Q_{cv}\}_R = \begin{Bmatrix} Q_1 \\ hAT_{\infty R} \end{Bmatrix} = \begin{Bmatrix} Q_1 \\ 0.02(100)(50) \end{Bmatrix}$$

$$= \begin{Bmatrix} Q_1 \\ 100 \end{Bmatrix}$$

Next, determine the convection matrix for the remaining free sides.

$$\{Q_{cv}\}_H = \frac{hPLT_{\infty H}}{2}\begin{Bmatrix} 1 \\ 1 \end{Bmatrix} = \frac{0.02(50)(20)(50)}{2}\begin{Bmatrix} 1 \\ 1 \end{Bmatrix}$$

$$= \begin{Bmatrix} 500 \\ 500 \end{Bmatrix}$$

Assemble the complete element heat flow matrix equation.

$$[[k_{cd}] + [k_{cv}]_R + [k_{cv}]_H]\{q_t\} = \{Q_{cv}\}_R + \{Q_{cv}\}_H$$

$$\begin{bmatrix} 26.7 & -16.7 \\ -16.7 & 28.7 \end{bmatrix}\begin{Bmatrix} T_1 \\ T_2 \end{Bmatrix} = \begin{Bmatrix} Q_1 + 500 \\ 600 \end{Bmatrix}$$

Substitute the (essential) boundary condition, $T_1 = 200$:

$$\begin{bmatrix} 1 & 0 \\ -16.7 & 28.7 \end{bmatrix}\begin{Bmatrix} T_1 \\ T_2 \end{Bmatrix} = \begin{Bmatrix} 200 \\ 600 \end{Bmatrix}$$

Solving for T_2 yields

$$-16.7(200) + 28.7T_2 = 600 \qquad T_2 = 137.3°C$$

3.13 ONE-DIMENSIONAL HEAT FLOW WITH CONVECTION

The heat flow Q_1 in the left face can be obtained from the "top" equation in the matrix equation.

$$26.7T_1 - - 16.7T_2 = Q_1 + 500$$
$$26.7(200) - 16.7(137.3) = Q_1 + 500 \qquad Q_1 = 2547 \text{ W}$$

The average heat flow in the element is calculated using Fourier's law:

$$q_1 = -k\frac{dT}{dx} = -k[B]\{q_t\} = -\frac{k}{L}[-1 \quad 1]\begin{Bmatrix} T_1 \\ T_2 \end{Bmatrix}$$

$$= -\frac{4}{20}[-1 \quad 1]\begin{Bmatrix} 200 \\ 137.3 \end{Bmatrix}$$

$$= 26.3 \text{ W/cm}^2 \text{ or a total heat flow of 2630 W}$$

Remember, the element we are using models a constant heat flow throughout the length of the element; a gradient is not possible.

Two-Element Model The two-element model is shown in Figure 3.22. First, calculate the conduction matrix for element 1. This element has half the length of the element in the previous single element model, so the conduction matrix entries are doubled.

$$[k_{cd}]_{①} = \begin{bmatrix} 40 & -40 \\ -40 & 40 \end{bmatrix}$$

There is no convection at the right face of element 1 since it joins element 2 and heat flow is by conduction.

$$[k_{cv}]_{R①} = [0]$$

From the single-element model, the convection matrix for the exposed sides is determined to be

$$[k_{cv}]_{H①} = \begin{bmatrix} 3.35 & 1.65 \\ 1.65 & 3.35 \end{bmatrix}$$

Since there is no convection at the local node 2 of element 1, we will not have a convection vector. However, we must include the nodal heat flow vector for heat conducted in node 1 and through node 2.

$$\{Q_t\}_{N①} = \begin{Bmatrix} Q_1 \\ -Q_2 \end{Bmatrix}$$

The convection vector for the free sides is, from the single-element model,

$$\{Q_{cv}\}_{H①} = \begin{Bmatrix} 250 \\ 250 \end{Bmatrix}$$

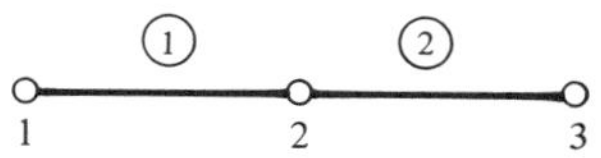

FIGURE 3.22 Two-element model.

The complete equation for element 1 becomes

$$\begin{bmatrix} 43.35 & -38.35 \\ -38.35 & 43.35 \end{bmatrix} \begin{Bmatrix} T_1 \\ T_2 \end{Bmatrix} = \begin{Bmatrix} Q_1 \\ -Q_2 \end{Bmatrix} + \begin{Bmatrix} 250 \\ 250 \end{Bmatrix}$$

Element 2 has convection at node 3 (local node 2) and conduction at global node 2. The matrices and vectors are as follows:

$$[k_{cd}]_{\textcircled{2}} = \begin{bmatrix} 40 & -40 \\ -40 & 40 \end{bmatrix}, \quad [k_{cv}]_{R\textcircled{2}} = \begin{bmatrix} 0 & 0 \\ 0 & 2 \end{bmatrix}$$

$$[k_{cv}]_{H\textcircled{2}} = \begin{bmatrix} 3.35 & 1.65 \\ 1.65 & 3.35 \end{bmatrix}, \quad \{Q_{cv}\}_{H\textcircled{2}} = \begin{Bmatrix} 250 \\ 250 \end{Bmatrix}$$

$$\{Q_{cv}\}_{R\textcircled{2}} = \begin{Bmatrix} Q_1 \\ hAT_{\infty R} \end{Bmatrix}_{local} = \begin{Bmatrix} Q_2 \\ hAT_{\infty R} \end{Bmatrix}_{global} = \begin{Bmatrix} Q_2 \\ 100 \end{Bmatrix}$$

The complete equation for element 2 becomes

$$\begin{bmatrix} 43.35 & -38.35 \\ -38.35 & 45.35 \end{bmatrix} \begin{Bmatrix} T_2 \\ T_3 \end{Bmatrix} = \begin{Bmatrix} Q_2 \\ 100 \end{Bmatrix} + \begin{Bmatrix} 250 \\ 250 \end{Bmatrix}$$

Assembling the equations for the two elements, we have

$$\begin{bmatrix} 43.35 & -38.35 & 0 \\ -38.35 & 86.70 & -38.35 \\ 0 & -38.35 & 45.35 \end{bmatrix} \begin{Bmatrix} T_1 \\ T_2 \\ T_3 \end{Bmatrix} = \begin{Bmatrix} Q_1 \\ 0 \\ 100 \end{Bmatrix} + \begin{Bmatrix} 250 \\ 500 \\ 250 \end{Bmatrix}$$

Substituting the boundary temperature, $T_1 = 200°C$, and solving gives us

$$\begin{Bmatrix} T_2 \\ T_3 \end{Bmatrix} = \begin{Bmatrix} 156 \\ 140 \end{Bmatrix} \quad °C$$

The heat flow, Q_1 into the fin can be calculated from the "top" equation in the system matrix equation.

$$43.35 T_1 - 38.35 T_2 = Q_1 + 250 \quad Q_1 = 2438 \text{ W}$$

The heat flow in element 1 is

$$q_{\textcircled{1}} = -k[B]\{q_t\} = \frac{-4}{10}[-1 \quad 1]\begin{Bmatrix} 200 \\ 156 \end{Bmatrix} = 17.6 \text{ W/cm}^2$$

or a total heat flow of 1760 W. The heat flow in element 2 is

$$q_{\textcircled{2}} = \frac{-4}{10}[-1 \quad 1]\begin{Bmatrix} 156 \\ 140 \end{Bmatrix} = 6.4 \text{ W/cm}^2$$

or a total heat flow of 640 W.

The reader is encouraged to refine this model further, calculate temperatures and heat flow, and examine convergence. □

3.14 PROGRAMMING NONZERO DIRICHLET (ESSENTIAL) BOUNDARY CONDITIONS

The solution of any problem requires the satisfaction of the natural and essential boundary conditions. In the three types of problems that we have looked at in some detail—structures, elasticity, and heat flow—the natural and essential boundary conditions have been identified as, respectively, boundary force and displacement for the structures and elasticity problem, boundary heat flow and temperature for the heat flow problem. We have mentioned that essential boundary conditions must be specified in order to have a unique solution. For the elasticity (and structures) problem, for example, sufficient boundary displacements must be specified to prevent rigid-body motion.

In all the structure and elasticity examples shown so far, the essential (displacement) boundary values have been zero. Substitution of these values into the assembled global force–displacement equation has permitted a reduction of the equation by removal of the rows and columns associated with those zero displacements. If nonzero boundary displacements are specified, a different solution procedure must be followed. The following is a description of two approaches (Zienkiewicz, 1971). The choice of method depends on the desired alteration of the "stiffness" matrix. Note that "stiffness" was placed in quotes because the method is a general procedure, applicable to specification of the essential boundary condition for elasticity or any other problem type.

For discussion purposes, we assume a typical equation of the following form:

$$[K]\{\phi\} = \{R\}$$

where we will have the essential nonzero boundary values specified for $\{\phi\}$.

The first procedure is recommended if there is no need to maintain the symmetry of the stiffness matrix.

1. If the value of ϕ_n is given at node n, replace the (n, n) entry in matrix $[K]$ with a 1, enter zeros in all other positions in the nth row, and enter the given value of ϕ_n in the nth row of the $\{R\}$ vector.

If this simple procedure sounds familiar, it is because it was introduced in Example 3.3. Note that although the symmetry of the matrix is lost, it does not affect the use of a general Gauss elimination routine.

The second technique is used if the symmetry of the matrix is to be maintained.

2. If the value of ϕ_n is given at node n as ϕ_n, multiply the (n, n) entry, $k_{n,n}$, of the $[K]$ matrix by a number G which is many orders of magnitude larger than any number in the matrix $[K]$. In addition,

replace the entry in row n of the vector $\{R\}$ with the product $(\phi_n)(G)(k_{n,n})$.

Note that this does not alter any of the off-diagonal terms in the matrix $[K]$, so symmetry is retained. The solution now follows using the Gauss or any other routine, and you can appreciate that with the number G large enough, all quantities in row n remain of negligible magnitude relative to the diagonal value. This second technique will be illustrated in Example 10.2.

3.15 CONCLUSION

In this chapter we have introduced the finite element method for the one-dimensional problem and have attempted to convey the message that this is a technique that has application to any problem for which a functional has been derived. Two problem types have been used, and simple examples have been used with the sole purpose of illustrating the basic steps underlying the method.

In the next chapter we expand our region to two dimensions. We focus initially on the elasticity problem, and later in the text bring in applications to other problems.

PROBLEMS

3.1 The steel member shown in Figure P3.1 has a cross-sectional area of 2 in.^{2}. Assuming a linear displacement field throughout the element, solve for the strain energy in terms of the displacements of the ends.

3.2 Assuming a linear displacement field throughout the element in Figure P3.2, solve for the nodal forces that are the energy equivalent of the distributed loading applied to the element.

$$T = 2 + 5X \quad \text{N/cm}$$

where X is the global coordinate.

3.3 Assuming a linear displacement field throughout the element shown in Figure P3.3, solve for the nodal forces that are the energy equivalent of the distributed loading applied to the element.

$$T = 10x^2 \quad \text{lb/in.}$$

where x is the local coordinate.

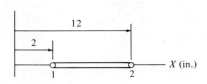

FIGURE P3.1

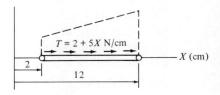

FIGURE P3.2

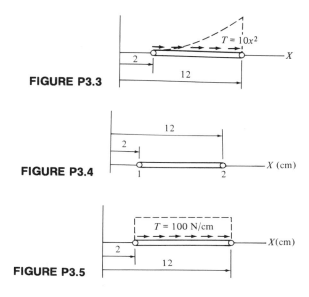

FIGURE P3.3

FIGURE P3.4

FIGURE P3.5

3.4 Assume the following displacement field throughout the element shown in Figure P3.4:
$$u = c_1 + c_2 X^2$$
where X is the global coordinate. Solve for the strain energy in the element in terms of the displacements of the ends of the element.

3.5 Assume the following displacement field throughout the element shown in Figure P3.5:
$$u = c_1 + c_2 x^2$$
where x is the local coordinate. Solve for the nodal forces that are the energy equivalent of the distributed loading applied to the element.

3.6 A steel rod is attached to rigid walls at each end and is subjected to the distributed loading $T(X)$ and the concentrated force of 4000 lb (Figure P3.6). Determine the stress distribution in the rod.

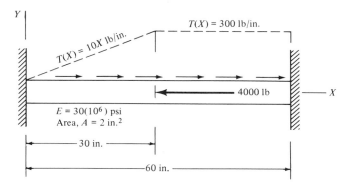

FIGURE P3.6

142 AN ELEMENT FOR THE ONE-DIMENSIONAL PROBLEM

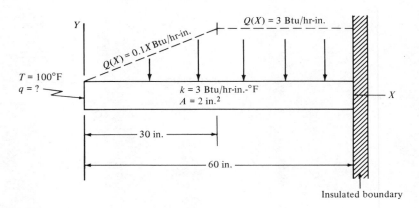

FIGURE P3.7

3.7 The bar shown in Figure P3.7 is subjected to a linearly varying and constant distributed heat source. The right end is against an insulated boundary, the left end is held at a temperature of 100°F. (Heat flow can occur into or out of the left end.) Determine the temperature distribution in the bar.

3.8 Derive the distributed load equivalent nodal load formula (3.15).

3.9 It has been proposed that the following displacement function be used for the uniaxial stress element (Figure P3.9):

$$u = \cos\frac{\pi x}{2L} u_1 + \left(1 - \cos\frac{\pi x}{2L}\right) u_2$$

Solve for the energy equivalent nodal loads for a uniformly distributed load, w, with units of force/length.

3.10 Solve Example 3.4 using four and then eight elements.

3.11 The bar shown in Figure P3.11 has a square cross section of 1 cm². All sides are insulated, but the ends are exposed. The ambient temperature at each end is 30°C, and the convection coefficient is 0.5 W/cm² · °C. If the center of the bar has a heat source of 50 W, determine the temperature distribution along the length of the bar. The conduction coefficient for the material is 8 W/cm · °C.

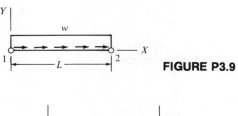

FIGURE P3.9

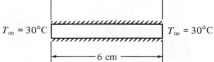

FIGURE P3.11

PROBLEMS

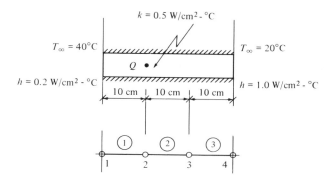

FIGURE P3.12

3.12 A bar of length equal to 30 cm has a square cross section of area equal to 4 cm². All sides are insulated except for the ends, which have heat transfer by convection with temperatures and coefficients as defined Figure P3.12. A heat source of 50 W is applied at the location 10 cm from the left end. Construct the finite element matrix equation for the three-element model. Solve for the temperatures at the four nodes.

CHAPTER 4

Development of the Constant-Strain Triangular Element

4.1 INTRODUCTION

The development of the one-dimensional stress element in Chapter 3 served to illustrate, with a minimum of complexity, the steps involved in constructing a continuum finite element model. In this chapter we develop a triangular element that can be (as is) used in the analysis of continua under plane strain or plane stress conditions. This particular triangular element will be characterized by having a constant strain (and stress) throughout.

In this chapter development of the triangle is detailed twice. First, the development will employ the global coordinates exclusively. Second, the local coordinate system will be used with results ultimately related to the global system. This method will clearly bring out the advantages and disadvantages of these coordinate systems.

After another look at the natural coordinate systems in Chapter 5, this triangle will be developed once again in Chapter 6 using the triangle natural coordinates.

DEVELOPMENT OF THE CONSTANT-STRAIN TRIANGULAR ELEMENT

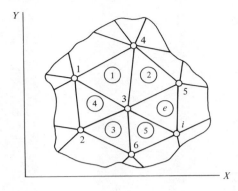

FIGURE 4.1 Continuum subdivided into triangular elements.

4.2 THE TRIANGLE ELEMENT: GLOBAL COORDINATES

Figure 4.1 shows a portion of a continuum subdivided into (solid) triangles. The global coordinate system is shown, and every node in the assemblage of triangles has a distinct global number. Each triangle is also identified with a number.

Figure 4.2 identifies the local numbering convention which is universally adopted. The nodes are always numbered in succession moving counterclockwise as shown. Each node of the triangle as well as all points in the interior of the element can undergo an X displacement, u, and a Y displacement, v.

We will now proceed through the five steps of the element development following the general format used in Chapter 3.

Step 1. Assumed Displacement Function. We assume displacement functions linear in X and Y as the simplest functions that yield piecewise continuous first derivatives over the entire domain of the continuum.

$$u(X,Y) = \alpha_1 + \alpha_2 X + \alpha_3 Y \quad (4.1a)$$
$$v(X,Y) = \beta_1 + \beta_2 X + \beta_3 Y \quad (4.1b)$$

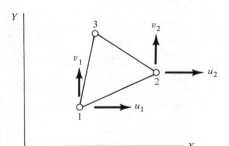

FIGURE 4.2 Local numbering convention.

4.2 THE TRIANGLE ELEMENT: GLOBAL COORDINATES

These equations involve the constants α_i and β_i (often referred to as *generalized coordinates*), which can be determined as follows. First, express the X displacement of each node using equation (4.1a):

$$u_1 = \alpha_1 + \alpha_2 X_1 + \alpha_3 Y_1$$
$$u_2 = \alpha_1 + \alpha_2 X_2 + \alpha_3 Y_2$$
$$u_3 = \alpha_1 + \alpha_2 X_3 + \alpha_3 Y_3$$

In matrix form,

$$\begin{Bmatrix} u_1 \\ u_2 \\ u_3 \end{Bmatrix} = \begin{bmatrix} 1 & X_1 & Y_1 \\ 1 & X_2 & Y_2 \\ 1 & X_3 & Y_3 \end{bmatrix} \begin{Bmatrix} \alpha_1 \\ \alpha_2 \\ \alpha_3 \end{Bmatrix}$$

or

$$\{q_1\} = [A_1]\{\alpha\} \qquad (4.2)$$

In a similar manner, using equation (4.1b), we have

$$\begin{Bmatrix} v_1 \\ v_2 \\ v_3 \end{Bmatrix} = \begin{bmatrix} 1 & X_1 & Y_1 \\ 1 & X_2 & Y_2 \\ 1 & X_3 & Y_3 \end{bmatrix} \begin{Bmatrix} \beta_1 \\ \beta_2 \\ \beta_3 \end{Bmatrix}$$

or

$$\{q_2\} = [A_1]\{\beta\} \qquad (4.3)$$

The generalized coordinates α_i and β_i are determined by premultiplying both sides of equations (4.2) and (4.3) by the inverse of the square matrix of nodal coordinates.

$$[A_1]^{-1} = \begin{bmatrix} 1 & X_1 & Y_1 \\ 1 & X_2 & Y_2 \\ 1 & X_3 & Y_3 \end{bmatrix}^{-1} = \frac{\text{adjoint of } [A_1]}{\text{determinant of } [A_1]}$$

$$= \frac{1}{\det} \begin{bmatrix} a_1 & a_2 & a_3 \\ b_1 & b_2 & b_3 \\ c_1 & c_2 & c_3 \end{bmatrix} \qquad (4.4)$$

where

$$a_1 = X_2 Y_3 - X_3 Y_2 \qquad a_2 = X_3 Y_1 - X_1 Y_3 \qquad a_3 = X_1 Y_2 - X_2 Y_1$$
$$b_1 = Y_2 - Y_3 \qquad b_2 = Y_3 - Y_1 \qquad b_3 = Y_1 - Y_2$$
$$c_1 = X_3 - X_2 \qquad c_2 = X_1 - X_3 \qquad c_3 = X_2 - X_1$$

$$\det = X_2 Y_3 - X_3 Y_2 + X_1(Y_2 - Y_3) + Y_1(X_3 - X_2)$$
$$= 2 \text{ (area of the triangle)}$$

From equations (4.2) and (4.3),

$$\{\alpha\} = [A_1]^{-1}\{q_1\}$$
$$\{\beta\} = [A_1]^{-1}\{q_2\}$$

Substituting into equation (4.1) yields

$$u = [1 \quad X \quad Y]\{\alpha\}$$
$$= [1 \quad X \quad Y][A_1]^{-1}\{q_1\} \qquad (4.5)$$
$$v = [1 \quad X \quad Y][A_1]^{-1}\{q_2\}$$

where we recall that

$$\{q_1\} = \begin{Bmatrix} u_1 \\ u_2 \\ u_3 \end{Bmatrix} \qquad \{q_2\} = \begin{Bmatrix} v_1 \\ v_2 \\ v_3 \end{Bmatrix}$$

Expansion of $[1 \quad X \quad Y][A_1]^{-1}$ yields

$$[1 \quad X \quad Y][A_1]^{-1}$$
$$= \frac{1}{\det}[a_1 + Xb_1 + Yc_1 \quad a_2 + Xb_2 + Yc_2 \quad a_3 + Xb_3 + Yc_3]$$
$$= [N_1 \quad N_2 \quad N_3]$$

Therefore, we can write the displacement in the interior of the element in terms of the displacement of the nodes.

$$u = [N_1 \quad N_2 \quad N_3]\begin{Bmatrix} u_1 \\ u_2 \\ u_3 \end{Bmatrix}$$

$$v = [N_1 \quad N_2 \quad N_3]\begin{Bmatrix} v_1 \\ v_2 \\ v_3 \end{Bmatrix}$$

or in matrix form,

$$\begin{Bmatrix} u \\ v \end{Bmatrix} = \begin{bmatrix} N_1 & 0 & N_2 & 0 & N_3 & 0 \\ 0 & N_1 & 0 & N_2 & 0 & N_3 \end{bmatrix} \begin{Bmatrix} u_1 \\ v_1 \\ u_2 \\ v_2 \\ u_3 \\ v_3 \end{Bmatrix}$$

In symbolic form,

$$\{u\} = [N]\{q\} \qquad (4.6)$$

This completes the determination of the element displacement function. Note that the displacement of any point in the interior of

4.2 THE TRIANGLE ELEMENT: GLOBAL COORDINATES

the triangle is in terms of the nodal displacements. These nodal displacements are the unknowns to be determined in a manner similar in theory to that presented for the uniaxial element in Chapter 3.

Step 2. Derivation of Strain and Strain Energy. We will devote our attention to developing the element to model the plane stress state, the thickness of the continuum being small in the z coordinate direction.

Referring to equation (2.5), we obtain the following definitions of the required strains:

$$\begin{Bmatrix} \epsilon_X \\ \epsilon_Y \\ \gamma_{XY} \end{Bmatrix} = \begin{bmatrix} \frac{\partial}{\partial X} & 0 \\ 0 & \frac{\partial}{\partial Y} \\ \frac{\partial}{\partial Y} & \frac{\partial}{\partial X} \end{bmatrix} \begin{Bmatrix} u \\ v \end{Bmatrix} = \begin{bmatrix} \frac{\partial}{\partial X} & 0 \\ 0 & \frac{\partial}{\partial Y} \\ \frac{\partial}{\partial Y} & \frac{\partial}{\partial X} \end{bmatrix} [N]\{q\}$$

$$= \frac{1}{\det} \begin{bmatrix} Y_2 - Y_3 & 0 & Y_3 - Y_1 & 0 & Y_1 - Y_2 & 0 \\ 0 & X_3 - X_2 & 0 & X_1 - X_3 & 0 & X_2 - X_1 \\ X_3 - X_2 & Y_2 - Y_3 & X_1 - X_3 & Y_3 - Y_1 & X_2 - X_1 & Y_1 - Y_2 \end{bmatrix} \begin{Bmatrix} u_1 \\ v_1 \\ u_2 \\ v_2 \\ u_3 \\ v_3 \end{Bmatrix}$$

(4.7)

where

$$\det = X_2 Y_3 - X_3 Y_2 + X_1(Y_2 - Y_3) + Y_1(X_3 - X_2)$$

In symbolic form,

$$\{\epsilon\} = [B]\{q\} \tag{4.8}$$

This now completes the derivation of the relationship that defines the strain at any point in the element in terms of the displacements of the nodes. Next, we derive the expression for the strain energy in the element.

Recall equation (2.16) for the strain energy in a continuum subjected to plane stress.

$$U = \int_\Omega \tfrac{1}{2} \{\epsilon\}^T [C]\{\epsilon\} \, d\Omega \tag{2.16}$$

The constitutive matrix $[C]$ is defined in equation (2.7). If we now substitute equation (4.8) into (2.16), we obtain the following:

$$U = \tfrac{1}{2} \int_\Omega \{q\}^T [B]^T [C][B]\{q\} \, d\Omega$$

Note that the differential volume equals $h \, dA$, where h is the thickness of

150 DEVELOPMENT OF THE CONSTANT-STRAIN TRIANGULAR ELEMENT

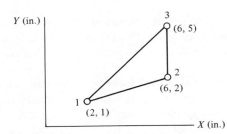

FIGURE 4.3 Single triangular element.

the element and dA is the differential area. Also, observe that matrices $\{q\}$, $[B]$, and $[C]$ are constant. This is obvious because matrix $[B]$ contains only the nodal coordinate values, and $[C]$ contains material constants. Realizing this, we find that the integral for the strain energy becomes

$$U = \tfrac{1}{2}\{q\}^T[B]^T[C][B]\{q\}h\int_A dA$$

Thus the strain energy in one element is given by

$$U = \tfrac{1}{2}\{q\}^T[B]^T[C][B]\{q\}hA \qquad (4.9)$$

where the element area equals $\tfrac{1}{2}$ det.

Example 4.1 A steel triangular element of thickness h is located in the global XY plane as shown in Figure 4.3. Solve for the strain in the triangle in terms of the nodal displacements.

Solution. Refer back to equation (4.8): $\{\epsilon\} = [B]\{q\}$, where the matrix $[B]$ is a 3×6 matrix of the spatial derivatives of the elements of the $[N]$ matrix as defined in equation (4.7). Substituting the coordinates of the nodes, we obtain

$$\det = 12$$

$$[B] = \begin{bmatrix} -3 & 0 & 4 & 0 & -1 & 0 \\ 0 & 0 & 0 & -4 & 0 & 4 \\ 0 & -3 & -4 & 4 & 4 & -1 \end{bmatrix} \frac{1}{12}$$

Thus the strain is defined by

$$\begin{Bmatrix} \epsilon_X \\ \epsilon_Y \\ \gamma_{XY} \end{Bmatrix} = \begin{bmatrix} -\tfrac{1}{4} & 0 & \tfrac{1}{3} & 0 & -\tfrac{1}{12} & 0 \\ 0 & 0 & 0 & -\tfrac{1}{3} & 0 & \tfrac{1}{3} \\ 0 & -\tfrac{1}{4} & -\tfrac{1}{3} & \tfrac{1}{3} & \tfrac{1}{3} & -\tfrac{1}{12} \end{bmatrix} \begin{Bmatrix} u_1 \\ v_1 \\ u_2 \\ v_2 \\ u_3 \\ v_3 \end{Bmatrix}$$

□

4.2 THE TRIANGLE ELEMENT: GLOBAL COORDINATES

Example 4.2 The strain energy in the element is determined using equation (4.9). This equation requires the matrix product $[B]^T[C][B]$. Evaluate this product for the element in the Example 4.1 and determine the strain energy in terms of the displacements of the nodes.

Solution. The matrix $[B]$ was determined in Example 4.1. The constitutive matrix $[C]$ is defined by equation (2.7):

$$[C] = \frac{E}{1 - \nu^2} \begin{bmatrix} 1 & \nu & 0 \\ \nu & 1 & 0 \\ 0 & 0 & \frac{1-\nu}{2} \end{bmatrix}$$

If we let $\phi = (1 - \nu)/2$,

$$[C][B] = \frac{E}{12(1-\nu^2)} \begin{bmatrix} -3 & 0 & 4 & -4\nu & -1 & 4\nu \\ -3\nu & 0 & 4\nu & -4 & -\nu & 4 \\ 0 & -3\phi & -4\phi & 4\phi & 4\phi & -\phi \end{bmatrix}$$

For steel, $E = 30(10^6)$ psi and $\nu = 0.3$. The matrix product $[B]^T[C][B]$ becomes

$$[B]^T[C][B] = \begin{bmatrix} 9 & 0 & -12 & 3.6 & 3 & -3.6 \\ & 3.15 & 4.2 & -4.2 & -4.2 & 1.05 \\ & & 21.6 & -10.4 & -9.6 & 6.2 \\ & & & 21.6 & 6.8 & -17.4 \\ & \text{symmetric} & & & 6.6 & -2.6 \\ & & & & & 16.35 \end{bmatrix} [2.2894(10^5)]$$

From equation (4.9) we obtain the strain energy by multiplying the matrix product above by $\frac{1}{2}$ (thickness, h) (element area) $= 3h$ and also by the nodal displacement vector $\{q\}$ and its transpose.

$$U = \{q\}^T \times \begin{bmatrix} 9 & 0 & -12 & 3.6 & 3 & -3.6 \\ & 3.15 & 4.2 & -4.2 & -4.2 & 1.05 \\ & & 21.6 & -10.4 & -9.6 & 6.2 \\ & & & 21.6 & 6.8 & -17.4 \\ & \text{symmetric} & & & 6.6 & -2.6 \\ & & & & & 16.35 \end{bmatrix} \{q\} 6.87(10^5) h \quad \text{in.-lb}$$

where

$$\{q\}^T = \begin{bmatrix} u_1 & v_1 & u_2 & v_2 & u_3 & v_3 \end{bmatrix} \qquad \square$$

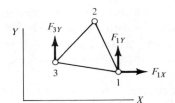

FIGURE 4.4 Examples of nodal forces.

Step 3. Derivation of Applied Load Energy Functions. Much of the following discussion of the derivation of force energy potential functions has been introduced in Sections 2.5 and 3.5. Reference to those articles may prove to be helpful as you proceed.

There are three types of element loading to be considered.

1. *Concentrated forces applied to nodes*: The potential energy function, V, for a force $\bar{F}$ having the components F_X and F_Y is defined by the expression (2.13).

$$V = -F_X u - F_Y v \tag{2.13}$$

For a force at node 1 with the components F_{1X} and F_{1Y} (Figure 4.4) the potential energy is

$$V_1 = -F_{1X} u_1 - F_{1Y} v_1$$

Extending this development to all force components acting at all nodes of the element, we obtain the complete scalar potential energy function, V_{NF}, for the nodal forces:

$$V_{\text{NF}} = -\{q\}^T \{Q\}_{\text{NF}} \tag{4.10}$$

where $\{Q\}_{\text{NF}}$ is the vector of element nodal forces.

$$\{Q\}_{\text{NF}}^T = [F_{1X} \quad F_{1Y} \quad F_{2X} \quad F_{2Y} \quad F_{3X} \quad F_{3Y}]$$

2. *Forces distributed along the edge of the element*: Distributed edge loading intensities are specified as force per unit length or force per unit area. The force per unit area at a particular point is called the *stress traction vector*, $\bar{T}$.

The force produced by the stress traction vector acting in the direction defined by the unit vector $\hat{n}$ on the area dA along one side of the element (Figure 4.5) is given by

$$\overline{dF} = T \, dA \, \hat{n}$$

The differential potential energy function, dV_T, for this differential force can be expressed as

$$dV_T = -(T_X \, dA) u - (T_Y \, dA) v$$

where u and v are the displacement components for the arbitrary point

4.2 THE TRIANGLE ELEMENT: GLOBAL COORDINATES

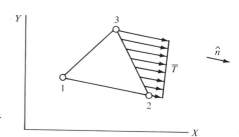

FIGURE 4.5 Example of distributed stress traction.

on the element boundary. Letting the thickness of the continuum (element) be h, this can be written as

$$V_T = -\int_S [u \quad v] \begin{Bmatrix} T_X \\ T_Y \end{Bmatrix} h \, dS$$

Substituting expression (4.6) gives us

$$V_T = -h \int_S \{q\}^T [N]^T \begin{Bmatrix} T_X \\ T_Y \end{Bmatrix} dS$$

Since the $\{q\}$ vector is independent of the integration variable, this can be written as

$$V_T = -h\{q\}^T \int_S [N]^T \begin{Bmatrix} T_X \\ T_Y \end{Bmatrix} dS \qquad (4.11)$$

Example 4.3 The triangular element of Example 4.1 is h units thick and has a uniform stress traction (force per unit area) on the side as shown in Figure 4.6. Solve for the potential energy function for this distributed loading in terms of the element node displacements.

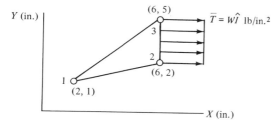

FIGURE 4.6 Uniform stress traction on one edge.

Solution. The solution will be obtained using expression (4.11):

$$V_T = -h\{q\}^T \int_S [N]^T \begin{Bmatrix} T_X \\ T_Y \end{Bmatrix} dS \qquad (4.11)$$

$$= -h\{q\}^T \int_S \begin{bmatrix} N_1 & 0 \\ 0 & N_1 \\ N_2 & 0 \\ 0 & N_2 \\ N_3 & 0 \\ 0 & N_3 \end{bmatrix} \begin{Bmatrix} W \\ 0 \end{Bmatrix} dY$$

$$= -Wh\{q\}^T \begin{Bmatrix} \int N_1 \, dY \\ 0 \\ \int N_2 \, dY \\ 0 \\ \int N_3 \, dY \\ 0 \end{Bmatrix}$$

where

$$\int_S N_1 \, dY = \frac{1}{\det} \int_2^5 (a_1 + Xb_1 + Yc_1) \, dY$$

$$= \frac{1}{12} \int_2^5 [18 + (6)(-3) + Y(0)] \, dY = 0$$

$$\int_S N_2 \, dY = \frac{1}{12} \int_2^5 (a_2 + Xb_2 + Yc_2) \, dY$$

$$= \frac{1}{12} \int_2^5 [-4 + (6)(4) + Y(-4)] \, dY = \tfrac{3}{2}$$

$$\int_S N_3 \, dY = \frac{1}{\det} \int_2^5 (a_3 + Xb_3 + Yc_3) \, dY$$

$$= \frac{1}{12} \int_2^5 [-2 + (6)(-1) + Y(4)] \, dY = \tfrac{3}{2}$$

In summary,

$$V_T = -\frac{3}{2} Wh \{q\}^T \begin{Bmatrix} 0 \\ 0 \\ 1 \\ 0 \\ 1 \\ 0 \end{Bmatrix}$$

It is very important to observe that the result is equivalent to two

4.2 THE TRIANGLE ELEMENT: GLOBAL COORDINATES

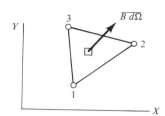

FIGURE 4.7 Differential body force.

concentrated forces each equal to half the resultant of the distributed loading and applied to nodes 2 and 3 in the X direction. □

3. *Body forces*: Forces that are exerted on each element of mass throughout the continuum (element) are called *body forces* (Figure 4.7). Gravitational loading is a common example of a body force. The units of the loading are force per unit volume.

Defining B_X as the body force in the X direction and B_Y as the body force in the Y direction, we can derive the differential potential function for these forces as

$$dV_{BF} = -(B_X \, d\Omega)u - (B_Y \, d\Omega)v$$
$$= -[u \quad v]\begin{Bmatrix} B_X \\ B_Y \end{Bmatrix} h \, dA$$

where u and v are the displacements of the arbitrary point in the interior of the element. Substituting expression (4.6) gives us

$$V_{BF} = -h\{q\}^T \int_A [N]^T \begin{Bmatrix} B_X \\ B_Y \end{Bmatrix} dA \qquad (4.12)$$

Example 4.4 The single constant-strain triangle shown in Figure 4.8 is of uniform thickness and uniform mass density ρ. Using equation (4.12), solve for the energy potential of the distributed weight (body force), assuming it to be acting in the negative Y direction.

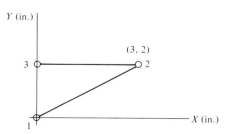

FIGURE 4.8 Constant-strain triangle of uniform density.

Solution. Beginning with equation (4.12), we have

$$V_{BF} = -h\{q\}^T \int_A [N]^T \begin{Bmatrix} B_X \\ B_Y \end{Bmatrix} dA$$

$$= -h\{q\}^T \int_A \begin{bmatrix} N_1 & 0 \\ 0 & N_1 \\ N_2 & 0 \\ 0 & N_2 \\ N_3 & 0 \\ 0 & N_3 \end{bmatrix} \begin{Bmatrix} 0 \\ -\rho g \end{Bmatrix} dA$$

where g is the gravitational acceleration.

$$V_{BF} = h\rho g \{q\}^T \begin{bmatrix} 0 \\ \int N_1 \, dA \\ 0 \\ \int N_2 \, dA \\ 0 \\ \int N_3 \, dA \end{bmatrix}$$

Evaluating the integrals gives us

$$\int_A N_1 \, dA = \frac{1}{\det} \int_A (a_1 + Xb_1 + Yc_1) \, dA$$

$$= \frac{1}{\det} \int_0^3 \int_{(2/3)X}^2 (a_1 + Xb_1 + Yc_1) \, dY \, dX$$

$$= \frac{1}{\det} \int_0^3 \left(a_1 Y + XYb_1 + \frac{Y^2}{2} c_1 \right) \Big|_{(2/3)X}^2 dX$$

$$= \frac{1}{\det} (3a_1 + 3b_1 + 4c_1)$$

where, referring back to equation (4.4), we have

$$a_1 = X_2 Y_3 - X_3 Y_2 = X_2 Y_3 = 6$$
$$b_1 = Y_2 - Y_3 = 0$$
$$c_1 = X_3 - X_2 = -X_2 = -3$$
$$\det = 2 \text{ (area of triangle)} = 6$$

Thus

$$\int N_1 \, dA = 1$$

Similarly, performing the other remaining two integrations yields

$$\int N_2 \, dA = 1 \quad \text{and} \quad \int N_3 \, dA = 1$$

4.2 THE TRIANGLE ELEMENT: GLOBAL COORDINATES

Note that the total weight of the triangle = $W = h\rho g$ (area) = $3h\rho g$, or

$$h\rho g = \frac{W}{3}$$

We may now write the body force potential energy as

$$V_{BF} = \{q\}^T \begin{Bmatrix} 0 \\ 1 \\ 0 \\ 1 \\ 0 \\ 1 \end{Bmatrix} \frac{W}{3}$$

Comparing the answer with the equation (4.10), we observe that the nodal equivalent loading consists of forces at each node acting in the negative Y direction and equal to one-third of the total weight. ☐

Step 4. Summation of the Energy Terms. The total energy in the single-element system consists of the element strain energy, U, and the loading energy potentials.

$$\Pi = U + V_{NF} + V_T + V_{BF}$$

$$= \tfrac{1}{2}\int_\Omega \{q\}^T[B]^T[C][B]\{q\}\, d\Omega - \{q\}^T \begin{Bmatrix} F_{1X} \\ F_{1Y} \\ F_{2X} \\ F_{2Y} \\ F_{3X} \\ F_{3Y} \end{Bmatrix}$$

$$- h\{q\}^T \int_S [N]^T \begin{Bmatrix} T_X \\ T_Y \end{Bmatrix} dS - h\{q\}^T \int_A [N]^T \begin{Bmatrix} B_X \\ B_Y \end{Bmatrix} dA$$

(4.13)

Step 5. Application of the Energy Minimization Principle. The total energy in the element is minimized relative to the nodal displacements. This can be accomplished by differentiating the scalar expression with respect to each and every one of the nodal displacements and setting each resulting equation equal to zero. This process yields six equations for this three-node element, and the equations are assembled into one matrix equation which becomes the force deflection equation for the single element.

This procedure was discussed in detail for the uniaxial element in Section 3.7. A review of that section may prove helpful.

The process of differentiating the energy with respect to each displacement and then assembling the resulting equations into matrix form is equivalent to differentiating with respect to the nodal vector $\{q\}$. The resulting matrix equation, which is the working force deflection equation for the element, is

$$\int_\Omega [B]^T[C][B]\,d\Omega \begin{Bmatrix} u_1 \\ v_1 \\ u_2 \\ v_2 \\ u_3 \\ v_3 \end{Bmatrix} = \begin{Bmatrix} F_{1X} \\ F_{1Y} \\ F_{2X} \\ F_{2Y} \\ F_{3X} \\ F_{3Y} \end{Bmatrix} + h\int_S [N]^T \begin{Bmatrix} T_X \\ T_Y \end{Bmatrix} dS$$

$$+ h\int_A [N]^T \begin{Bmatrix} B_X \\ B_Y \end{Bmatrix} dA \quad (4.14a)$$

or

$$[k]\{q\} = \{Q\}_{NF} + \{Q\}_T + \{Q\}_{BF} \quad (4.14b)$$

where

- $[k]$ = element stiffness matrix
- $\{q\}$ = element nodal displacement vector
- $\{Q\}_{NF}$ = vector of forces applied to the element nodes
- $\{Q\}_T$ = force vector resulting from distributed edge loading (surface traction)
- $\{Q\}_{BF}$ = force vector resulting from the distributed body force

Compare equations (4.14) with equations (3.18). With the single exception of the body force vector in (4.14), they are very similar.

This completes the development of the stiffness property (force–displacement equation) of the triangle using the global reference coordinates. We will now look at a simple example that illustrates the steps required for the solution of the displacements of the nodes of a single triangle. It should be emphasized that this example will illustrate the force–displacement response for the particular element that is a *constant strain* element. Do not get the idea that a triangular continuum can be modeled successfully with one element. Your background in stress analysis would lead you to suspect a considerable variation in strain (and stress) in the continuum of this example.

Example 4.5 Solve for the nodal displacements of the triangle shown in Figure 4.9, subjected to the distributed loading. This is the same triangle of Example 4.1 with the loading of Example 4.3. The material is steel, and it is h units thick.

Solution. The solution will use the single-element force–deflection equation (4.14). Return to Example 4.2 to obtain the $[B]^T[C][B]$ matrix product for this example. This matrix product can be

4.2 THE TRIANGLE ELEMENT: GLOBAL COORDINATES

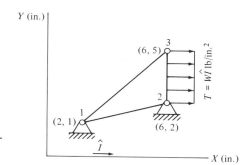

FIGURE 4.9 Triangle with uniform stress traction on one face.

substituted into equation (4.14) on the left side; the right side will include the nodal forces exerted by the supports and the result of the distributed loading on the side 2–3. The element load vector for the distributed load was determined in Example 4.3. Substituting these values, we have

$$[k]\{q\} = \{Q\}_{NF} + \{Q\}_T$$

$$[B]^T[C][B]h\int_A dA\{q\} = \begin{Bmatrix} F_{1X} \\ F_{1Y} \\ F_{2X} \\ F_{2Y} \\ 0 \\ 0 \end{Bmatrix} + \frac{3}{2}Wh\begin{Bmatrix} 0 \\ 0 \\ 1 \\ 0 \\ 1 \\ 0 \end{Bmatrix}$$

Integration of the left side amounts simply to determining the area of the triangle. This can be evaluated with ease using the expression for the determinant of matrix $[A_1]$ of equation (4.4), area $= (\frac{1}{2})\det = 6$. The element stiffness matrix now becomes

$$[k]_{XY} = 1.374(10^6)h \begin{bmatrix} 9.00 & 0. & -12.0 & 3.60 & 3.00 & -3.60 \\ & 3.15 & 4.20 & -4.20 & -4.20 & 1.05 \\ & & 21.6 & -10.4 & -9.60 & 6.20 \\ & & & 21.6 & 6.80 & -17.4 \\ & \text{symmetric} & & & 6.60 & -2.60 \\ & & & & & 16.4 \end{bmatrix} \text{lb/in.}$$

Next, substitute the displacement boundary conditions, row-and-column-reduce the equation, and solve for the displacements.

$$\begin{bmatrix} k_{55} & k_{56} \\ k_{65} & k_{66} \end{bmatrix}\begin{Bmatrix} u_3 \\ v_3 \end{Bmatrix} = \frac{3}{2}Wh\begin{Bmatrix} 1 \\ 0 \end{Bmatrix}$$

$$1.374(10)^6 h \begin{bmatrix} 6.60 & -2.60 \\ -2.60 & 16.4 \end{bmatrix}\begin{Bmatrix} u_3 \\ v_3 \end{Bmatrix} = \frac{3}{2}Wh\begin{Bmatrix} 1 \\ 0 \end{Bmatrix}$$

The resulting displacements are
$$u_3 = 1.76(10^{-7})W \quad \text{in.}$$
$$v_3 = 2.80(10^{-8})W \quad \text{in.}$$

From equation (4.8) (and Example 4.1), we find the strain to be
$$\epsilon_X = -1.46(10^{-8})W$$
$$\epsilon_Y = 0.93(10^{-8})W$$
$$\gamma_{XY} = 5.6(10^{-8})W$$

From equation (2.7) the stresses are
$$\sigma_X = -0.39W \text{ psi}$$
$$\sigma_Y = 0.16W \text{ psi}$$
$$\tau_{XY} = 0.65W \text{ psi}$$

Observe that this element yields strain and stress, which is a constant throughout the element. This is a result of our development of an element model using a linear displacement relationship. □

4.3 THE TRIANGLE ELEMENT: LOCAL COORDINATES

The local coordinates used for the triangle are most often oriented as shown in Figure 4.10. The origin is at local node 1; the x axis passes through local node 2.

We shall now proceed through the development of the constant strain element force–displacement equations using the local coordinate system. The procedure will follow the same steps used in the previous development using global coordinates.

Step 1. The Assumed Displacement Function. Again, assume a linear function, but this time in terms of local x and y coordinates.

$$u(x, y) = \alpha_1 + \alpha_2 x + \alpha_3 y \tag{4.15a}$$
$$v(x, y) = \beta_1 + \beta_2 x + \beta_3 y \tag{4.15b}$$

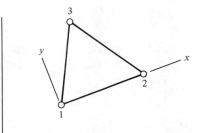

FIGURE 4.10 Triangle local coordinates.

4.3 THE TRIANGLE ELEMENT: LOCAL COORDINATES

Substitute the coordinates and displacements of the nodes. This will yield

$$\begin{Bmatrix} u_1 \\ u_2 \\ u_3 \end{Bmatrix} = \begin{bmatrix} 1 & 0 & 0 \\ 1 & x_2 & 0 \\ 1 & x_3 & y_3 \end{bmatrix} \begin{Bmatrix} \alpha_1 \\ \alpha_2 \\ \alpha_3 \end{Bmatrix}$$

or

$$\{q_1\} = [A_1]\{\alpha\} \tag{4.16}$$

Similarly,

$$\begin{Bmatrix} v_1 \\ v_2 \\ v_3 \end{Bmatrix} = \{q\} = [A_1]\{\beta\} \tag{4.17}$$

Solving for $[A_1]^{-1}$ gives us

$$[A_1]^{-1} = \frac{1}{\det} \begin{bmatrix} a_1 & a_2 & a_3 \\ b_1 & b_2 & b_3 \\ c_1 & c_2 & c_3 \end{bmatrix} \tag{4.18}$$

where

$$a_1 = x_2 y_3 \qquad a_2 = 0 \qquad a_3 = 0$$
$$b_1 = -y_3 \qquad b_2 = y_3 \qquad b_3 = 0$$
$$c_1 = x_3 - x_2 \qquad c_2 = -x_3 \qquad c_3 = x_2$$
$$\det = x_2 y_3 = 2(\text{triangle area})$$

Compare the entries in this local $[A_1]^{-1}$ matrix with the entries in the global $[A_1]^{-1}$ matrix of equation (4.7). Observe the greater simplicity together with reduced possibility of round-off errors arising from differences of products of large numbers of similar magnitude in the global matrix. From equations (4.16) and (4.17),

$$\{\alpha\} = [A_1]^{-1}\{q_1\}$$
$$\{\beta\} = [A_1]^{-1}\{q_2\}$$

Substituting into equations (4.15) yields

$$u = \begin{bmatrix} 1 & x & y \end{bmatrix} [A_1]^{-1}\{q_1\}$$
$$v = \begin{bmatrix} 1 & x & y \end{bmatrix} [A_1]^{-1}\{q_2\} \tag{4.19}$$

This yields

$$\begin{Bmatrix} u \\ v \end{Bmatrix} = \begin{bmatrix} N_1 & 0 & N_2 & 0 & N_3 & 0 \\ 0 & N_1 & 0 & N_2 & 0 & N_3 \end{bmatrix} \begin{Bmatrix} u_1 \\ v_1 \\ u_2 \\ v_2 \\ u_3 \\ v_3 \end{Bmatrix}_{xy}$$

where

$$N_1 = \frac{y_3(x_2 - x) + y(x_3 - x_2)}{x_2 y_3}$$

$$N_2 = \frac{xy_3 - yx_3}{x_2 y_3}$$

$$N_3 = \frac{yx_2}{x_2 y_3}$$

In symbolic form

$$\{u\} = [N]\{q\} \tag{4.20}$$

This completes the development of the displacement function in terms of the nodal displacements. Remember, all coordinates and displacements are in the local coordinate directions.

Step 2. Derivation of Strain and Strain Energy. Here we have to be careful. The strain that we ultimately will want will be relative to the global coordinate system. We are going to derive the strain first relative to the local system by differentiating the local displacements with respect to the local coordinate directions. The strain energy will be developed in terms of these local quantities and ultimately a coordinate transformation (rotation) will be performed on the force–displacement equation to have it referenced to the global system. The process will be quite analogous to the transformation performed on the bar element in Section 1.4.

Applying the strain definition to equation (4.20) yields

$$\begin{Bmatrix} \epsilon_x \\ \epsilon_y \\ \gamma_{xy} \end{Bmatrix} = \begin{bmatrix} \frac{\partial}{\partial x} & 0 \\ 0 & \frac{\partial}{\partial y} \\ \frac{\partial}{\partial y} & \frac{\partial}{\partial x} \end{bmatrix} \begin{Bmatrix} u \\ v \end{Bmatrix} = \begin{bmatrix} \frac{\partial}{\partial x} & 0 \\ 0 & \frac{\partial}{\partial y} \\ \frac{\partial}{\partial y} & \frac{\partial}{\partial x} \end{bmatrix} [N]\{q\}$$

$$= \frac{1}{x_2 y_3} \begin{bmatrix} -y_3 & 0 & y_3 & 0 & 0 & 0 \\ 0 & x_3 - x_2 & 0 & -x_3 & 0 & x_2 \\ x_3 - x_2 & -y_3 & -x_3 & y_3 & x_2 & 0 \end{bmatrix} \begin{Bmatrix} u_1 \\ v_1 \\ u_2 \\ v_2 \\ u_3 \\ v_3 \end{Bmatrix}$$

or

$$\{\epsilon\} = [B]\{q\} \tag{4.21}$$

Compare this $[B]$ matrix for the local coordinate system with the $[B]$ matrix of equation (4.8) for the global coordinate system. The local matrix is of simpler form with less tendency for round-off error.

4.3 THE TRIANGLE ELEMENT: LOCAL COORDINATES

From equation (2.16) for the strain energy, we can derive the following expression for the strain energy of the element. Note the similarity between the relationships for global and local reference system development. The symbolic expressions are identical. The entries in the matrices are certainly different, however.

$$U = \tfrac{1}{2}\{q\}^T[B]^T[C][B]\{q\}h\int_A dA \qquad (4.22)$$

Example 4.6 The steel triangular element of Example 4.1 is shown again in Figure 4.11. It is of thickness h, and the global node coordinates are given.

Solve for the strain in the triangle in terms of the nodal displacements and referenced to the local coordinate system.

Solution. Recall equation (4.21):

$$\{\epsilon\} = [B]\{q\}$$

The local nodal coordinates are

$$x_1 = 0 \qquad y_1 = 0$$
$$x_2 = \sqrt{17} \qquad y_2 = 0$$
$$x_3 = \frac{20}{\sqrt{17}} \qquad y_3 = \frac{12}{\sqrt{17}}$$

Thus the local $[B]$ matrix is

$$[B] = \frac{1}{(12)(\sqrt{17})}\begin{bmatrix} -12 & 0 & 12 & 0 & 0 & 0 \\ 0 & 3 & 0 & -20 & 0 & 17 \\ 3 & -12 & -20 & 12 & 17 & 0 \end{bmatrix}$$

where the det $= 12$. Substituting into equation (4.21) gives us

$$\begin{Bmatrix} \epsilon_x \\ \epsilon_y \\ \gamma_{xy} \end{Bmatrix} = \frac{1}{\sqrt{17}}\begin{bmatrix} -1 & 0 & 1 & 0 & 0 & 0 \\ 0 & \tfrac{1}{4} & 0 & -\tfrac{5}{3} & 0 & \tfrac{17}{12} \\ \tfrac{1}{4} & -1 & -\tfrac{5}{3} & 1 & \tfrac{17}{12} & 0 \end{bmatrix}\begin{Bmatrix} u_1 \\ v_1 \\ u_2 \\ v_2 \\ u_3 \\ v_3 \end{Bmatrix}_{xy}$$

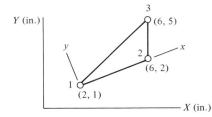

FIGURE 4.11 Triangle with global and local coordinates.

Compare this result with the strain determined in Example 4.1. The strain values are not the same because the local coordinate system is rotated relative to the global coordinates. □

This completes the derivation of the strain in terms of the nodal displacements. It should be reemphasized that the strain is relative to the local xy system and the displacements are also in the local xy directions.

Example 4.7 The strain energy in the element is determined using the equation (4.9). This equation requires the matrix product $[B]^T[C][B]$. Evaluate this product using the local coordinate system for the element in Example 4.6.

Solution. Using the $[C]$ matrix for the plane stress state, and the $[B]$ matrix from Example 4.6, we have

$$[B]^T[C][B]$$

$$=134670\begin{bmatrix} 14.7 & -2.34 & -16.5 & 8.46 & 1.78 & -6.12 \\ & 5.94 & 9.48 & -11.0 & -7.14 & 5.10 \\ & & 28.4 & -15.6 & -11.9 & 6.12 \\ & & & 45.0 & 7.14 & -34.0 \\ & \text{symmetric} & & & 10.1 & 0 \\ & & & & & 28.9 \end{bmatrix}$$
□

Step 3. Derivation of Applied Load Energy Functions. At this point we could go back to step 3 in Section 4.2. The relationships derived there are directly applicable to the local system; the only change we have to make is to refer all displacement and force components to the local coordinate system.

Step 4. Summation of the Energy Terms. Again, equation (4.13) can be rewritten here as long as we change all global coordinates to the local coordinates. The symbolic form of the equations is identical.

Step 5. Application of the Energy Minimization Principle. Summarizing, we write the final form of the equation as

$$\int_\Omega [B]^T[C][B]\,d\Omega \begin{Bmatrix} u_1 \\ v_1 \\ u_2 \\ v_2 \\ u_3 \\ v_3 \end{Bmatrix}_{xy} = \begin{Bmatrix} F_{1x} \\ F_{1y} \\ F_{2x} \\ F_{2y} \\ F_{3x} \\ F_{3y} \end{Bmatrix} + h\int_S [N]^T \begin{Bmatrix} T_x \\ T_y \end{Bmatrix} ds$$

$$+ h\int_A [N]^T \begin{Bmatrix} B_x \\ B_y \end{Bmatrix} dA \quad (4.23a)$$

4.4 LOCAL-TO-GLOBAL COORDINATE TRANSFORMATION

or symbolically,

$$[k]_{xy}\{q\}_{xy} = \{Q\}_{NF_{xy}} + \{Q\}_{T_{xy}} + \{Q\}_{BF_{xy}} \quad (4.23b)$$

4.4 LOCAL-TO-GLOBAL COORDINATE TRANSFORMATION

It is obvious that if we have more than one element modeling a continuum, we have more than one set of local coordinate axes. We have seen that there are advantages of simplicity and round-off error reduction when the local axes are used. For these reasons we choose the local axes for the model development.

In construction of the master global stiffness matrix for the complete system, we know from our work in Chapter 1 that the global matrix is formed by summing the stiffness matrices of all the elements. This can be done only when the element stiffness matrices are all developed with reference to a common coordinate system.

Return to Section 1.4 for a review of this transformation development for the two force bar element in a plane. The coordinate transformation relationships for our current element, the triangle, will be identical symbolically. However, where the bar rotation matrix $[A_2]$ was a 4×4, the triangle rotation matrix is a 6×6. For example, the transformation of the nodal force vector would be accomplished as follows [compare with equation (1.10)]:

$$\begin{Bmatrix} F_{1X} \\ F_{1Y} \\ F_{2X} \\ F_{2Y} \\ F_{3X} \\ F_{3Y} \end{Bmatrix} = \begin{bmatrix} [A] & 0 & 0 \\ 0 & [A] & 0 \\ 0 & 0 & [A] \end{bmatrix} \begin{Bmatrix} F_{1x} \\ F_{1y} \\ F_{2x} \\ F_{2y} \\ F_{3x} \\ F_{3y} \end{Bmatrix}$$

or

$$\{Q\}_{XY} = [A_3]\{Q\}_{xy}$$

where

$$[A] = \begin{bmatrix} \cos\theta & -\sin\theta \\ \sin\theta & \cos\theta \end{bmatrix}$$

The development of the transformation relationship to transform the stiffness matrix from a local system to the global system follows exactly the discussion in Section 1.4. The only difference is the substitution of the rotation matrix $[A_3]$ for $[A_2]$. We may then write the transformation as

$$[k]_{XY} = [A_3][k]_{xy}[A_3]^T \quad (4.24)$$

Having calculated the element stiffness matrix using the local system and then transformed the matrix to the global system using equation

(4.24), equation (4.14) is normally used because applied loads are generally given relative to the global coordinate system.

$$[k]_{XY}\{q\}_{XY} = \{Q\}_{NF_{XY}} + \{Q\}_{T_{XY}} + \{Q\}_{BF_{XY}} \quad (4.14)$$

Example 4.8 In Example 4.7 we calculated the matrix product $[B]^T[C][B]$ relative to the local coordinate system for the triangle of Example 4.6.

(a) Determine the stiffness matrix of the element relative to the local coordinate system.
(b) Transform the stiffness matrix to the global coordinate system. Compare the result with the stiffness matrix constructed for Example 4.5.

Solution. (a) From equation (4.23),

$$[k]_{xy} = [B]^T[C][B]_{xy}hA \quad \text{where } A = 6$$

$$= 808{,}020h \begin{bmatrix} 14.7 & -2.34 & -16.5 & 8.46 & 1.78 & -6.12 \\ & 5.94 & 9.48 & -11.0 & -7.14 & 5.10 \\ & & 28.4 & -15.6 & -11.9 & 6.12 \\ & & & 45.0 & 7.14 & -34.0 \\ & \text{symmetric} & & & 10.1 & 0. \\ & & & & & 28.9 \end{bmatrix}$$

(b) From equation (4.24)

$$[k]_{XY} = [A_3][k]_{xy}[A_3]^T$$

The rotation matrix $[A_3]$ is

$$[A_3] = \frac{1}{\sqrt{17}} \begin{bmatrix} 4 & -1 & 0 & 0 & 0 & 0 \\ 1 & 4 & 0 & 0 & 0 & 0 \\ 0 & 0 & 4 & -1 & 0 & 0 \\ 0 & 0 & 1 & 4 & 0 & 0 \\ 0 & 0 & 0 & 0 & 4 & -1 \\ 0 & 0 & 0 & 0 & 1 & 4 \end{bmatrix}$$

The result of the matrix product is

$$[k]_{XY}$$

$$= 1.374(10^6)h \begin{bmatrix} 9.00 & 0. & -12.0 & 3.6 & 3.00 & -3.60 \\ & 3.15 & 4.20 & -4.20 & -4.20 & 1.05 \\ & & 21.6 & -10.4 & -9.60 & 6.20 \\ & & & 21.6 & 6.80 & -17.4 \\ & \text{symmetric} & & & 6.60 & -2.60 \\ & & & & & 16.4 \end{bmatrix}$$

4.5 ASSEMBLING THE ELEMENTS

The assembling of the elements in a continuum is based on the minimization of the total energy (the functional) for the entire domain. The process was presented in detail in an example involving the one-dimensional element in Section 3.9; we will simply outline the theory for this element.

From expressions (4.13) and (4.14), we can write the expression for the energy of one element as follows:

$$\Pi_e = \frac{1}{2}\{q\}_e^T[k]_e\{q\}_e - \{q\}_e^T\{Q\}_{\text{NF}_e} - \{q\}_e^T\{Q\}_{T_e} - \{q\}_e^T\{Q\}_{\text{BF}_e}$$

The total energy of the domain (assemblage of elements) would involve the following:

$$\Pi = \boxed{\begin{array}{c}\text{total strain}\\\text{energy}\end{array}} + \boxed{\begin{array}{c}\text{boundary force}\\\text{potential energy}\end{array}}$$

$$+ \boxed{\begin{array}{c}\text{interior force}\\\text{potential energy}\end{array}}$$

Examining each term, we observe the following:

$$\boxed{\begin{array}{c}\text{total strain}\\\text{energy}\end{array}} = \boxed{\begin{array}{c}\text{the sum of all element}\\\text{strain energies}\end{array}}$$

$$\boxed{\begin{array}{c}\text{boundary force}\\\text{potential energy}\end{array}} = \boxed{\begin{array}{c}\text{potential energies of}\\\text{all domain nodal and}\\\text{distributed surface}\\\text{tractions on domain}\\\text{boundary}\end{array}}$$

$$\boxed{\begin{array}{c}\text{interior force}\\\text{potential energy}\end{array}} = \boxed{\begin{array}{c}\text{potential energies of}\\\text{all interior nodal}\\\text{and body forces}\end{array}}$$

The total domain energy can be written symbolically as

$$\Pi = \{r\}^T \sum_{e=1}^{m} [k]_e \{r\} - \{r\}^T \{R\}_{NF} - \{r\}^T \{R\}_T - \{r\}^T \{R\}_{BF}$$

where for a region containing n nodes and m elements

$\{r\}$ = $2n \times 1$ nodal displacement vector
   = $[u_1 \quad v_1 \quad u_2 \quad ,\ldots, \quad u_n \quad v_n]^T$
$\{R\}_{NF}$ = $2n \times 1$ vector of all nodal forces (including zeros)
$\{R\}_T$ = vector of node forces derived from distributed surface tractions
$\{R\}_{BF}$ = vector of node forces derived from the distributed body forces

Application of the Rayleigh–Ritz procedure to the total energy yields the force–displacement equation for the assemblage of elements:

$$[K]\{r\} = \{R\}_{NF} + \{R\}_T + \{R\}_{BF} \qquad (4.25)$$

where [K] is the assembled stiffness matrix.

The process of assembly of the individual element stiffness matrices into the global assemblage stiffness matrix follows directly the procedures discussed in Section 1.4, in particular in the solution of Example 1.5. The only difference is the number of nodes involved per element and thus the dimension of the matrices. Consider the two-element system shown in Figure 4.12.

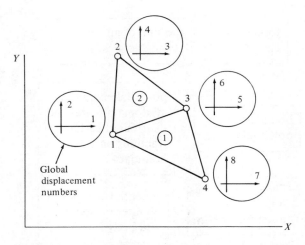

FIGURE 4.12 Assemblage of two triangular elements.

4.5 ASSEMBLING THE ELEMENTS

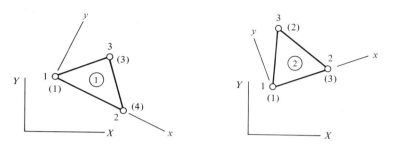

FIGURE 4.13 Individual elements with local coordinates, local node numbers, i, and global node numbers, (N).

1. Assign global displacement numbers to node i according to:

   $$\text{displacement number for } X\text{-direction displacement} = (2)(\text{global node number}) - 1$$

   $$\text{displacement number for } Y\text{-direction displacement} = (2)(\text{global node number})$$

2. Construct the local–global node map. (Figure 4.13).

Element Number	Global Node Number of:		
	Local Node 1	Local Node 2	Local Node 3
1	1	4	3
2	1	3	2

3. The element ① stiffness matrix undergoes transformation to global XY system and has its entries addressed according to the global displacement numbers.

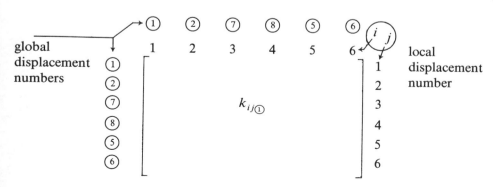

DEVELOPMENT OF THE CONSTANT-STRAIN TRIANGULAR ELEMENT

4. The element ② stiffness matrix is transformed to the global system and has entries addressed according to the global displacement numbers.

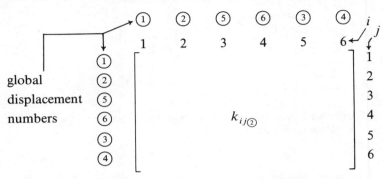

5. Each element stiffness matrix entry is added to the global assemblage stiffness matrix. The rows and columns of the assemblage stiffness matrix are numbered in the ascending order of the global displacement numbers. The correct location of each element matrix entry is defined by the global displacement numbers of the row and column of that entry.

$$[K] = \begin{array}{c} \text{global} \\ \text{displacement} \\ \text{numbers} \end{array} \begin{array}{c} 1 \\ 2 \\ 3 \\ 4 \\ 5 \\ 6 \\ 7 \\ 8 \end{array} \begin{bmatrix} & 1 & 2 & 3 & 4 & 5 & 6 & 7 & 8 \\ & & & & K_{ij} & & & \\ & & \text{assemblage stiffness} & & \\ & & \text{matrix} & & \end{bmatrix}$$

For example,

$$K_{11} = k_{11①} + k_{11②}$$

$$K_{33} = k_{55②}$$

$$K_{56} = k_{56①} + k_{34②}$$

$$K_{87} = k_{43①}$$

Observe the notation. Lower case $[k]$ represents a particular element stiffness matrix. Upper case $[K]$ represents the global assemblage stiffness matrix.

Example 4.9 A square region is divided into two triangular elements (Figure 4.14). Assume a uniform thickness of 1 in., an elastic modulus

4.5 ASSEMBLING THE ELEMENTS

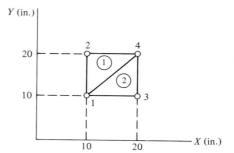

FIGURE 4.14 Two-element system.

of $30(10^6)$ psi, and a Poisson's ratio of 0.5 (for computation convenience).

The element nodes have been numbered, and the element local–global node numbering map is as follows:

Element Number	Global Node Number of: Local Node		
	1	2	3
1	1	4	2
2	1	3	4

The element stiffness matrices have been calculated relative to the global XY reference and are as follows:

$$[k_1] = 10^6 \begin{bmatrix} 5 & 0 & 0 & -5 & -5 & 5 \\ & 20 & -10 & 0 & 10 & -20 \\ & & 20 & 0 & -20 & 10 \\ & & & 5 & 5 & -5 \\ & \text{symmetric} & & & 25 & -15 \\ & & & & & 25 \end{bmatrix}$$

$$[k_2] = 10^6 \begin{bmatrix} 20 & 0 & -20 & 10 & 0 & -10 \\ & 5 & 5 & -5 & -5 & 0 \\ & & 25 & -15 & -5 & 10 \\ & & & 25 & 5 & -20 \\ & \text{symmetric} & & & 5 & 0 \\ & & & & & 20 \end{bmatrix}$$

Determine the global assemblage stiffness matrix.

172   DEVELOPMENT OF THE CONSTANT-STRAIN TRIANGULAR ELEMENT

Solution. The first step is to identify the rows and columns of each matrix with the global displacement numbers.

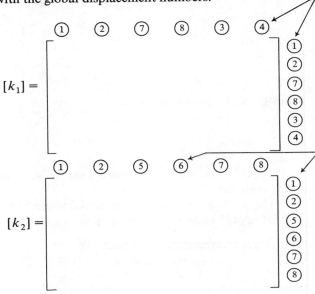

The second step involves placing the individual element entries into the 8 × 8 assemblage matrix in positions defined by the displacement number of the row and column of the element entry. Multiple element entries having the same "address" are summed, for example (all values to be multiplied by 10^6):

$$K_{11} = 5 + 20 = 25 \qquad K_{17} = 0 + 0 = 0$$
$$K_{13} = -5 \qquad K_{18} = -5 + (-10) = -15$$

The resultant assemblage stiffness matrix is

$$[K] = 10^6$$

①	②	③	④	⑤	⑥	⑦	⑧	
25	0	−5	5	−20	10	0	−15	①
	25	10	−20	5	−5	−15	0	②
		25	+15	0	0	+20	5	③
			25	0	0	10	−5	④
				25	−15	−5	10	⑤
	symmetric				25	5	−20	⑥
						25	0	⑦
							25	⑧

← global displacement numbers

□

4.5 ASSEMBLING THE ELEMENTS

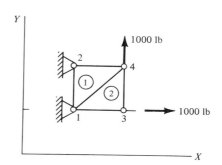

FIGURE 4.15 Boundary conditions on assembly of two elements.

Example 4.10 The nodes 1 and 2 of the two-element system of Example 4.9 are fixed in space. Forces are applied to nodes 3 and 4 as defined in Figure 4.15.

(a) Construct the nodal force vector, $\{R\}_{NF}$, and displacement vector, $\{r\}$.
(b) Write the reduced force–displacement matrix equation and solve for strain and stress in the elements.

Solution. (a) For the force vector, the forces are known at nodes 3 and 4. Nodes 1 and 2 have the unknown reactions. On the other hand, displacements are known to be zero at nodes 1 and 2 and are unknown at nodes 3 and 4.

Notation reminder: Single-element load and displacement vectors are $\{Q\}$ and $\{q\}$. Assemblage load and displacement vectors are $\{R\}$ and $\{r\}$.

The two vectors are

$$\{R\}_{NF} = \begin{Bmatrix} R_{1X} \\ R_{1Y} \\ R_{2X} \\ R_{2Y} \\ 1000 \\ 0 \\ 0 \\ 1000 \end{Bmatrix} \quad \{r\} = \begin{Bmatrix} 0 \\ 0 \\ 0 \\ 0 \\ u_3 \\ v_3 \\ u_4 \\ v_4 \end{Bmatrix}$$

(b) The assemblage matrix equation is $[K]\{r\} = \{R\}_{NF}$. To reduce the equation, we remove the rows associated with the unknown force components and columns associated with the zero displacement values. This yields the following:

$$10^6 \begin{bmatrix} 25 & -15 & -5 & 10 \\ & 25 & 5 & -20 \\ & & 25 & 0 \\ & & & 25 \end{bmatrix} \begin{Bmatrix} u_3 \\ v_3 \\ u_4 \\ v_4 \end{Bmatrix} = \begin{Bmatrix} 1000 \\ 0 \\ 0 \\ 1000 \end{Bmatrix}$$

Solution of this equation yields the following displacement values:

Node Number	u (in.)	v (in.)
3	$7.74\text{E}-05$	$15.8\text{E}-05$
4	$-1.62\text{E}-05$	$13.5\text{E}-05$

Equation (4.8) can be used to obtain the strain in the elements. For element 1:

$$\{\epsilon\} = [B]\{q\}$$

where the q for element 1 is the displacement vector for global nodes 1, 4, and 2.

$$\{q\}_1 = \begin{Bmatrix} 0 \\ 0 \\ -1.62\text{E}-05 \\ 13.5\text{E}-05 \\ 0 \\ 0 \end{Bmatrix} \text{ in.}$$

Thus

$$\{\epsilon\}_1 = \begin{bmatrix} 0 & 0 & 0.1 & 0 & -0.1 & 0 \\ 0 & -0.1 & 0 & 0 & 0 & 0.1 \\ -0.1 & 0 & 0 & 0.1 & 0.1 & -0.1 \end{bmatrix} \begin{Bmatrix} 0 \\ 0 \\ -1.62 \\ 13.5 \\ 0 \\ 0 \end{Bmatrix} 10^{-5}$$

$$= \begin{Bmatrix} -1.62 \\ 0 \\ 13.5 \end{Bmatrix} 10^{-6}$$

The stress in element 1 is obtained from equation (2.7):

$$\begin{Bmatrix} \sigma_X \\ \sigma_Y \\ \tau_{XY} \end{Bmatrix} = \frac{E}{1-\nu^2} \begin{bmatrix} 1 & \nu & 0 \\ \nu & 1 & 0 \\ 0 & 0 & \frac{1-\nu}{2} \end{bmatrix} \begin{Bmatrix} \epsilon_X \\ \epsilon_Y \\ \gamma_{XY} \end{Bmatrix} \quad (2.7)$$

$$\begin{Bmatrix} \sigma_X \\ \sigma_Y \\ \tau_{XY} \end{Bmatrix}_1 = \frac{30(10^6)}{1-0.5^2} \begin{bmatrix} 1 & 0.5 & 0 \\ 0.5 & 1 & 0 \\ 0 & 0 & \frac{1-0.5}{2} \end{bmatrix} \begin{Bmatrix} -1.62 \\ 0 \\ 13.5 \end{Bmatrix} 10^{-6}$$

$$= \begin{Bmatrix} -65 \\ -32 \\ 135 \end{Bmatrix} \text{ psi}$$

Following the same procedure for element 2, using

$$\{q\}_2 = \begin{Bmatrix} 0 \\ 0 \\ 7.74\text{E} - 05 \\ 15.8\text{E} - 05 \\ -1.62\text{E} - 05 \\ 13.5\text{E} - 05 \end{Bmatrix}$$

$$\begin{Bmatrix} \sigma_X \\ \sigma_Y \\ \tau_{XY} \end{Bmatrix}_2 = \begin{Bmatrix} 264 \\ 65 \\ 65 \end{Bmatrix} \text{ psi}$$

□

You should pay close attention to the results of this "stress analysis." The results are clearly absurd if interpreted to be the stress distribution in a continuum loaded as defined. Remember, the element used is a constant strain (stress) element. There can be no variation of strain (or stress) across the element. Clearly, for the continuum of this example, there would exist a very significant strain gradient. Compare the results of this analysis with that which you would predict using strength-of-materials relationships (Mc/I and F/A, etc.).

The obvious conclusion is that many more elements have to be used in an analysis, and this will require the computer. Chapter 6 will involve more discussion of applications and will include an elementary program for plane analysis.

PROBLEMS

4.1 Solve for the area of the triangle in Figure P4.1 in terms of the global X, Y coordinates of the nodes. Then show that the resulting expression is equivalent to $(\frac{1}{2})$ det.

4.2 Derive the constants a_1 and det of (4.4).

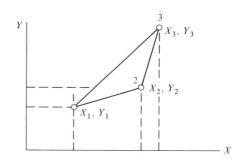

FIGURE P4.1 Arbitrary triangle in XY plane.

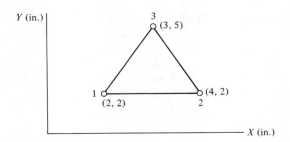

FIGURE P4.3 Global coordinates of a triangle.

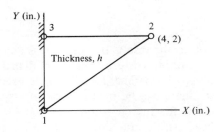

FIGURE P4.5 Global coordinates of triangle with one free node.

4.3 Calculate the magnitude of the entries in the $[N]$ matrix of (4.6) at nodes 1, 2, and 3 and also at the centroid of the triangle of Figure P4.3.

4.4 If the displacement vector for the nodes of the triangle of Figure P4.3 is

$$\{q\} = [0 \quad 0 \quad 1 \quad 2 \quad 0 \quad 1]^T (10^{-3}) \quad \text{in.}$$

solve for the displacement of the centroid of the triangle.

4.5 Solve for the strain vector $\{\epsilon\}$ of the triangle of Figure P4.5 if the nodal displacement vector is

$$\{q\} = [0 \quad 0 \quad 2 \quad -2 \quad 0 \quad 0]^T (10^{-3}) \quad \text{in.}$$

4.6 The stress traction T_X varies linearly on side 2–3 of the triangle of Figure P4.6 according to

$$T_X = C_1 + C_2 Y$$

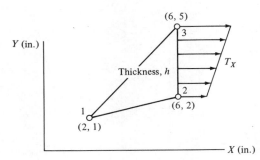

FIGURE P4.6 Triangle side with linear traction.

PROBLEMS

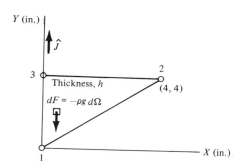

FIGURE P4.7 Triangle with linearly distributed body force.

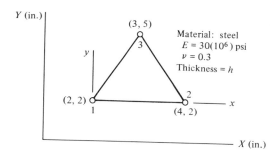

FIGURE P4.8 Global coordinates of the triangle.

Solve for the potential energy function for this distributed loading in terms of the element node displacements.

4.7 The three-node triangle in Figure P4.7 has a mass density that varies linearly according to

$$\rho = \rho_0 Y$$

Assuming the weight to act in the negative Y direction, solve for the weight energy potential function in terms of the element node displacements.

4.8 The global coordinates are given for the nodes of the triangle in Figure P4.8. Solve for the local reference $[N]$ matrix defined in (4.20) for nodes 1, 2, and 3 and the centroid.

4.9 For the triangle of Figure P4.8, solve for the element stiffness matrix using:
(a) The global XY reference.
(b) The local xy reference. Compare the two matrices.

4.10 Figure P4.10 defines the global coordinates of the triangle nodes.
(a) Using the global coordinate $[B]$ matrix of (4.8), solve for the stiffness matrix relative to the global XY reference.
(b) Using the local coordinate $[B]$ matrix of (4.22), solve for the stiffness matrix relative to the local xy coordinate axes.
(c) Transform the local coordinate reference stiffness matrix to the global reference using the transformation defined by (4.24) and check to see if, as it should, the transformed matrix agrees with the result in part (a).

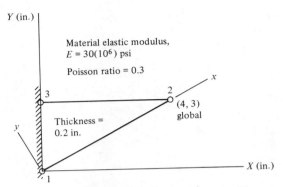

FIGURE P4.10 Triangle with local and global reference coordinate systems.

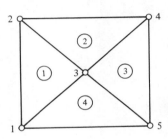

FIGURE P4.11 Assemblage of four elements with node and element numbering.

4.11 An assemblage of 4 elements is shown in Figure P4.11. The elements and nodes have been numbered randomly. The following is the global–local node numbering map for the elements:

Element Number	Global Node Number of:		
	Local Node 1	Local Node 2	Local Node 3
1	2	1	3
2	2	3	4
3	3	5	4
4	1	5	3

Assemble the 10×10 global stiffness matrix in terms of the element stiffness entries $k_{ij\circledcirc}$. For example,

$$K_{12} = k_{34\circled{1}} + k_{12\circled{4}}$$

As a minimum, solve for K_{17}, K_{56}, K_{66}, and $K_{6,10}$.

4.12 An assemblage of five triangular elements is shown in Figure P4.12. The nodes and elements have been numbered. The global–local node numbering map is as

PROBLEMS

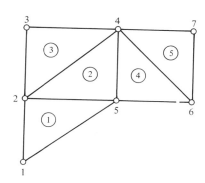

FIGURE P4.12 Assemblage of five elements with node and element numbering.

follows:

Element Number	Global Number of:		
	Local Node 1	Local Node 2	Local Node 2
1	1	5	2
2	2	5	4
3	2	4	3
4	4	5	6
5	4	6	7

Assemble the 14–14 global stiffness matrix in terms of the element stiffness entries $k_{ij\text{\textcircled{e}}}$. For example,

$$K_{11} = k_{11\text{\textcircled{1}}}$$
$$K_{12,12} = k_{66\text{\textcircled{4}}} + k_{44\text{\textcircled{5}}}$$
$$K_{12,13} = k_{45\text{\textcircled{5}}}$$

As a minimum, solve for K_{44}, K_{79}, and $K_{8,11}$.

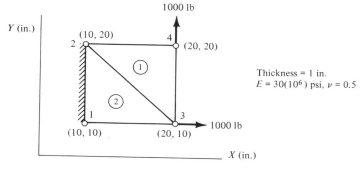

FIGURE P4.14 Region with two triangles.

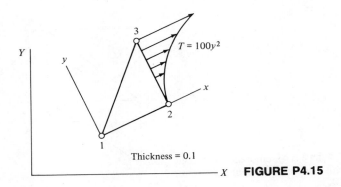

FIGURE P4.15

4.13 If the triangle of Problem 4.10 has uniform mass density ρ with the weight acting in the negative Y direction, solve for the deflection, in terms of ρ, of the free node 2 and the strain and stress in the triangle.

4.14 The square region of Examples 4.9 and 4.10 is divided into two triangles with the diagonal the opposite way (Figure P4.14).

Solve for the deflection of the free nodes 3 and 4 and determine the strain and stress in the two triangles. Compare with the results from Example 4.10.

4.15 Consider the region shown in Figure P4.15. The distributed loading, $T(y) = 100y^2$, on the side $x = 2$, which happens to be local side 1, is expressed in terms of the local coordinate variable and acts in the local coordinate x direction.

Solve for the local direction energy equivalent nodal force components acting at each node of the triangle.

CHAPTER 5

Interpolation Formulas and Numerical Integration

5.1 INTRODUCTION

In Chapters 3 and 4 we saw that one of the steps in the derivation of the element force–deflection equations involved obtaining an expression for the displacement of any point in the interior of the element as a linear combination of the displacements of each and every node. [Refer to equations (3.4) and (4.6).] Obtaining these displacement relations required a formal solution of some simultaneous equations, the number being equal to the number of nodes in the element. This process was not difficult for a two or three-node element, but the effort becomes quite objectionable as the number of nodes increases beyond three. To address this difficulty, we will introduce a technique for developing interpolation formulas.

Also in these same chapters we saw that obtaining the stiffness matrix required the integration of a matrix over the area (or length for the two-node element) of the element. We discovered, however, that the matrix product was independent of the integration coordinates for the axial and triangular constant-strain element, and the integration simply involved a length or area determination. This simplicity exists only for the very simple elements. As we study more complex elements, we will find that the [B] matrix will contain the integration variable. This results in a rather complicated integral. Rather than try to evaluate the integral in closed form, we will perform the integration numerically.

5.2 INTERPOLATION FORMULA: ONE INDEPENDENT VARIABLE

Consider the two-dimensional plot of the variables ϕ and θ shown in Figure 5.1. Assume that values of the dependent variable ϕ are associated with a number of arbitrarily spaced values of the independent variable θ. We wish to construct a function that will allow interpolation between the known values.

One approach to developing this function would be to pass a polynomial of degree $n - 1$ through the n known points. For example, assume that three (data) points are given. Assume the polynomial $\phi = C_0 + C_1\theta + C_2\theta^2$, or in matrix form,

$$\phi = \begin{bmatrix} 1 & \theta & \theta^2 \end{bmatrix} \begin{Bmatrix} C_0 \\ C_1 \\ C_2 \end{Bmatrix} \tag{5.1}$$

Next, write three equations using the data points:

$$\begin{Bmatrix} \phi_1 \\ \phi_2 \\ \phi_3 \end{Bmatrix} = \begin{bmatrix} 1 & \theta_1 & \theta_1^2 \\ 1 & \theta_2 & \theta_2^2 \\ 1 & \theta_3 & \theta_3^2 \end{bmatrix} \begin{Bmatrix} C_0 \\ C_1 \\ C_2 \end{Bmatrix}$$

Solving for the coefficient vector and substituting back into equation (5.1) yields the desired function. This procedure should look familiar. We recognize that we are faced with increasing complexity as the number of data points (and the degree of the polynomial) increases, so we will look for an alternative procedure.

Lagrange's interpolation formula accomplishes the desired curve fitting without having to solve a set of simultaneous equations. The formula to fit n points is

$$\phi = N_1\phi_1 + N_2\phi_2 + N_3\phi_3 + \cdots + N_n\phi_n \tag{5.2}$$

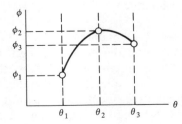

FIGURE 5.1 Plot of two variables.

5.2 INTERPOLATION FORMULA: ONE INDEPENDENT VARIABLE

The N_i coefficients are defined as

$$N_1 = \frac{(\theta - \theta_2)(\theta - \theta_3) \cdots (\theta - \theta_n)}{(\theta_1 - \theta_2)(\theta_1 - \theta_3) \cdots (\theta_1 - \theta_n)}$$

$$N_2 = \frac{(\theta - \theta_1)(\theta - \theta_3) \cdots (\theta - \theta_n)}{(\theta_2 - \theta_1)(\theta_2 - \theta_3) \cdots (\theta_2 - \theta_n)}$$

$$N_i = \frac{(\theta - \theta_1)(\theta - \theta_2) \cdots (\theta - \theta_{i-1})(\theta - \theta_{i+1}) \cdots (\theta - \theta_n)}{(\theta_i - \theta_1)(\theta_i - \theta_2) \cdots (\theta_i - \theta_{i-1})(\theta_i - \theta_{i+1}) \cdots (\theta_i - \theta_n)} \quad (5.3)$$

It is very important to recognize the following characteristics of the interpolation formula.

1. The independent variable, θ, need not be evenly spaced.
2. Each N_i coefficient is a polynomial of degree $n - 1$ in the independent variable.
3. $N_i = 1$ for $\theta = \theta_i$ and $N_i = 0$ for $\theta = \theta_j$ where $j \neq i$. This is an absolute requirement of an interpolation formula.

One additional comment. The polynomial approximation interpolation formula will yield the correct values of ϕ at intermediate values of θ only if the function being modeled (the "true" curve) is the same polynomial. This often is not the case, but the approximation can be made to yield results with an acceptable magnitude of error.

Example 5.1 Given the tabular values of v and t, derive a function for v with t the independent variable. Use the interpolation formula, equation (5.2).

Data Point	t	v
1	1	2
2	2	4
3	5	7

Solution

$$v = N_1 v_1 + N_2 v_2 + N_3 v_3$$

$$N_1 = \frac{(t - t_2)(t - t_3)}{(t_1 - t_2)(t_1 - t_3)} = \frac{(t - 2)(t - 5)}{(-1)(-4)} \qquad N_1 = \frac{(t - 2)(t - 5)}{4}$$

$$N_2 = \frac{(t - t_1)(t - t_3)}{(t_2 - t_1)(t_2 - t_3)} = \frac{(t - 1)(t - 5)}{(1)(-3)} \qquad N_2 = -\frac{(t - 1)(t - 5)}{3}$$

$$N_3 = \frac{(t - t_1)(t - t_2)}{(t_3 - t_1)(t_3 - t_2)} = \frac{(t - 1)(t - 2)}{(4)(3)} \qquad N_3 = \frac{(t - 1)(t - 2)}{12}$$

The final equation becomes

$$v = \frac{(t-2)(t-5)}{4}2 - \frac{(t-1)(t-5)}{3}4 + \frac{(t-1)(t-2)}{12}7$$

Solving for v at $t = 4$, just to try it out, we get

$$v = \frac{(2)(-1)}{4}2 - \frac{(3)(-1)}{3}4 + \frac{(3)(2)}{12}7$$

$$v(t=4) = 6.5 \qquad \square$$

5.3 INTERPOLATION FORMULA: TWO INDEPENDENT VARIABLES

Consider the situation where the dependent variable is a function of two variables x and y. We are given the value of the dependent variable, ϕ, for specific values of x and y as shown in the Figure 5.2. The points plotted in the figure are assumed to lie on a surface, and we wish to be able to interpolate other values on the surface without knowing the exact equation of the surface. We will approximate the surface function using the interpolation formula.

The procedure for constructing the formula for this case can be explained as follows:

1. Hold the y variable constant at $y = y_1$ and derive the interpolation formula using the three values of ϕ given for $y = y_1$.

$$\phi_I(x, y_1) = N_1(x)\phi_1 + N_2(x)\phi_2 + N_3(x)\phi_3$$

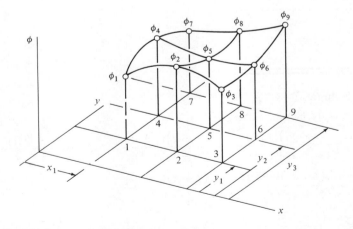

FIGURE 5.2 Plot of three variables.

5.3 INTERPOLATION FORMULA: TWO INDEPENDENT VARIABLES

where

$$N_i = \frac{(x-x_1)\cdots(x-x_{i-1})(x-x_{i+1})\cdots(x-x_n)}{(x_i-x_1)\cdots(x_i-x_{i-1})(x_i-x_{i+1})\cdots(x_i-x_n)}$$

2. Hold y constant at $y = y_2$ and derive a second formula using the three given values of ϕ along that line of constant y.

$$\phi_{II}(x, y_2) = N_4(x)\phi_4 + N_5(x)\phi_5 + N_6(x)\phi_6$$

where N_i is defined in step 1.

3. Hold y constant at $y = y_3$ and derive a third formula.

$$\phi_{III}(x, y_3) = N_7(x)\phi_7 + N_8(x)\phi_8 + N_9(x)\phi_9$$

4. The next step is to develop the interpolation formula for the three functions of x to define the dependence of each function on the y variable. We might imagine a plot of these functions in the ϕ–y plane as shown in Figure 5.3. We wish to define the y dependence of the continuum of the functions of the x variable. Constructing the interpolation formula, we obtain

$$\phi(x, y) = N_1(y)\phi_I(x, y_1) + N_2(y)\phi_{II}(x, y_2) + N_3(y)\phi_{III}(x, y_3)$$

where

$$N_i(y) = \frac{(y-y_1)\cdots(y-y_{i-1})(y-y_{i+1})\cdots(y-y_n)}{(y_i-y_1)\cdots(y_i-y_{i-1})(y_i-y_{i+1})\cdots(y_i-y_n)}$$

5. Substitution of the formulas developed in steps 1 through 3 into the formula written in step 4 yields the two-dimensional interpolation formula.

Before making the substitution suggested in step 5, we will impose a restriction on the placement of the data points in the xy plane of Figure 5.2. This will simplify the resulting formula without diminishing its effectiveness for our purposes. We will require that all data values project onto lines parallel to the coordinate axes. Thus $x_1 = x_4 = x_7$, and so on. This will cause $N_1(x) = N_4(x) = N_7(x)$, $N_2(x) = N_5(x) = N_8(x)$, and $N_3(x) = N_6(x) = N_9(x)$. With this restriction, we write the final form of

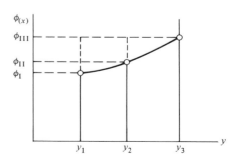

FIGURE 5.3 Plot of $\phi(x)$ versus y.

the interpolation formula as

$$\phi(x, y) = N_1(y)N_1(x)\phi_1 + N_1(y)N_2(x)\phi_2 + N_1(y)N_3(x)\phi_3$$
$$+ N_2(y)N_1(x)\phi_4 + N_2(y)N_2(x)\phi_5 + N_2(y)N_3(x)\phi_6$$
$$+ N_3(y)N_1(x)\phi_7 + N_3(y)N_2(x)\phi_8 + N_3(y)N_3(x)\phi_9 \quad (5.4)$$

where

$$N_1(s) = \frac{(s - s_2)(s - s_3)}{(s_1 - s_2)(s_1 - s_3)} \qquad s = x \text{ or } y$$

$$N_2(s) = \frac{(s - s_1)(s - s_3)}{(s_2 - s_1)(s_2 - s_3)}$$

$$N_3(s) = \frac{(s - s_1)(s - s_2)}{(s_3 - s_1)(s_3 - s_2)}$$

The interpolation formula, equation (5.4), was derived using nine data points distributed in the particular manner defined in Figure 5.2. If the procedure is understood, you should be able to derive a formula for differing number and pattern of data points.

Example 5.2 Given the data values ϕ_1, ϕ_2, ϕ_3, and ϕ_4, solve for the interpolation formula for the four-point rectangular pattern (Figure 5.4).

Solution. The interpolation formula is

$$\phi(x, y) = N_1(y)N_1(x)\phi_1 + N_1(y)N_2(x)\phi_2$$
$$+ N_2(y)N_1(x)\phi_4 + N_2(y)N_2(x)\phi_3$$

with

$$N_1(y) = \frac{y - y_4}{y_1 - y_4} \qquad N_2(y) = \frac{y - y_1}{y_4 - y_1}$$

$$N_1(y) = \frac{y - h}{-h} \qquad N_2(y) = \frac{y}{h}$$

$$N_1(x) = \frac{x - x_2}{x_1 - x_2} \qquad N_2(x) = \frac{x - x_1}{x_2 - x_1}$$

$$N_1(x) = \frac{x - b}{-b} \qquad N_2(x) = \frac{x}{b}$$

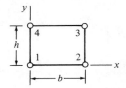

FIGURE 5.4 Arbitrary rectangle with sides parallel to coordinate axes.

Thus we have

$$\phi(x, y) = \frac{(y - h)(x - b)}{hb}\phi_1 - \frac{(y - h)x}{hb}\phi_2$$
$$- \frac{y(x - b)}{hb}\phi_4 + \frac{xy}{hb}\phi_3$$

As a check on the formula, put in the coordinates of the data points. For example, $\phi(b, 0) = \phi_2$. □

It should be pointed out that the polynomial coefficients of the data points can be obtained formally using the Lagrange formula as we have done in this discussion. However, they may also be obtained by inspection or trial and error. Regardless of the method employed, the one overriding requirement that must be satisfied is that the polynomial coefficient of a data point must equal unity when the independent variables correspond to that particular data point, while all other coefficients must equal zero.

5.4 INTERPOLATION APPLIED TO A THREE-NODE TRIANGLE ELEMENT

Referring to Figure 5.5, we would like to write an interpolation formula for any scalar variable in terms of the coordinate variables. This is rather confusing, because the pattern of the data points 1, 2, and 3 does not conform to anything like that in our previous discussion.

Let us try to write the Lagrange polynomials by inspection keeping in mind their "zero-unity" requirement. For the scalar dependent variable ϕ, we write the formula

$$\phi(X, Y) = N_1(X, Y)\phi_1 + N_2(X, Y)\phi_2 + N_3(X, Y)\phi_3$$

$$N_1 = \frac{(X - X_2)(X - X_3)(Y - Y_2)(Y - Y_3)}{(X_1 - X_2)(X_1 - X_3)(Y_1 - Y_2)(Y_1 - Y_3)} \quad \text{by inspection}$$

$$N_2 = \frac{(X - X_1)(X - X_3)(Y - Y_1)(Y - Y_3)}{(X_2 - X_1)(X_2 - X_3)(Y_2 - Y_1)(Y_2 - Y_3)} \quad \text{etc.}$$

These polynomial coefficients are quite cumbersome, and not of much

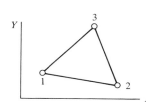

FIGURE 5.5 Three-node triangle.

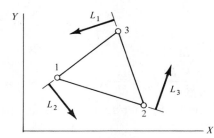

FIGURE 5.6 Triangle natural coordinates.

practical value for our development. Imagine if the $\phi(X, Y)$ variable is replaced with the displacement in the X direction $u(X, Y)$.

$$u(X, Y) = N_1 u_1 + N_2 u_2 + N_3 u_3$$

The strain in the element in the X direction is defined as $\partial u / \partial X$, which will involve $\partial N_i / \partial X$. The resulting expression for strain will be seen to be a function of differing powers of X and Y. This will not be a suitable strain function for modeling a continuum.

An alternative interpolation scheme that does work nicely and has been used extensively involves the application of triangle natural coordinates (Figure 5.6). It may be helpful to review the definition of these coordinates in Section 2.2.

The interpolation formula written in terms of the natural coordinates would be

$$\phi(L_1, L_2, L_3) = L_1 \phi_1 + L_2 \phi_2 + L_3 \phi_3$$

Note that substitution of the nodal coordinates verifies the required characteristics of the coefficients. For example,

$$\phi(1, 0, 0) = \phi_1 \qquad \phi(0, 1, 0) = \phi_2 \qquad \text{etc.}$$

In the next look at the triangular element as a finite element model, in Chapter 6, we will use the interpolation for shape as well as displacement. We will discuss them fully at that time.

$$u = L_1 u_1 + L_2 u_2 + L_3 u_3, \qquad v = L_1 v_1 + L_2 v_2 + L_3 v_3$$
$$X = L_1 X_1 + L_2 X_2 + L_3 X_3, \qquad Y = L_1 Y_1 + L_2 Y_2 + L_3 Y_3$$

Example 5.3 The temperature in the triangular continuum shown in Figure 5.7 is not uniform throughout. At nodes 1, 2, and 3 the temperature is known to be T_1, T_2, and T_3, respectively. Assuming a linear variation of temperature, solve for the temperature at a point p in the triangle located one-third of the way from side 1 and one-fourth of the way from side 3.

Solution. The natural coordinates of the point p in the triangle are

$$L_1 = \tfrac{1}{3} \qquad L_3 = \tfrac{1}{4} \qquad L_2 = 1 - L_1 - L_3 = \tfrac{5}{12}$$

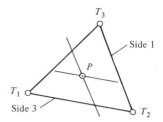

FIGURE 5.7 Triangle node temperatures.

The temperature at point p is thus

$$T_p = L_1 T_1 + L_2 T_2 + L_3 T_3$$
$$= \frac{T_1}{3} + \frac{5T_2}{12} + \frac{T_3}{4} \qquad \square$$

5.5 INTEGRATION BY GAUSS QUADRATURE

In our previous discussion of element development, we saw that the element strain energy, which produced the stiffness property, and the distributed surface traction and body force potential energy, which produced equivalent nodal forces, all required integration over a line or area. The uniaxial and triangular constant strain elements yielded functions that were very easily integrated. This will not be the case for the higher-order elements. The integral expressions are so complex that they will be evaluated numerically.

There are many techniques of numerical integration. We will touch very briefly on the Gauss method because it has found great acceptance in finite element work. Our discussion will introduce the basic principle and outline the method of application for situations of specific interest in our finite element development.

Our goal is to evaluate the magnitude of a definite integral, ideally without error, without having to perform the integration and substitute the limits. Consider the evaluation of the following integral (Figure 5.8):

$$I = \int_{-1}^{1} f(x)\, dx$$

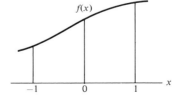

FIGURE 5.8 Plot of arbitrary function.

An approximation of this integral would be the product of the function $f(x)$ value at the midpoint and the "width" defined by the limits.

$$I_a = 2f(0) \tag{5.5}$$

Another approximation would be the product of an average value of $f(x)$ and the limit dimension.

$$I_a = 2\left[\frac{f(-1) + f(1)}{2}\right] \quad I_a = f(-1) + f(1) \tag{5.6}$$

Both of these approximations would yield the correct result for the integral if $f(x)$ is a linear function. Functions of higher degree could not be evaluated accurately using these formulas.

A word should be said about the particular limits of integration used in the example. In the development of higher-order elements, natural coordinate systems are used almost exclusively. The limits on the integration variables for these coordinates are 0 to 1 or -1 to $+1$. Our experience with natural coordinates at this point is limited to the triangle, where we recall the limits on these natural coordinates to be 0 and 1. Thus, as we continue this discussion using examples having limits of unity, we will be using limits directly applicable to the natural coordinates of an element. To extend the numerical integration formulas to other applications having different limits, we have only to normalize the coordinate variable (introduce a change of variable which has a maximum magnitude of unity).

Returning to the integral approximation, we saw that the integral of a linear function, $f(x) = C_0 + C_1 x$, between limits of ± 1 could be evaluated by summing products of constants and particular function values. Recalling (5.5) and (5.6), we have

$$\int_{-1}^{1} (C_0 + C_1 x)\, dx = 2f(0) = 2C_0 \quad \text{from (5.5)}$$

$$= f(-1) + f(1) = C_0 - C_1 + C_0 + C_1 = 2C_0 \quad \text{from (5.6)}$$

To generalize these simple formulas, we could write

$$I = \int f(x)\, dx = W_1 f(x_1) + W_2 f(x_2) = \sum_{i=1}^{2} W_i f(x_i)$$

The constant W_i is called a *weighting factor*.

Let us extend this concept to the integration of a function of higher degree, a cubic polynomial for example. The exact integral of the general cubic is

$$I = \int_{-1}^{1} (C_0 + C_1 x + C_2 x^2 + C_3 x^3)\, dx = 2C_0 + \tfrac{2}{3} C_2 \tag{5.7}$$

Note that when integrating between symmetric limits, the integrals of odd powers of x equal zero. Thus the integral of the complete polynomial

5.5 INTEGRATION BY GAUSS QUADRATURE

of degree 3 yields two summed terms. We ask the question: Is it possible to evaluate the integral using the same number of summed weighted function values? Let us select a "two-point" formula for this example. The numerically approximated integral is

$$I_a = W_1 f(x_1) + W_2 f(x_2)$$
$$= W_1(C_0 + C_1 x_1 + C_2 x_1^2 + C_3 x_1^3)$$
$$+ W_2(C_0 + C_1 x_2 + C_2 x_2^2 + C_3 x_2^3) \quad (5.8)$$

The difference (error) between (5.8) and (5.7) is

$$E = I_a - I$$
$$= (W_1 + W_2 - 2)C_0 + (x_1 W_1 + x_2 W_2)C_1 + (W_1 x_1^2 + W_2 x_2^2 - \tfrac{2}{3})C_2$$
$$+ (W_1 x_1^3 + W_2 x_2^3)C_3 \quad (5.9)$$

If we set the error equal to zero, and if the coefficients of the polynomial variable are arbitrary, we can write the following four independent equations:

$$\begin{aligned} W_1 + W_2 - 2 &= 0 \\ x_1 W_1 + x_2 W_2 &= 0 \\ W_1 x_1^2 + W_2 x_2^2 &= \tfrac{2}{3} \\ W_1 x_1^3 + W_2 x_2^3 &= 0 \end{aligned} \quad (5.10)$$

Here we have four equations for four unknowns: W_1, W_2, x_1, and x_2. Solving these equations, we obtain

$$W_1 = W_2 = 1$$
$$x_1 = -x_2 = -\frac{1}{\sqrt{3}}$$

Table 5.1 lists some abscissa locations and associated weighting factors for exact numerical integration of polynomials of varying degree. It must be emphasized that exact results are obtained for integration of polynomials. If a nonpolynomial function is integrated, an error will result, the magnitude of which depends on how well the polynomial matches the function.

Example 5.4 Estimate the integral $I = \int_{-1}^{1} e^x \, dx$ using the two-point and then the three-point quadrature formulas.

Solution. Using the two-point formula, we have

$$I \approx \sum_{i=1}^{2} W_i f(x_i) = (1) e^{(-0.57735)} + (1) e^{(0.57735)}$$
$$= 0.56138 + 1.78131$$
$$I \approx 2.3427$$

TABLE 5.1 Gauss Quadrature Data

$$\int_{-1}^{+1} f(x)\,dx = \sum_{i=1}^{n} W_i f(x_i)$$

Polynomial Degree	Number of Points[a]	Locations, x_i	Weighting Factors, W_i
1	1	$x_1 = 0$	$W_1 = 2$
≤ 3	2	$x_1 = 0.5773503$	$W_1 = 1$
		$x_2 = 0.5773503$	$W_2 = 1$
$= 5$	3	$x_1 = -0.7745967$	$W_1 = 0.5555556$
		$x_2 = 0$	$W_2 = 0.8888889$
		$x_3 = 0.7745967$	$W_3 = 0.5555556$

[a] n points yield exact results for polynomials of degree $\leq 2n - 1$.

Using the three-point formula, we have

$$I \approx \sum_{i=1}^{3} W_i f(x_i) = 0.55556 e^{(-0.774597)}$$

$$+ 0.88889 e^{(0)} + 0.55556 e^{(0.774597)}$$
$$= 0.25605 + 0.88889 + 1.20540$$
$$I \approx 2.3503$$

These results can be compared to the exact value:

$$I = \int_{-1}^{1} e^x\,dx = e^x \Big|_{-1}^{1} = 2.35040$$

The three-point formula gives the better approximation. Note that the quadrature, based on polynomial functions, will not yield the exact value for this function. We have to decide on "good enough." ∎

Example 5.5 Given the coordinates of three points which conform to some unknown function, estimate the integral

$$I = \int_{-1}^{1} f(x)\,dx$$

x	$f(x)$
1	1.5
0	2.0
-1	1.0

5.5 INTEGRATION BY GAUSS QUADRATURE

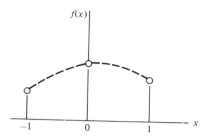

FIGURE 5.9 Assumed quadratic function.

Solution. Since only three points are given, we assume that the function $f(x)$ can be represented by a quadratic polynomial (Figure 5.9). We will then estimate the integral using the two-point formula. First, however, we must interpolate to find the values at the Gauss points $x = \pm 0.577\ldots$.

The interpolation formula can be derived easily (see Example 5.1).

$$f(x) = N_1 f(x_1) + N_2 f(x_2) + N_3 f(x_3)$$

$$= \frac{(x - x_2)(x - x_3)}{(x_1 - x_2)(x_1 - x_3)} f(x_1)$$

$$+ \frac{(x - x_1)(x - x_3)}{(x_2 - x_1)(x_2 - x_3)} f(x_2) + \text{etc.}$$

$$= \frac{x(x-1)}{2}(1.0) + \frac{(x+1)(x-1)}{-1}(2.0) + \frac{(x+1)x}{2}(1.5)$$

$$f(-0.57735) = 1.6057$$

$$f(0.57735) = 1.8943$$

Now using the two-point formula gives us

$$I = \int_{-1}^{1} f(x)\, dx = \sum_{i=1}^{2} W_i f(x_i) = 1.6057 + 1.8943$$

$$= 3.500$$

This would be the exact result if the points given did obey a quadratic function. Of course, we have no way of knowing this with the information given. □

We must now extend the integration to two variables. This is accomplished without much difficulty for a square region.

$$I = \int_{-1}^{1} \int_{-1}^{1} f(x, y)\, dx\, dy$$

If we assume a three-point formula and integrate first in the y direction,

INTERPOLATION FORMULAS AND NUMERICAL INTEGRATION

we have

$$I_{a'} = \int_{-1}^{1} [W_1 f(x, y_1) + W_2 f(x, y_2) + W_3 f(x, y_3)] \, dx$$
$$= \int_{-1}^{1} g(x) \, dx$$

Next, integrating in the x direction, we have

$$\begin{aligned}I_a = & W_1[W_1 f(x_1, y_1) + W_2 f(x_1, y_2) + W_3 f(x_1, y_3)] \\ & + W_2[W_1 f(x_2, y_1) + W_2 f(x_2, y_2) + W_3 f(x_2, y_3)] \\ & + W_3[W_1 f(x_3, y_1) + W_2 f(x_3, y_2) + W_3 f(x_3, y_3)]\end{aligned}$$

This can be written in a general form for an n-point approximation formula:

$$I_a = \sum_{i=1}^{n} \sum_{j=1}^{n} W_i W_j f(x_i, y_j) \tag{5.11}$$

Example 5.6 Evaluate the integral over the square region (Figure 5.10).

$$I = \int_{-1}^{1} \int_{-1}^{1} x^2 y \, dx \, dy$$

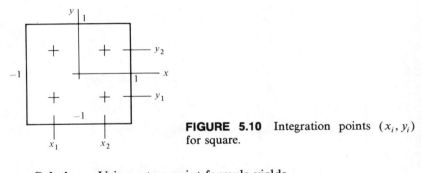

FIGURE 5.10 Integration points (x_i, y_i) for square.

Solution. Using a two-point formula yields

$$I_a = W_1 W_1 x_1^2 y_1 + W_1 W_2 x_1^2 y_2 + W_2 W_1 x_2^2 y_1 + W_2 W_2 x_2^2 y_2$$

Selecting the integration point coordinates and weighting factors from Table 5.1 gives us

$$\begin{aligned}I_a = & (1)(1)(-0.577^2)(-0.577) + (1)(1)(-0.577^2)(0.577) \\ & + (1)(1)(0.577^2)(-0.577) + (1)(1)(0.577^2)(0.577) \\ = & 0\end{aligned}$$ □

We will employ these quadrature formulas when we develop the four-node quadrilateral element in Chapter 7, the six-node triangle, and the eight-node quadrilateral in Chapter 8.

5.6 CHANGE OF INTEGRATION VARIABLE: THE JACOBIAN DETERMINANT

In our previous discussion of the development of the constant strain uniaxial and triangular element, the mathematical formulation was solely in terms of the rectangular Cartesian local or global reference system. As we encounter higher-level elements, we are going to have a mix of global and natural coordinate systems, and transformations will be made from one to the other. One very important transformation involves the change of differential variable for the double (or triple) integral. Since our discussion is focusing on the two-dimensional stress state, we will discuss the transformation of a differential element of area of thickness h.

The area of a differential element in terms of the rectangular Cartesian coordinates is $dx\,dy$ (Figure 5.11). The volume of the element for uniform thickness is simply $h\,dx\,dy$. This can be derived from the triple scalar product of three vectors defining the sides of the element.

$$d\Omega = (dx\,\hat{i} \times dy\,\hat{j}) \cdot h\hat{k}$$

$$= \begin{vmatrix} 0 & 0 & h \\ dx & 0 & 0 \\ 0 & dy & 0 \end{vmatrix} \qquad (5.12)$$

Introduce a second coordinate system, not necessarily orthogonal (Figure 5.12), and let the transformation equations for the two coordinate

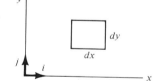

FIGURE 5.11 Differential area.

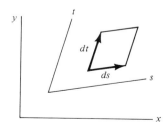

FIGURE 5.12 Differential area with two coordinate systems.

systems be
$$x = x(s,t)$$
$$y = y(s,t)$$

Referring to the differential element with sides parallel to the st coordinate axes, we can express a differential change in the s variable only as a vector with x and y components:

$$dx = \frac{\partial x}{\partial s} ds \qquad dy = \frac{\partial y}{\partial s} ds$$

$$\overline{ds} = dx\,\hat{i} + dy\,\hat{j}$$

$$= \hat{i}\frac{\partial x}{\partial s} ds + \hat{j}\frac{\partial y}{\partial s} ds$$

Similarly, a differential change in the t variable can be expressed as another vector:

$$dx = \frac{\partial x}{\partial t} dt \qquad dy = \frac{\partial y}{\partial t} dt$$

$$\overline{dt} = dx\,\hat{i} + dy\,\hat{j}$$

$$= \hat{i}\frac{\partial x}{\partial t} dt + \hat{j}\frac{\partial y}{\partial t} dt$$

The volume of the element with sides ds, dt, and h is

$$d\Omega = \begin{vmatrix} 0 & 0 & h \\ \frac{\partial x}{\partial s} ds & \frac{\partial y}{\partial s} ds & 0 \\ \frac{\partial x}{\partial t} dt & \frac{\partial y}{\partial t} dt & 0 \end{vmatrix} \tag{5.13}$$

$$= h \begin{vmatrix} \frac{\partial x}{\partial s} & \frac{\partial y}{\partial s} \\ \frac{\partial x}{\partial t} & \frac{\partial y}{\partial t} \end{vmatrix} ds\,dt$$

or
$$d\Omega = h|J|\,ds\,dt \tag{5.14}$$

where $|J|$ is the Jacobian determinant of the transformation matrix for the two coordinate systems. The transformation of a double integral now takes the form

$$\int\int_A f(x,y)\,dx\,dy = \int\int_A g(s,t)\,dx\,dy$$

$$= \int\int_{A^*} g(s,t)|J|\,ds\,dt \tag{5.15}$$

5.6 CHANGE OF INTEGRATION VARIABLE: THE JACOBIAN DETERMINANT

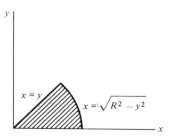

FIGURE 5.13 Rectangular and polar coordinate axes and integration region.

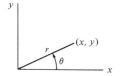

Example 5.7 Perform the indicated change of variable for the integral over the region.

$$\int_0^{R/\sqrt{2}} \int_y^{\sqrt{R^2-y^2}} (x^2 + y^2) \, dx \, dy = \int \int_{A^*} g(r,\theta) |J| \, dr \, d\theta$$

where $x = r\cos\theta$, $y = r\sin\theta$. See Figure 5.13.

Solution. $x^2 + y^2 = r^2$; thus $g(r, \theta) = r^2$.

$$|J| = \begin{vmatrix} \dfrac{\partial x}{\partial r} & \dfrac{\partial y}{\partial r} \\ \dfrac{\partial x}{\partial \theta} & \dfrac{\partial y}{\partial \theta} \end{vmatrix} = \begin{vmatrix} \cos\theta & \sin\theta \\ -r\sin\theta & r\cos\theta \end{vmatrix} = r$$

Thus

$$\int\int_A (x^2 + y^2) \, dx \, dy = \int_0^{\pi/4} \int_0^R r^3 \, dr \, d\theta \qquad \square$$

Example 5.8 Perform the indicated change of variable and integration using the Gauss quadrature with the two-point formula

$$I = \int_0^{10} \int_2^6 y^2 \, dx \, dy = \int_{-1}^1 \int_{-1}^1 f(s,t) \, ds \, dt$$

See Figure 5.14.

Solution. The transformation equations for the two parallel orthogonal coordinate system are

$$s = \frac{x-4}{2} \qquad t = \frac{y-5}{5}$$

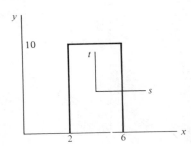

FIGURE 5.14 Region of integration and coordinate axes.

for limits of ± 1 on s and t, or
$$y = 5(t + 1) \qquad x = 2s + 4$$
From (5.14),
$$dx\,dy = |J|\,ds\,dt = \begin{vmatrix} 2 & 0 \\ 0 & 5 \end{vmatrix} ds\,dt = 10\,ds\,dt$$

Substituting for y and $dx\,dy$ in the original form of the integral, we have
$$I = \int_{-1}^{1} \int_{-1}^{1} 25(t + 1)^2 (10)\,ds\,dt = 250 \int_{-1}^{1} \int_{-1}^{1} (t + 1)^2\,ds\,dt$$

Using the two-point formula gives us
$$I = 2\{250[(-0.57735 + 1)^2 + (0.57735 + 1)^2]\}$$
$$= 1333.3 \qquad \square$$

PROBLEMS

5.1 Derive an interpolation formula for the data points given in Figure P5.1. Using the formula, calculate and plot values for $x = 1.5, 2.5,$ and 3.5.

5.2 Derive an interpolation formula for any scalar variable ϕ in terms of ϕ at each corner node and coordinates x and y. The coordinate axis is to be at the geometric center of the square which has sides of length 2 (Figure P5.2).

5.3 Solve the integral exactly and then evaluate the integral numerically using the two-point and the three-point formulas.
$$I = \int_{-1}^{1} (3 + 3x^2)\,dx$$

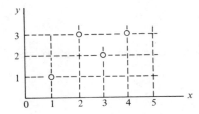

FIGURE P5.1

PROBLEMS

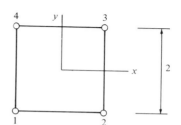

FIGURE P5.2

5.4 Solve the integral exactly and then evaluate the integral numerically using the two-point and then the three-point formula.

$$I = \int_{-1}^{1} \int_{-1}^{1} (8x^3 + 5x^2y^3)\, dx\, dy$$

5.5 Evaluate the integral.

$$I = \int_{-1}^{1} \int_{-1}^{1} \frac{xy^2}{x-2}\, dx\, dy$$

5.6 Evaluate the area under the curve that passes through the three points defined in the table.

Point	x	y
1	1.0	1.0
2	2.0	2.0
3	3.0	4.0

5.7 Evaluate the area under the curve that passes through the four points defined in the table.

Point	x	y
1	-1.0	0
2	-0.33	1.0
3	0.33	3.0
4	1.0	2.5

5.8 Evaluate the integral over the region using first the two-point formula and then the three-point formula.

$$I = \int_{-1}^{1} \int_{-1}^{1} \frac{e^x}{\cos y}\, dx\, dy$$

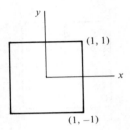

FIGURE P5.10

5.9 Change the integration variable so that the limits are ± 1 and evaluate the integral using the one-point, two-point, and three-point formulas.

$$I = \int_0^{\pi/2} \cos x \, dx$$

5.10 Use the two-point formula to numerically estimate the value of the integral of the function over the unit square (Figure P5.10).

$$I = \int_A x^2 e^{(x/y)^2} \, dA$$

CHAPTER 6

Isoparametric Three-Node Triangular Element

6.1 INTRODUCTION

We will have one more look at the development of the three-node triangular element. The method of development will involve natural coordinates and interpolation functions. Using the three-node triangle to introduce this approach is beneficial because the development is not complex, and second, it allows us to make a comparison with the two methods previously discussed in Chapter 4.

A review of Section 4.2 at this time would be helpful to bring back the details of the development of the triangle using the global coordinate variables in the assumed linear displacement function. The only substantial differences between that development and this new development of the triangle include the introduction of the natural coordinates and the use of the interpolation formula for displacement and position of any point in the interior of the element.

6.2 DEVELOPMENT OF THE STIFFNESS MATRIX

Step 1. The Assumed Displacement Function. The triangle with the natural and global coordinates is shown in Figure 6.1. The first step in the development of the stiffness matrix is, as before, to assume a displacement function. As discussed in Section 5.4, the interpolation function for displacement in terms of the nodal displacements is

$$u = L_1 u_1 + L_2 u_2 + L_3 u_3 \tag{6.1}$$

We recall from Section 2.2 that the natural coordinates are related through the expression

$$L_1 + L_2 + L_3 = 1$$

Substitution into equation (6.1) gives us

$$u = L_1 u_1 + L_2 u_2 + (1 - L_1 - L_2) u_3 \tag{6.2a}$$

In like manner,

$$v = L_1 v_1 + L_2 v_2 + (1 - L_1 - L_2) v_3 \tag{6.2b}$$

The next step we take is simple, but it has a profound impact on the development of the element mathematics. It is proposed that the shape of the element (spatial coordinates) be described using the same interpolation equation coefficients. We write the following expressions for the global position coordinates of any point in the element:

$$X = L_1 X_1 + L_2 X_2 + L_3 X_3 \tag{6.3a}$$

or

$$X = L_1 X_1 + L_2 X_2 + (1 - L_1 - L_2) X_3$$

and

$$Y = L_1 Y_1 + L_2 Y_2 + (1 - L_1 - L_2) Y_3 \tag{6.3b}$$

Definition. *Elements that use the same interpolation coefficients for the spatial coordinates (shape) and displacement interpolation formulas are called* isoparametric elements.

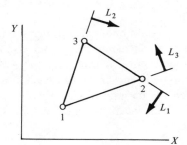

FIGURE 6.1 Three-node triangle with natural and global coordinates.

6.2 DEVELOPMENT OF THE STIFFNESS MATRIX

Step 2. Derivation of Strain. Recalling the definitions for the strain components, equations (2.3) and (2.4), we understand that the strain requires the differentiation of the displacement functions relative to the global X and Y coordinates. The assumed function for the displacements, equation (6.1), are in terms of the natural coordinates, not the global. We must initiate a change of variable.

Note that from equations (6.1) and (6.2) we can perform the following differentiations:

$$\frac{\partial u}{\partial L_1}, \quad \frac{\partial u}{\partial L_2}, \quad \ldots, \quad \frac{\partial X}{\partial L_1}, \quad \frac{\partial X}{\partial L_2}, \quad \ldots$$

Treating L_1 as the independent variable, we can write

$$\frac{\partial u}{\partial L_1} = \frac{\partial u}{\partial X}\frac{\partial X}{\partial L_1} + \frac{\partial u}{\partial Y}\frac{\partial Y}{\partial L_1}$$

Also, treating L_2 as the independent variable gives us

$$\frac{\partial u}{\partial L_2} = \frac{\partial u}{\partial X}\frac{\partial X}{\partial L_2} + \frac{\partial u}{\partial Y}\frac{\partial Y}{\partial L_2}$$

or in matrix form,

$$\begin{bmatrix} \frac{\partial X}{\partial L_1} & \frac{\partial Y}{\partial L_1} \\ \frac{\partial X}{\partial L_2} & \frac{\partial Y}{\partial L_2} \end{bmatrix} \begin{Bmatrix} \frac{\partial u}{\partial X} \\ \frac{\partial u}{\partial Y} \end{Bmatrix} = \begin{Bmatrix} \frac{\partial u}{\partial L_1} \\ \frac{\partial u}{\partial L_2} \end{Bmatrix} \quad (6.4)$$

Now, substituting the derivatives of equations (6.2) and (6.3) yields

$$\begin{bmatrix} X_1 - X_3 & Y_1 - Y_3 \\ X_2 - X_3 & Y_2 - Y_3 \end{bmatrix} \begin{Bmatrix} \frac{\partial u}{\partial X} \\ \frac{\partial u}{\partial Y} \end{Bmatrix} = \begin{Bmatrix} u_1 - u_3 \\ u_2 - u_3 \end{Bmatrix} \quad (6.5)$$

Solving for the global coordinate derivatives from (6.5), we have

$$\begin{Bmatrix} \frac{\partial u}{\partial X} \\ \frac{\partial u}{\partial Y} \end{Bmatrix} = \frac{1}{|J|} \begin{bmatrix} Y_2 - Y_3 & Y_3 - Y_1 \\ X_3 - X_2 & X_1 - X_3 \end{bmatrix} \begin{Bmatrix} u_1 - u_3 \\ u_2 - u_3 \end{Bmatrix}$$

where $|J|$ is the Jacobian determinant of the transformation matrix for the two coordinate systems. Now substitute the following:

$$\begin{Bmatrix} u_1 - u_3 \\ u_2 - u_3 \end{Bmatrix} = \begin{bmatrix} 1 & 0 & -1 \\ 0 & 1 & -1 \end{bmatrix} \begin{Bmatrix} u_1 \\ u_2 \\ u_3 \end{Bmatrix}$$

The resultant form of the expression is

$$\left\{\begin{array}{c}\dfrac{\partial u}{\partial X}\\ \dfrac{\partial u}{\partial Y}\end{array}\right\} = \dfrac{1}{|J|}\begin{bmatrix} Y_2 - Y_3 & Y_3 - Y_1 & Y_1 - Y_2 \\ X_3 - X_2 & X_1 - X_3 & X_2 - X_1 \end{bmatrix}\left\{\begin{array}{c}u_1\\u_2\\u_3\end{array}\right\} \quad (6.6)$$

where

$$|J| = (X_1 - X_3)(Y_2 - Y_3) - (X_2 - X_3)(Y_1 - Y_3)$$

In like manner, we are able to find global derivatives of v,

$$\left\{\begin{array}{c}\dfrac{\partial v}{\partial X}\\ \dfrac{\partial v}{\partial Y}\end{array}\right\} = \dfrac{1}{|J|}\begin{bmatrix} Y_2 - Y_3 & Y_3 - Y_1 & Y_1 - Y_2 \\ X_3 - X_2 & X_1 - X_3 & X_2 - X_1 \end{bmatrix}\left\{\begin{array}{c}v_1\\v_2\\v_3\end{array}\right\} \quad (6.7)$$

Combining equations (6.6) and (6.7), we have

$$\left\{\begin{array}{c}\dfrac{\partial u}{\partial X}\\ \dfrac{\partial v}{\partial Y}\\ \dfrac{\partial u}{\partial Y} + \dfrac{\partial v}{\partial X}\end{array}\right\}$$

$$= \dfrac{1}{|J|}\begin{bmatrix} Y_2 - Y_3 & 0 & Y_3 - Y_1 & 0 & Y_1 - Y_2 & 0 \\ 0 & X_3 - X_2 & 0 & X_1 - X_3 & 0 & X_2 - X_1 \\ X_3 - X_2 & Y_2 - Y_3 & X_1 - X_3 & Y_3 - Y_1 & X_2 - X_1 & Y_1 - Y_2 \end{bmatrix}$$

$$\times \left\{\begin{array}{c}u_1\\v_1\\u_2\\v_2\\u_3\\v_3\end{array}\right\} \quad (6.8)$$

In symbolic form,

$$\{\epsilon\} = [B]\{q\}$$

Comparison of equation (6.8) with equation (4.7) reveals that they are the same with the exception of the factor $1/\det$ in (4.7) as opposed to $1/|J|$ in (6.8). Examination of the expressions for det and $|J|$ reveals that they are identical, and this means that the $[B]$ matrix of (6.8) is the same as the $[B]$ matrix of (4.7). One final reminder, the Jacobian determinant $|J|$ being equal to det is thus also equal to twice the area of the three-node triangle.

6.3 DETERMINING THE EQUIVALENT NODAL FORCES

In general, derivation of the $[B]$ matrix is simpler using the natural coordinate system. The real comparison of effort is seen when the matrix is derived for a higher-order element. The methods of Chapter 4 are conceptually simple but generate horrible detail in the higher-order elements.

Having derived the strain function, it would be a straightforward substitution into equation (4.9) to obtain the strain energy in the element. Symbolically, the strain energy formula for elements with $[B]$ and $[C]$ matrices independent of the coordinates is

$$U = \tfrac{1}{2}\{q\}^T[B]^T[C][B]\{q\}hA \tag{4.9}$$

Subsequent addition of the element force energy potentials and application of the minimization of total energy (Section 4.2, steps 4 and 5) would yield the same symbolic form for the element stiffness:

$$[k] = [B]^T[C][B]hA$$

The stiffness matrix of each element in an assemblage is easily calculated by substituting the coordinates in the $[B]$ matrix, selecting the correct constitutive matrix and evaluating the product above. This stiffness matrix will be identical to that derived directly with the global reference coordinates in Chapter 4.

It should be emphasized, perhaps, that the stiffness matrix is referenced directly to the global reference system. Thus no coordinate rotation need by applied. If this is not clearly understood, you must go back and review the derivation. Observe that the derivation for the strain is referenced, by a change of variable, to the global XY coordinate system.

6.3 DETERMINING THE EQUIVALENT NODAL FORCES

The discussion in Section 4.2, step 3, concerning the applied load energy functions should be reviewed at this time. The final form of the energy functions defining energy potentials of the applied element loads is seen in equation (4.13). The subsequent application of the minimization of energy yielded the following expressions for the nodal load vectors [refer to equation (4.14)]:

Element nodal force vector:

$$\{Q\}_{NF} = \begin{Bmatrix} F_{1X} \\ F_{1Y} \\ F_{2X} \\ F_{2Y} \\ F_{3X} \\ F_{3Y} \end{Bmatrix} \tag{6.9}$$

Distributed edge loading vector:

$$\{Q\}_T = h \int_S [N]^T \begin{Bmatrix} T_X \\ T_Y \end{Bmatrix} ds \qquad (6.10)$$

Distributed body force vector:

$$\{Q\}_{BF} = h \int_A [N]^T \begin{Bmatrix} B_X \\ B_Y \end{Bmatrix} dA \qquad (6.11)$$

where $\{Q\}$ is the 6×1 equivalent node force component vector, T_X and T_Y are the X and Y components of the stress traction on one side of the triangle, B_X and B_Y are the X and Y components of the distributed body force, ds is the differential length of the side, and dA is the differential element of area in the triangle.

These relationships for the nodal force vectors, although derived in Chapter 4, are completely general and applicable to any two-dimensional element with three nodes. The important difference in application for different coordinate systems will be seen in the form of the $[N]$ matrix.

We can rewrite equations (6.2) in the following form:

$$\begin{Bmatrix} u \\ v \end{Bmatrix} = \begin{bmatrix} L_1 & 0 & L_2 & 0 & L_3 & 0 \\ 0 & L_1 & 0 & L_2 & 0 & L_3 \end{bmatrix} \begin{Bmatrix} u_1 \\ v_1 \\ u_2 \\ v_2 \\ u_3 \\ v_3 \end{Bmatrix} \qquad (6.12)$$

where $L_3 = 1 - L_1 - L_2$, or symbolically,

$$\{u\} = [N]\{q\}$$

The entries in the $[N]$ matrix are the interpolation formula coefficients, and these coefficients are also used in the interpolation formulas for the position coordinates of any point in the element. These entries are commonly called *shape functions*. Substitution of the shape function matrix $[N]$ into the equivalent node force relationship, equation (6.10) yields

$$\{Q\}_T = h \int_s \begin{bmatrix} L_1 & 0 \\ 0 & L_1 \\ L_2 & 0 \\ 0 & L_2 \\ 1 - L_1 - L_2 & 0 \\ 0 & 1 - L_1 - L_2 \end{bmatrix} \begin{Bmatrix} T_X \\ T_Y \end{Bmatrix} ds \quad (6.10a)$$

A similar substitution can be made in equation (6.11).

Equation (6.10) yields element node forces which are equivalent to the distributed edge traction. These equivalent forces are to be applied just like an externally applied load. You must appreciate that the relationship

6.3 DETERMINING THE EQUIVALENT NODAL FORCES

provides the X and Y global coordinate components. If it is not clear why, review the derivation in detail.

Example 6.1 Solve for the equivalent nodal forces resulting from the distributed stress traction, T_X (force/area), on one side of the triangle shown in Figure 6.2. The element thickness $= h$.

Solution. We are concerned with evaluating the equivalent nodal force vector for the stress distribution on side 1. For this case, $L_1 = 0$. Also, for the isoparametric element,

$$X = L_1 X_1 + L_2 X_2 + (1 - L_1 - L_2) X_3$$

and

$$Y = L_1 Y_1 + L_2 Y_2 + (1 - L_1 - L_2) Y_3$$

Therefore,

$$T_X = 10Y = 10[L_2(Y_2 - Y_3) + Y_3] = 10[L_2(2 - 5) + 5]$$
$$= 50 - 30 L_2$$

and

$$T_Y = 0$$

Also,

$$ds = \sqrt{dX^2 + dY^2} = |dY| = |(Y_2 - Y_3)| dL_2$$
$$= |(2 - 5)| dL_2 = 3 dL_2$$

Substituting these relations into (6.10), we have

$$\{Q\}_T = \begin{Bmatrix} F_{1X} \\ F_{1Y} \\ F_{2X} \\ F_{2Y} \\ F_{3X} \\ F_{3Y} \end{Bmatrix} = h \int_0^1 \begin{bmatrix} 0 & 0 \\ 0 & 0 \\ L_2 & 0 \\ 0 & L_2 \\ 1 - L_2 & 0 \\ 0 & 1 - L_2 \end{bmatrix} \begin{Bmatrix} 50 - 30 L_2 \\ 0 \end{Bmatrix} (3 dL_2)$$

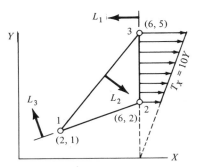

FIGURE 6.2 Linear stress traction on one side of triangle.

Expanding the matrix product yields

$$\{Q\}_T = h \int_0^1 \begin{Bmatrix} 0 \\ 0 \\ L_2(50 - 30L_2)(3) \\ 0 \\ (1 - L_2)(50 - 30L_2)(3) \\ 0 \end{Bmatrix} dL_2$$

Integrating gives us

$$\{Q\}_T = h \begin{Bmatrix} 0 \\ 0 \\ 75L_2^2 - 30L_2^3 \big|_0^1 \\ 0 \\ 150L_2 - 120L_2^2 + 30L_2^3 \big|_0^1 \\ 0 \end{Bmatrix}$$

Substituting the limits, we obtain

$$\{Q\}_T = \begin{Bmatrix} F_{1X} \\ F_{1Y} \\ F_{2X} \\ F_{2Y} \\ F_{3X} \\ F_{2Y} \end{Bmatrix} = h \begin{Bmatrix} 0 \\ 0 \\ 45 \\ 0 \\ 60 \\ 0 \end{Bmatrix} \quad \text{(force units)}$$

This equivalent loading is shown in Figure 6.3. □

In Section 6.4, two formulas are presented which simplify the integration of functions that are expressed in terms of the triangle natural coordinates. One formula is for the integration along the side of the triangle; the second is for the integration over the area. Illustrative examples detail the determination of the equivalent nodal loads for the

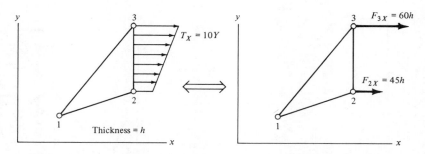

FIGURE 6.3 Node forces equivalent to the distributed edge loading.

6.4 INTEGRATION FORMULAS FOR TRIANGLE NATURAL COORDINATES

Example 6.1 required transformation of the differential variable, dY, as well as the traction component, T_X, from the physical coordinates (X, Y) to the natural coordinates (L_i). Two formulas that can be helpful in integrating natural coordinates over a specified region in real space are as follows:

For integration along a side of the triangle;

$$\int_S L_i^\alpha L_j^\beta \, ds = \frac{\alpha! \beta!}{(\alpha + \beta + 1)!} S \qquad i, j = 1, 2, \text{ or } 3 \qquad (6.13)$$

where S is the length of the side.

For integration over the area,

$$\int_A L_1^\alpha L_2^\beta L_3^\gamma \, dA = \frac{\alpha! \beta! \gamma!}{(\alpha + \beta + \gamma + 2)!} 2A \qquad (6.14)$$

where A is the area of the triangular region. A derivation of formula (6.14) is given in Davies (1980, Appendix 2).

Example 6.2 Perform the integration for the nodal force vector of the distributed loading in Example 6.1.

Solution. From Example 6.1 with $L_1 = 0$ and $[N]$ in terms of L_2 and L_3:

$$\{Q\}_T = h \int_S [N]^T \begin{Bmatrix} T_X \\ T_Y \end{Bmatrix} ds = h \int_S \begin{bmatrix} 0 & 0 \\ 0 & 0 \\ L_2 & 0 \\ 0 & L_2 \\ L_3 & 0 \\ 0 & L_3 \end{bmatrix} \begin{Bmatrix} 50 - 30L_2 \\ 0 \end{Bmatrix} dY$$

$$= h \int_S \begin{Bmatrix} 0 \\ 0 \\ (50L_2 - 30L_2^2) \, dY \\ 0 \\ (50L_3 - 30L_2 L_3) \, dY \\ 0 \end{Bmatrix}$$

Integrating each term, we have

$$\int_S (50L_2 - 30L_2^2)\, dY = \int_S 50L_2\, dY - \int_S 30L_2^2\, dY$$

Use formula (6.13) with the length of the side equal to 3.

$$\int_S 50L_2\, dY = 50\frac{0!\,1!}{2!}(3) = 75$$

$$\int_S 30L_2^2\, dY = 30\frac{0!\,2!}{3!}(3) = 30$$

Thus

$$\int_S (50L_2 - 30L_2^2)\, dY = 75 - 30 = 45$$

Also,

$$\int_S (50L_3 - 30L_2 L_3)\, dY = 50\frac{1!}{2!}(3) - 30\frac{1!\,1!}{3!}(3)$$

$$= 75 - 15 = 60$$

Therefore, we arrive at the same result obtained in Example 6.1. □

Example 6.3 The triangle of Example 6.1 has a mass per unit volume, ρ. Solve for the equivalent element nodal forces (see Figure 6.4).

Solution. From (6.11),

$$\{Q\}_{BF} = h\int_A [N]^T \begin{Bmatrix} B_X \\ B_Y \end{Bmatrix} dA$$

where the $[N]$ matrix is defined in equation (6.12). Thus we have

$$\{Q\}_{BF} = h\int_A \begin{bmatrix} L_1 & 0 \\ 0 & L_1 \\ L_2 & 0 \\ 0 & L_2 \\ L_3 & 0 \\ 0 & L_3 \end{bmatrix} \begin{Bmatrix} 0 \\ -\rho g \end{Bmatrix} dA = -g\rho h\int_A \begin{Bmatrix} 0 \\ L_1 \\ 0 \\ L_2 \\ 0 \\ L_3 \end{Bmatrix} dA$$

Applying formula (6.14) yields

$$\int_A L_1\, dA = \frac{1!\,0!\,0!}{3!}2A = \frac{1}{3}A \quad \text{where } A \text{ is the triangle area}$$

$$\int_A L_2\, dA = \frac{0!\,1!\,0!}{3!}2A = \frac{1}{3}A.$$

6.4 INTEGRATION FORMULAS FOR TRIANGLE NATURAL COORDINATES

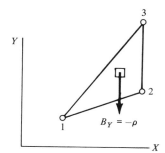

FIGURE 6.4 Body force direction and magnitude.

and

$$\int_A L_3 \, dA = \tfrac{1}{3} A$$

The final result is

$$\{Q\}_{BF} = -g\rho hA \begin{Bmatrix} 0 \\ \tfrac{1}{3} \\ 0 \\ \tfrac{1}{3} \\ 0 \\ \tfrac{1}{3} \end{Bmatrix}$$

Since the weight of the element is $g\rho hA$, the equivalent nodal forces are each one-third of the weight acting in the direction of the gravitational field. (Compare this solution procedure with Example 4.4.) □

Example 6.4 The side 3 ($L_3 = 0$) of the triangle shown in Figure 6.5 has a distributed edge traction which varies according to

$$T_X = CL_2^2 \quad \text{(force/area)}$$

Solve for the equivalent nodal forces. Element thickness = h.

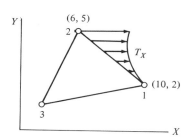

FIGURE 6.5 Distributed edge loading on side 3.

Solution. From (6.10),

$$\{Q\}_T = h\int_S [N]^T \begin{Bmatrix} T_X \\ T_Y \end{Bmatrix} ds$$

$$= h\int_S \begin{bmatrix} L_1 & 0 \\ 0 & L_1 \\ L_2 & 0 \\ 0 & L_2 \\ 0 & 0 \\ 0 & 0 \end{bmatrix} \begin{Bmatrix} CL_2^2 \\ 0 \end{Bmatrix} ds$$

$$= Ch \int_S \begin{Bmatrix} L_1 L_2^2 \\ 0 \\ L_2^3 \\ 0 \\ 0 \\ 0 \end{Bmatrix} ds$$

Using formula (6.13) with the length of side 3 as 5 in., we obtain

$$\int_S L_1 L_2^2 \, ds = \frac{1!\,2!}{4!}(5) = \frac{5}{12}$$

$$\int_S L_2^3 \, ds = \frac{0!\,3!}{4!}(5) = \frac{5}{4}$$

or

$$\{Q\}_T = \frac{5Ch}{12}\begin{Bmatrix} 1 \\ 0 \\ 3 \\ 0 \\ 0 \\ 0 \end{Bmatrix} \quad \text{(force units)}$$

Thus if we had a distributed loading that was quadratic with a maximum value, for example, of $C = 100$, we would have equivalent nodal forces applied to the element nodes as shown in Figure 6.6. □

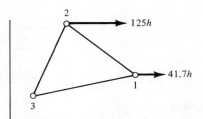

FIGURE 6.6 Equivalent node forces for $C = 100$.

6.4 INTEGRATION FORMULAS FOR TRIANGLE NATURAL COORDINATES

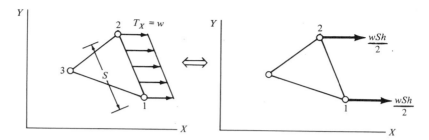

FIGURE 6.7 Equivalent loads for uniform stress traction, T_x force/area, on side of length S and element thickness h.

Two of the most common distributed edge tractions are the uniform stress and the linearly varying stress. The equivalent nodal loads for these distributions are simple to derive and easily remembered (see Figures 6.7 and 6.8). For a uniform stress on side 3 of length S and element thickness h:

$$\{Q\}_T = wSh \begin{Bmatrix} \frac{1}{2} \\ 0 \\ \frac{1}{2} \\ 0 \\ 0 \\ 0 \end{Bmatrix}$$

where wSh is the resultant force of the distributed stress traction.

For a linearly varying traction on side 3 (length S and element thickness h),

$$\{Q\}_T = \tfrac{1}{2} wSh \begin{Bmatrix} \frac{1}{3} \\ 0 \\ \frac{2}{3} \\ 0 \\ 0 \\ 0 \end{Bmatrix}$$

where $\tfrac{1}{2} wSh$ is the resultant force of the distributed traction. The reader is encouraged to derive these equivalent loads.

Recalling the discussion of equivalent nodal loads in Section 3.5, we present an alternative method of calculating the nodal loads with the following formula, which was derived for the linear displacement function and the linear load distribution (see Figure 6.9). This form is more convenient for computer implementation.

$$\begin{Bmatrix} F_{1X} \\ F_{2X} \end{Bmatrix} = h \frac{S}{6} \begin{bmatrix} 2 & 1 \\ 1 & 2 \end{bmatrix} \begin{Bmatrix} T_X(1) \\ T_X(2) \end{Bmatrix} \tag{6.15}$$

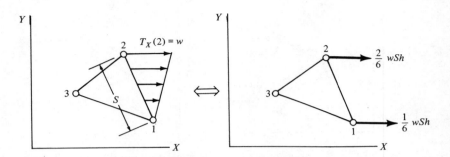

FIGURE 6.8 Equivalent loads for linear edge stress traction.

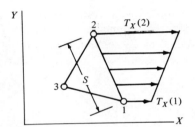

FIGURE 6.9 General linear distribution.

6.5 CONCLUDING REMARKS ON THE ELEMENT DEVELOPMENT

The derivation of the basic force–displacement relationship for the three-node triangle was presented in Chapter 4. The relationship derived is in symbolic form,

$$[k]\{q\} = \{Q\}_{\text{NF}} + \{Q\}_T + \{Q\}_{\text{BF}} \qquad (4.14)$$

where

$$[k] = [B]^T [C][B] hA$$

and the force vectors are defined in (6.9), (6.10), and (6.11). In Chapter 4 the matrix $[B]$ was developed in two ways: once using a linear displacement function in terms of the global coordinates, followed by the use of a displacement function in terms of the local coordinate system. In this chapter we have derived the $[B]$ matrix by using the triangle natural coordinates. We have seen that the development is less work when the natural coordinates are used, and this is the reason for using them. The end result, as was pointed out in Section 6.2, is a $[B]$ matrix that is identical to that derived in Chapter 4 using the global coordinates. Once the $[B]$ matrix has been derived, the natural coordinates are not used to calculate the element stiffness matrix. It should be emphasized that when using the $[B]$ matrix in equation (6.8), coordinate rotations are not necessary because all strain values were derived in terms of the global coordinate system.

In Section 6.4 it was shown that the natural coordinates can simplify the determination of the equivalent nodal forces. You should appreciate, however, that once you have found the equivalent nodal forces for common edge distributions (uniform or linear are the most common), there is no need to repeat the formal determination.

We have reached the point where we should become involved with an application of the triangular element to a "real" problem. As noted in Chapter 4, however, a system involving any more than two elements becomes quite impossible for a solution by hand; the computer becomes a necessity. A program written in Applesoft BASIC is listed in Appendix 3, and this will be used for the examples in the next section.

6.6 THE COMPUTER PROGRAM AND APPLICATION EXAMPLES

Refer to Appendix 3 and study the program listing. You will perhaps note many similarities between this program and the structures program in Appendix 2. Very similar logic has been used in the assembly of the global stiffness matrix and force vector. Also, the global stiffness matrix for the triangular elements has been assembled in band matrix form and the same Gaussian elimination routine employed for solving the matrix equation. Both programs have the restriction that node displacement boundary conditions must be specified in terms of global X and Y components, and all forces must be applied directly to the nodes.

The program was written to present simple applications of basic algorithms without a particular emphasis on speed or efficient memory usage. Many changes can be made to improve these characteristics. For example, much memory can be recovered by removing all REM lines. The reader is referred to Bathe and Wilson (1976) for more discussion on computer implementation of the finite element method.

Before we proceed to tackle the stress analysis of some arbitrary problem with, perhaps, no previous knowledge of the correct result, we must perform some tests to ascertain a sense of the accuracy of the method and to gain some appreciation of the characteristics and limitations of the technique.

Example 6.5 Consider the square plate shown in Figure 6.10, which is 8 in. on a side and $\frac{1}{8}$ in. thick (area = unity). Establish a uniform stress traction as shown, and determine the stress in each element of a four-element model.

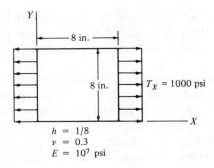

FIGURE 6.10 Square plate with uniform stress traction and defined material constants.

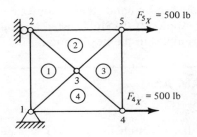

FIGURE 6.11 Equivalent node forces, element and node numbers, and boundary supports.

Solution. The plate is divided into four elements as shown in Figure 6.11, boundary conditions are applied to prevent rigid-body displacement, and the equivalent nodal forces are calculated [using (6.15)] and applied to the free nodes.

The computer printout of the stresses in each element follows; we are encouraged to note that all elements indicate the correct magnitude of the uniform stress field throughout the plate.

```
*****************************************
**                                     **
**              STRESS                 **
**                                     **
*****************************************
```

ELEMENT	SXX	SYY	SXY
1	999.999	-1E-03	0
2	999.999	-1E-03	0
3	999.999	-1E-03	0
4	999.999	-1E-03	0

Example 6.6 For the square plate of Example 6.5, establish a uniform shear stress traction (Figure 6.12), and determine the stress in each element of the same four-element model.

6.6 THE COMPUTER PROGRAM AND APPLICATION EXAMPLES

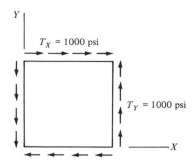

FIGURE 6.12 Plate of Example 6.5 loaded in pure shear.

Solution. Boundary conditions are applied to prevent rigid-body displacement and the equivalent nodal forces are calculated [using (6.15)] and applied to the free nodes. A plot, using the distortion plot program, of the original shape [Figure 6.13(a)] and the (exaggerated) shape under load [Figure 6.13(b)] is shown, and the deformation is what is expected of pure shear loading.

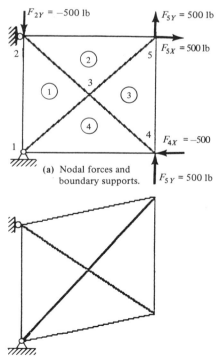

FIGURE 6.13 Plate in pure shear. (b) Distortion of the plate.

The computer printout of the stresses in each element follows. All elements indicate the same (correct) uniform shear stress throughout the plate.

```
*****************************************
**                                     **
**             STRESS                  **
**                                     **
*****************************************

ELEMENT     SXX           SYY           SXY
   1         0             0            1000
   2       -1E-03        -1E-03         1000
   3         0             0            1000
   4         0             0            1000
```

□

Example 6.7 Continuing with the same plate as Example 6.5, superimpose the two loadings of uniform longitudinal and uniform shear stress (Figure 6.14) and determine the resulting stress in each element of a 16-element model.

Solution. The first task is the determination of the equivalent nodel forces for each node of the model. Taking elements 13 and 15, for example, the uniformly distributed shear and normal stress tractions would yield the equivalent nodal forces [calculated using (6.15)] as shown in Figure 6.15.

In Figure 6.16, the model is shown with the nodal forces on the free nodes along with the rigid-body motion constraints. In addition, a computer-generated plot of the distortion of the plate indicates a distorted shape, which would be as expected for the superimposed tractions. The plate exhibits both axial and angular distortion. Although not evident by eye, perhaps, the Poisson contraction in the Y direction is indicated.

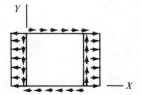

FIGURE 6.14 Superposition of the uniform shear and uniform normal stress tractions.

6.6 THE COMPUTER PROGRAM AND APPLICATION EXAMPLES

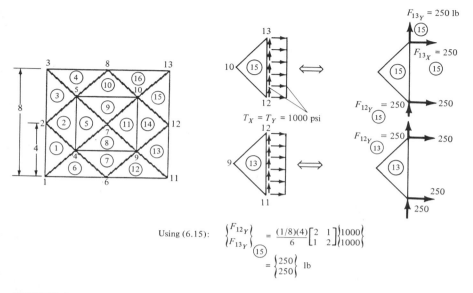

FIGURE 6.15 Element pattern, numbering, and calculation example for equivalent nodal forces.

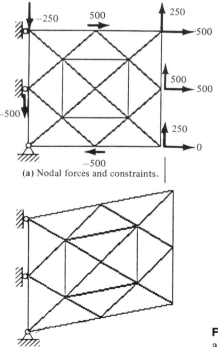

FIGURE 6.16 Applied nodal forces and node constraints and resulting distortion.

```
*****************************************
**                                     **
**              STRESS                 **
**                                     **
*****************************************

ELEMENT      SXX            SYY              SXY

   1       999.999        -1E-03           999.999
   2       1000           -1E-03           999.999
   3       1000           -1E-03           999.999
   4       1000           -1E-03           999.999
   5       1000           -1E-03           999.999
   6       999.999        -1E-03           999.999
   7       999.999         0               999.999
   8       999.999        -1E-03           999.999
   9       1000            0               999.999
  10       1000            0               999.999
  11       999.999        -1E-03           999.999
  12       999.999        -1E-03           999.999
  13       999.999        -1E-03           1000
  14       1000            0               999.999
  15       999.999         0               999.999
  16       1000            0               999.999
```
◻

We should note in reviewing these examples that this imposed stress state is one that establishes a constant state of stress throughout the region. We will next examine what happens when a loading is applied that forces a stress gradient throughout the plate.

Example 6.8 The same square plate of the last three examples is loaded with a linearly varying stress traction (Figure 6.17) which is statically equivalent to a couple of 8000 in.-lb. Determine the stress distribution in the elements at an interior cross section.

Solution. The equivalent nodal forces are determined using any of the relationships given in Section 6.4, equation (6.15), for example. For the two boundary elements 13 and 15, we obtain the forces shown in Figure 6.18.

In Figure 6.19, we show the model with the applied nodal loads and the boundary constraints. Note that there is no force applied to the midside node because the individual element forces canceled at node 12 (refer to Figure 6.18).

The deformation of the plate looks reasonable, as shown in the Figure 6.19. The upper elements elongate and the lower elements

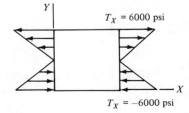

FIGURE 6.17 Linear stress traction on edge of square plate.

6.6 THE COMPUTER PROGRAM AND APPLICATION EXAMPLES

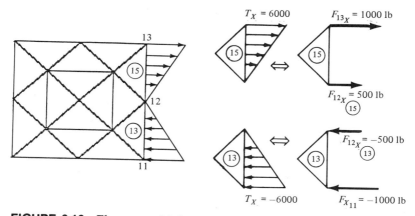

FIGURE 6.18 Element nodal forces equivalent to the distributed stress traction.

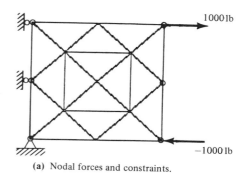

(a) Nodal forces and constraints.

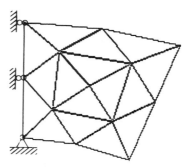

FIGURE 6.19 Applied nodal forces and node constraints and resulting distortion.

(b) Computer-plotted distortion.

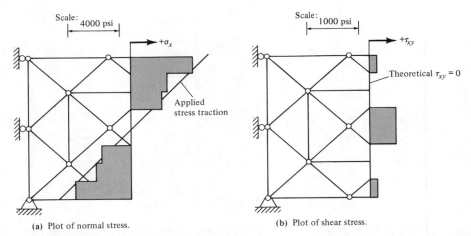

FIGURE 6.20 Plot of element normal and shear stresses on cross section of square plate loaded with a couple.

contract. The boundary side rotates and remains reasonably straight, in keeping with small deformation theory of beams in pure bending.

In Figure 6.20 we show plots of the normal and shear stress at an interior plane. Interpretation of the results requires that you recall that the triangle element we are using is a constant strain element. A strain (and stress) field that varies throughout the region is difficult to model accurately with an element that represents no variation of the strain and stress. Nevertheless, the results of this load case look rather good, although certainly not as good as the previous uniform stress field examples. □

For the next example, we examine the results for a cantilever beam loaded with an end force. With this example we will be able to see how well the element can model the flexure stress and the shear stress at an interior cross section. We compare the deflection and stress results with small deflection beam theory.

Example 6.9 A very short cantilever beam loaded with a single concentrated force is defined in Figure 6.21. Using several finite element mesh patterns, examine the dependence of the displacement of point A and the stress at point B on the degree of refinement of the model. Finally, plot the distribution of the flexure and shear stresses on a cross section.

Solution. Four triangle element mesh patterns were designed, and they are shown in Figure 6.22. Note that succeeding refinements contain all the nodes of a previous pattern.

6.6 THE COMPUTER PROGRAM AND APPLICATION EXAMPLES

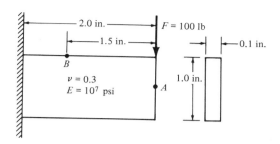

FIGURE 6.21 Cantilever beam with single force at the end.

In Figure 6.23 is a plot of the Y displacement of point A obtained for each of the four triangle mesh patterns shown in Figure 6.22. It is noted that the result from the 45-node model is very nearly equal to the simple beam theory prediction, but the slope of the curve indicates that further refinement would yield a deflection larger than the simple beam theory value. This is to be expected for a short beam, where shear distortion begins to contribute significantly to the beam deflection. When this shear distortion is included in the beam deflection equation, the theoretical prediction is 25% greater than our most refined finite element model. We will find that the application of "better" elements to this cantilever beam problem in Chapters 7 and 8 will show a convergence toward this higher deflection value of 0.0040.

The rate of convergence of the stress in the beam can be studied by observing the change in the stress vector for each element having a preselected point as a common node. In Figure 6.24 the stress vector for elements having point B as a common node is identified for the three models that have B as a node. As the model is refined, the flexure stress, σ_X, increases in each element, the other stress components, σ_Y and τ_{XY}, vary rather erratically.

We note that for all models there is a large variation of the magnitude of the stress components for adjacent elements, and further refinement should be made to reduce these differences. In an attempt to chart the convergence of the stress, the flexure stress at node B was evaluated as the average of the σ_X values for each element having B as a node. The results for this small sampling is shown in Figure 6.25. The plot indicates very slow convergence to the theoretical result; the slope is small, and the 69-element model produced a value approximately 22% below the simple beam theory.

In Figure 6.26, the flexure stress, σ_X, and the shear stress have been plotted at the cross section 1.38 in. from the applied load. The general trend of the flexure stress is good, but the results are lower than the theoretical prediction. The shear stress is not good; the magnitudes

224 ISOPARAMETRIC THREE-NODE TRIANGULAR ELEMENT

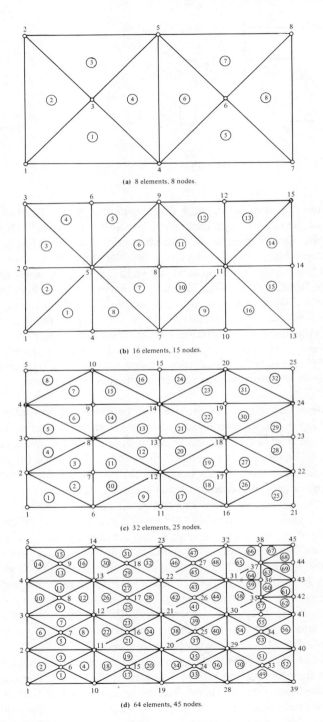

(a) 8 elements, 8 nodes.

(b) 16 elements, 15 nodes.

(c) 32 elements, 25 nodes.

(d) 64 elements, 45 nodes.

FIGURE 6.22 Element mesh patterns used in cantilever beam analysis.

6.6 THE COMPUTER PROGRAM AND APPLICATION EXAMPLES

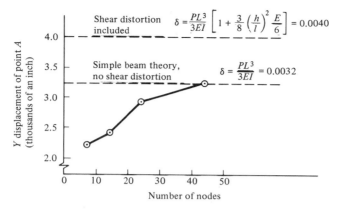

FIGURE 6.23 Dependence of displacement of point A on the degree of mesh refinement. (The formula of "shear distortion included" is from S. P. Timoshenko, *Strength of Materials*, Part I, Article 67, New York: D. Van Nostrand Co., Inc., 1930.)

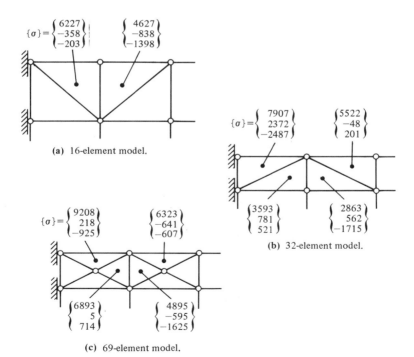

FIGURE 6.24 Stress vector for elements having point B as a common node.

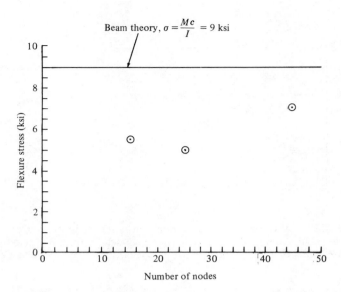

FIGURE 6.25 Dependence of average stress at node B on the mesh pattern.

differ greatly in adjacent elements. The conclusion is obvious that the model will have to be refined further or a better element will have to be developed. The latter choice is the topic of study in Chapters 7 and 8. ◻

Perhaps a comment should be made regarding the appropriateness of the simple beam flexure stress relationship My/I for short beams that undergo some shear distortion. In the derivation of this relationship, it was assumed that cross sections of the beam would remain plane during bending. In Figure 6.27 we show the computer plot of the undeformed and the deformed beam. There is perhaps a slight indication of distortion of the cross section, but it is very small. From this result we would expect that the linear strain and stress variation along the cross section can be assumed and the simple beam flexure stress formula can be used.

We should mention, in addition, that in cases where the beam bending causes a greater distortion of the cross section, if the shearing force is constant along the length of the beam, the distortion of all sections will be the same. This means that the linear strain gradient assumption is not violated and the simple beam equation is satisfactory.

The next example is presented to illustrate the importance of correct modeling procedures in regions near applied loads. In many situations, the regions of maximum stress may be far removed from the point of load application and the applied load need be correct only in an average sense. This is the situation encountered in elementary strength of materi-

6.6 THE COMPUTER PROGRAM AND APPLICATION EXAMPLES

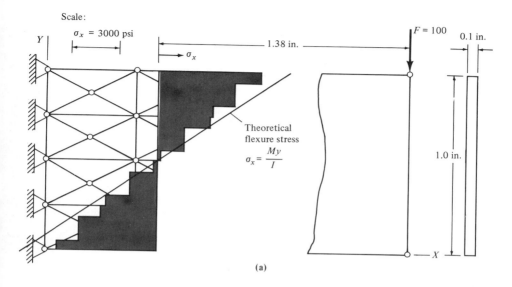

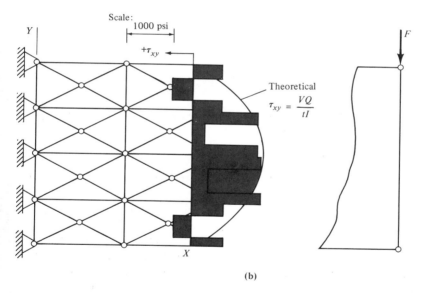

FIGURE 6.26 Flexure stress, σ_x, and shear stress, τ_{xy}, on cross section 1.38 in. from load.

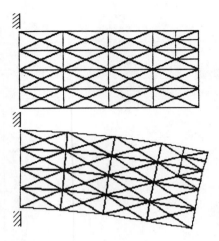

FIGURE 6.27 Computer-generated plot of the deformed and undeformed cantilever beam of Example 6.9.

als applications. In "real" problems, however, the region of maximum stress intensity may be very close to the point of load application. In these cases the load must be applied to the model as it is applied to the real continuum and the mesh refinement must be adequate to reflect the strain gradients. We will investigate the effect of loading and its region of influence by applying a self-equilibrated (zero resultant) force at the end of the cantilever beam of the previous example.

Example 6.10 The cantilever beam shown in Figure 6.28 has a zero resultant force system applied to the beam. Solve for the stress in the beam.

Solution.* The 69-element model of the beam illustrated in Figure 6.22(d) was used for this analysis. The results are shown in Figure 6.29. Note that for this case, the stress decays to less than 5% of the maximum within a radius 1.5 times the extent of the boundary length containing the applied load. This indicates that any nonzero but statically equivalent loads applied over this same region would yield, for all practical purposes, the same result beyond the significant decay region. Within the decay region, however, the results will depend on the exact manner of the distribution of the boundary loads. ☐

---

*In 1853, Barre de Saint-Venant presented his solution for torsion in long prismatic bars. He postulated that variations in the distribution of statically equivalent end loads must have little effect on the twisted bar except near the ends. He went on to say that the influence of forces in equilibrium acting on a small portion of a body extends very little beyond the parts on which they act. This declaration became the basis for what is now known as the Saint-Venant principle, and the mathematical clarification of the principle has been the subject of serious consideration since its enunciation.

6.6 THE COMPUTER PROGRAM AND APPLICATION EXAMPLES   229

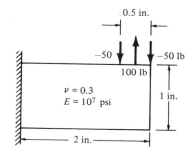

FIGURE 6.28 Beam with loading having a zero resultant force.

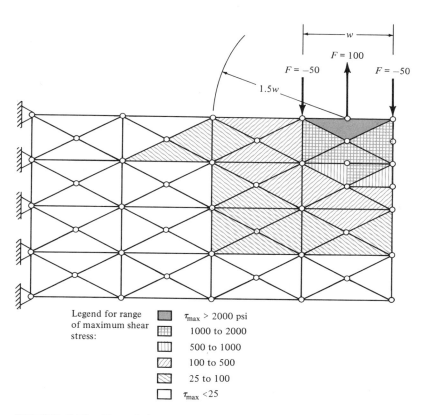

FIGURE 6.29 Plot of the ranges of the maximum shear stress in the decay region caused by the self-equilibrated loading.

6.7 DATA REDUCTION FOR BEST RESULTS

It would be of value to review the discussion of the results of the one-dimensional problem in Section 3.11. Recall that the one-dimensional element and the triangle element both use linear displacement functions. The example problem in Chapter 3 clearly illustrated the following characteristics of the constant-strain element.

1. The strain and stress are constant throughout an element.
2. The error in satisfaction of the equilibrium equations is in proportion to the magnitude of the discontinuity of the strain across element boundaries.
3. The element yields the best estimate of stress (or strain) near the midpoint of the element. The node locations have the greatest error.

Based on the experience with the one-dimensional element, we might expect the following solution characteristics when using the constant-strain triangle:

1. The element stress value will best represent the stress for a point located somewhere in the interior of the element (location unknown, of course).
2. The element boundary location is the region of greatest error for the element stress value.

In evaluating the results of an analysis using the three-node triangle, a good data reduction technique involves the averaging of element values

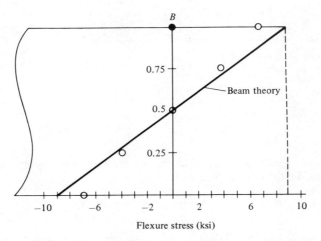

FIGURE 6.30 Comparison of node average flexure stress with elementary theory (section 1.5 in. from end load) for beam of Example 6.9.

at common nodes. This implies that the most reliable values will occur at interior nodes (which are surrounded by elements); the least reliable values will occur at the region boundary.

In Figure 6.30 is shown a plot of the element average stresses at the nodes on the beam cross section containing node B, which is identified in Example 6.9. The plot clearly shows that the interior node average stresses are in better agreement with the beam theory prediction than is the boundary node B.

6.8 ELEMENT AND NODE NUMBERING

The elements in the descretized region can be numbered sequentially in an arbitrary manner. In theory, the nodes can be numbered arbitrarily also, but for an important practical reason some caution should be applied.

Consider a triangular element that has two different node numbering schemes as shown in Figure 6.31. If you recall the technique for assembling the global stiffness matrix for the entire system of elements, you remember that the entries in the element stiffness matrix are placed into the assembled global matrix at addresses that depend on the global node numbers.

For example, element (a) would have its local entry (1, 1) placed in global position (9, 9), and the local entry (1, 6) would be placed in global position (9, 14). For element (b), however, local entry (1, 1) would be in global (9, 9), but local (1, 6) would go in global (9, 40). You should recognize that the greater the difference in the magnitude of the global node numbers for an element, the more "scattered" the entries in the global stiffness matrix. Recall the assembled stiffness matrix of the one-dimensional example in Section 3.10. When the nodes were numbered sequentially, the maximum difference between the node numbers was 1, and this placed all nonzero global stiffness matrix entries in a

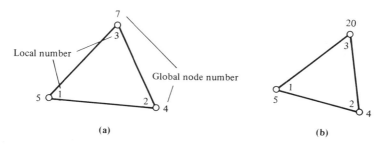

FIGURE 6.31 Triangle with two different global node numberings.

"band" of three diagonal rows. A random numbering of the nodes would have increased the width of this band.

There are benefits in having the nonzero stiffness matrix entries close to the diagonal of the matrix. The first benefit is in the area of computer storage requirements. The stiffness matrix is completely defined with storage of the band of nonzero entries. Actually, remembering that the matrix is symmetric, only the main diagonal and half of the band requires storage. The smaller the bandwidth, the fewer the numbers that have to be stored. A second benefit is in computational time. Special equation solvers have been developed for band storage of the matrix; considerable time can be saved if they are employed. The programs in the Appendixes make use of a simple band storage and equation solver routine. More elaborate (memory efficient) can be found in the literature.

6.9 TAKING ADVANTAGE OF LOADING AND REGION SYMMETRY

The analyst should be alert to situations of symmetry that allow reduction in the size of the region to be analyzed. Refer to Figure 6.32, which shows a tensile plate with a square hole in the center and subjected to a uniform load distribution. There is obvious loading and part geometry symmetry in this situation. Recognizing that points on a line along the Y

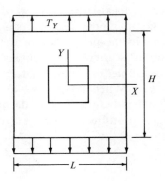

FIGURE 6.32 Symmetric plate with symmetric load.

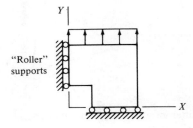

FIGURE 6.33 Boundary conditions on reduced part.

6.9 TAKING ADVANTAGE OF LOADING AND REGION SYMMETRY   233

axis will not displace in the X direction and that points along the X axis will not displace in the Y direction, we can restrict our analysis to one-fourth of the plate, which is supplied with boundary support conditions that will allow the correct displacements of points on the axes of symmetry as shown in Figure 6.33.

Example 6.11 The rectangular thin plate shown in Figure 6.34 has a centrally located hole. Assuming the application of a uniformly distributed stress traction on the two edges as indicated, solve for the maximum stress in the plate and compare the result with published data on stress concentration factors.

Solution. If we recognize the loading and geometric symmetry, the region can be reduced to one-fourth that of the total plate. A triangle mesh pattern is drawn in the region, the constraint boundary conditions are applied, and the work-equivalent nodal forces are applied to the nodes. This is shown in Figure 6.35. Note the boundary constraints. The nodes on the edge parallel to the Y axis must be free to displace in the Y direction, but not in the X; the nodes on the edge parallel to the X axis must be free to move in the X direction but not the Y.

In Figure 6.36 the tensile stress distribution across the section through the center of the hole has been plotted from the results of the 40-element model. The plot clearly indicates the large strain gradient near the hole, and the ratio of the maximum stress of 34 ksi to the

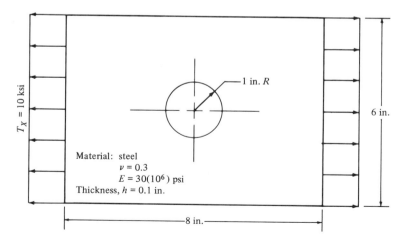

FIGURE 6.34 Square thin plate with centrally located hole and uniform stress traction.

234   ISOPARAMETRIC THREE-NODE TRIANGULAR ELEMENT

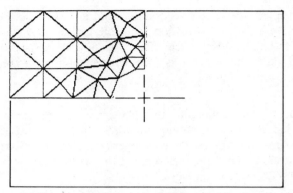

(a) Element mesh for analysis of one-fourth of the plate.

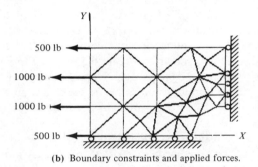

(b) Boundary constraints and applied forces.

FIGURE 6.35 Finite element mesh pattern and boundary constraints.

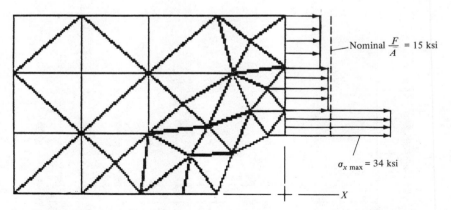

FIGURE 6.36 Plot of normal stress, σ_x, at cross section through center of hole.

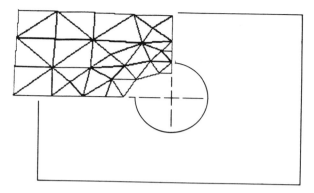

FIGURE 6.37 Distortion of the quarter-plate model.

nominal stress of 15 ksi yields a stress concentration factor of 2.27. This agrees remarkably well with a published value of 2.35 in view of the fact that the mesh is rather coarse. Finally, the distortion of the quarter-plate model is shown (much exaggerated, to be sure) in Figure 6.37). □

PROBLEMS

6.1 Node 3 of the three-node triangle shown in Figure P6.1 has the following displacement:

$$u_3 = 12(10^{-4}) \text{ in.}$$

$$v_3 = -15(10^{-4}) \text{ in.}$$

The other two nodes have zero displacement. Solve for the displacement of the area centroid of the triangle.

6.2 Solve for the X and Y coordinates of the point in a triangle (Figure P6.2) located by the natural coordinates:

$$L_1 = \tfrac{1}{2} \qquad L_2 = \tfrac{1}{3}$$

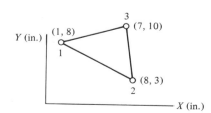

FIGURE P6.1

236 ISOPARAMETRIC THREE-NODE TRIANGULAR ELEMENT

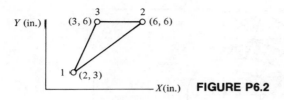

FIGURE P6.2

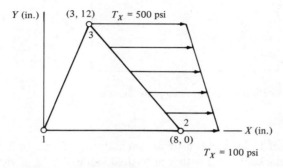

FIGURE P6.7 Stress traction on side 1.

6.3 Solve for the displacement of the point specified in Problem 6.2 for the following nodal displacements:

$$u_1 = 2(10^{-3}) \text{ in.}$$
$$u_2 = -2(10^{-3}) \text{ in.}$$
$$u_3 = 5(10^{-3}) \text{ in.}$$

6.4 Solve for the strain vector $\{\epsilon\}$ for the triangle in Problem 6.1.
6.5 Assuming plane stress, solve for the stress vector $\{\sigma\}$ for the triangle in Problem 6.1.
6.6 Beginning with the equations (6.2) and (6.3), derive equation (6.7).
6.7 Solve for the element nodal force vector $\{Q\}_T$ equivalent to the distributed stress traction on side 1 of the triangle in Figure P6.7. The triangle is 0.5 in. thick.

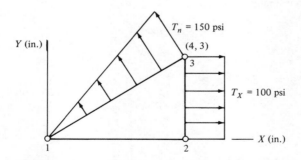

FIGURE P6.8 Element with stress traction on two sides.

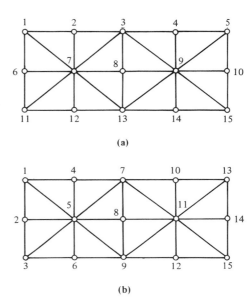

FIGURE P6.11 Two different global numbering schemes.

6.8 For the element shown in Figure P6.8, solve for the element nodal force vector $\{Q\}_T$ equivalent to the distributed stress tractions normal to sides 1 and 2. The element thickness is 0.2 in.

6.9 The mass density of a triangle of thickness h varies linearly with the natural coordinate according to

$$\rho = \rho_0 + \rho_1 L_1$$

Assume that gravity acts in the negative Y direction and solve for the equivalent element nodal body force vector $\{Q\}_{BF}$.

6.10 Using the computer program in Appendix 2, refine the mesh of the tensile plate in Example 6.11 and obtain a better estimate of the maximum stress in the part.

6.11 Two different node numbering schemes are illustrated in Figure P6.11. Considering the entries in the global stiffness matrix, which of the two-node numbering schemes would put the nonzero entries closest to the diagonal of the matrix?

Discuss the advantage of such a matrix (called a "banded" matrix) with regard to solving the set of equations.

CHAPTER 7

Isoparametric Four-Node Quadrilateral Element

7.1 INTRODUCTION

The general expression for the stiffness matrix of any plane finite element of thickness h was defined in Section 4.2 to be

$$[k] = h \int_A [B]^T [C][B] \, dA$$

In the three different developments of the three-node triangle in Chapters 4 and 6, we found that the matrices $[B]$ and $[C]$ were constant and the integration simply amounted to the multiplication of the triple matrix product by the area of the element. In this chapter we are going to study the development of the popular and practical four-node quadrilateral. The matrix $[B]$ is not a constant for this element.

We will encounter three new applications of previously discussed analytical tools:

1. Natural coordinates of the quadrilateral.
2. Interpolation formula coefficients (shape functions) for this element.
3. Numerical integration of the $[B]^T[C][B]$ product over the area of the element.

As we saw in Chapter 6, the major effort in the development of the element is concentrated in the determination of the matrix $[B]$. We will detail the derivation of the $[B]$ matrix for this element, discuss the numerical integration of the matrix product to obtain the stiffness matrix, and show once again how to transform distributed loadings into energy equivalent forces applied to the nodes.

7.2 THE ELEMENT NATURAL COORDINATES

An arbitrary four-node quadrilateral has been drawn in the global XY plane in Figure 7.1. The nodes have been numbered in sequence counterclockwise. This will be the standard local node numbering scheme.

A nonorthogonal intersecting pair of lines has been superimposed on the quadrilateral. We are going to define these lines as the natural coordinate s, t axes. We will define the side joining nodes 1 and 2 as having a constant t coordinate of -1. The side joining nodes 2 and 3 has a constant s coordinate of 1, the side joining 3 and 4 has a coordinate t of 1, and the remaining side joining 4 and 1 has a coordinate s of -1.

Instead of plotting the quadrilateral on the XY plane, it could be drawn relative to a set of orthogonal s, t natural coordinate axes. This plot is shown in Figure 7.2. The natural coordinates of the nodes have been identified.

Compare this natural coordinate system with the triangle natural coordinate system. Note that we can define the location of a point in the

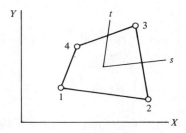

FIGURE 7.1 Standard local node numbering and coordinate axes of quadrilateral.

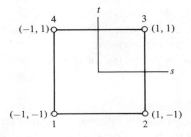

FIGURE 7.2 Natural coordinates of the quadrilateral.

7.3 THE INTERPOLATION FORMULA

element relative to the natural systems, but we need to develop the transformation from the natural system to the global reference. At this point we do not have the transformation relationship between the two coordinate systems.

7.3 THE INTERPOLATION FORMULA

The quadrilateral of our discussion is to be isoparametric. We will write interpolation formulas for displacements and global coordinates, and the coefficients (shape functions) will be the same for each set of equations. The independent variables for all equations will be the natural s, t coordinates. These equations will be as follows:

For displacements in the global X and Y directions:

$$u(s,t) = N_1 u_1 + N_2 u_2 + N_3 u_3 + N_4 u_4$$
$$v(s,t) = N_1 v_1 + N_2 v_2 + N_3 v_3 + N_4 v_4 \qquad (7.1)$$

For global coordinates:

$$X(s,t) = N_1 X_1 + N_2 X_2 + N_3 X_3 + N_4 X_4$$
$$Y(s,t) = N_1 Y_1 + N_2 Y_2 + N_3 Y_3 + N_4 Y_4 \qquad (7.2)$$

where the interpolation polynomial coefficients (shape functions) for two independent variables and a four-node region are defined as (review Section 5.3 if necessary)

$$N_1 = \frac{(1-s)(1-t)}{4} \qquad N_2 = \frac{(1+s)(1-t)}{4}$$

$$N_3 = \frac{(1+s)(1+t)}{4} \qquad N_4 = \frac{(1-s)(1+t)}{4}$$

Example 7.1 Given the global coordinates of the four nodes of an element (Figure 7.3), solve, using the interpolation formula (7.2), for

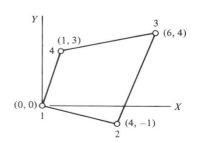

FIGURE 7.3 A Particular quadrilateral with global coordinates for each node.

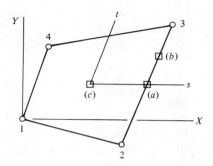

FIGURE 7.4 Scale drawing of quadrilateral and s, t coordinate axes.

the global coordinates of the points in the element with natural coordinates defined in the following table.

Point	s	t
a	1.0	0
b	1.0	0.5
c	0	0

Solution. Using equation (7.2), we have
$$X = N_1 X_1 + N_2 X_2 + N_3 X_3 + N_4 X_4$$
$$Y = N_1 Y_1 + N_2 Y_2 + N_3 Y_3 + N_4 Y_4$$

Point	s	t	N_1	N_2	N_3	N_4	X	Y
a	1.0	0	0	$\frac{1}{2}$	$\frac{1}{2}$	0	5.0	1.5
b	1.0	0.5	0	$\frac{1}{4}$	$\frac{3}{4}$	0	5.5	2.75
c	0	0	$\frac{1}{4}$	$\frac{1}{4}$	$\frac{1}{4}$	$\frac{1}{4}$	2.75	1.5

A plot of the three points is produced on the scaled drawing, Figure 7.4. Note that point c is the origin of the natural coordinates, and coordinate axis s passes through point a. The t coordinate axis passes through the midpoint of the line 3–4. □

7.4 DEVELOPMENT OF THE ELEMENT STRAIN–DISPLACEMENT MATRIX

Having specified the displacement function for the element [equations (7.1)], we can proceed with the derivation of the matrix $[B]$, which defines the strain in terms of the node displacements. This derivation will be very similar to that for the triangle in Chapter 6.

7.4 DEVELOPMENT OF THE ELEMENT STRAIN-DISPLACEMENT MATRIX

The determination of strain requires partial differentiation of the displacement functions with respect to the global coordinates. Our displacement functions (7.1) are written in terms of the natural coordinates. However, with equations (7.2) we can solve for the required derivatives as follows. Treating s as the independent variable,

$$\frac{\partial u}{\partial s} = \frac{\partial u}{\partial X}\frac{\partial X}{\partial s} + \frac{\partial u}{\partial Y}\frac{\partial Y}{\partial s}$$

and with t as the independent variable,

$$\frac{\partial u}{\partial t} = \frac{\partial u}{\partial X}\frac{\partial X}{\partial t} + \frac{\partial u}{\partial Y}\frac{\partial Y}{\partial t}$$

or in matrix form,

$$\begin{bmatrix} \frac{\partial X}{\partial s} & \frac{\partial Y}{\partial s} \\ \frac{\partial X}{\partial t} & \frac{\partial Y}{\partial t} \end{bmatrix} \begin{Bmatrix} \frac{\partial u}{\partial X} \\ \frac{\partial u}{\partial Y} \end{Bmatrix} = \begin{Bmatrix} \frac{\partial u}{\partial s} \\ \frac{\partial u}{\partial t} \end{Bmatrix} \quad (7.3)$$

Solving for the spatial derivative terms, we have

$$\begin{Bmatrix} \frac{\partial u}{\partial X} \\ \frac{\partial u}{\partial Y} \end{Bmatrix} = \frac{1}{|J|} \begin{bmatrix} \frac{\partial Y}{\partial t} & -\frac{\partial Y}{\partial s} \\ -\frac{\partial X}{\partial t} & \frac{\partial X}{\partial s} \end{bmatrix} \begin{Bmatrix} \frac{\partial u}{\partial s} \\ \frac{\partial u}{\partial t} \end{Bmatrix} \quad (7.4)$$

where $|J|$ is the Jacobian determinant,

$$|J| = \frac{\partial Y}{\partial t}\frac{\partial X}{\partial s} - \frac{\partial Y}{\partial s}\frac{\partial X}{\partial t}$$

Performing the differentiation of equations (7.2) yields

$$\frac{\partial X}{\partial s} = \frac{\partial N_1}{\partial s}X_1 + \frac{\partial N_2}{\partial s}X_2 + \frac{\partial N_3}{\partial s}X_3 + \frac{\partial N_4}{\partial s}X_4$$

$$\frac{\partial X}{\partial s} = \sum_{i=1}^{4} \frac{\partial N_i}{\partial s}X_i$$

Similarly,

$$\frac{\partial Y}{\partial s} = \sum_{i=1}^{4} \frac{\partial N_i}{\partial s}Y_i$$

$$\frac{\partial X}{\partial t} = \sum_{i=1}^{4} \frac{\partial N_i}{\partial t}X_i \qquad \frac{\partial Y}{\partial t} = \sum_{i=1}^{4} \frac{\partial N_i}{\partial t}Y_i \quad (7.5)$$

The derivatives of the shape functions are

$$\frac{\partial N_1}{\partial s} = \frac{-1}{4}(1-t) \qquad \frac{\partial N_1}{\partial t} = \frac{-1}{4}(1-s)$$
$$\frac{\partial N_2}{\partial s} = \frac{1}{4}(1-t) \qquad \frac{\partial N_2}{\partial t} = \frac{-1}{4}(1+s)$$
$$\frac{\partial N_3}{\partial s} = \frac{1}{4}(1+t) \qquad \frac{\partial N_3}{\partial t} = \frac{1}{4}(1+s) \qquad (7.6)$$
$$\frac{\partial N_4}{\partial s} = \frac{-1}{4}(1+t) \qquad \frac{\partial N_4}{\partial t} = \frac{1}{4}(1-s)$$

We can now develop an expression for the Jacobian determinant.

$$|J| = \frac{\partial Y}{\partial t}\frac{\partial X}{\partial s} - \frac{\partial Y}{\partial s}\frac{\partial X}{\partial t}$$
$$= \sum_{i=1}^{4} \frac{\partial N_i}{\partial t} Y_i \sum_{i}^{4} \frac{\partial N_i}{\partial s} X_i - \sum_{i}^{4} \frac{\partial N_i}{\partial s} Y_i \sum_{i}^{4} \frac{\partial N_i}{\partial t} X_i$$
$$= \sum_{i}^{4}\sum_{j}^{4} \left\{ Y_i \left[\frac{\partial N_i}{\partial t}\frac{\partial N_j}{\partial s} - \frac{\partial N_i}{\partial s}\frac{\partial N_j}{\partial t} \right] X_j \right\} \qquad (7.7)$$

This can be written as a matrix product,

$$|J| = [Y_1 \quad Y_2 \quad Y_3 \quad Y_4][a]\begin{Bmatrix} X_1 \\ X_2 \\ X_3 \\ X_4 \end{Bmatrix} \qquad (7.8a)$$

where

$$a_{ij} = \frac{\partial N_i}{\partial t}\frac{\partial N_j}{\partial s} - \frac{\partial N_i}{\partial s}\frac{\partial N_j}{\partial t} \qquad (7.8b)$$

Solving for the $[a]$ matrix entries yields

$$a_{11} = \frac{\partial N_1}{\partial t}\frac{\partial N_1}{\partial s} - \frac{\partial N_1}{\partial s}\frac{\partial N_1}{\partial t} = 0$$

$$a_{12} = \frac{\partial N_1}{\partial t}\frac{\partial N_2}{\partial s} - \frac{\partial N_1}{\partial s}\frac{\partial N_2}{\partial t}$$
$$= -\frac{1}{16}[(1-s)(1-t) + (1-t)(1+s)] = -\frac{1-t}{8}$$

$$a_{13} = \frac{\partial N_1}{\partial t}\frac{\partial N_3}{\partial s} - \frac{\partial N_1}{\partial s}\frac{\partial N_3}{\partial t} = -\frac{t-s}{8}$$

$$a_{14} = \frac{\partial N_1}{\partial t}\frac{\partial N_4}{\partial s} - \frac{\partial N_1}{\partial s}\frac{\partial N_4}{\partial t} = -\frac{s-1}{8} \qquad \text{etc.}$$

7.4 DEVELOPMENT OF THE ELEMENT STRAIN–DISPLACEMENT MATRIX

Thus we have for $[a]$,

$$[a] = \frac{-1}{8} \begin{bmatrix} 0 & 1-t & t-s & s-1 \\ t-1 & 0 & s+1 & -(t+s) \\ s-t & -(s+1) & 0 & t+1 \\ 1-s & s+t & -(1+t) & 0 \end{bmatrix}$$

(7.9)

Returning to equation (7.4), we have

$$\frac{\partial u}{\partial X} = \frac{1}{|J|} \left[\frac{\partial Y}{\partial t} \frac{\partial u}{\partial s} - \frac{\partial Y}{\partial s} \frac{\partial u}{\partial t} \right]$$

whereas from (7.1),

$$\frac{\partial u}{\partial s} = \sum_{i=1}^{4} \frac{\partial N_i}{\partial s} u_i \qquad \frac{\partial u}{\partial t} = \sum_{i=1}^{4} \frac{\partial N_i}{\partial t} u_i$$

and using equations (7.5), we can obtain the following:

$$\frac{\partial u}{\partial X} = \frac{1}{|J|} \sum_{i=1}^{4} \sum_{j=1}^{4} \left\{ Y_i \left[\frac{\partial N_i}{\partial t} \frac{\partial N_j}{\partial s} - \frac{\partial N_i}{\partial s} \frac{\partial N_j}{\partial t} \right] u_j \right\}$$

or in matrix form,

$$\frac{\partial u}{\partial X} = \frac{1}{|J|} [Y_1 \quad Y_2 \quad Y_3 \quad Y_4][a] \begin{Bmatrix} u_1 \\ u_2 \\ u_3 \\ u_4 \end{Bmatrix} = \epsilon_X \quad (7.10)$$

Return again to equation (7.4).

$$\frac{\partial u}{\partial Y} = \frac{-1}{|J|} \left| \frac{\partial X}{\partial t} \frac{\partial u}{\partial s} - \frac{\partial X}{\partial s} \frac{\partial u}{\partial t} \right|$$

$$= \frac{-1}{|J|} \sum_{i=1}^{4} \sum_{j=1}^{4} \left\{ X_i \left[\frac{\partial N_i}{\partial t} \frac{\partial N_j}{\partial s} - \frac{\partial N_i}{\partial s} \frac{\partial N_j}{\partial t} \right] u_j \right\}$$

In matrix form,

$$\frac{\partial u}{\partial Y} = \frac{-1}{|J|} [X_1 \quad X_2 \quad X_3 \quad X_4][a] \begin{Bmatrix} u_1 \\ u_2 \\ u_3 \\ u_4 \end{Bmatrix} \quad (7.11)$$

In summary, equation (7.10) defines the strain ϵ_X, and equation (7.11) defines one term of the shear strain function. We now have to repeat the preceding steps, the only change being the substitution of the Y displacement, v, for the X displacement, u. Doing this would yield the following relationships:

$$\left\{\begin{array}{c}\frac{\partial v}{\partial X}\\ \frac{\partial v}{\partial Y}\end{array}\right\} = \frac{1}{|J|}\begin{bmatrix}\frac{\partial Y}{\partial t} & -\frac{\partial Y}{\partial s}\\ -\frac{\partial X}{\partial t} & \frac{\partial X}{\partial s}\end{bmatrix}\left\{\begin{array}{c}\frac{\partial v}{\partial s}\\ \frac{\partial v}{\partial t}\end{array}\right\} \quad (7.4a)$$

$$\frac{\partial v}{\partial X} = \frac{1}{|J|}\left[\frac{\partial Y}{\partial t}\frac{\partial v}{\partial s} - \frac{\partial Y}{\partial s}\frac{\partial v}{\partial t}\right]$$

$$= \frac{1}{|J|}[Y_1 \quad Y_2 \quad Y_3 \quad Y_4][a]\left\{\begin{array}{c}v_1\\ v_2\\ v_3\\ v_4\end{array}\right\} \quad (7.10a)$$

where $|J|$ is defined by (7.7) and $[a]$ by (7.9).

Also, we obtain the expression for the strain ϵ_Y:

$$\frac{\partial v}{\partial Y} = \frac{-1}{|J|}\left[\frac{\partial X}{\partial t}\frac{\partial v}{\partial s} - \frac{\partial X}{\partial s}\frac{\partial v}{\partial t}\right]$$

$$= \frac{-1}{|J|}[X_1 \quad X_2 \quad X_3 \quad X_4][a]\left\{\begin{array}{c}v_1\\ v_2\\ v_3\\ v_4\end{array}\right\} = \epsilon_Y \quad (7.11a)$$

Now that we have the expressions for the derivatives of the displacements relative to the global coordinates, the next task is to assemble the relationships to give us the strain vector. Recall that shear strain is defined as

$$\gamma_{XY} = \frac{\partial u}{\partial Y} + \frac{\partial v}{\partial X}$$

This is the sum of (7.11) and (7.10a).

$$\{\epsilon\} = \left\{\begin{array}{c}\epsilon_X\\ \epsilon_Y\\ \gamma_{XY}\end{array}\right\} = [B]\left\{\begin{array}{c}u_1\\ v_1\\ u_2\\ \vdots\\ u_4\\ v_4\end{array}\right\} = [B]\{q\} \quad (7.12)$$

From equation (7.10) we identify B_{1j} as follows:

$$B_{1,(2j-1)} = \frac{1}{|J|}\sum_{i=1}^{4} Y_i a_{ij} \quad j = 1,2,3,4 \quad (7.13a)$$

$$B_{1j} = 0 \quad j = 2,4,6,8 \quad (7.13b)$$

7.4 DEVELOPMENT OF THE ELEMENT STRAIN-DISPLACEMENT MATRIX

From equation (7.11a),

$$B_{2,2j} = -\frac{1}{|J|} \sum_{i=1}^{4} X_i a_{ij} \quad j = 1, 2, 3, 4 \quad (7.13c)$$

$$B_{2j} = 0 \quad j = 1, 3, 5, 7 \quad (7.13d)$$

From equations (7.11) and (7.10a),

$$B_{3j} = B_{2,(j+1)} \quad j = 1, 3, 5, 7 \quad (7.13e)$$

$$= B_{1,(j-1)} \quad j = 2, 4, 6, 8 \quad (7.13f)$$

Expanding these summation terms and substituting the simplified notation

$$X_{mn} = X_m - X_n$$
$$Y_{mn} = Y_m - Y_n,$$

the entries in the $[B]$ matrix are

$$B_{11} = B_{32} = \frac{1}{8|J|}(Y_{24} + sY_{43} + tY_{32})$$

$$B_{13} = B_{34} = \frac{1}{8|J|}(Y_{31} + sY_{34} + tY_{14})$$

$$B_{15} = B_{36} = \frac{1}{8|J|}(Y_{42} + sY_{12} + tY_{41})$$

$$B_{17} = B_{38} = \frac{1}{8|J|}(Y_{13} + sY_{21} + tY_{23}) \quad (7.14)$$

$$B_{22} = B_{31} = \frac{1}{8|J|}(X_{42} + sX_{34} + tX_{23})$$

$$B_{24} = B_{33} = \frac{1}{8|J|}(X_{13} + sX_{43} + tX_{41})$$

$$B_{26} = B_{35} = \frac{1}{8|J|}(X_{24} + sX_{21} + tX_{14})$$

$$B_{28} = B_{37} = \frac{1}{8|J|}(X_{31} + sX_{12} + tX_{32})$$

and the Jacobian determinant is

$$|J| = \frac{1}{8}[X_1 \quad X_2 \quad X_3 \quad X_4]$$

$$\times \begin{bmatrix} 0 & 1-t & t-s & s-1 \\ t-1 & 0 & s+1 & -(t+s) \\ s-t & -(s+1) & 0 & t+1 \\ 1-s & s+t & -(1+t) & 0 \end{bmatrix} \begin{Bmatrix} Y_1 \\ Y_2 \\ Y_3 \\ Y_4 \end{Bmatrix}$$

$$(7.15)$$

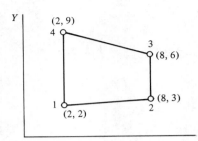

FIGURE 7.5 Single quadrilateral element.

Example 7.2 Solve for the [B] matrix of the element shown in Figure 7.5.

Solution. Referring to equation (7.12) and the definitions of the nonzero entries in the [B] matrix, (7.14), we substitute coordinate values and obtain the following:

$$B_{11} = B_{32} = \frac{1}{8|J|}(-6 + 3s + 3t) \qquad B_{22} = B_{31} = \frac{1}{8|J|}(-6 + 6s)$$

$$B_{13} = B_{34} = \frac{1}{8|J|}(4 - 3s - 7t) \qquad B_{24} = B_{33} = \frac{1}{8|J|}(-6 - 6s)$$

$$B_{15} = B_{36} = \frac{1}{8|J|}(6 - s + 7t) \qquad B_{26} = B_{35} = \frac{1}{8|J|}(6 + 6s)$$

$$B_{17} = B_{38} = \frac{1}{8|J|}(-4 + s - 3t) \qquad B_{28} = B_{37} = \frac{1}{8|J|}(6 - 6s)$$

The Jacobian determinant is

$$|J| = \frac{1}{8}[2 \quad 2] \times \begin{bmatrix} 0 & 1-t & t-s & s-1 \\ t-1 & 0 & s+1 & -(t+s) \\ s-t & -(s+1) & 0 & t+1 \\ 1-s & s+t & -(1+t) & 0 \end{bmatrix} \begin{Bmatrix} 2 \\ 3 \\ 6 \\ 9 \end{Bmatrix}$$

In contrast to the three-node triangle, the entries in the [B] matrix are not constant but are a function of the natural coordinates of a point in the element. Consider, for example, the [B] matrix for the node 1 location. At node 1, $s = -1$ and $t = -1$. Substituting these coordinate values into the expression above for the Jacobian determinant gives us

$$|J| = \frac{1}{8}[2 \quad 8 \quad 8 \quad 2] \begin{bmatrix} 0 & 2 & 0 & -2 \\ -2 & 0 & 0 & 2 \\ 0 & 0 & 0 & 0 \\ 2 & -2 & 0 & 0 \end{bmatrix} \begin{Bmatrix} 2 \\ 3 \\ 6 \\ 9 \end{Bmatrix} = 10.5$$

7.4 DEVELOPMENT OF THE ELEMENT STRAIN–DISPLACEMENT MATRIX

Substituting into the B_{ij} expressions, we have

$$B_{11} = B_{32} = \frac{1}{8|J|}(-6 + 3s + 3t) = \frac{1}{8(10.5)}(-6 - 3 - 3) = -0.143 \quad \text{etc.}$$

The $[B]$ matrix for node 1 becomes

$$[B]_{-1,-1} = \begin{bmatrix} -0.143 & 0 & 0.167 & 0 & 0 & 0 & -0.024 & 0 \\ 0 & -0.143 & 0 & 0 & 0 & 0 & 0 & 0.143 \\ -0.143 & -0.143 & 0 & 0.167 & 0 & 0 & 0.143 & -0.024 \end{bmatrix}$$

Similarly, for the node 2 location we would have

$|J| = 4.5$

$$[B]_{1,-1} = \begin{bmatrix} -0.167 & 0 & 0.222 & 0 & -0.056 & 0 & 0 & 0 \\ 0 & 0 & 0 & -0.333 & 0 & 0.333 & 0 & 0 \\ 0 & -0.167 & -0.333 & 0.222 & 0.333 & -0.056 & 0 & 0 \end{bmatrix}$$

□

Example 7.3 The four-node element shown in Figure 7.6 is fixed in space at nodes 1 and 4. Solve for the strains ϵ_X, ϵ_Y, and γ_{XY} at node 1 and the "midpoint" of the element as a function of the displacements u and v of nodes 2 and 3.

Solution. The strain in the element is defined by (7.12): $\{\epsilon\} = [B]\{q\}$. From the boundary conditions we know that $u_1 = v_1 = u_4 = v_4 = 0$. Taking the $[B]$ matrix for node 1 from Example 7.2 and discarding the entries that multiply the zero displacements, we have

$$\begin{Bmatrix} \epsilon_X \\ \epsilon_Y \\ \gamma_{XY} \end{Bmatrix}_{-1,-1} = \begin{bmatrix} 0.167 & 0 & 0 & 0 \\ 0 & 0 & 0 & 0 \\ 0 & 0 & 0.167 & 0 \end{bmatrix} \begin{Bmatrix} u_2 \\ v_2 \\ u_3 \\ v_3 \end{Bmatrix}$$

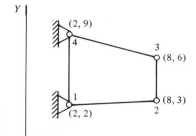

FIGURE 7.6 Single element with boundary conditions of zero displacement at nodes 1 and 4.

At the midpoint of the element, $s = t = 0$. Substituting these values for the Jacobian determinant and the B_{ij} entries, we have

$$|J| = 7.5$$

$$[B]_{0,0} = \begin{bmatrix} -0.100 & 0 & 0.067 & 0 & 0.100 & 0 & -0.067 & 0 \\ 0 & -0.100 & 0 & -0.100 & 0 & 0.100 & 0 & 0.100 \\ -0.100 & -0.100 & -0.100 & 0.067 & 0.100 & 0.100 & 0.100 & -0.067 \end{bmatrix}$$

Retaining only the $[B]$ entries that multiply the free node displacements yields

$$\begin{Bmatrix} \epsilon_X \\ \epsilon_Y \\ \gamma_{XY} \end{Bmatrix}_{0,0} = \begin{bmatrix} 0.067 & 0 & 0.100 & 0 \\ 0 & -0.100 & 0 & 0.100 \\ -0.100 & 0.067 & 0.100 & 0.100 \end{bmatrix} \begin{Bmatrix} u_2 \\ v_2 \\ u_3 \\ v_3 \end{Bmatrix}$$

$\square$

7.5 THE ELEMENT STIFFNESS MATRIX

Recall once again that we sum strain and potential energies for the element.

$$\Pi = \tfrac{1}{2} \int_\Omega \{\epsilon\}^T [C] \{\epsilon\} \, d\Omega + \sum_i V_i$$

where $\{\epsilon\}$ is known in terms of the nodal displacement vector $\{q\}$ [equation (7.12)], and the force potentials are derived for the body forces, surface tractions, and joint loads in terms of the nodal displacement vector and shape functions, N_i. Substitution of the $[B]\{q\}$ product for $\{\epsilon\}$ and differentiation of Π with respect to $\{q\}$ and setting equal to zero yields the force displacement relation for the element. The stiffness matrix is found to be (as we have seen before).

$$[k] = \int_\Omega [B]^T [C] [B] \, d\Omega$$

The determination of the stiffness matrix $[k]$ requires integration of the matrix product over the volume, $h \, dA$, of the element. There are two "difficulties" confronting us:

1. The $[B]$ matrix has entries that involve ratios of functions of the natural coordinates s and t.
2. The differential area dA can easily be expressed as $dX \, dY$, but this integration variable is not the same as the $[B]$ matrix variable, so a change of variable must be undertaken.

7.5 THE ELEMENT STIFFNESS MATRIX

We have already discussed the means of dealing with these two problems. First, the integration will be done numerically, and second, the coordinates of the differential area will be changed from the physical X and Y to the natural s and t by application of the Jacobian determinant of the transformation equations for the two coordinate systems. The numerical double-integration and change-of-variable techniques should be reviewed in Sections 5.5 and 5.6, respectively. Note that the discussion in Section 5.6 is directly related to the change of variables from X, Y to s, t; the same symbols are used. The result of the derivation, equation (5.14), is as follows:

$$\text{differential volume} = (\text{thickness})(\text{differential area})$$
$$d\Omega = h\, dX\, dY = h|J|\, ds\, dt$$

where

$$|J| = \frac{\partial X}{\partial s}\frac{\partial Y}{\partial t} - \frac{\partial X}{\partial t}\frac{\partial Y}{\partial s}$$

This is the same determinant defined in equation (7.7). Therefore, changing the integration variable, we obtain

$$[k] = h\int_A [B]^T[C][B]\, dX\, dY = h\int_{A^*} [B]^T[C][B]|J|\, ds\, dt \quad (7.16)$$

Next, we can integrate this product over the area by using Gauss quadrature (Figure 7.7). We will elect to use the two-point formula; it is easy to use and it yields good results for the four-node element (Bathe & Wilson (1976)). The application of the formula yields the following. [refer to Section 5.5 for a review; in particular, see expression (5.11)]:

$$\int_{A^*} [B]^T[C][B]|J|\, ds\, dt$$
$$\approx W_1 W_1 [B(s_1, t_1)]^T [C][B(s_1, t_1)]|J(s_1, t_1)|$$
$$+ W_1 W_2 [B(s_1, t_2)]^T [C][B(s_1, t_2)]|J(s_1, t_2)|$$
$$+ W_2 W_1 [B(s_2, t_1)]^T [C][B(s_2, t_1)]|J(s_2, t_1)|$$
$$+ W_2 W_2 [B(s_2, t_2)]^T [C][B(s_2, t_2)]|J(s_2, t_2)|$$

$$(7.17)$$

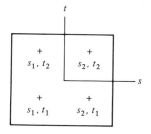

FIGURE 7.7 Integration points for Gauss quadrature.

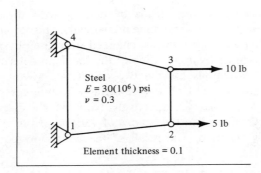

FIGURE 7.8 Supported element with forces on nodes 2 and 3.

where

$$s_1 = -0.5773503 \quad t_1 = -0.5773503 \quad W_1 = W_2 = 1.0$$
$$s_2 = -s_1 \quad t_2 = -t_1$$

The process amounts to evaluating the product at four points and summing. The result when multiplied by the thickness is the stiffness matrix of the element (review Example 5.6).

Example 7.4 Continuing with the element of Example 7.3, apply the forces at the nodes 2 and 3 as shown in Figure 7.8. Solve for the displacements of nodes 2 and 3 and the strains and stresses at node 1.

Solution. The first step will be the determination of the stiffness matrix for the element. This will be accomplished using equation (7.17). The evaluation of the matrix product $h[B]^T[C][B]|J|$ at the four locations in the element, calculated by computer, are listed below.

1. At $s = -0.577, \quad t = -0.577$:

$$|J| = 9.23$$

$$h[B]^T[C][B]|J| = \begin{bmatrix} 674{,}674 & 324{,}843 & -469{,}229 & -140{,}461 & -180{,}779 & -87{,}042 & -24{,}668 & -97{,}341 \\ & 674{,}674 & -107{,}961 & -46{,}724 & -87{,}042 & -180{,}779 & -129{,}842 & -447{,}173 \\ & & 545{,}527 & -89{,}888 & 125{,}729 & 28{,}927 & -202{,}029 & 168{,}919 \\ & & & 222{,}420 & 37{,}636 & 12{,}519 & 192{,}711 & -188{,}216 \\ & & & & 48{,}439 & 23{,}322 & 6{,}609 & 26{,}082 \\ & & \text{symmetric} & & & 48{,}439 & 34{,}790 & 119{,}819 \\ & & & & & & 220{,}086 & -97{,}662 \\ & & & & & & & 515{,}568 \end{bmatrix}$$

7.5 THE ELEMENT STIFFNESS MATRIX

2. At $s = 0.577$, $\quad t = -0.577$:

$$|J| = 5.77$$

$h[B]^T[C][B]|J|$

$$= \begin{bmatrix} 341{,}600 & 88{,}323 & -263{,}063 & 102{,}124 & -149{,}026 & -163{,}084 & 70{,}487 & -27{,}365 \\ & 169{,}955 & 134{,}624 & 96{,}005 & -186{,}876 & -240{,}237 & -36{,}073 & -25{,}725 \\ & & 635{,}478 & -346{,}625 & -202{,}141 & 119{,}122 & -170{,}276 & 92{,}877 \\ & & & 924{,}332 & 151{,}622 & -772{,}664 & 92{,}877 & -247{,}675 \\ & & & & 297{,}002 & 75{,}879 & 54{,}163 & -40{,}628 \\ & \text{symmetric} & & & & 805{,}865 & -31{,}919 & 207{,}034 \\ & & & & & & 45{,}625 & -24{,}887 \\ & & & & & & & 66{,}364 \end{bmatrix}$$

3. At $s = 0.577$, $\quad t = 0.577$:

$$|J| = 5.77$$

$h[B]^T[C][B]|J|$

$$= \begin{bmatrix} 77{,}531 & 37{,}329 & 115{,}181 & 78{,}357 & -289{,}350 & -139{,}317 & 96{,}637 & 23{,}629 \\ & 77{,}531 & 87{,}066 & 228{,}390 & -139{,}317 & -289{,}350 & 14{,}920 & -16{,}572 \\ & & 308{,}055 & 97{,}432 & -429{,}862 & -324{,}935 & 6{,}625 & 140{,}436 \\ & & & 809{,}734 & -292{,}435 & -852{,}367 & 116{,}644 & -185{,}759 \\ & & & & 1{,}079{,}868 & 519{,}936 & -360{,}658 & -88{,}186 \\ & \text{symmetric} & & & & 1{,}079{,}868 & -55{,}686 & 61{,}847 \\ & & & & & & 257{,}393 & -75{,}880 \\ & & & & & & & 140{,}483 \end{bmatrix}$$

4. At $s = -0.577$, $\quad t = 0.577$:

$$|J| = 9.23$$

$h[B]^T[C][B]|J|$

$$= \begin{bmatrix} 375{,}780 & 205{,}942 & -9{,}729 & -5{,}778 & -402{,}360 & -221{,}726 & 36{,}307 & 21{,}560 \\ & 570{,}061 & 2{,}931 & 114{,}101 & -197{,}934 & -258{,}332 & -10{,}941 & -425{,}831 \\ & & 28{,}505 & -15{,}549 & 87{,}607 & -45{,}411 & -106{,}385 & 58{,}027 \\ & & & 41{,}462 & -36{,}703 & -824 & 58{,}027 & -154{,}741 \\ & & & & 641{,}706 & 97{,}661 & -326{,}954 & 136{,}974 \\ & \text{symmetric} & & & & 256{,}082 & 169{,}474 & 3{,}072 \\ & & & & & & 397{,}030 & -216{,}563 \\ & & & & & & & 577{,}499 \end{bmatrix}$$

□

Summing the four matrix products yields the stiffness matrix $[k]$ for the element:

$$[k] = \begin{bmatrix} 1{,}469{,}586 & 656{,}438 & -626{,}838 & 34{,}243 & -1{,}021{,}514 \\ & 1{,}492{,}222 & 116{,}661 & 391{,}773 & -611{,}167 \\ & & 1{,}517{,}567 & -354{,}628 & -418{,}666 \\ & & & 1{,}997{,}949 & -139{,}878 \\ & & & & 2{,}067{,}017 \\ & \text{symmetric} & & & \end{bmatrix}$$

$$\begin{bmatrix} -611{,}167 & 178{,}764 & -79{,}516 \\ -968{,}697 & -161{,}934 & -915{,}300 \\ -222{,}296 & -472{,}064 & 460{,}261 \\ -1{,}613{,}334 & 460{,}261 & -776{,}390 \\ 716{,}800 & -626{,}838 & 34{,}243 \\ 2{,}190{,}257 & 116{,}661 & 391{,}773 \\ & 920{,}136 & -414{,}990 \\ & & 1{,}299{,}914 \end{bmatrix}$$

Writing the force–displacement equation, equation (4.14), with nodal forces only, we obtain

$$[k]\{q\} = \{Q\}_{\text{NF}}$$

Substituting the boundary conditions yields

$$[k]\begin{Bmatrix} 0 \\ 0 \\ u_2 \\ v_2 \\ u_3 \\ v_3 \\ 0 \\ 0 \end{Bmatrix} = \begin{Bmatrix} R_{1X} \\ R_{1Y} \\ 5 \\ 0 \\ 10 \\ 0 \\ R_{4X} \\ R_{4Y} \end{Bmatrix}$$

Reducing the size of the matrix equation by removing columns and rows 1, 2, 7, and 8, we have

$$10^6 \begin{bmatrix} 1.52 & -0.355 & -0.419 & -0.222 \\ & 2.00 & -0.140 & -1.61 \\ & & 2.07 & 0.717 \\ & \text{symmetric} & & 2.19 \end{bmatrix} \begin{Bmatrix} u_2 \\ v_2 \\ u_3 \\ v_3 \end{Bmatrix} = \begin{Bmatrix} 5 \\ 0 \\ 10 \\ 0 \end{Bmatrix}$$

7.5 THE ELEMENT STIFFNESS MATRIX

Solving for the displacements gives us

$$\begin{Bmatrix} u_2 \\ v_2 \\ u_3 \\ v_3 \end{Bmatrix} = \begin{Bmatrix} 4.8 \\ 0 \\ 6.4 \\ -1.6 \end{Bmatrix} (10^{-6}) \text{ in.}$$

The strain is obtained from equation (7.12). The $[B]$ matrix for this element at node 1 was derived in Example 7.3.

$$\begin{Bmatrix} \epsilon_X \\ \epsilon_Y \\ \gamma_{XY} \end{Bmatrix}_{(-1,-1)} = 0.1667 \begin{bmatrix} 1 & 0 & 0 & 0 \\ 0 & 0 & 0 & 0 \\ 0 & 1 & 0 & 0 \end{bmatrix} \begin{Bmatrix} u_2 \\ v_2 \\ u_3 \\ v_3 \end{Bmatrix} = \begin{Bmatrix} 0.1667 u_2 \\ 0 \\ 0.1667 v_2 \end{Bmatrix}$$

$$= \begin{Bmatrix} 0.8 \\ 0 \\ 0 \end{Bmatrix} (10^{-6})$$

The stress is obtained from equation (2.7) for plane stress.

$$\begin{Bmatrix} \sigma_X \\ \sigma_Y \\ \tau_{XY} \end{Bmatrix}_{(-1,-1)} = \frac{E}{1-\nu^2} \begin{bmatrix} 1 & \nu & 0 \\ \nu & 1 & 0 \\ 0 & 0 & \frac{1-\nu}{2} \end{bmatrix} \begin{Bmatrix} \epsilon_X \\ \epsilon_Y \\ \gamma_{XY} \end{Bmatrix}$$

$$= \frac{30(10^6)}{1-0.3^2} \begin{bmatrix} 1 & 0.3 & 0 \\ 0.3 & 1 & 0 \\ 0 & 0 & 0.35 \end{bmatrix} \begin{Bmatrix} 0.8(10^{-6}) \\ 0 \\ 0 \end{Bmatrix}$$

$$= \begin{Bmatrix} 26.7 \\ 7.91 \\ 0 \end{Bmatrix} \text{psi}$$

Out of curiosity, you may wish to try solving for the stress at another point in the element. In Example 7.3 we solved for the $[B]$ matrix at the midpoint ($s = t = 0$) of the element also. Performing the matrix products, you should arrive at the following:

$$\begin{Bmatrix} \epsilon_X \\ \epsilon_Y \\ \gamma_{XY} \end{Bmatrix}_{(0,0)} = \begin{Bmatrix} 0.96 \\ -0.16 \\ 0 \end{Bmatrix} (10^{-6})$$

$$\begin{Bmatrix} \sigma_X \\ \sigma_Y \\ \tau_{XY} \end{Bmatrix}_{(0,0)} = \begin{Bmatrix} 30. \\ 4.2 \\ 0 \end{Bmatrix} \text{psi} \qquad \square$$

7.6 DISTRIBUTED SURFACE TRACTIONS: EQUIVALENT NODAL FORCES

Reviewing for a moment, we recall that the potential energy functions for all applied loads are determined and added to the strain energy to obtain the total energy potential of the loaded and deformed continuum. The general derivation of these force energy potentials for node, surface, and body forces can be reviewed in Section 4.2. Subsequently, the total energy is differentiated with respect to the displacement vector and the resulting equation is our now very familiar force–displacement equation. An example of this equation is equation (4.14). The right side of the equation defines the nodal forces accounting for directly applied nodal forces, $\{Q\}_{NF}$, distributed loading equivalent forces, $\{Q\}_T$, and equivalent body forces, $\{Q\}_{BF}$. As an example showing the steps leading to the development of equivalent nodal forces for a particular distributed loading, we will work out the details of a surface traction distributed on one side of the quadrilateral.

Assume the distributed loading on one side of the element as shown in Figure 7.9. The tractions are given in component form in the global X and Y directions.

The general expression for the equivalent nodal loads for this distributed surface traction is [from (4.14) or (6.10)]

$$\{Q\}_T = h \int_l [N]^T \begin{Bmatrix} T_X \\ T_Y \end{Bmatrix} dl \tag{7.18}$$

where the symbol l is used for the length of the side to avoid confusing the symbol s with the natural coordinate.

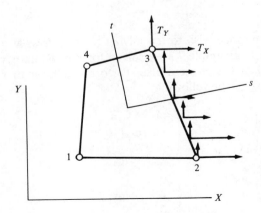

FIGURE 7.9 Distributed stress traction on one side of the element.

7.6 DISTRIBUTED SURFACE TRACTIONS: EQUIVALENT NODAL FORCES

For the four-node quadrilateral.

$$[N] = \begin{bmatrix} N_1 & 0 & N_2 & 0 & N_3 & 0 & N_4 & 0 \\ 0 & N_1 & 0 & N_2 & 0 & N_3 & 0 & N_4 \end{bmatrix}$$

and

$$N_1 = \frac{(1-s)(1-t)}{4} \qquad N_2 = \frac{(1+s)(1-t)}{4}$$

$$N_3 = \frac{(1+s)(1+t)}{4} \qquad N_4 = \frac{(1-s)(1+t)}{4}$$

For the particular side on which we have placed the distributed load, $s = 1$. Thus we have

$$N_1 = 0 \qquad N_2 = \frac{1-t}{2}$$

$$N_3 = \frac{1+t}{2} \qquad N_4 = 0$$

We will assume that the distributed loading is a function of the position along the edge of the element instead of being a function of the X and Y coordinates. This is a reasonable assumption, since loadings are generally defined locally on the continuum. For a general derivation, we will assume a quadratic variation of the traction in the X direction on one face. The result will be applicable to a traction in any direction on any face.

Assume the following distributed loading function:

$$T_X = C_1 + C_2 t + C_3 t^2 \qquad T_Y = 0$$

where t is the natural coordinate. The differential length of the side is

$$dl = (dX^2 + dY^2)^{1/2}$$

$$dX = \frac{\partial X}{\partial s} ds + \frac{\partial X}{\partial t} dt = \sum_{i=1}^{4} \frac{\partial N_i}{\partial s} X_i \, ds + \sum_{i=1}^{4} \frac{\partial N_i}{\partial t} X_i \, dt$$

For the side involved, $ds = 0$ and $s = 1$, N_1 and $N_4 = 0$. Thus we have

$$dX = \left(\frac{\partial N_2}{\partial t} X_2 + \frac{\partial N_3}{\partial t} X_3 \right) dt$$

$$= \left(\frac{-X_2}{2} + \frac{X_3}{2} \right) dt$$

or

$$dX = \frac{X_3 - X_2}{2} dt$$

Similarly for dY,

$$dY = \frac{\partial Y}{\partial s} ds + \frac{\partial Y}{\partial t} dt \quad \text{where } ds = 0$$

$$= \left(\frac{\partial N_2}{\partial t} Y_2 + \frac{\partial N_3}{\partial t} Y_3\right) dt$$

$$= \frac{Y_3 - Y_2}{2} dt$$

Thus we have the result for dl,

$$dl = \frac{\sqrt{(X_3 - X_2)^2 + (Y_3 - Y_2)^2}}{2} dt$$

Observe that the radical term is the length of the side. Let this term be replaced by the symbol L.

$$dl = \frac{L}{2} dt$$

Now expand equation (7.18):

$$\begin{Bmatrix} F_{1X} \\ F_{1Y} \\ F_{2X} \\ F_{2Y} \\ F_{3X} \\ F_{3Y} \\ F_{4X} \\ F_{4Y} \end{Bmatrix} = h\frac{L}{2} \begin{Bmatrix} 0 \\ 0 \\ \int_{-1}^{1} N_2 T_X \, dt \\ 0 \\ \int_{-1}^{1} N_3 T_X \, dt \\ 0 \\ 0 \\ 0 \end{Bmatrix}$$

Integrating each nonzero term gives us

$$\int_{-1}^{1} N_2 T_X \, dt = \tfrac{1}{2} \int_{-1}^{1} (1-t)(C_1 + C_2 t + C_3 t^2) \, dt$$

$$= C_1 + \tfrac{1}{3}(C_3 - C_2)$$

Also,

$$\int_{-1}^{1} N_3 T_X \, dt = \tfrac{1}{2} \int_{-1}^{1} (1+t)(C_1 + C_2 t + C_3 t^2) \, dt$$

$$= C_1 + \tfrac{1}{3}(C_2 + C_3)$$

The resulting matrix equation for the nonzero nodal forces becomes

$$\begin{Bmatrix} F_{2X} \\ F_{3X} \end{Bmatrix} = \frac{hL}{6} \begin{bmatrix} 3 & -1 & 1 \\ 3 & 1 & 1 \end{bmatrix} \begin{Bmatrix} C_1 \\ C_2 \\ C_3 \end{Bmatrix} \qquad (7.19)$$

7.6 DISTRIBUTED SURFACE TRACTIONS: EQUIVALENT NODAL FORCES

To solve for the C_i constants, we substitute values of the stress traction at three locations on the side.

$$T_X(t = 1) = C_1 + C_2 + C_3$$
$$T_X(t = 0) = C_1$$
$$T_X(t = -1) = C_1 - C_2 + C_3$$

In matrix form,

$$\begin{Bmatrix} T_X(1) \\ T_X(0) \\ T_X(-1) \end{Bmatrix} = \begin{bmatrix} 1 & 1 & 1 \\ 1 & 0 & 0 \\ 1 & -1 & 1 \end{bmatrix} \begin{Bmatrix} C_1 \\ C_2 \\ C_3 \end{Bmatrix} \qquad (7.20)$$

Inverting the matrix and solving for the vector of constants gives us

$$\begin{Bmatrix} C_1 \\ C_2 \\ C_3 \end{Bmatrix} = \frac{1}{2} \begin{bmatrix} 0 & 2 & 0 \\ 1 & 0 & -1 \\ 1 & -2 & 1 \end{bmatrix} \begin{Bmatrix} T_X(1) \\ T_X(0) \\ T_X(-1) \end{Bmatrix} \qquad (7.21)$$

Substituting (7.21) into (7.19) yields

$$\begin{Bmatrix} F_{2X} \\ F_{3X} \end{Bmatrix} = \frac{hL}{6} \begin{bmatrix} 0 & 2 & 1 \\ 1 & 2 & 0 \end{bmatrix} \begin{Bmatrix} T_X(1) \\ T_X(0) \\ T_X(-1) \end{Bmatrix} \qquad (7.22)$$

This formula can be put into a form that may be easier to use if we reorder the traction values so that their order in the vector is in the same sequence of position along the side of the element as the nodal force vector components. Reordering the traction vector, we have the following formula for the equivalent nodal forces for a *quadratic traction distribution*:

$$\begin{Bmatrix} F_{(\text{node } i)X} \\ F_{(\text{node } j)X} \end{Bmatrix} = \frac{hL}{6} \begin{bmatrix} 1 & 2 & 0 \\ 0 & 2 & 1 \end{bmatrix} \begin{Bmatrix} T_X(\text{node } i) \\ T_X(\text{midpoint}) \\ T_X(\text{node } j) \end{Bmatrix} \qquad (7.23a)$$

Of course, the other components would obey the same relation.

$$\begin{Bmatrix} F_{(\text{node } i)Y} \\ F_{(\text{node } j)Y} \end{Bmatrix} = \frac{hL}{6} \begin{bmatrix} 1 & 2 & 0 \\ 0 & 2 & 1 \end{bmatrix} \begin{Bmatrix} T_Y(\text{node } i) \\ T_Y(\text{midpoint}) \\ T_Y(\text{node } j) \end{Bmatrix} \qquad (7.23b)$$

This is a general formula for any loading up to a quadratic form. The most common loading for this element would be a linear distribution. For this case we can derive a simpler formula by replacing the midpoint traction as follows:

$$\begin{Bmatrix} T_X(\text{node } i) \\ T_X(\text{midpoint}) \\ T_X(\text{node } j) \end{Bmatrix} = \begin{Bmatrix} T_X(\text{node } i) \\ \frac{1}{2}(T_X(\text{node } i) + T_X(\text{node } j)) \\ T_X(\text{node } j) \end{Bmatrix}$$

$$= \begin{bmatrix} 1 & 0 \\ \frac{1}{2} & \frac{1}{2} \\ 0 & 1 \end{bmatrix} \begin{Bmatrix} T_X(\text{node } i) \\ T_X(\text{node } j) \end{Bmatrix}$$

Substitution into (7.23) yields the formula for a *linear distribution*:

$$\begin{Bmatrix} F_{(\text{node } i)X} \\ F_{(\text{node } j)X} \end{Bmatrix} = \frac{hL}{6} \begin{bmatrix} 2 & 1 \\ 1 & 2 \end{bmatrix} \begin{Bmatrix} T_X(\text{node } i) \\ T_X(\text{node } j) \end{Bmatrix} \quad (7.24)$$

Obviously, the X direction can be replaced with any direction, and the formula holds for any face of the element.

Compare this equivalent nodal force vector formula (7.24) with the nodal force vector formula (3.15) for the linear distribution on the one-dimensional element. They are equivalent. In addition, this formula can be used for the three-node triangle.

The advantage of having this formula is that it may be included in a computer program so that the distributed loading information can be the input data. This saves having to calculate equivalent loads by hand.

Example 7.5 For the element shown in Figure 7.10, replace the quadratic traction distribution with the energy equivalent nodal forces.

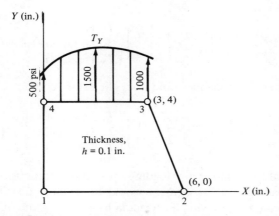

FIGURE 7.10 Quadratic traction distribution on side 3-4.

7.6 DISTRIBUTED SURFACE TRACTIONS: EQUIVALENT NODAL FORCES

Solution. Substituting into (7.23b), we have

$$\begin{Bmatrix} F_{(\text{node } i)Y} \\ F_{(\text{node } j)Y} \end{Bmatrix} = \frac{hL}{6} \begin{bmatrix} 1 & 2 & 0 \\ 0 & 2 & 1 \end{bmatrix} \begin{Bmatrix} T_Y \text{ (node } i) \\ T_Y \text{ (midpoint)} \\ T_Y \text{ (node } j) \end{Bmatrix}$$

$$\begin{Bmatrix} F_{4Y} \\ F_{3Y} \end{Bmatrix} = \frac{0.1(3)}{6} \begin{bmatrix} 1 & 2 & 0 \\ 0 & 2 & 1 \end{bmatrix} \begin{Bmatrix} 500 \\ 1500 \\ 1000 \end{Bmatrix} = \begin{Bmatrix} 175 \\ 200 \end{Bmatrix} \text{ lb} \quad \square$$

Example 7.6 For the element shown in Figure 7.11, replace the distributed loading with energy equivalent nodal forces.

Solution. The length of side 2–3 is 5 in. Substituting into equation (7.24) yields

$$\begin{Bmatrix} F_{2X} \\ F_{3X} \end{Bmatrix} = \frac{0.1(5)}{6} \begin{bmatrix} 2 & 1 \\ 1 & 2 \end{bmatrix} \begin{Bmatrix} 1000 \\ 1000 \end{Bmatrix} = \begin{Bmatrix} 250 \\ 250 \end{Bmatrix} \text{ lb} \quad \square$$

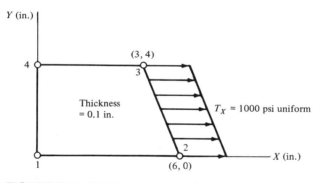

FIGURE 7.11 Uniform stress traction on one face.

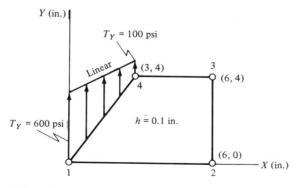

FIGURE 7.12 Nonuniform stress traction on one face of element.

Example 7.7 For the element shown in Figure 7.12, replace the distributed loading with energy equivalent nodal forces.

Solution. The length of side 1–4 is 5 in. Substituting into (7.7) with the Y coordinate direction replaced by X gives us

$$\begin{Bmatrix} F_{1Y} \\ F_{4Y} \end{Bmatrix} = \frac{hL}{6}\begin{bmatrix} 2 & 1 \\ 1 & 2 \end{bmatrix}\begin{Bmatrix} T_Y(1) \\ T_Y(4) \end{Bmatrix}$$

$$= \frac{0.1(5)}{6}\begin{bmatrix} 2 & 1 \\ 1 & 2 \end{bmatrix}\begin{Bmatrix} 600 \\ 100 \end{Bmatrix}$$

$$= \begin{Bmatrix} F_{1Y} \\ F_{4Y} \end{Bmatrix} = \begin{Bmatrix} 108.33 \\ 66.67 \end{Bmatrix} \text{ lb} \qquad \square$$

7.7 DISTRIBUTED BODY FORCES: EQUIVALENT NODAL FORCES

The equivalent nodal forces for the distributed body force in the four-node quadrilateral is defined in symbolic form in the general force–displacement equation (4.14). In symbolic form the formula is identical to that used for the triangle (6.11) or any other plane two-dimensional element:

$$\{Q\}_{\text{BF}} = h\int_A [N]^T \begin{Bmatrix} B_X \\ B_Y \end{Bmatrix} dX\, dY \qquad (6.11)$$

Since the matrix $[N]^T$ is in terms of the natural coordinates s and t, we will change the variable of integration to the natural coordinates.

$$\{Q\}_{\text{BF}} = h\int_{-1}^{1}\int_{-1}^{1} [N]^T \begin{Bmatrix} B_X \\ B_Y \end{Bmatrix} |J|\, ds\, dt$$

As an alternative to expanding the product of the matrix $[N]^T$, the body force vector and the Jacobian $|J|$ and then integrating each term, the expression is complicated enough to warrant approximating the integration with Gauss quadrature. The two-point formula is sufficient to yield good results for this four-node element, so referring back to the area integration approximation formula (5.11), we will obtain the equivalent nodal forces using the following summation:

$$\{Q\}_{\text{BF}} \approx h\sum_{i=1}^{2}\sum_{j=1}^{2} W_i W_j [N(s_i, t_j)]^T \begin{Bmatrix} B_X(s_i, t_j) \\ B_Y(s_i, t_j) \end{Bmatrix} |J(s_i, t_j)| \qquad (7.25)$$

For the two-point formula, the weighting functions are unity, and the Gauss points are located at s and t equal to ± 0.57735.

7.7 DISTRIBUTED BODY FORCES: EQUIVALENT NODAL FORCES

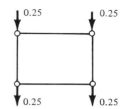

FIGURE 7.13 Fraction of total weight distributed to element nodes of a rectangle.

A program has been written for the evaluation of this summation for the case of a uniform body force distribution. It is given in Appendix 6. Using this program, you can verify that the distribution of the nodal forces is dependent on the shape of the quadrilateral.

Example 7.8
Using the computer program in Appendix 6, solve for the fraction of the total weight of the element that should be applied to each node of the rectangle of uniform density and arbitrary dimensions.

Solution. The nodal force at each node equals one-fourth of the total weight of the element (Figure 7.13). ☐

Example 7.9
Solve for the fraction of the total weight of the element to be applied to each node if the shape is a parallelogram.

Solution. The nodal forces are one-fourth of the total weight for any parallelogram (Figure 7.14). ☐

Example 7.10
Determine the fraction of the total weight to be applied to the nodes of the trapezoid and the general quadrilateral.

Solution. The weight distributions are shown in Figure 7.15. It is interesting to note that as one side length goes to zero, the

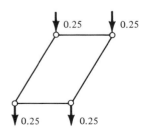

FIGURE 7.14 Fraction of total weight distributed to nodes of parallelogram.

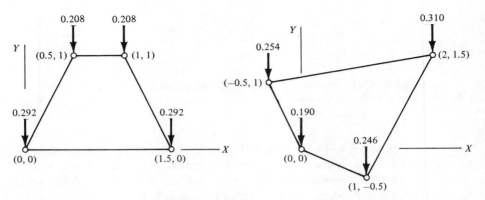

FIGURE 7.15 Body force distributions for trapezoid and arbitrary quadrilateral of uniform density.

distribution approaches that of the three-node triangle, one-third at each of the three nodes. □

7.8 ASSEMBLING THE ELEMENTS AND SOLVING FOR STRESSES

The force–displacement equation for the single quadrilateral element is, symbolically, the same as other elements studied so far:

$$[k]\{q\} = \{Q\}_{\text{NF}} + \{Q\}_T + \{Q\}_{\text{BF}}$$

where

$[k]$ = element stiffness matrix

$\{q\}$ = element nodal displacement vector

$\{Q\}$ = nodal force, surface traction force, and body force vectors

For a continuum modeled with many elements, the total system equation is represented symbolically as

$$[K]\{r\} = \{R\}_{\text{NF}} + \{R\}_T + \{R\}_{\text{BF}} \qquad (4.5.1)$$

where

$[K]$ = assembled global stiffness matrix

$\{r\}$ = complete global node displacement vector

$\{R\}$ = nodal, boundary traction, and body force vectors for loads applied to the continuum

In this chapter we have developed the quadrilateral element stiffness matrix and demonstrated how to obtain the equivalent nodal forces.

7.8 ASSEMBLING THE ELEMENTS AND SOLVING FOR STRESSES

Assembling the element equations follows the procedures described in Chapter 3 for the one-dimensional element and Chapters 4 and 6 for the three-node triangle. For a system containing n nodes:

1. The algorithm for the assembly of the $2n \times 2n$ global stiffness matrix is identical to that of the triangle with the one exception that the local stiffness matrix of the quadrilateral is an 8×8 as opposed to the triangle 6×6 dimension.
2. The $2n \times 1$ nodal force vector $\{R\}_{NF}$ is the vector of all nodal loads, including zero, applied to the continuum.
3. The $2n \times 1$ surface traction vector $\{R\}_T$ is the sum of all element equivalent nodal loads (review the example in Section 3.10 and Example 6.11).
4. The $2n \times 1$ body force vector $\{R\}_{BF}$ is the sum of all element equivalent nodal loads.

The concern discussed in Section 6.7 about element node numbering to achieve a small bandwidth stiffness matrix is the same for this and, for that matter, all elements.

In Example 7.4 it was made clear that after solving for the element nodal displacements, the strain and stress calculations were to be made for specific locations in the element. The question arises as to the location in the element for the most accurate stress value. We recall that the constant strain one-dimensional example in Chapter 3 yielded the best result near the center of the element, the worst result at the nodes. We might expect that the quadrilateral would have the "best" value at some location inside the element and the next examples will show that the best location is at the element center. Good results can also be achieved by averaging the element values at interior nodes. Element averages for boundary nodes are not very satisfactory.

Example 7.11 The square plate in Figure 7.16 has a linearly distributed stress traction equivalent to a couple of 8000 in.-lb. (This is the same problem as in Example 6.8.) Model this problem with a single quadrilateral element and solve for the resulting stress distributions on the element boundary.

Solution. Using (7.24), the distributed traction is replaced with nodal forces as indicated in Figure 7.16(b), and the boundary constraints allow displacement of node 4 in the Y direction.

The stress results are obtained for several locations on each side of the element; these values are plotted in Figure 7.17.

Observe the following. First, the normal stress σ_X varies linearly in agreement with the theory, but this single element model yields a value 25% below the theoretical value. This looks very promising, and

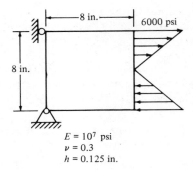

$E = 10^7$ psi
$\nu = 0.3$
$h = 0.125$ in.

(a) Square plate with distributed stress traction.

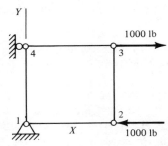

(b) Equivalent nodal forces.

FIGURE 7.16 Nodal forces equivalent to the distributed stress tractions.

an improvement in accuracy would be expected with mesh refinement. The normal stress σ_Y and the shear stress τ_{XY} should both be zero everywhere in the region and on the boundary. We note that these stresses oscillate about the center of the element, and we should expect that mesh refinement would reduce this error.

The reason for the error in the stresses σ_Y and τ_{XY} can be rationalized by observing the nature of the distortion of the element. As shown in the Figure 7.18(a), the sides of the four-node quadrilateral remain straight during deformation so that angles formed by intersecting edges change. This causes shear strain.

For pure bending, the angles of the intersecting edges do not change, as illustrated in Figure 7.18(b). The simple four-node element cannot model this condition. The procedures to remedy this error include mesh refinement to reduce the angle change for each element and/or development of a new, more sophisticated element. In this chapter we examine the effect of mesh refinement; in the next chapter we discuss a more sophisticated element.

It should be noted that the stresses calculated at the center of this single element model are all zero, the theoretically correct value. As

7.8 ASSEMBLING THE ELEMENTS AND SOLVING FOR STRESSES

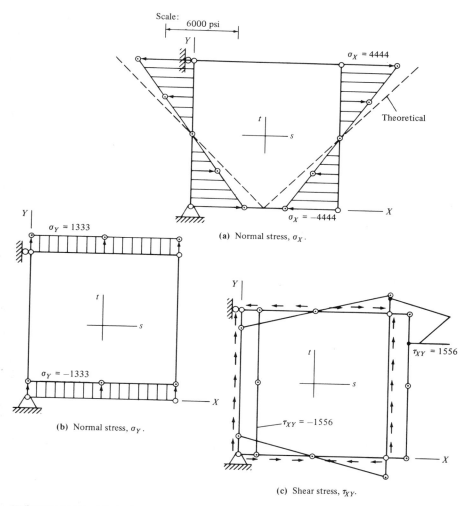

FIGURE 7.17 Variation of the stresses on the boundaries of the single-element model.

we illustrate in the next example, the best procedure in using this element is to accept values calculated at the center of the element ($s = 0 = t$) and to obtain stresses in desired locations by mesh refinement. □

Example 7.12 Solve the problem of Example 7.11 using four-elements (Figure 7.19). Examine the stresses at individual element

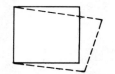

(a) Element deformation.

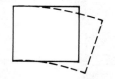

(b) Pure bending deformation.

FIGURE 7.18 Quadrilateral deformation with and without shear strain.

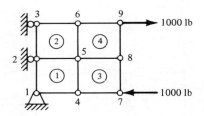

FIGURE 7.19 Four-element mesh.

nodes and centers, and identify the location of the most accurate results.

Solution. The element mesh is defined in Figure 7.19, and the results of the stress analysis are shown in Table 7.1. In the column under "Element number" there is the element number identification with the location by global node number listed in ascending local number order. The first label, however, is the element center location, defined as zero.

Examining the results in Table 7.1, we note that all locations but the element centers have significant error in the stresses σ_Y and τ_{XY}; they should all be zero. Since we observed in Example 7.11 that these stresses oscillated about element centers, perhaps good results could be achieved by averaging the values at each node. Table 7.2 lists the nodal averages, together with the theoretical values in parentheses, and we note that some values improve. The σ_X stress is within 11% of the correct value. However, with the exception of interior node 5, all nodal values have a very large error in one of the stresses σ_Y or τ_{XY}.

7.8 ASSEMBLING THE ELEMENTS AND SOLVING FOR STRESSES

TABLE 7.1 Stress Results for Example 7.12

Element Number	Sigma X (psi)	Sigma Y (psi)	Tau XY (psi)
1 − 0	−2.67E3	1.39E − 6	8.19E − 7
1 − 1	−5.61E3	−8.82E2	−1.02E3
1 − 4	−5.61E3	−8.82E2	1.02E3
1 − 5	2.64E2	8.82E2	1.02E3
1 − 2	2.64E2	8.82E2	−1.02E3
2 − 0	2.67E3	−1.33E − 6	1.69E − 8
2 − 2	−2.64E2	−8.82E2	−1.02E3
2 − 5	−2.64E2	−8.82E2	1.02E3
2 − 6	5.61E3	8.82E2	1.02E3
2 − 3	5.61E3	8.82E2	−1.02E3
3 − 0	−2.67E3	0.00	2.51E − 6
3 − 4	−5.61E3	−8.82E2	−1.02E3
3 − 7	−5.61E3	−8.82E2	1.02E3
3 − 8	2.64E2	8.82E2	1.02E3
3 − 5	2.64E2	8.82E2	−1.02E3
4 − 0	2.67E3	2.36E − 6	8.74E − 7
4 − 5	−2.64E2	−8.82E2	−1.02E3
4 − 8	−2.64E2	−8.82E2	1.02E3
4 − 9	5.61E3	8.82E2	1.02E3
4 − 6	5.61E3	8.82E2	−1.02E3

Table 7.3 summarizes the stresses at the element centers. The stress σ_X is within 11%, and the stresses σ_Y and τ_{XY} are the correct values.

The conclusion from the study of this example is that the best results for the four-node quadrilateral are obtained at the element center. Results throughout a region have to be obtained by mesh refinement and careful extrapolation. ◻

Example 7.13 Refine the mesh of the Example 7.12 as shown in Figure 7.20 and plot the stress on a cross section. Compare with the theoretical value.

Solution. The element mesh is defined in Figure 7.20, and Table 7.4 lists the stress results at the element centers. Note the reduction

TABLE 7.2 Nodal Average Stresses[a]

Node Number	Sigma X	Sigma Y	Tau XY
1	−5610	−882	−1020
	(−6000)	(0)	(0)
2	0	0	−1020
	(0)	(0)	(0)
3	5610	882	−1020
	(6000)	(0)	(0)
4	−5610	−882	0
	(−6000)	(0)	(0)
5	0	0	0
	(0)	(0)	(0)
6	5610	882	0
	(6000)	(0)	(0)
7	−5610	−882	1020
	(−6000)	(0)	(0)
8	0	0	1020
	(0)	(0)	(0)
9	5610	882	1020
	(6000)	(0)	(0)

[a] Theoretical values are given in parentheses.

TABLE 7.3 Element Center Stresses[a]

Element Number	Sigma X	Sigma Y	Tau XY
1	−2670	0	0
	(−3000)	(0)	(0)
2	2670	0	0
	(3000)	(0)	(0)
3	−2670	0	0
	(−3000)	(0)	(0)
4	2670	0	0
	(3000)	(0)	(0)

[a] Theoretical values are given in parentheses.

7.8 ASSEMBLING THE ELEMENTS AND SOLVING FOR STRESSES

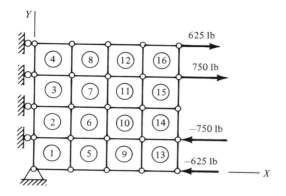

FIGURE 7.20 Sixteen-element model of the square plate.

TABLE 7.4 Stress at Element Center

Element Number	Sigma X (psi)	Sigma Y (psi)	Tau XY (psi)
1	−4369	−1	0
2	−1456	−1	−1
3	1455	0	−1
4	4368	0	0
5	−4368	1	−2
6	−1457	2	1
7	1456	−3	1
8	4367	−2	−2
9	−4367	11	−13
10	−1466	11	12
11	1465	−14	12
12	4366	−12	−13
13	−4361	−14	−32
14	−1506	−17	31
15	1505	16	31
16	4360	13	−32

of the initially small error in the σ_Y and τ_{XY} stresses and the improvement in the linearity of σ_X in elements away from the applied load boundary.

Figure 7.21 compares the σ_X stress with the theoretical value. The error has been reduced from 11% for the 4-element model to 3% for this 16-element model. ☐

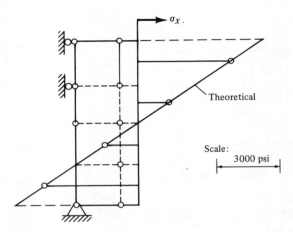

FIGURE 7.21 Plot of the element center value of the σ_X stress on cross section through centers of elements 5, 6, 7, and 8.

7.9 COMPARING THE FOUR-NODE QUADRILATERAL AND THE THREE-NODE TRIANGLE

The four-node quadrilateral is based on a displacement function which includes a quadratic term. For the X displacement, for example,

$$u = c_1 + c_2 s + c_3 t + c_4 st$$

where u is the displacement and s and t are the natural coordinates. This function ensures that the sides of the element will remain straight during displacement, satisfying interelement compatibility, and it also allows a certain degree of strain gradient modeling capability. Note that if the element is rectangular and s and t correspond to coordinates X and Y, respectively, then the strain

$$\epsilon_X = c_2 + c_4 Y$$

is independent of X. This implies that for a constant Y coordinate, the element is unable to model a gradient of this strain value in the X direction. In a similar manner, it may be shown that the strain ϵ_Y is independent of Y, whereas the shear strain varies linearly with both X and Y.

Although the four-node quadrilateral has greater capability than the three-node triangle, its inability to model strain gradients fully in all directions is a severe limitation. In Chapter 8 we introduce a higher-order

7.9 FOUR-NODE QUADRILATERAL AND THREE-NODE TRIANGLE

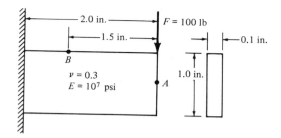

FIGURE 7.22 Cantilever beam with single force at the end.

triangle and quadrilateral that have greater capability and are very popular in finite element analysis of real problems.

In the next example, comparison of a four-node quadrilateral and three-node triangle analysis is made using the cantilever beam problem of Example 6.9.

Example 7.14 Using the quadrilateral element, determine the displacement of point A and the stress at point B in the beam shown in Figure 7.22. Compare the convergence characteristics of the quadrilateral with the triangle solution of this problem in Example 6.9.

Solution. Three different mesh patterns were used to study the convergence character of the quadrilateral; they are shown in Figure 7.23. The vertical displacement of the endpoint A as determined by

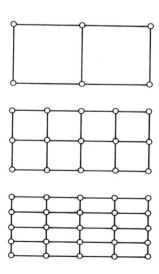

FIGURE 7.23 Models for deflection and stress analysis.

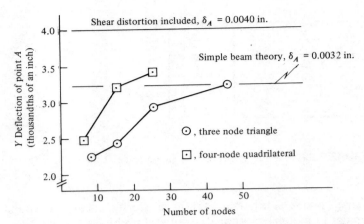

FIGURE 7.24 Convergence comparison for three-node triangle and four-node quadrilateral.

each of these models is shown in Figure 7.24. The quadrilateral models are more flexible than the triangle element models. Note that the 25-node model displacement exceeds the simple beam theory prediction because this short beam has a shear distortion that is not included in the simple beam formula.

The result for the stress at the boundary point B is shown in Figure 7.25, and the improvement of the quadrilateral over the triangle is quite apparent. Since the triangle results had been obtained by

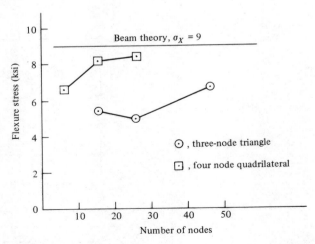

FIGURE 7.25 Comparison of results of three-node triangle and four-node quadrilateral. Values are node averages.

7.9 FOUR-NODE QUADRILATERAL AND THREE-NODE TRIANGLE

averaging element values at a common node, for comparison the quadrilateral values shown are also element averages, despite our earlier recognition that the nodal values are not the most accurate. Results may be improved perhaps by a more elaborate averaging and extrapolation scheme, but the best approach is to have enough elements (sufficient refinement) so that simple averaging will yield good results. □

The inability of the quadrilateral to model the strain and stress gradient in the X direction is illustrated in Figure 7.26. Note that there is a negligible difference in the σ_X stress component in an element along a line of constant Y. There is a large change in this component, however, at the common nodes of adjacent elements. This is always an indication of the need for further mesh refinement.

The other stress components, as well, indicate a large change from one element to another at common boundary locations. Except for the element center, the magnitudes of the σ_Y and τ_{XY} stress components are far from correct; they should be zero on the boundary of the beam. The stress vector at the center gives a reasonably good estimate of the correct stress in the beam. One more example will be used to illustrate the accuracy obtainable at this location in the element.

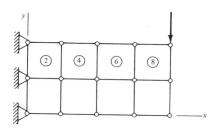

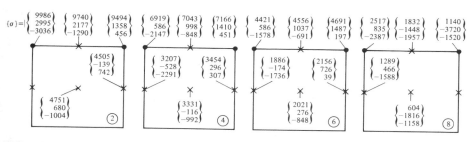

FIGURE 7.26 Stress vector at selected points on boundary and at the center of adjacent elements in the eight-element model.

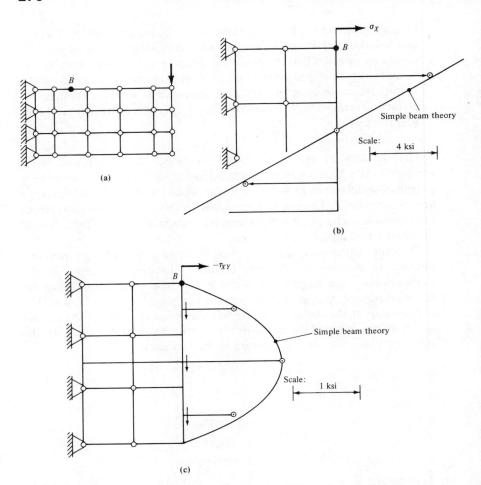

FIGURE 7.27 Fifteen-element model and plots of flexure and shear stresses.

The final model is shown in Figure 7.27(a). The model and the resulting flexure and shear stresses are plotted on the cross section containing point B. All values were calculated at the center of the elements; no averaging was involved. Note in Figure 7.27 the amazingly good agreement with theoretical values calculated with formulas for simple (long and slender) beams. This is discussed further in Chapter 8.

7.10 TRIANGLE SUBASSEMBLY AND CONDENSATION

Before leaving the discussion of the quadrilateral element, it would perhaps be of interest to discuss briefly an alternative method of constructing an element of this shape. Shown in Figure 7.28 is a quadrilateral

7.10 TRIANGLE SUBASSEMBLY AND CONDENSATION

made by assembling four three-node triangles. Although the boundary looks like a quadrilateral, we in fact have four constant-strain triangles. The characteristic of the strain throughout this quadrilateral is not the same as that of the isoparametric four-node element.

The extra node inside the element, now viewed as a quadrilateral, is undesirable because it increases the number of equations to be solved by two per element. The equations associated with the extra node inside the element can be removed, however, by a process referred to as *condensation*.

For simplicity, let us assume that nodal forces are to be applied only to the quadrilateral corner nodes; no externally applied force will be applied to the inner node. We have the following equation derived from the assemblage of the four triangle elements:

$$[K]\begin{Bmatrix} u_1 \\ v_1 \\ \vdots \\ u_5 \\ v_5 \end{Bmatrix} = \begin{Bmatrix} R_{1X} \\ R_{1Y} \\ \vdots \\ R_{5X} = 0 \\ R_{5Y} = 0 \end{Bmatrix}$$

The next step is to partition the matrix equation as follows:

$$\begin{bmatrix} K_{11} & | & K_{12} \\ \hline K_{21} & | & K_{22} \end{bmatrix} \begin{Bmatrix} r_1 \\ r_2 \end{Bmatrix} = \begin{Bmatrix} R_1 \\ R_2 \end{Bmatrix}$$

where

$[K_{11}]$ is an 8×8 matrix

$[K_{12}]$ is an 8×2 matrix

$[K_{21}]$ is an 2×8 matrix

$[K_{22}]$ is an 2×2 matrix

$\{r_1\} = \begin{bmatrix} u_1 & v_1 & \cdots & u_4 & v_4 \end{bmatrix}^T \qquad \{r_2\} = \begin{bmatrix} u_5 & v_5 \end{bmatrix}^T$

$\{R_1\} = \begin{bmatrix} R_{1X} & R_{1Y} & \cdots & R_{4X} & R_{4Y} \end{bmatrix}^T$

$\{R_2\} = \begin{bmatrix} 0 & 0 \end{bmatrix}^T$

From the partitioned matrix equation, we can write two equations:

$$[K_{11}]\{r_1\} + [K_{12}]\{r_2\} = \{R_1\} \qquad (7.26)$$

$$[K_{21}]\{r_1\} + [K_{22}]\{r_2\} = \begin{Bmatrix} 0 \\ 0 \end{Bmatrix} \qquad (7.27)$$

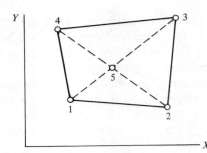

FIGURE 7.28 Quadrilateral made by assembling four triangles.

If we solve for $\{r_2\}$, with (7.27), we obtain

$$\{r_2\} = -[K_{22}]^{-1}[K_{21}]\{r_1\}$$

and substituting this into matrix equation (7.26) yields the following:

$$[K_{11}]\{r_1\} - [K_{12}][K_{22}]^{-1}[K_{21}]\{r_1\} = \{R_1\}$$

We thus have the condensed stiffness matrix for the quadrilateral:

$$[K_c] = [K_{11}] - [K_{12}][K_{22}]^{-1}[K_{21}] \qquad (7.28)$$

Quadrilaterals with this stiffness matrix will yield displacements identical to that which would be obtained with the triangles. The payoff is the reduction in the size of the matrix equation.

PROBLEMS

7.1 Calculate the value of the Jacobian determinant at the origin of the natural coordinates and at node 3 of the quadrilateral shown in Figure P7.1.

7.2 Write a computer program to calculate the Jacobian determinant at any point in a quadrilateral element.

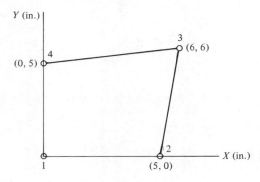

FIGURE P7.1 Quadrilateral.

PROBLEMS

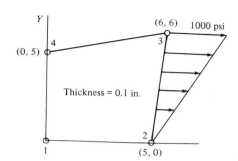

FIGURE P7.7 Linearly distributed stress traction.

7.3 Determine the $[B]$ matrix for nodes 1 and 3 of the quadrilateral shown in Figure P7.1.

7.4 Write a computer program to determine the $[B]$ matrix at any point in a quadrilateral element.

7.5 Write a computer program to determine the stiffness matrix of a quadrilateral element.

7.6 Verify, by calculating, the strain and stress values at $s = t = 0$ given at the end of Example 7.4.

7.7 The traction nodal force vector $\{Q\}_T$ can be obtained by using the derived formulas (7.23) or (7.24). Alternatively, the vector $\{Q\}_T$ can be obtained by numerically integrating (7.18).

For the linearly distributed loading in Figure P7.7, derive the traction nodal force vector $\{Q\}_T$ numerically using the two-point quadrature formula.

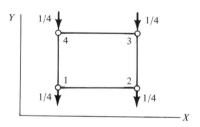

(a) Fraction of weight to each node for rectangle.

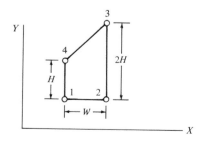

FIGURE P7.8   (b) Quadrilateral.

Compare the result with the value obtained using formula (7.24).
Hint: From (7.18),

$$\{Q\}_T = h\int_S [N]^T \begin{Bmatrix} T_X \\ 0 \end{Bmatrix} dl$$

$$= \frac{hL}{2} \int_{-1}^{1} [N]^T \begin{Bmatrix} T_X \\ 0 \end{Bmatrix} dt$$

Use the two-point quadrature formula to evaluate each component.

7.8 Assuming that gravity acts in the negative Y direction, the body force vector $\{Q\}_{BF}$ for a *rectangle* of uniform thickness and density consists of an even allocation of the weight to each node of the element, as shown in Figure P7.8(a). If the quadrilateral is not a rectangle, the distribution to each node not be equal.

Solve for the nodal body force component to be applied to node 1 for the quadrilateral shown in Figure P7.8(b).

Hint: From (6.11),

$$\{Q\}_{BF} = h\int_A [N]^T \begin{Bmatrix} 0 \\ -\rho \end{Bmatrix} dX\,dY$$

$$= h\int_{A^*} [N]^T \begin{Bmatrix} 0 \\ -\rho \end{Bmatrix} |J|\,ds\,dt$$

Evaluate the component using the two-point quadrature formula.

7.9 Project. Develop a stress analysis computer program that employs the quadrilateral element.

CHAPTER 8

Higher-Order Elements and Additional Capabilities

8.1 ISOPARAMETRIC SIX-NODE TRIANGLE STIFFNESS MATRIX

A triangle with corner and midside nodes is shown in Figure 8.1(a) as it might appear in the global XY plane. The sides of the triangle can be curved or straight. In Figure 8.1(b) the triangle is shown as it is defined in terms of the natural triangular coordinates. The sides are always straight when referenced to the natural coordinate system.

For the isoparametric element, displacement and position coordinates are defined with interpolation formulas that have common coefficients.

$$u = \sum_{i=1}^{6} N_i u_i \quad v = \sum_{i=1}^{6} N_i v_i$$

$$X = \sum_{i=1}^{6} N_i X_i \quad Y = \sum_{i=1}^{6} N_i Y_i \quad (8.1)$$

where

$$N_1 = L_1(2L_1 - 1) \quad N_2 = L_2(2L_2 - 1) \quad N_3 = L_3(2L_3 - 1)$$
$$N_4 = 4L_1 L_2 \quad N_5 = 4L_2 L_3 \quad N_6 = 4L_1 L_3 \quad (8.2)$$

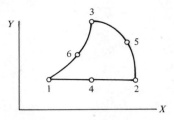

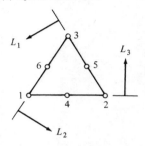

(a) Spatial coordinates.

(b) Natural coordinates.   **FIGURE 8.1** Isoparametric triangle.

and it should be remembered that $L_1 + L_2 + L_3 = 1$. Treating L_1 and L_2 as the independent variables, $L_3 = 1 - L_1 - L_2$. Substituting into (8.2) gives us

$$N_1 = L_1(2L_1 - 1) \qquad N_2 = L_2(2L_2 - 1)$$
$$N_3 = L_1(2L_1 - 3) + L_2(2L_2 - 3) + 4L_1L_2 + 1$$
$$N_4 = 4L_1L_2 \qquad (8.3)$$
$$N_5 = 4L_2(1 - L_2) - 4L_2L_1 \qquad N_6 = 4L_1(1 - L_1) - 4L_2L_1$$

Now that we have defined the interpolation coefficients for the displacement and shape functions, the next step will be to define the strain components. Following the procedures of Section 6.2 for the three-node triangle, we proceed to obtain the matrix $[B]$.

$$\epsilon_X = \frac{\partial u}{\partial X} = \frac{1}{|J|}\left[\frac{\partial Y}{\partial L_2}\frac{\partial u}{\partial L_1} - \frac{\partial Y}{\partial L_1}\frac{\partial u}{\partial L_2}\right] \qquad (8.4a)$$

$$\epsilon_Y = \frac{\partial v}{\partial Y} = \frac{-1}{|J|}\left[\frac{\partial X}{\partial L_2}\frac{\partial v}{\partial L_1} - \frac{\partial X}{\partial L_1}\frac{\partial v}{\partial L_2}\right] \qquad (8.4b)$$

$$\gamma_{XY} = \frac{\partial u}{\partial Y} + \frac{\partial v}{\partial X}$$
$$= \frac{-1}{|J|}\left[\frac{\partial X}{\partial L_2}\frac{\partial u}{\partial L_1} - \frac{\partial X}{\partial L_1}\frac{\partial u}{\partial L_2} + \frac{\partial Y}{\partial L_2}\frac{\partial v}{\partial L_1} - \frac{\partial Y}{\partial L_1}\frac{\partial v}{\partial L_2}\right] \qquad (8.4c)$$

$$|J| = \frac{\partial Y}{\partial L_2}\frac{\partial X}{\partial L_1} - \frac{\partial Y}{\partial L_1}\frac{\partial X}{\partial L_2} \qquad (8.4d)$$

8.1 ISOPARAMETRIC SIX-NODE TRIANGLE STIFFNESS MATRIX

From equations (8.1),

$$\frac{\partial u}{\partial L_1} = \sum_{i=1}^{6} \frac{\partial N_i}{\partial L_1} u_i \qquad \frac{\partial u}{\partial L_2} = \sum_{i=1}^{6} \frac{\partial N_i}{\partial L_2} u_i \qquad \text{etc.}$$

Thus we can obtain the following:

$$\epsilon_X = \frac{1}{|J|} \left(\sum_{i=1}^{6} \frac{\partial N_i}{\partial L_2} Y_i \sum_{i=1}^{6} \frac{\partial N_i}{\partial L_1} u_i - \sum_{i=1}^{6} \frac{\partial N_i}{\partial L_1} Y_i \sum_{i=1}^{6} \frac{\partial N_i}{\partial L_2} u_i \right)$$

$$= \frac{1}{|J|} \sum_{i=1}^{6} \sum_{j=1}^{6} \left[Y_i \frac{\partial N_i}{\partial L_2} \frac{\partial N_j}{\partial L_1} u_j - Y_i \frac{\partial N_i}{\partial L_1} \frac{\partial N_j}{\partial L_2} u_j \right]$$

or, in matrix form:

$$\epsilon_X = \frac{1}{|J|} [Y_1 \quad Y_2 \quad \cdots \quad Y_6][b] \begin{Bmatrix} u_1 \\ u_2 \\ \vdots \\ u_6 \end{Bmatrix} \qquad (8.5)$$

where

$$b_{ij} = \frac{\partial N_i}{\partial L_2} \frac{\partial N_j}{\partial L_1} - \frac{\partial N_i}{\partial L_1} \frac{\partial N_j}{\partial L_2} \qquad (8.6)$$

Similarly, we can derive the following:

$$\epsilon_Y = \frac{-1}{|J|} [X_1 \quad X_2 \quad \cdots \quad X_6][b] \begin{Bmatrix} v_1 \\ v_2 \\ \vdots \\ v_6 \end{Bmatrix} \qquad (8.7)$$

and

$$\gamma_{XY} = \frac{-1}{|J|} [X_1 \quad X_2 \quad \cdots \quad X_6][b] \begin{Bmatrix} u_1 \\ u_2 \\ \vdots \\ u_6 \end{Bmatrix}$$

$$+ \frac{1}{|J|} [Y_1 \quad Y_2 \quad \cdots \quad Y_6][b] \begin{Bmatrix} v_1 \\ v_2 \\ \vdots \\ v_6 \end{Bmatrix} \qquad (8.8)$$

We can now define the strain vector in matrix form:

$$\begin{Bmatrix} \epsilon_X \\ \epsilon_Y \\ \gamma_{XY} \end{Bmatrix} = [B] \begin{Bmatrix} u_1 \\ v_1 \\ u_2 \\ \vdots \\ u_6 \\ v_6 \end{Bmatrix} = [B]\{q\} \qquad (8.9)$$

The elements of the $[B]$ matrix are

$$B_{1,(2j-1)} = \frac{1}{|J|} \sum_i^6 Y_i b_{ij} \qquad j = 1, 2, \ldots, 6 \qquad (8.10a)$$

$$B_{2,(2j)} = \frac{-1}{|J|} \sum_{i=1}^6 X_i b_{ij} \qquad j = 1, 2, \ldots, 6 \qquad (8.10b)$$

$$B_{3,(2j-1)} = B_{2,(2j)} \qquad j = 1, 2, \ldots, 6 \qquad (8.10c)$$

$$B_{3,(2j)} = B_{1,(2j-1)} \qquad j = 1, 2, \ldots, 6 \qquad (8.10d)$$

$$|J| = -\begin{bmatrix} X_1 & X_2 & \cdots & X_6 \end{bmatrix} [b] \begin{Bmatrix} Y_1 \\ \vdots \\ Y_6 \end{Bmatrix} \qquad (8.10e)$$

When the b_{ij} values are determined, it is seen that they are functions of the natural coordinates. Refer to Appendix 5 for the definition of the b_{ij} terms as functions of the derivatives of the shape functions, N_i.

The next step is the determination of the stiffness matrix. As with the quadrilateral element, this will be determined using numerical integra-

TABLE 8.1 Quadrature Formulas for Triangular Areas

$$I = \int_A f(L_1, L_2) \, dA = \sum_i W_i f_i + e$$

Order	Sampling Points (triangular coordinates)		Weighting Factor, W_i
Linear		(a) $\frac{1}{3}, \frac{1}{3}, \frac{1}{3}$	$\frac{1}{2}$
Quadratic		(a) $\frac{1}{2}, \frac{1}{2}, 0$	$\frac{1}{6}$
		(b) $0, \frac{1}{2}, \frac{1}{2}$	$\frac{1}{6}$
		(c) $\frac{1}{2}, 0, \frac{1}{2}$	$\frac{1}{6}$

tion. The stiffness matrix is defined as

$$[k] = h \int [B]^T [C][B]|J|\, dL_1\, dL_2$$

We will perform the integration using Gauss quadrature over a triangular region. The sampling points at which the matrix product is determined and the corresponding weighting factors are given in the Table 8.1. Experience has shown that the quadratic formula gives good results for the six-node triangle.

$$[k] = h \sum_{j=a}^{c} W_j \left[B(L_{1_j}, L_{2_j}) \right]^T [C] \left[B(L_{1_j}, L_{2_j}) \right] |J(L_{1_j}, L_{2_j})|$$

(8.11)

8.2 EQUIVALENT NODAL FORCES: SURFACE TRACTION

Distributed surface tractions are replaced with equivalent nodal forces according to the now very familiar relationship

$$\{Q\}_T = h \int_S [N]^T \begin{Bmatrix} T_X \\ T_Y \end{Bmatrix} ds \qquad (6.10)$$

We will use the integration formula for the triangular region and natural coordinates:

$$\int_S L_i^\alpha L_j^\beta\, ds = \frac{\alpha!\,\beta!}{(\alpha + \beta + 1)!} S \qquad (6.13)$$

where S is the length of the side.

Example 8.1 The six-node triangle shown in Figure 8.2 has a uniformly distributed surface traction along side 1 ($L_1 = 0$). The traction is in the global X direction. Solve for the equivalent nodal forces.

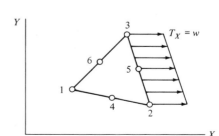

FIGURE 8.2 Uniform traction on side 1.

Solution. For $L_1 = 0$, $N_1 = 0 = N_4 = N_6$. Substituting into (6.10), we have

$$\{Q\}_T = h\int_S \begin{bmatrix} 0 & 0 \\ 0 & 0 \\ N_2 & 0 \\ 0 & N_2 \\ N_3 & 0 \\ 0 & N_3 \\ 0 & 0 \\ 0 & 0 \\ N_5 & 0 \\ 0 & N_5 \\ 0 & 0 \\ 0 & 0 \end{bmatrix} \begin{Bmatrix} T_X \\ 0 \end{Bmatrix} ds = h\int_S \begin{Bmatrix} 0 \\ 0 \\ N_2 T_X \\ 0 \\ N_3 T_X \\ 0 \\ 0 \\ 0 \\ N_5 T_X \\ 0 \\ 0 \\ 0 \end{Bmatrix} ds$$

integrating each term gives us

$$\int_S N_2 T_X \, ds = \int_S (2L_2^2 - L_2) w \, ds = w\left(\int_S 2L_2 \, ds - \int_S L_2 \, ds\right)$$

$$= w\left(2\frac{2!}{3!}S - \frac{1!}{2!}S\right) \quad \text{(where } S \text{ is the length of side 1)}$$

$$= wS\left(\frac{2}{3} - \frac{1}{2}\right)$$

$$h\int_S N_2 T_X \, ds = \frac{wSh}{6} = F_{2X}$$

Performing the integration of the other two terms yields

$$h\int_S N_3 T_X \, ds = \frac{wSh}{6} = F_{3X}$$

$$h\int_S N_5 T_X \, ds = \frac{2wSh}{3} = F_{5X}$$

The equivalent forces are shown in Figure 8.3(a). Performing a similar derivation for a general linear traction, we would have the equivalent nodal forces as indicated in Figure 8.3(b).

The equivalent nodal forces for any traction up through a quadratic distribution may be determined conveniently with the following matrix product:

$$\begin{Bmatrix} F_a \\ F_b \\ F_c \end{Bmatrix} = \frac{hS}{30} \begin{bmatrix} 4 & 2 & -1 \\ 2 & 16 & 2 \\ -1 & 2 & 4 \end{bmatrix} \begin{Bmatrix} T_a \\ T_b \\ T_c \end{Bmatrix} \quad (8.12)$$

The symbols are defined in Figure 8.4. □

8.2 EQUIVALENT NODAL FORCES: SURFACE TRACTION

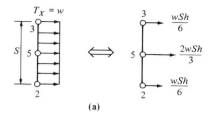

(a)

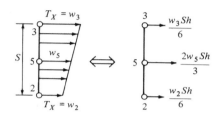

(b)

FIGURE 8.3 Equivalent nodal forces of side 1 of element of thickness h.

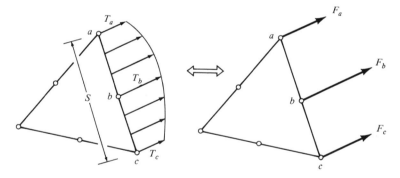

FIGURE 8.4 Equivalent nodal forces for surface traction distributed on side of a six-node isoparametric triangle of thickness h.

Example 8.2 Repeat the problem used in Examples 6.8 and 7.11 of the square plate with the stress traction couple loading. Use two six-node triangles with the node numbering and nodal forces, from (8.12), as shown in Figure 8.5.

Solve for the stress components at the nodes of each triangle. Compare the results with the 16-element model of Example 6.8.

Solution. The numerical analysis of this problem was performed with the commercially available structures and finite element program

288 HIGHER-ORDER ELEMENTS AND ADDITIONAL CAPABILITIES

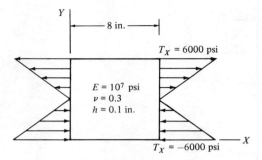

(a) Square plate with couple loading.

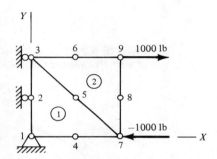

(b) Element and node numbering, boundary constraints and node forces.

FIGURE 8.5 Equivalent nodal forces and node constraints.

called ICES STRUDL-II (Integrated Civil Engineering Systems Structural Design Language). A sample listing of the required program input and resulting output can be found in Appendix 4.

For this particular problem, the stress results are as follows:

ELEMENT	NODE	SIGMAX	SIGMAY	TAUXY
1	1	-0.6000E 4	-0.2686E-2	0.4478E-3
1	7	-0.6000E 4	-0.7324E-3	0.1074E-1
1	3	0.6000E 4	-0.1221E-2	-0.3134E-2
2	3	0.6000E 4	-0.2197E-2	-0.4478E-3
2	7	-0.6000E 4	-0.7080E-2	-0.3582E-2
2	9	0.6000E 4	-0.1196E-1	0.0000E 0

A comparison of these results with the results for Example 6.8 reveals the superiority of the six-node triangle over the three-node triangle. Here with two six-node elements and a total of nine-nodes, we have much better results than obtained with the Example 6.8 model having 16 three-node elements and 13 nodes. ☐

8.2 EQUIVALENT NODAL FORCES: SURFACE TRACTION

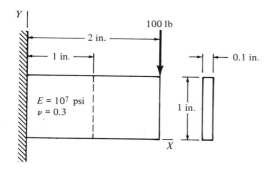

FIGURE 8.6 Section of interest in the short cantilever beam.

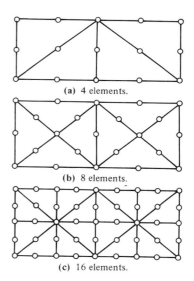

FIGURE 8.7 Six-node triangle models of the beam.

Example 8.3 Referring once again to the short-beam problem, investigate the stress distribution along the cross section at $X = 1$ in Figure 8.6 as predicted by simple models of six-node triangular elements.

Solution. Three different element mesh patterns were selected as shown in Figure 8.7. In Figure 8.8, the flexure and shear stress results for each model are compared to the simple beam theory prediction. The finite element values were arrived at by averaging the element values at common nodes.

We observe that the flexure stress agrees very well with the beam theory for all models. The shear stress, however, is considerably

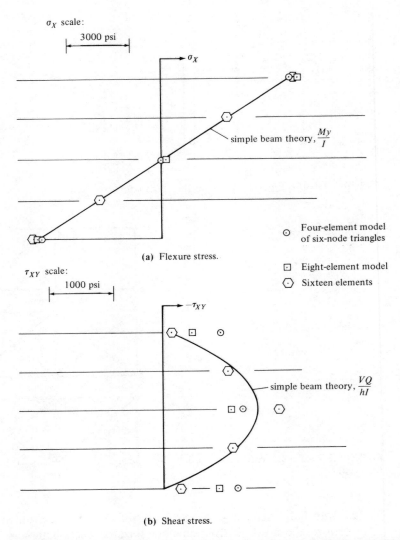

FIGURE 8.8 Comparison of results of the finite element models with the simple beam theory at section $X = 1$. All values are nodal averages of element values at common nodes.

altered with each model, and further refinement is required to achieve an acceptable result.

The ability of the six-node element to model a strain gradient is indicated in Figure 8.9. Compare these results with those of the four-node quadrilateral in Figure 7.26. The gradient of the flexure stress in each element along the length of the beam is greatly

8.3 ISOPARAMETRIC EIGHT-NODE QUADRILATERAL STIFFNESS MATRIX

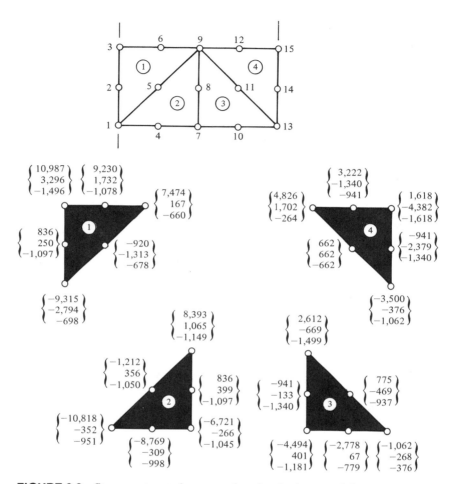

FIGURE 8.9 Stress vector at element nodes of each element of the four-element model.

improved. The large differences noted at common nodes of adjacent elements is, again, an indication of the need for further refinement. □

8.3 ISOPARAMETRIC EIGHT-NODE QUADRILATERAL STIFFNESS MATRIX

The quadrilateral with the local node numbering convention is shown in Figure 8.10 in real XY space and also in terms of the natural coordinates.

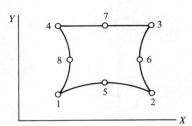

(a) Spatial coordinates.

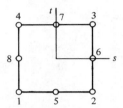

(b) Natural coordinates.   **FIGURE 8.10** Isoparametric quadrilateral.

The interpolation formulas for the displacement and spatial coordinates are

$$u = \sum_{i=1}^{8} N_i u_i \qquad v = \sum_{i=1}^{8} N_i v_i$$
$$X = \sum_{i=1}^{8} N_i X_i \qquad Y = \sum_{i=1}^{8} N_i Y_i \qquad (8.13)$$

where

$$N_1 = -\tfrac{1}{4}(1-s)(1-t)(1+s+t) \qquad N_2 = -\tfrac{1}{4}(1+s)(1-t)(1-s+t)$$
$$N_3 = -\tfrac{1}{4}(1+s)(1+t)(1-s-t) \qquad N_4 = -\tfrac{1}{4}(1-s)(1+t)(1+s-t)$$
$$N_5 = \tfrac{1}{2}(1-s^2)(1-t) \qquad\qquad N_6 = \tfrac{1}{2}(1+s)(1-t^2) \qquad (8.14)$$
$$N_7 = \tfrac{1}{2}(1-s^2)(1+t) \qquad\qquad N_8 = \tfrac{1}{2}(1-s)(1-t^2)$$

Following the same procedures of Section 7.2 for the four-node isoparametric quadrilateral element development, we now define strain and develop the matrix $[B]$. Using the equations (7.4) and (7.4a), we can define the strain as follows:

$$\epsilon_X = \frac{\partial u}{\partial X} = \frac{1}{|J|}\left[\frac{\partial Y}{\partial t}\frac{\partial u}{\partial s} - \frac{\partial Y}{\partial s}\frac{\partial u}{\partial t}\right] \qquad (8.15a)$$

$$\epsilon_Y = \frac{\partial v}{\partial Y} = \frac{-1}{|J|}\left[\frac{\partial X}{\partial t}\frac{\partial v}{\partial s} - \frac{\partial X}{\partial s}\frac{\partial v}{\partial t}\right] \qquad (8.15b)$$

$$\gamma_{XY} = \frac{\partial u}{\partial Y} + \frac{\partial v}{\partial X}$$
$$= \frac{-1}{|J|}\left[\frac{\partial X}{\partial t}\frac{\partial u}{\partial s} - \frac{\partial X}{\partial s}\frac{\partial u}{\partial t}\right] + \frac{1}{|J|}\left[\frac{\partial Y}{\partial t}\frac{\partial v}{\partial s} - \frac{\partial Y}{\partial s}\frac{\partial v}{\partial t}\right] \qquad (8.15c)$$

8.3 ISOPARAMETRIC EIGHT-NODE QUADRILATERAL STIFFNESS MATRIX

The partial derivatives of coordinates and displacements are obtained from the interpolation formulas, equations (8.13). For example,

$$\frac{\partial u}{\partial s} = \sum_{i=1}^{8} \frac{\partial N_i}{\partial s} u_i \qquad \frac{\partial X}{\partial t} = \sum_{i=1}^{8} \frac{\partial N_i}{\partial t} X_i \qquad \text{etc.}$$

The expressions for the strains can take the following forms:

$$\epsilon_X = \frac{1}{|J|} [Y_1 \quad Y_2 \quad \cdots \quad Y_8][g] \begin{Bmatrix} u_1 \\ u_2 \\ \vdots \\ u_8 \end{Bmatrix} \qquad (8.16)$$

$$\epsilon_Y = \frac{-1}{|J|} [X_1 \quad X_2 \quad \cdots \quad X_8][g] \begin{Bmatrix} v_1 \\ v_2 \\ \vdots \\ v_8 \end{Bmatrix} \qquad (8.17)$$

$$\gamma_{XY} = \frac{-1}{|J|} [X_1 \quad X_2 \quad \cdots \quad X_8][g] \begin{Bmatrix} u_1 \\ u_2 \\ \vdots \\ u_8 \end{Bmatrix}$$

$$+ \frac{1}{|J|} [Y_1 \quad Y_2 \quad \cdots \quad Y_8][g] \begin{Bmatrix} v_1 \\ v_2 \\ \vdots \\ v_8 \end{Bmatrix} \qquad (8.18)$$

where

$$|J| = -[X_1 \quad X_2 \quad \cdots \quad X_8][g] \begin{Bmatrix} Y_1 \\ Y_2 \\ \vdots \\ Y_8 \end{Bmatrix} \qquad (8.19)$$

and

$$g_{ij} = \frac{\partial N_i}{\partial t} \frac{\partial N_j}{\partial s} - \frac{\partial N_i}{\partial s} \frac{\partial N_j}{\partial t} \qquad (8.20)$$

(See Appendix 5 for derivatives of N_i.)

We can now define the strain in matrix form:

$$\begin{Bmatrix} \epsilon_X \\ \epsilon_Y \\ \gamma_{XY} \end{Bmatrix} = [B] \begin{Bmatrix} u_1 \\ v_1 \\ u_2 \\ v_2 \\ \vdots \\ u_8 \\ v_8 \end{Bmatrix} \qquad (8.21)$$

The elements of the matrix $[B]$ are

$$B_{1,(2j-1)} = \frac{1}{|J|} \sum_{i=1}^{8} Y_i g_{ij} \qquad j = 1, 2, \ldots, 8$$

$$B_{2,(2j)} = \frac{1}{|J|} \sum_{i=1}^{8} X_i g_{ij} \qquad j = 1, 2, \ldots, 8$$

$$B_{3,(2j-1)} = B_{2,(2j)} \qquad j = 1, 2, \ldots, 8$$

$$B_{3,(2j)} = B_{1,(2j-1)} \qquad j = 1, 2, \ldots, 8$$

Refer to Appendix 5 for derivatives of the shape functions.

The entries in the $[B]$ matrix are recognized to be functions of the natural coordinates s and t. In the determination of the stiffness matrix, a numerical procedure will be used to evaluate the integral:

$$[k] = h \int [B]^T [C][B]|J|\, ds\, dt$$

The numerical integration of the four-node quadrilateral employed the two-point Gauss quadrature formulas. The eight-node quadrilateral could employ the two-point formula, but the higher order of the element suggests that the three-point formula would be advisable (Bathe and Wilson, 1976). Calculation experience has shown significant difference in results between the two-point and the three-point formulas for this element, but negligible difference between the three-point and four-point formulas. The coordinates and weighting factors for the two and three-point formulas are summarized in Table 8.2.

$$[k] \approx h \sum_{i=1}^{3} \sum_{j=1}^{3} \left\{ W_i W_j [B(s_i, t_j)]^T [C][B(s_i, t_j)]|J(s_i, t_j)| \right\} \qquad (8.22)$$

TABLE 8.2 Quadrature Formulas for the Unit Square

$\int f(s,t)\, ds\, dt = \sum_i \sum_j W_i W_j f(s_i, t_j) + e$			
	Sampling Point Location		
Formula	i	s_i, t_i	Weighting Factor, W_i
Two-point	1	-0.577350	1.0
	2	0.577350	1.0
Three-point	1	-0.774597	0.555556
	2	0	0.888889
	3	0.774597	0.555556

8.4 EQUIVALENT NODAL FORCES

The equivalent nodal forces for distributed edge tractions can be determined through explicit evaluation of the integral in equation (6.10):

$$\{Q\}_T = h \int_S [N]^T \begin{Bmatrix} T_X \\ T_Y \end{Bmatrix} ds \qquad (6.10)$$

If we consider a quadratic stress traction distribution along the side $s = 1$ [see Figure 8.10(b)], for example:

$$T_X = C_0 + C_1 t + C_2 t^2$$
$$= T_6 + \frac{T_3 - T_2}{2} t + \frac{T_3 - 2T_6 + T_2}{2} t^2$$

Substitution into

$$\{Q\}_T = h \int_{-1}^{1} [N]^T \begin{Bmatrix} T_X \\ 0 \end{Bmatrix} \frac{S}{2} dt$$

and integrating yields the following:

$$\{Q\}_T = \frac{hS}{30} \begin{Bmatrix} 0 \\ 4T_2 + 2T_6 - T_3 \\ -T_2 + 2T_6 + 4T_3 \\ 0 \\ 0 \\ 2T_2 + 16T_6 + 2T_3 \\ 0 \\ 0 \end{Bmatrix}$$

Placing this into a more convenient form and generalizing to account for stress traction in the X or Y direction on any face of the element, we arrive at the same formula presented for the six-node triangle:

$$\begin{Bmatrix} F_a \\ F_b \\ F_c \end{Bmatrix} = \frac{hS}{30} \begin{bmatrix} 4 & 2 & -1 \\ 2 & 16 & 2 \\ -1 & 2 & 4 \end{bmatrix} \begin{Bmatrix} T_a \\ T_b \\ T_c \end{Bmatrix} \qquad (8.12)$$

where the symbols are defined in Figure 8.11.

The equivalent nodal forces for body force distributions can be determined by integrating (6.11), but perhaps an appealing alternative would be numerical integration. We would thus evaluate the following:

$$\{Q\}_{BF} = h \int_A [N]^T \begin{Bmatrix} B_X \\ B_Y \end{Bmatrix} dA$$

$$= h \sum_{i=1}^{3} \sum_{j=1}^{3} \left\{ W_i W_j [N(s_i, t_j)]^T \begin{Bmatrix} B_X(s_i, t_j) \\ B_Y(s_i, t_j) \end{Bmatrix} |J(s_i, t_j)| \right\}$$

$$(8.23)$$

296   HIGHER-ORDER ELEMENTS AND ADDITIONAL CAPABILITIES

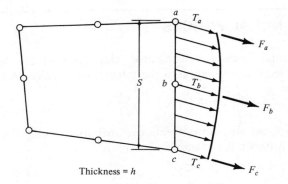

FIGURE 8.11 Equivalent nodal forces and nodal traction values.

Example 8.4 Once again, use the square plate with couple loading problem, and investigate the stress distribution in a single eight-node quadrilateral element. Compare the results with the four-node quadrilateral results of Examples 7.11 and 7.12.

Solution. The distributed stress traction is replaced with equivalent nodal forces using the formula (8.12), and the node numbering and boundary constraints are defined in Figure 8.12.

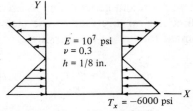

(a) Square plate with couple loading.

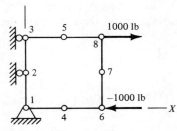

(b) Node numbering of the element.

FIGURE 8.12 Equivalent nodal forces and boundary constraints.

8.4 EQUIVALENT NODAL FORCES

The calculations are performed with the STRUDL program, and the results are as follows:

ELEMENT	NODE	SIGMAX	SIGMAY	TAUXY
1	1	-0.6000E 4	0.1978E-1	0.9851E-2
	6	-0.6000E 4	-0.5859E-2	0.3369E-2
	8	0.6000E 4	0.2930E-2	0.6268E-2
	3	0.6000E 4	0.1074E-1	-0.4478E-2
	4	-0.6000E 4	-0.1709E-2	0.6268E-2
	7	-0.5117E-2	-0.1535E-2	0.1903E-2
	5	0.6000E 4	0.7324E-3	0.8955E-3
	2	0.7157E-2	0.1610E-1	0.2798E-2

A comparison of these results with the results for Example 7.11 reveals the superiority of the eight-node quadrilateral over the four-node quadrilateral. The error in all stress components is negligible for the eight-node element. This single eight-node-element model also gives much better results than the four four-node-element model of Example 7.12. This element, like the six-node triangle, proves to be a much more efficient element than the corresponding linear displacement element. ☐

Example 8.5 Using the short-beam problem again, investigate the stress distribution along the cross section at $X = 1$ in Figure 8.13 as predicted by simple models of eight-node quadrilateral elements.

Solution. We will employ two models, a two-element and a four-element model, as shown in Figure 8.14. The analysis was performed with the STRUDL program and the node average results are plotted in Figure 8.15.

Note that both models give a very good correlation of the flexure stress with the simple beam theory; the shear stress results are significantly different between models and the theory. The four-element model results, however, begin to approach the correct value of zero at the beam top and bottom surface. Further refinement is called for

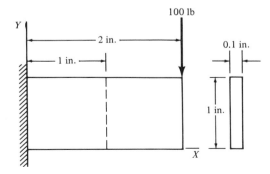

FIGURE 8.13 Section of interest in cantilever beam.

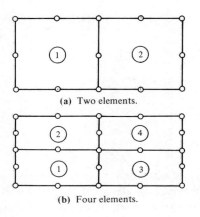

FIGURE 8.14 Eight-node quadrilateral models.

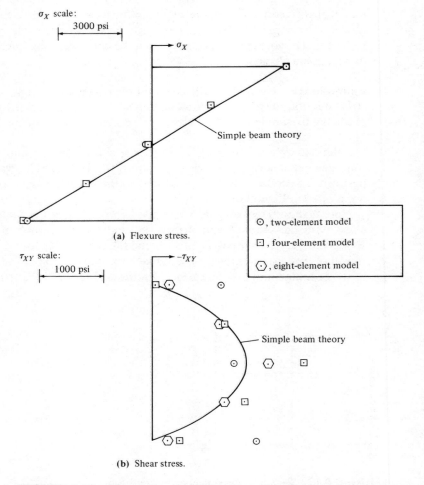

FIGURE 8.15 Comparison of results of models with simple beam theory at section $X = 1$. All values are nodal average values.

8.5 USING HIGHER-ORDER ELEMENTS

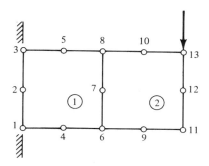

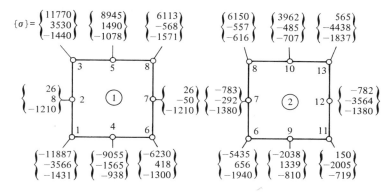

FIGURE 8.16 Stress vector at the nodes of each element in the two-element model of the cantilever beam.

to obtain good results; values for an eight-element model are also shown.

Figure 8.16 is presented to show how the stress vector varies from node to node in an element and to indicate the relatively small difference in magnitudes of the stress components, with the possible exception of shear stress, between elements at a common node. Compare these results with those shown in Figure 7.26 for the four-node quadrilateral; the eight-node element provides a much greater variation in the X and Y directions. ☐

8.5 USING HIGHER-ORDER ELEMENTS

Regardless of the element used, the basic force–displacement equation is

$$[K]\{r\} = \{R\}_{\text{NF}} + \{R\}_T + \{R\}_{\text{BF}} \qquad (4.25)$$

The procedures for assembling the global stiffness matrix and the force vectors should not require reiteration. One point that should be discussed, however, is the possibility of assembling the equation with a mixture of types of elements.

When using more than one type of element, care must be exercised in maintaining compatibility at the boundaries of the elements. The linear elements, the three-node triangle and the four-node quadrilateral, may be joined because their sides remain straight during a deformation. Similarly, the quadratic elements, the six-node triangle and the eight-node quadrilateral, may be joined because their sides conform to a quadratic displacement function. On the other hand, the linear elements must not be joined with the quadratic elements unless steps are taken to account for the lack of compatibility.

It is possible to formulate a mesh that has a mix of elements of varying order. However, the procedure will not be covered in this book. The reader is referred to Cook (1974) for a discussion on the "degrading" of elements.

8.6 CONVERGENCE COMPARISONS OF FOUR ELEMENTS: AN EXAMPLE

In Chapters 6 through 8, we have developed the finite element solution of the two-dimensional elasticity problem using four different elements, the three- and six-node triangles and the four- and eight-node quadrilaterals. Examples in each chapter have been used to point out the solution behavior of each element for particular problems, and it would be worthwhile to make a final comparison of the effectiveness of each element in the solution of a problem. As a problem for comparison, we will use the short beam which has been used repeatedly in these chapters.

The problem is to find the deflection of point A and the stress at point B, the two points identified in Figure 8.17.

The results of the analysis using the element models from Examples 6.9, 7.14, 8.3, and 8.5 are summarized in Figures 8.18 and 8.19.

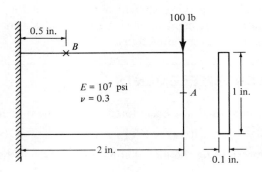

FIGURE 8.17 Short cantilever beam used for convergence comparisons.

8.6 CONVERGENCE COMPARISONS OF FOUR ELEMENTS: AN EXAMPLE

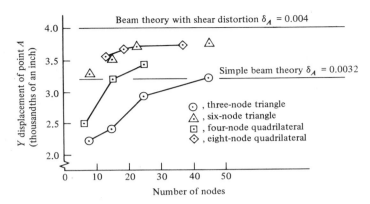

FIGURE 8.18 Convergence of the Y-displacement of point A for four different element types.

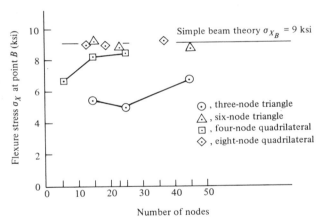

FIGURE 8.19 Convergence of the flexure stress, σ_X, at point B for four different element types. Values are nodal averages for three- and four-node elements; nodal values for six- and eight-node elements.

The graphs indicate very well the improved modeling capability of the higher-order elements. The six-node triangle and eight-node quadrilateral converge very quickly to the theoretical stress at point B. The deflection at point A is within 7% of the theoretical prediction for the eight-node quadrilateral. The plot shows a very slow rate of convergence from this point on, but with further refinement, a higher value could be obtained.

In general, but not always, the convergence will be monotonic if for one type of element, the succeeding more refined mesh patterns are derived by subdividing the original patterns (Zienkiewicz, 1971). For a test of convergence, therefore, refined patterns should contain all the nodes of previous coarser mesh patterns. In practice this is a bit of a problem to accomplish because just a few subdivisions can create an excessive number of nodes. For the elasticity problem, one generally sees the stiffness of the system reduced as more nodes are added even though the original nodes are not retained in their exact position, so a plot of results similar to that shown in Figure 8.18 should define the progress.

In practice, one should make convergence tests but also, if possible, use a comparison analysis. Test the model on a similar problem for which results are known, or can be determined, either theoretically or better yet, by experimentation.

8.7 SKEW BOUNDARY CONDITIONS

Consider the three types of boundary conditions shown in Figure 8.20.

1. At node 1, the displacement is zero in the X and Y directions.
2. At node 2, the displacement is zero in the Y direction only.
3. At node 3, the displacement is zero in the y direction only of the rotated xy coordinate system.

Before application of the boundary conditions, the force–displacement relationships would be in the following form with all quantities referenced to the global XY coordinate system.

$$[K]_{XY} \begin{Bmatrix} u_1 \\ v_1 \\ u_2 \\ v_2 \\ \vdots \\ u_n \\ v_n \end{Bmatrix}_{XY} = \begin{Bmatrix} F_{1X} \\ F_{1Y} \\ F_{2X} \\ F_{2Y} \\ \vdots \\ F_{nX} \\ F_{nY} \end{Bmatrix} \qquad (8.24)$$

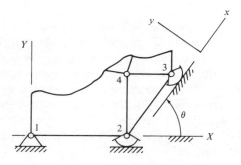

FIGURE 8.20 Examples of displacement boundary conditions.

8.7 SKEW BOUNDARY CONDITIONS

Substitution of boundary conditions for cases 1 and 2 is a simple matter:

$$u_{1X} = 0 = v_{1Y} = v_{2Y}$$

For case 3, however, we have to deal with components of displacement and force at node 3 in terms of the rotated xy reference.

Recall equation (1.9), which relates the components of a vector expressed in terms of two different coordinate systems. Applying the equation to the displacement and force vectors for node 3, we have the following:

$$\begin{Bmatrix} u_{3X} \\ v_{3Y} \end{Bmatrix} = \begin{bmatrix} \cos\theta & -\sin\theta \\ \sin\theta & \cos\theta \end{bmatrix} \begin{Bmatrix} u_{3x} \\ v_{3y} \end{Bmatrix} = [A] \begin{Bmatrix} u_{3x} \\ v_{3y} \end{Bmatrix}$$

and

$$\begin{Bmatrix} F_{3X} \\ F_{3Y} \end{Bmatrix} = [A] \begin{Bmatrix} F_{3x} \\ F_{3y} \end{Bmatrix}$$

We can now substitute these expressions into the displacement and force vectors of equation (8.24).

$$\begin{Bmatrix} u_1 \\ v_1 \\ u_2 \\ v_2 \\ u_3 \\ v_3 \\ \vdots \\ u_n \\ v_n \end{Bmatrix}_{XY} = \begin{bmatrix} 1 & 0 & 0 & 0 & 0 & \cdots & & & 0 \\ 0 & 1 & 0 & 0 & 0 & & & 0 & 0 \\ 0 & 0 & 1 & 0 & 0 & & & 0 & 0 \\ 0 & 0 & 0 & 1 & 0 & & & 0 & 0 \\ 0 & 0 & 0 & 0 & \cos\theta & -\sin\theta & & 0 & 0 \\ 0 & 0 & 0 & 0 & \sin\theta & \cos\theta & & 0 & 0 \\ \vdots & & & & & & & & \vdots \\ 0 & 0 & 0 & 0 & 0 & 0 & & 1 & 0 \\ 0 & 0 & 0 & 0 & 0 & 0 & 0 & 0 & 1 \end{bmatrix} \begin{Bmatrix} u_{1X} \\ v_{1Y} \\ u_{2X} \\ v_{2Y} \\ u_{3x} \\ v_{3y} \\ \vdots \\ u_{nX} \\ v_{nY} \end{Bmatrix}$$

or

$$\{r\}_{XY} = [IA]\{r\}_{XY, xy} \qquad (8.25)$$

We note that the vector $\{r\}_{XY,xy}$ contains displacement components referenced to the two different coordinate systems. Doing the same for the force vector gives us

$$\{R\}_{XY} = [IA]\{R\}_{XY,xy} \quad (8.26)$$

Substituting (8.25) and (8.26) into (8.24) yields

$$[K]_{XY}[IA]\{r\}_{XY,xy} = [IA]\{R\}_{XY,xy}$$

or

$$[IA]^T[K]_{XY}[IA]\{r\}_{XY,xy} = \{R\}_{XY,xy}$$

This can be written in terms of the "partially rotated" stiffness matrix as

$$[K]_{XY,xy}\{r\}_{XY,xy} = \{R\}_{XY,xy} \quad (8.27)$$

Now, substitution of the boundary conditions followed by row and column reduction of the stiffness matrix is a straightforward process. Note that for the example shown in the Figure 8.20, reduction and solution of (8.27) would yield X and Y components of displacements of all nodes except node 3. For node 3 the displacements would be in the x direction. Be certain to reapply equation (8.25) or equivalent to transform all displacement components to the XY reference before applying the $[B]$ matrix for individual elements to obtain strain.

The method just described presents the general concept, but from a practical standpoint it involves procedures wasteful of time and storage space. In application, it is preferable to perform the transformation on the elements involved one at a time and then assemble the element rotated stiffness matrix into the global assemblage matrix.

8.8 TEMPERATURE-INDUCED STRAIN

A temperature change, ΔT, of a material will cause expansion or contraction of all linear dimensions of the body. The resulting strain vector, for plane strain, is defined as follows:

$$\{\epsilon_T\} = \begin{Bmatrix} \epsilon_{X_T} \\ \epsilon_{Y_T} \\ \gamma_{XY_T} \end{Bmatrix} = \begin{Bmatrix} \alpha \Delta T \\ \alpha \Delta T \\ 0 \end{Bmatrix} \quad (8.28)$$

where α is the material coefficient of thermal expansion. The total strain $\{\epsilon\}$ in the continuum is the sum of the strain $\{\epsilon_F\}$ due to the applied loads (elastic strain) and the temperature-induced strain $\{\epsilon_T\}$:

$$\{\epsilon\} = \{\epsilon_F\} + \{\epsilon_T\} \quad (8.29)$$

The stress in the continuum is thus defined in terms of the elastic strain $\{\epsilon_F\}$, the total strain $\{\epsilon\}$, and the temperature-induced strain $\{\epsilon_T\}$ as follows:

$$\{\sigma\} = [C]\{\epsilon_F\} \quad (8.30)$$

$$\{\sigma\} = [C](\{\epsilon\} - \{\epsilon_T\}) \quad (8.31)$$

8.8 TEMPERATURE-INDUCED STRAIN

The strain energy in an element is defined by equation (2.15):

$$U = \tfrac{1}{2} \int_\Omega \{\epsilon\}^T \{\sigma\} \, d\Omega \tag{2.15}$$

where the strain energy must be based on the elastic strain $\{\epsilon_F\}$. Thus

$$U = \tfrac{1}{2} \int_\Omega (\{\epsilon\} - \{\epsilon_T\})^T ([C](\{\epsilon\} - \{\epsilon_T\})) \, d\Omega$$

$$= \tfrac{1}{2} \int_\Omega \left[(\{\epsilon\} - \{\epsilon_T\})^T ([C]\{\epsilon\} - [C]\{\epsilon_T\}) \right] d\Omega$$

$$= \tfrac{1}{2} \int_\Omega \left[\{\epsilon\}^T [C]\{\epsilon\} - \{\epsilon\}^T [C]\{\epsilon_T\} - \{\epsilon_T\}^T [C]\{\epsilon\} \right.$$
$$\left. + \{\epsilon_T\}^T [C]\{\epsilon_T\} \right] d\Omega$$

Substituting $\{\epsilon\} = [B]\{q\}$ into the equation above and adding all the force potential terms, we apply the energy minimization principle (review Section 4.3) and obtain the following equation:

$$\int_\Omega [B]^T [C][B] \, d\Omega \, (q) = \int_\Omega [B]^T [C]\{\epsilon_T\} \, d\Omega$$
$$+ \{Q\}_{NF} + \{Q\}_T + \{Q\}_{BF} \tag{8.32}$$

The first term on the right side represents the "loading" resulting from the temperature. The three remaining terms are the familiar nodal, surface traction, and body force contributions. The temperature nodal load vector is

$$\{Q\}_{temp} = \int_\Omega [B]^T [C]\{\epsilon_T\} \, d\Omega = h \int_A [B]^T [C]\{\epsilon_T\} \, dA \tag{8.33}$$

The evaluation of the vector can be done explicitly or by numerical methods.

Example 8.6 A single three-node constant strain element is supported against rigid-body motion (Figure 8.21). No real forces are applied to the element, but the element is heated so that it undergoes a uniform temperature increase of 100°F.

Solve for the temperature nodal load vector. Solve for the nodal displacements using equation (8.32), and then solve for the stress in the element.

306 HIGHER-ORDER ELEMENTS AND ADDITIONAL CAPABILITIES

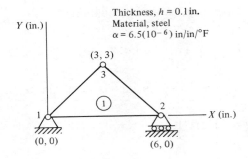

FIGURE 8.21 Single constant-strain element.

Solution. From (8.33),

$$\{Q\}_{\text{temp}} = h \int_A [B]^T [C] \{\epsilon_T\} \, dA = h \int_A [B]^T [C] \begin{Bmatrix} \epsilon_X \\ \epsilon_Y \\ \epsilon_{XY} \end{Bmatrix} dA$$

$$= Ah[B]^T[C] \begin{Bmatrix} \alpha \Delta T \\ \alpha \Delta T \\ 0 \end{Bmatrix}$$

The $[B]$ matrix for the constant-strain triangle is defined in (6.8). Substituting the nodal coordinates and using the assumption of plane stress for the constitutive matrix $[C]$ gives us

$$\{Q\}_{\text{temp}} = 9(0.1) \begin{bmatrix} -0.1667 & 0 & -0.1667 \\ 0 & -0.1667 & -0.1667 \\ 0.1667 & 0 & -0.1667 \\ 0 & -0.1667 & 0.1667 \\ 0 & 0 & 0.3333 \\ 0 & 0.3333 & 0 \end{bmatrix}$$

$$\times \begin{bmatrix} 1 & 0.3 & 0 \\ 0.3 & 1 & 0 \\ 0 & 0 & 0.35 \end{bmatrix} \begin{Bmatrix} 1 \\ 1 \\ 0 \end{Bmatrix} \frac{6.5(10^{-4})}{0.91} 30(10^6)$$

$$= \begin{Bmatrix} -4179 \\ -4179 \\ 4179 \\ -4179 \\ 0 \\ 8358 \end{Bmatrix} \text{lb} = \begin{Bmatrix} F_{1X} \\ F_{1Y} \\ F_{2X} \\ F_{2Y} \\ F_{3X} \\ F_{3Y} \end{Bmatrix}$$

The "fictitious" temperature loads are placed on the element as shown in Figure 8.22. The displacements of the nodes of this single

8.8 TEMPERATURE-INDUCED STRAIN

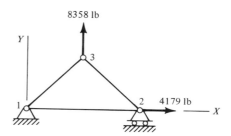

FIGURE 8.22 Temperature nodal forces.

element are found using the computer program in Appendix 3:

NODE NUMBER	XDISPL	YDISPL
1	0	0
2	3.9003E–03	0
3	1.9501E–03	1.9502E–03

Substituting these displacements into the strain–displacement relation, we have

$$\{\epsilon\} = [B]\{q\}$$

$$= \begin{Bmatrix} 6.5(10^{-4}) \\ 6.5(10^{-4}) \\ 0 \end{Bmatrix} = \begin{Bmatrix} \alpha \Delta T \\ \alpha \Delta T \\ 0 \end{Bmatrix}$$

Substituting this total strain vector into equation (8.32) yields a stress vector of zero.

$$\{\sigma\} = [C](\{\epsilon\} - \{\epsilon_T\})$$

$$= [C]\left(\begin{Bmatrix} 6.5(10^{-4}) \\ 6.5(10^{-4}) \\ 0 \end{Bmatrix} - \begin{Bmatrix} 6.5(10^{-6})(100) \\ 6.5(10^{-6})(100) \\ 0 \end{Bmatrix}\right)$$

$$= \begin{Bmatrix} 0 \\ 0 \\ 0 \end{Bmatrix}$$

We understand this to be correct because there are no real forces acting on the element. On the other hand, had the nodes, all nodes, been completely constrained in the X and Y directions, the 100°F temperature rise would have given rise to a compressive stress in the X and Y directions equal to 27,860 psi. Note that in this situation, the total strain vector would be zero. □

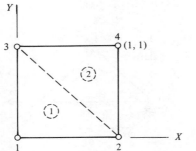

FIGURE 8.23 Unit square plate of thickness $h = 0.1$.

Example 8.7 The square plate (Figure 8.23) undergoes nonuniform heating so that the temperature of the corners increases as follows:

Corner Node	Temperature Increase
1	100
2	80
3	60
4	10

The plate is divided into two three-node triangle elements, and we wish to calculate the stress approximation in each triangle resulting from the temperature increase. Assume the following values for material constants:

$$\text{Poisson's ratio} = \tfrac{1}{3}$$

$$\text{thermal expansion coefficient} = 10\text{E}{-}6$$

$$\text{elastic modulus} = 30\text{E}6$$

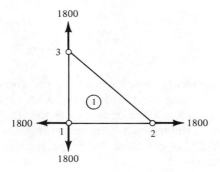

FIGURE 8.24 Thermal forces on element 1.

8.8 TEMPERATURE-INDUCED STRAIN

Solution. The thermal "load" vector is calculated for each element using the element temperature as the average of the nodal temperatures. For element 1 (Figure 8.24) the average temperature is 80. The load vector for element 1 is calculated as follows:

$$\{Q\}_{\text{temp}} = Ah[B]^T[C]\begin{Bmatrix}1\\1\\0\end{Bmatrix}\alpha\Delta T$$

$$\{Q\}_{\text{temp}①} = \frac{1}{2}(0.1)\begin{bmatrix}-1 & 0 & -1\\0 & -1 & -1\\1 & 0 & 0\\0 & 0 & 1\\0 & 0 & 1\\0 & 1 & 0\end{bmatrix}$$

$$\times \begin{bmatrix}1 & \frac{1}{3} & 0\\\frac{1}{3} & 1 & 0\\0 & 0 & \frac{1}{3}\end{bmatrix}\begin{Bmatrix}1\\1\\0\end{Bmatrix}\frac{30\text{E}6(10\text{E}-6)(80)}{0.89}$$

or

$$\{Q\}_{\text{temp}①} = \begin{Bmatrix}-1800\\-1800\\1800\\0\\0\\1800\end{Bmatrix}$$

Similarly for element 2 (Figure 8.25), using an average temperature of 50, we find that

$$\{Q\}_{\text{temp}②} = \frac{1}{2}(0.1)\begin{bmatrix}0 & 0 & -1\\0 & -1 & 0\\1 & 0 & 1\\0 & 1 & 1\\-1 & 0 & 0\\0 & 0 & -1\end{bmatrix}\begin{bmatrix}1 & \frac{1}{3} & 0\\\frac{1}{3} & 1 & 0\\0 & 0 & \frac{1}{3}\end{bmatrix}\begin{Bmatrix}1\\1\\0\end{Bmatrix}$$

$$\times \frac{30\text{E}6(10\text{E}-6)(50)}{0.89}$$

or

$$\{Q\}_{\text{temp}②} = \begin{Bmatrix}0\\-1125\\1125\\1125\\-1125\\0\end{Bmatrix}$$

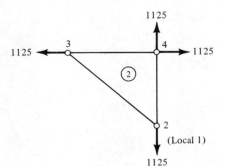

FIGURE 8.25 Thermal forces on element 2.

The assembled global thermal load vector is

$$\{R\}_{\text{temp}} = \begin{Bmatrix} -1800 \\ -1800 \\ 1800 \\ -1125 \\ -1125 \\ 1800 \\ 1125 \\ 1125 \end{Bmatrix}$$

Next, execute the finite element program in Appendix 3 with the thermal load vector as the applied load and apply boundary conditions for free thermal expansion but no rigid-body motion (Figure 8.26).

The program yields the following node displacements and element TOTAL strain values.

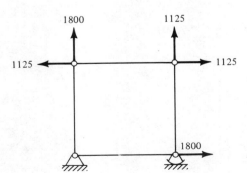

FIGURE 8.26 Thermal forces and boundary support.

```
******************************************
**                                      **
**           DISPLACEMENT               **
**                                      **
******************************************
  NODE        XDISPL          YDISPL
   1         0                0
   2         7.4999975E-04    0
   3         1.999995E-04     7.4999975E-04
   4         7.4999975E-04    5.5000025E-04


******************************************
**                                      **
**             STRAIN                   **
**                                      **
******************************************
  ELEMENT     EXX            EYY              EXY

   1         -5.01E-05      -5.01E-05         1.999E-04
   2          5E-05          5E-05           -2E-04
```

Results of Plane Stress Analysis

From the total strain for each element we must subtract the thermal strain in order to have the elastic strain. The thermal strains for the elements are

$$\{\epsilon_T\}_① = \alpha \Delta T \begin{Bmatrix} 1 \\ 1 \\ 0 \end{Bmatrix} = (10\text{E}-6)(80) \begin{Bmatrix} 1 \\ 1 \\ 0 \end{Bmatrix}$$

$$\{\epsilon_T\}_② = (10\text{E}-6)(50) \begin{Bmatrix} 1 \\ 1 \\ 0 \end{Bmatrix}$$

Calculating the stress from (8.31) yields

$$\{\sigma\} = [C](\{\epsilon\} - \{\epsilon_T\})$$

This yields the following results:

$$\{\sigma\}_1 = \begin{Bmatrix} -2250 \\ -2250 \\ 2250 \end{Bmatrix} \quad \{\sigma\}_2 = \begin{Bmatrix} 2250 \\ 2250 \\ -2250 \end{Bmatrix}$$

We note that even with the free expansion boundary conditions, the nonuniform temperature increase in the square resulted in stresses occurring in each element. We should view the stress magnitudes qualitatively only for such a crude model. ☐

PROBLEMS

8.1 Calculate the values of the entries in the six-node triangle $[N]$ matrix for the centroid location.

8.2 Solve for the nodal force vector $\{Q\}_{BF}$ equivalent to the weight of an isoparametric six-node triangle of uniform thickness h and mass density ρ. Assume that gravity acts in the negative Y direction.

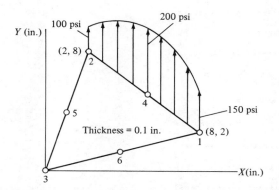

FIGURE P8.3 Distributed T_Y on side 3 of six-node triangle.

8.3 Determine the element nodal force vector $\{Q\}_T$ for the distributed stress traction shown in Figure P8.3.

8.4 Solve for the 1,1 entry in the $[B]$ matrix for the node 1 position of the six-node triangle shown in Figure P8.3.

8.5 Write a computer program to evaluate the $[B]$ matrix for any point in the six-node triangle.

8.6 *Project.* Write a computer program for stress analysis using the six-node triangle.

8.7 Calculate the entries of the eight-node quadrilateral $[N]$ matrix at the location given by $s = 0 = t$.

8.8 Show that for side $t = 1$ of the eight-node quadrilateral the following transformation relationship is correct:

$$dX = \frac{X_3 - X_4}{2} ds$$

8.9 Prove that a uniformly distributed load on a side of the eight-node quadrilateral is allocated to the nodes as shown in Figure P8.9.

Note: This distribution is the same as that for the six-node triangle.

8.10 Show that the weight of an eight-node quadrilateral of uniform thickness and density is distributed to the nodes as shown in Figure P8.10.

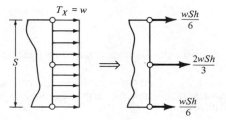

FIGURE P8.9 Distributed load allocation to nodes.

PROBLEMS

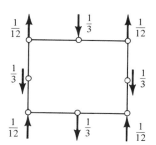

FIGURE P8.10 Allocation of weight for uniform thickness and density.

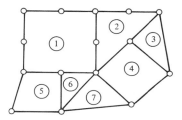

FIGURE P8.13 Assemblage of compatible and incompatible elements.

8.11 *Project.* Write a computer for stress analysis using the eight-node quadrilateral.

8.12 Discuss the compatibility of a six-node triangle adjacent to an eight-node quadrilateral.

8.13 Discuss the compatibility of the elements in the assemblage shown in Figure P8.13.

8.14 *Project.* Write a computer program for stress analysis of a continuum with the discretization including both the three-node triangle and four-node quadrilateral.

8.15 *Project.* Write a computer program for stress analysis of a continuum with the element mesh containing both the six-node triangle and eight-node quadrilateral.

8.16 Referring to the Example 8.6, show that the three-node triangle element yields a compressive stress of 27,860 psi for the temperature increase of 100°F if the nodes are constrained as shown in Figure P8.16.

8.17 Write a computer program to calculate the temperature load vector $\{Q\}_{temp}$ for the three-node triangle.

8.18 Solve for the temperature nodal load vector for the four-node quadrilateral shown in Figure P8.18 for a temperature increase of 100°F.

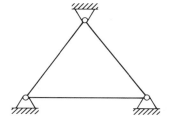

FIGURE P8.16 Triangle with all nodes constrained.

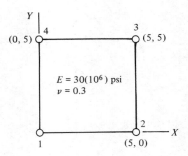

FIGURE P8.18 Coordinates of nodes of quadrilateral.

Hint:

$$\{Q\}_{\text{temp}} = h \int_A [B]^T [C] \{\epsilon_T\} \, dA$$

$$= h \int_{A^*} [B]^T [C] \{\epsilon_T\} |J| \, ds \, dt$$

Solve numerically using two-point quadrature.

8.19 Write a program to calculate the temperature load vector for the four-node quadrilateral.

8.20 Fill in all the steps leading to the derivation of equation (8.32).

CHAPTER 9

The Galerkin and Other Weighted Residual Methods

9.1 INTRODUCTION

A mathematical foundation for the finite element method was presented in Sections 2.10 through 2.14. If, by variational methods, a functional can be established corresponding to the governing differential equations and boundary conditions defining a problem, then the solution of the problem centers on finding the dependent variables that yield a stationary value of the functional. A major attribute of this functional formulation is its usefulness in obtaining approximate solutions using, for example, the well-established Rayleigh–Ritz method.

As described in Chapter 2, the finite element method is based on the concept of mathematically discretizing the region of the functional and using the Rayleigh–Ritz procedure with simple approximation functions. In that discussion you were cautioned not to become discouraged if you had little or no background in variational calculus. Given the functional, the main goal was to show how to use it. In addition, you were promised an alternative procedure for formulating the finite element equations.

In this chapter we examine that alternative approach. This approach can be and is used as an alternative method or in situations where the functional cannot be found by the variational method. The method is based on the Galerkin weighted residual method of approximating solutions to differential equations. We will see that for the problems described in this book, the variational method and the Galerkin method yield the same system equations.

In this discussion, several weighted residual methods are described and compared to broaden your understanding of the techniques.

9.2 APPROXIMATION TECHNIQUES FOR DIFFERENTIAL EQUATIONS: WEIGHTED RESIDUAL METHODS

Consider a differential operator D acting on an unknown function u to produce a function p:

$$D(u) = p \qquad (9.1)$$

For example, if $D = (d^2/dx^2) + 1$, the differential equation would be

$$\frac{d^2u}{dx^2} + u = p$$

The general solution to (9.1) is the function u satisfying that differential equation throughout a specified domain. A unique solution must, in addition, assume the specified values on the boundary of the domain (the "essential" boundary conditions).

We desire a method of obtaining a solution to a differential equation that will satisfy the boundary conditions exactly without necessarily satisfying the differential equation in the domain except in some average sense. The approximation schemes we will introduce are called *weighted residual methods*.

The weighted residual methods involve the "making up" of an assumed (approximated) solution function that satisfies the boundary conditions exactly and contains arbitrary (initially unknown) parameters that can be adjusted to minimize the integral over the domain of the error of the assumed solution in the differential equation. The assumed solution function is to consist of linearly independent functions, commonly called *basis functions*, chosen from a complete set of functions. The assumed solution could take the following form:

$$\tilde{u} = \sum a_i G_i \qquad i = 1, 2, \ldots, n \qquad (9.2)$$

where

a_i = adjustable coefficients

G_i = independent functions of the independent variable

n = range equal to the number of coefficients

9.2 APPROXIMATION TECHNIQUES FOR DIFFERENTIAL EQUATIONS

Substitution of the approximate solution function (9.2) into (9.1) will not yield the equality. The difference between the left side and the right side is the error, E (or residual), of the approximation.

$$E(x) = D\tilde{u} - p \neq 0$$

The weighted residual methods require the integral of the error E times a weighting function W_i to be zero, that is,

$$\int_X W_i E \, dx = 0 \qquad i = 1, 2, \ldots, n \qquad (9.3)$$

where the number of weighting functions is equal to the number of adjustable coefficients, a_i, in the assumed solution. If the weighting functions are properly chosen, this yields n independent algebraic equations for the n unknown coefficients. The various methods we introduce are characterized by the particular weighting function used in the integral.

Collocation Method. The collocation method uses a family of Dirac delta functions for the weighting functions:

$$\int_X \delta(x - x_i) E \, dx = 0 \qquad i = 1, 2, \ldots, n \qquad (9.4a)$$

This is equivalent to forcing the residual, E, to be zero at n points in the region:

$$E(x_i) = 0 \qquad i = 1, 2, \ldots, n \qquad (9.4b)$$

The solution of the n equations yields the n a_i coefficients.

Subdomain Method. For this method the weighting function is equal to unity, and simultaneous equations are generated by integrating the residual over separate regions of the domain.

$$\int_{X_i} E(x) \, dx = 0 \qquad i = 1, 2, \ldots, n \qquad (9.5)$$

The number of regions, of course, must equal the number of a_i coefficients.

Least-Squares Method. For this method the weighting functions are the individual function coefficients of the a_i parameters in the error function E.

$$W_i = \frac{\partial E}{\partial a_i} \qquad i = 1, 2, \ldots, n \qquad (9.6a)$$

Note that this partial derivative simply "picks out" the assumed solution basis function of which a_i is the coefficient. Thus

$$\int_X \frac{\partial E}{\partial a_i} E \, dx = 0 \qquad i = 1, 2, \ldots, n \qquad (9.6b)$$

This is equivalent to minimizing the integral of the square of the error.

$$\frac{\partial}{\partial a_i} \int_X E^2 \, dx = 0 \qquad i = 1, 2, \ldots, n \qquad (9.6c)$$

The reason for this equivalent relation is

$$\frac{\partial}{\partial a_i} E^2 = 2E \frac{\partial E}{\partial a_i}$$

Galerkin Method. The Galerkin method employs weighting functions that are the individual function coefficients of the a_i parameters in the assumed trial function, $\tilde{u}$ defined in (9.2).

$$W_i = \frac{\partial \tilde{u}}{\partial a_i} = G_i \qquad i = 1, 2, \ldots, n \qquad (9.7a)$$

or

$$\int_X G_i E \, dx = 0 \qquad i = 1, 2, \ldots, n \qquad (9.7b)$$

Example 9.1 Given the equation

$$\frac{d^2 u}{dx^2} + u = 1$$

with the boundary conditions: $u(0) = 1$ and $u(1) = 0$, derive the approximate solution function using each of the weighted residual methods and compare results with the exact solution.

Solution. The exact solution of this problem is

$$u = 1 - 1.1884 \sin x$$

An approximate solution function satisfying the boundary values is proposed:

$$\tilde{u} = \sum a_i G_i = a_0 + a_1 x + a_2 x^2 + \cdots + a_n x^n$$

To illustrate the methods with a minimum of effort, we will assume only three unknown coefficients:

$$\tilde{u} = a_0 + a_1 x + a_2 x^2$$

Substituting the boundary conditions, we obtain

$$a_0 = 1$$
$$a_1 = -(1 + a_2)$$

9.2 APPROXIMATION TECHNIQUES FOR DIFFERENTIAL EQUATIONS

The assumed solution function can now be expressed with one remaining undetermined coefficient, a_2:

$$\tilde{u} = 1 - x + a_2(x^2 - x)$$

Substituting this assumed solution into the differential equation, we can solve for the residual function.

$$E = \frac{d^2\tilde{u}}{dx^2} + \tilde{u} - 1$$

$$= a_2(2 - x + x^2) - x \qquad (9.8)$$

Collocation Method From (9.4b),

$$E(x_i) = 0 \qquad i = 1$$

We will select the midpoint of the region for the zero error. Substituting $x = 0.5$ into (9.8) and equating to zero gives us

$$E(0.5) = a_2(2 - \tfrac{1}{2} + \tfrac{1}{4}) - \tfrac{1}{2} = 0$$

From which $a_2 = \tfrac{2}{7}$ and the approximate solution has the form

$$\tilde{u} = 1 - x + \tfrac{2}{7}(x^2 - x)$$

Subdomain Method From (9.5),

$$\int_{X_i} E(x)\,dx = 0 \qquad i = 1$$

The integration will extend over the complete region, since we need only one equation for the one unknown coefficient. Substituting the error function (9.8) yields

$$\int_0^1 \left[a_2(2 - x + x^2) - x\right] dx = 0$$

from which $a_2 = \tfrac{3}{11}$ and the solution function is

$$\tilde{u} = 1 - x + \tfrac{3}{11}(x^2 - x)$$

Least-Squares Method From (9.6a) and (9.8),

$$W_i = W_1 = \frac{\partial E}{\partial a_2} = 2 - x + x^2$$

From (9.6b),

$$\int_0^1 \frac{\partial E}{\partial a_2} E\,dx = \int_0^1 (2 - x + x^2)\left[a_2(2 - x + x^2) - x\right] dx = 0$$

Integrating and solving for a_2 yields $a_2 = \tfrac{55}{202}$ and the function

$$\tilde{u} = 1 - x + \tfrac{55}{202}(x^2 - x)$$

TABLE 9.1 Comparison of Approximate Solutions

x	Exact	Collocation	Subdomain	Least Squares	Galerkin
0.2	0.7639	0.7542	0.7564	0.7564	0.7556
0.4	0.5372	0.5314	0.5345	0.5347	0.5333
0.6	0.3290	0.3314	0.3345	0.3346	0.3333
0.8	0.1475	0.1543	0.1564	0.1564	0.1555

Galerkin Method From (9.7),

$$W_i = W_1 = \frac{\partial \tilde{u}}{\partial a_2} = x^2 - x$$

$$\int_0^1 \frac{\partial \tilde{u}}{\partial a_2} E\, dx = (x^2 - x)\left[a_2(2 - x + x^2) - x\right] dx = 0$$

Integrating and solving for a_2, we obtain the solution $a_2 = \frac{5}{18}$ and

$$\tilde{u} = 1 - x + \tfrac{5}{18}(x^2 - x)$$

The Table 9.1 contains calculated values at even increments of x for the exact solution and each approximation. Figure 9.1 shows the nature of the distribution of the difference between the exact and approximate values. □

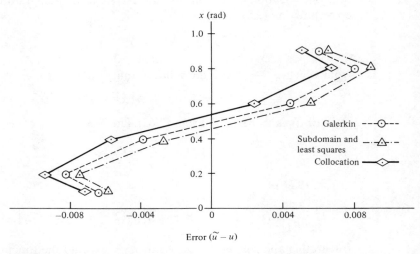

FIGURE 9.1 Comparison of error of the four approximation methods.

9.2 APPROXIMATION TECHNIQUES FOR DIFFERENTIAL EQUATIONS

Example 9.2 Derive an approximate solution for the differential equation of Example 9.1 using the Galerkin method with two constants.

Solution. The trial solution function that satisfies the essential boundary conditions is

$$\tilde{u} = (x-1)(-1 + a_1 x + a_2 x^2)$$
$$= 1 - x + a_1 x(x-1) + a_2 x^2(x-1)$$

The weighting functions for the Galerkin method are

$$W_1 = \frac{\partial \tilde{u}}{\partial a_1} = x(x-1) \qquad W_2 = \frac{\partial \tilde{u}}{\partial u_2} = x^2(x-1)$$

Substituting the trial function into the differential equation to obtain the error function yields

$$E = a_1(2 - x + x^2) + a_2(-2 + 6x - x^2 + x^3) - x$$

Substituting into (9.7b) for $i = 1$,

$$\int_0^1 W_1 E \, dx = 0$$

we obtain the following equation for a_1 and a_2:

$$18a_1 + 9a_2 = 5$$

Also, for $i = 2$,

$$\int_0^1 W_2 E \, dx = 0$$

and we obtain the second equation for a_1 and a_2:

$$63a_1 + 52a_2 = 21$$

Solving the two equations for a_1 and a_2 gives us

$$a_1 = \tfrac{71}{369}$$
$$a_2 = \tfrac{21}{123}$$

The resulting trial solution is, therefore,

$$\tilde{u} = -(x-1) + \tfrac{71}{369}x(x-1) + \tfrac{21}{123}x^2(x-1)$$

Table 9.2 contains the comparison of the exact solution with the one- and two-constant Galerkin solutions. The improvement in the solution is very evident.

The exact solution of the differential equation was a sine function. An "accurate" approximate solution over the complete period of the

TABLE 9.2 Comparison of Two Galerkin Solutions

x	Exact	One Constant (Example 9.1)	Two Constants (Example 9.2)
0.2	0.7639	0.7556	0.7638
0.4	0.5372	0.5333	0.5374
0.6	0.3290	0.3333	0.3292
0.8	0.1475	0.1555	0.1474

function would require a polynomial trial function of high degree. This is the motivation for subdivision of the region. By dividing the region up into small increments and taking care to ensure compatibility of the functions in adjacent subdivisions (elements), the solution can be approximated very well with simple trial functions of low degree. □

9.3 COMPATIBILITY REQUIREMENTS

If we are to work with small elements within a region, it is conceivable that each element will have a solution function different from the others. If we are to assemble these elements in a mathematical sense, there must be some compatibility requirements placed on the functions of the adjacent elements.

Return to our example differential equation in terms of the trial solution $\tilde{u}$. The residual for one particular element e is

$$E^e = \frac{d^2\tilde{u}^e}{dx^2} + \tilde{u}^e - 1$$

Using one of the weighting methods, Galerkin, for example, we have

$$\int_{X_e} WE^e\,dx = \int_{X_e} W\frac{d^2\tilde{u}^e}{dx^2}\,dx + \int_{X_e} \tilde{u}^e W\,dx - \int_{X_e} W\,dx = 0$$

We want the error, E, to be averaged to zero not only for each element, but also for the entire region. This will certainly require that the integrals be finite throughout each element and also across all element boundaries.

Let us assume that we have trial functions $\tilde{u}^1$ and $\tilde{u}^2$ for two adjacent elements, and that these functions are continuous in the manner defined in Figure 9.2(a). The first and second derivatives are defined in the Figure 9.2(b) and (c), respectively. Clearly, element functions that have a discontinuity in the first derivative are not suitable if the integration over both elements involves the second derivative of the function.

In order to have a well-defined integral of the second derivative of the element trial functions, the functions must possess a continuous first

9.4 REDUCTION OF ORDER

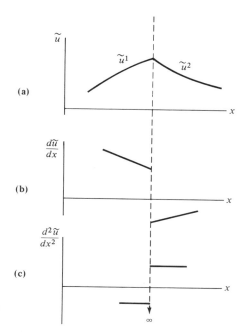

FIGURE 9.2 Behavior of function at element boundary.

derivative across element boundaries. For a polynomial function, for example, this would require at least a quadratic polynomial (which was adopted in Example 9.1). One way of reducing the level of differentiability requirement on the trial function is to reduce, if possible, the order of the derivative of the trial function in the integral. This idea is discussed in the next section.

9.4 REDUCTION OF ORDER

To put forth in simple terms the basic principle of order reduction, let us work with the same differential equation of Example 9.1:

$$\frac{d^2u}{dx^2} + u = 1$$

Multiply each term by a Galerkin weighting factor and integrate over the region:

$$\int_X W \frac{d^2u}{dx^2} \, dx + \int_X Wu \, dx = \int_X W \, dx$$

Integration (by parts) of the first term on the left yields

$$W \frac{du}{dx} \bigg|_{x_1}^{x_2} - \int_X \frac{dW}{dx} \frac{du}{dx} \, dx + \int_X Wu \, dx = \int_X W \, dx \tag{9.9}$$

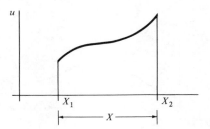

FIGURE 9.3 Definition of variables.

where X_1 and X_2 are the coordinates at the endpoints (boundary) of the region, and X defines the total region (Figure 9.3).

The first term on the left,

$$W \left. \frac{du}{dx} \right|_{X_1}^{X_2}$$

involves the weighting function and du/dx at the boundary. The function du/dx at $x = X_1$ and $x = X_2$ is called a *natural boundary condition*, and its value may be specified. If it is specified at $x = X_1$ or $x = X_2$, then the corresponding W function (at the boundary) must not be zero. If du/dx is not specified, and if the trial function, $\tilde{u}$, satisfies the essential boundary conditions, u, at $x = X_1$ and X_2, then the weighting functions will automatically be zero at the boundaries. Recall in Examples 9.1 and 9.2, where the essential boundary values were specified [$u(0) = 1$, $u(1) = 0$], that all the Galerkin weighting functions were zero at the boundaries. Observe, therefore, that it is not possible to specify both essential and natural boundary conditions at the same point. Note also that for a unique solution of this differential equation, at least one essential boundary condition must be specified.

The second term on the left of (9.9),

$$\int_X \frac{dW}{dx} \frac{du}{dx} \, dx$$

involves the functions u and W, which must have finite-valued first derivatives. This integral can accommodate a finite number of jumps in the integrand and we wish to select the lowest-degree basis functions defined on the subdomains that meet this condition. Hence we select piecewise linear functions.

9.5 FINITE ELEMENT SOLUTION OF THE DIFFERENTIAL EQUATION

In Section 9.2 it was pointed out that the quadratic trial function used in Example 9.1 was adequate for a small region, but as the region grew in size, the function could not possibly approximate the correct solution. In

9.5 FINITE ELEMENT SOLUTION OF THE DIFFERENTIAL EQUATION

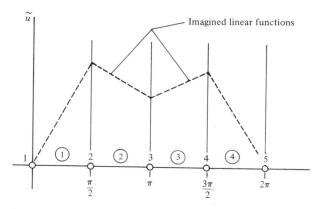

FIGURE 9.4 Four elements in the defined region with imagined piecewise linear functions.

this section we wish to show how a large region can be discretized and a "reasonable" solution obtained using a trial function of lesser degree. To illustrate the method, the same differential equation will be solved:

$$\frac{d^2u}{dx^2} + u = 1$$

with (different) boundary conditions:

$$u(0) = 0$$
$$u(2\pi) = 0$$

(The exact solution is $u = 1 - \cos x$.)

Following is the solution procedure.

1. *Discretize the region.* In Figure 9.4 the region has been divided into four elements and the elements and the nodes have been numbered (a familiar step by now).

2. *Establish the trial function.* In Figure 9.5 is shown the general element e and the labeled nodes i and j. For element trial function compatibility, a general linear function is selected based on the discus-

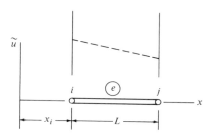

FIGURE 9.5 The general element.

sion in Section 9.4:
$$\tilde{u} = a_1 + a_2 x$$

Rather than formulate the problem in terms of the arbitrary constants a_1 and a_2, we prefer to recast this linear trial function in terms of the values of the dependent function at nodes i and j. This will attach physical significance to the function constants. Also, we will observe that the resulting trial function will be identical to the function developed in Chapter 3 and also in Chapter 5 using the Lagrange interpolation polynomials.

We proceed in a manner identical to that in Chapter 3 to convert the arbitrary constants a_1 and a_2 into the dependent function values at nodes i and j:

$$\left. \begin{array}{l} \tilde{u} = u_i = a_1 + a_2 x_i \\ \tilde{u} = u_j = a_1 + a_2 x_j \end{array} \right\} \qquad a_1 = \frac{x_j u_i - x_i u_j}{x_j - x_i} \qquad a_2 = \frac{-u_i + u_j}{x_j - x_i}$$

or we can write

$$\tilde{u} = \begin{bmatrix} \dfrac{x_j - x}{x_j - x_i} & \dfrac{x - x_i}{x_j - x_i} \end{bmatrix} \begin{Bmatrix} u_i \\ u_j \end{Bmatrix} = [N_1 \quad N_2]\{q\} \qquad (9.10)$$

Compare (9.10) with (3.4). The trial function constants now are the nodal values of the dependent variable $\tilde{u}$, and the N_i terms are the familiar interpolation, or shape, functions.

3. *Substitute the trial functions into the governing equation.* For our particular problem, the governing equation that will yield the solution of the trial function constants is given by (9.9). Writing this equation in terms of the discretized region, we have

$$W \left. \frac{du}{dx} \right|_{X_1}^{X_2} - \sum_{e=1}^{4} \int_{X_e} \frac{dW}{dx} \frac{d\tilde{u}}{dx} dx + \sum_{e=1}^{4} \int_{X_e} W\tilde{u}\, dx - \sum_{e=1}^{4} \int_{X_e} W\, dx = 0$$
(9.11)

There are two weighting functions for each element. Recall the definition of the weighting function for the Galerkin method:

$$W_i = \frac{\partial \tilde{u}}{\partial a_i} \qquad (9.7a)$$

Since we have written the trial function $\tilde{u}$ with the constants a_1 and a_2 replaced by the nodal constant values u_i and u_j, we redefine the weighting functions as

$$W_1 = \frac{\partial \tilde{u}}{\partial u_i} = N_1 \qquad W_2 = \frac{\partial \tilde{u}}{\partial u_j} = N_2$$

Thus, from equation (9.10),

$$W_1 = N_1 = \frac{x_j - x}{x_j - x_i} \qquad W_2 = \frac{x - x_i}{x_j - x_i} = N_2$$

9.5 FINITE ELEMENT SOLUTION OF THE DIFFERENTIAL EQUATION

Note also that

$$\frac{dW_1}{dx} = \frac{dN_1}{dx} = \frac{-1}{x_j - x_i} \qquad \frac{dW_2}{dx} = \frac{dN_2}{dx} = \frac{1}{x_j - x_i}$$

Writing (9.11) in matrix form for one element with nodes at x_i and x_j, we obtain

$$\left\{\begin{array}{c} N_1 \\ N_2 \end{array}\right\} \frac{du}{dx}\bigg|_{x_i}^{x_j} - \int_{x_i}^{x_j} \left\{\begin{array}{c} \frac{dN_1}{dx} \\ \frac{dN_2}{dx} \end{array}\right\} \left[\begin{array}{cc} \frac{dN_1}{dx} & \frac{dN_2}{dx} \end{array}\right] \left\{\begin{array}{c} u_i \\ u_j \end{array}\right\} dx$$

$$+ \int_{x_i}^{x_j} \left\{\begin{array}{c} N_1 \\ N_2 \end{array}\right\} [N_1 \quad N_2] \left\{\begin{array}{c} u_i \\ u_j \end{array}\right\} dx - \int_{x_i}^{x_j} \left\{\begin{array}{c} N_1 \\ N_2 \end{array}\right\} dx = \{0\} \quad (9.12)$$

The first term on the left involves the natural boundary values, du/dx, at each end node of the element. Expanding this boundary term, we have

$$\left\{\begin{array}{c} N_1 \frac{du}{dx}\big|_{x_j} \\ N_2 \frac{du}{dx}\big|_{x_j} \end{array}\right\} - \left\{\begin{array}{c} N_1 \frac{du}{dx}\big|_{x_i} \\ N_2 \frac{du}{dx}\big|_{x_i} \end{array}\right\} = \left\{\begin{array}{c} 0 \\ \frac{du}{dx}\big|_{x_j} \end{array}\right\} - \left\{\begin{array}{c} \frac{du}{dx}\big|_{x_i} \\ 0 \end{array}\right\}$$

$$= \left\{\begin{array}{c} -\frac{du}{dx}\big|_{x_i} \\ \frac{du}{dx}\big|_{x_j} \end{array}\right\} \quad (9.13)$$

Now, when the element equations are assembled according to (9.11), the equations for the elements must combine in such a manner that only the boundary terms for element nodes on the region boundary will contribute; all other terms for interior nodes will be zero. This implies that the boundary terms for elements at common interior nodes cancel one another. This is the case for the exact solution, where we have a function that has a continuous first derivative throughout the region. For the approximate solution, however, we may not have a continuous first derivative. The linear function for example, implies a different first derivative for each element. Thus interelement continuity of the first derivative is achieved only in an approximate sense.

The second term on the left side of (9.12) can be written as

$$\int_{x_i}^{x_j} \left[\begin{array}{cc} \frac{dN_1}{dx}\frac{dN_1}{dx} & \frac{dN_1}{dx}\frac{dN_2}{dx} \\ \frac{dN_2}{dx}\frac{dN_1}{dx} & \frac{dN_2}{dx}\frac{dN_2}{dx} \end{array}\right] \left\{\begin{array}{c} u_i \\ u_j \end{array}\right\} dx = \frac{1}{L_e}\left[\begin{array}{cc} 1 & -1 \\ -1 & 1 \end{array}\right]\left\{\begin{array}{c} u_i \\ u_j \end{array}\right\}$$

(9.14)

THE GALERKIN AND OTHER WEIGHTED RESIDUAL METHODS

where $L_e = x_j - x_i$. For the third term,

$$\int_{x_i}^{x_j} \begin{Bmatrix} N_1 \\ N_2 \end{Bmatrix} [N_1 \quad N_2] \begin{Bmatrix} u_i \\ u_j \end{Bmatrix} dx = \int_{x_i}^{x_j} \begin{bmatrix} N_1 N_1 & N_1 N_2 \\ N_2 N_1 & N_2 N_2 \end{bmatrix} \begin{Bmatrix} u_i \\ u_j \end{Bmatrix} dx$$

$$= \frac{x_j^3 - x_i^3 - 3x_i x_j L_e}{6 L_e^2} \begin{bmatrix} 2 & 1 \\ 1 & 2 \end{bmatrix} \begin{Bmatrix} u_i \\ u_j \end{Bmatrix}$$

(9.15)

The fourth term becomes

$$\int_{x_i}^{x_j} \begin{Bmatrix} N_1 \\ N_2 \end{Bmatrix} dx = \frac{x_j^2 + x_i^2 - 2 x_j x_i}{2 L_e} \begin{Bmatrix} 1 \\ 1 \end{Bmatrix}$$

(9.16)

Summarizing, we can express (9.12) as

$$\begin{Bmatrix} -\dfrac{du}{dx}(x_i) \\ \dfrac{du}{dx}(x_j) \end{Bmatrix} - \frac{1}{L_e} \begin{bmatrix} 1 & -1 \\ -1 & 1 \end{bmatrix} \begin{Bmatrix} u_i \\ u_j \end{Bmatrix} + \frac{x_j^3 - x_i^3 - 3x_i x_j L_e}{6 L_e^2} \begin{bmatrix} 2 & 1 \\ 1 & 2 \end{bmatrix} \begin{Bmatrix} u_i \\ u_j \end{Bmatrix}$$

$$- \frac{x_j^2 + x_i^2 - 2 x_j x_i}{2 L_e} \begin{Bmatrix} 1 \\ 1 \end{Bmatrix} = \begin{Bmatrix} 0 \\ 0 \end{Bmatrix}$$

(9.17)

We can now apply equation (9.17) to each element. Assuming equal-length elements $L_e = L$ for simplicity, we have:

Element 1: $i = 1$, $j = 2$, $x_1 = 0$, and $x_2 = L$

$$\begin{Bmatrix} -\dfrac{du}{dx}\Big|_{x=0} \\ \dfrac{du}{dx}\Big|_{x=L} \end{Bmatrix} - \frac{1}{L} \begin{bmatrix} 1 & -1 \\ -1 & 1 \end{bmatrix} \begin{Bmatrix} u_1 \\ u_2 \end{Bmatrix} + \frac{L}{6} \begin{bmatrix} 2 & 1 \\ 1 & 2 \end{bmatrix} \begin{Bmatrix} u_1 \\ u_2 \end{Bmatrix}$$

$$- \frac{L}{2} \begin{Bmatrix} 1 \\ 1 \end{Bmatrix} = \begin{Bmatrix} 0 \\ 0 \end{Bmatrix}$$

Element 2: $i = 2$, $j = 3$, $x_2 = L$, and $x_3 = 2L$

$$\begin{Bmatrix} -\dfrac{du}{dx}\Big|_{x=L} \\ \dfrac{du}{dx}\Big|_{x=2L} \end{Bmatrix} - \frac{1}{L} \begin{bmatrix} 1 & -1 \\ -1 & 1 \end{bmatrix} \begin{Bmatrix} u_2 \\ u_3 \end{Bmatrix} + \frac{L}{6} \begin{bmatrix} 2 & 1 \\ 1 & 2 \end{bmatrix} \begin{Bmatrix} u_2 \\ u_3 \end{Bmatrix}$$

$$- \frac{L}{2} \begin{Bmatrix} 1 \\ 1 \end{Bmatrix} = \begin{Bmatrix} 0 \\ 0 \end{Bmatrix} \quad \text{etc.}$$

Recalling the discussion following expression (9.13), we recognize that

9.5 FINITE ELEMENT SOLUTION OF THE DIFFERENTIAL EQUATION

$du/dx|_{x=L}$ for element 1 is assumed to be equal to $du/dx|_{X=L}$ for element 2. If we use the technique employed in Section 1.2 of adding zeros to increase the size of the matrices so that each equation has the same nodal displacement vector and each vector term has the same dimension as the displacement vector, then we can sum all the element equations term by term. The resulting assembly of the equations is

$$\begin{Bmatrix} -\dfrac{du}{dx}\bigg|_{x=0} \\ 0 \\ 0 \\ 0 \\ \dfrac{du}{dx}\bigg|_{x=4L} \end{Bmatrix} - \dfrac{1}{L}\begin{bmatrix} 1 & -1 & 0 & 0 & 0 \\ -1 & 2 & -1 & 0 & 0 \\ 0 & -1 & 2 & -1 & 0 \\ 0 & 0 & -1 & 2 & -1 \\ 0 & 0 & 0 & -1 & 1 \end{bmatrix}\begin{Bmatrix} u_1 \\ u_2 \\ u_3 \\ u_4 \\ u_5 \end{Bmatrix}$$

$$+ \dfrac{L}{6}\begin{bmatrix} 2 & 1 & 0 & 0 & 0 \\ 1 & 4 & 1 & 0 & 0 \\ 0 & 1 & 4 & 1 & 0 \\ 0 & 0 & 1 & 4 & 1 \\ 0 & 0 & 0 & 1 & 2 \end{bmatrix}\begin{Bmatrix} u_1 \\ u_2 \\ u_3 \\ u_4 \\ u_5 \end{Bmatrix} - \dfrac{L}{2}\begin{Bmatrix} 1 \\ 2 \\ 2 \\ 2 \\ 1 \end{Bmatrix} = \begin{Bmatrix} 0 \\ 0 \\ 0 \\ 0 \\ 0 \end{Bmatrix}$$

Now, if $u_1 = 0$ and $u_5 = 0$ (stated essential boundary conditions), the matrix equation can be reduced by removing rows and columns 1 and 5. Note that in the process, the natural boundary conditions at nodes 1 and 5 are removed from consideration. We thus have

$$\dfrac{-1}{L}\begin{bmatrix} 2 & -1 & 0 \\ -1 & 2 & -1 \\ 0 & -1 & 2 \end{bmatrix}\begin{Bmatrix} u_2 \\ u_3 \\ u_4 \end{Bmatrix} + \dfrac{L}{6}\begin{bmatrix} 4 & 1 & 0 \\ 1 & 4 & 1 \\ 0 & 1 & 4 \end{bmatrix}\begin{Bmatrix} u_2 \\ u_3 \\ u_4 \end{Bmatrix} - \dfrac{L}{2}\begin{Bmatrix} 2 \\ 2 \\ 2 \end{Bmatrix} = \begin{Bmatrix} 0 \\ 0 \\ 0 \end{Bmatrix}$$

Multiplying each term by $6L$, substituting the element common length, $L = \pi/2$, and summing the first two terms on the left yields

$$\begin{bmatrix} -2.1304 & 8.4674 & 0 \\ & -2.1304 & 8.4674 \\ \text{symmetric} & & -2.1304 \end{bmatrix}\begin{Bmatrix} u_2 \\ u_3 \\ u_4 \end{Bmatrix} = 14.804\begin{Bmatrix} 1 \\ 1 \\ 1 \end{Bmatrix}$$

Solving gives us

$$\begin{Bmatrix} u_2 \\ u_3 \\ u_4 \end{Bmatrix} = \begin{Bmatrix} 1.130 \\ 2.033 \\ 1.130 \end{Bmatrix}$$

For comparison purposes, the problem was solved once again with eight elements of equal length $L = \pi/4$. Table 9.3 gives the finite element results together with the exact values.

TABLE 9.3 Comparison of Finite Element Results

X	Exact	Four Elements	Eight Elements
$\pi/4$	0.293	—	0.332
$2\pi/4$	1.000	1.130	1.038
$3\pi/4$	1.707	—	1.722
$4\pi/4$	2.000	2.033	2.003
$5\pi/4$	1.707	—	1.722
$6\pi/4$	1.000	1.130	1.038
$7\pi/4$	0.293	—	0.332

In summary, the Galerkin method of weighted residuals has been used to develop the finite element procedure for obtaining solutions at discrete points in the continuum. This is in contrast to using the Galerkin method to obtain a continuous function approximation as in Examples 9.1 and 9.2.

We should recall that values of the dependent variable at intermediate points in the region may be calculated using the interpolation formula

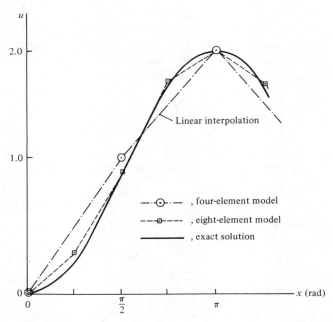

FIGURE 9.6 Comparison of results for two solutions and effect of linear interpolation for intermediate values.

(9.10). The error of these intermediate values, however, may far exceed the error of the nodal values.

Figure 9.6 shows a plot of the exact solution with the results of the two finite element solutions. Straight lines, which represent the interpolation function, have been drawn between the nodal values, and the error magnification at intermediate points is readily apparent.

Improved accuracy can be achieved by employing either a larger number of elements or a higher-order element capable of modeling a nonlinear solution.

9.6 JUMP CONDITIONS IN SYSTEM PROPERTIES

As one more example, we will look at a solution involving our familiar one-dimensional elasticity problem. As an added feature, however, we will allow the region cross-sectional area, A, and the material elastic modulus, E, to be a variable. The governing differential equation of equilibrium for this case is

$$\frac{d}{dx}\left(AE\frac{du}{dx}\right) + T(x) = 0 \qquad (9.18)$$

Multiply by the weighting factor and integrate.

$$\int_X W_i \frac{d}{dx}\left(AE\frac{du}{dx}\right) dx + \int_X W_i T(x)\, dx = 0$$

Integrate the first term by parts:

$$W_i AE \frac{du}{dx}\bigg|_{x=0}^{x=L} - \int_X \frac{dW_i}{dx} AE \frac{du}{dx}\, dx + \int_X W_i T(x)\, dx = 0 \quad (9.19)$$

We now proceed with the familiar steps of:

1. Discretizing the region and summing element integrals.
2. Assuming the linear displacement function:

$$u = [N_1 \quad N_2]\begin{Bmatrix} u_i \\ u_j \end{Bmatrix}$$

3. Substituting the shape functions, N_i, for the weighting functions, W_i.

For the general element equation we arrive at the following:

$$\begin{Bmatrix} -AE\frac{du}{dx}\bigg|_{x=x_i} \\ AE\frac{du}{dx}\bigg|_{x=x_j} \end{Bmatrix} - \int_{X_e} AE \begin{Bmatrix} \frac{dN_1}{dx} \\ \frac{dN_2}{dx} \end{Bmatrix} \begin{bmatrix} \frac{dN_1}{dx} & \frac{dN_2}{dx} \end{bmatrix} \begin{Bmatrix} u_i \\ u_j \end{Bmatrix} dx$$

$$+ \int_{X_e} \begin{Bmatrix} N_1 \\ N_2 \end{Bmatrix} T(x)\, dx = \{0\} \qquad (9.20)$$

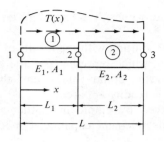

FIGURE 9.7 Two-element region with varying area and elastic modulus.

Assume now that the area A and the elastic modulus E are constant for the element (but not for the entire region), as shown in the Figure 9.7.

Performing the integration on the second term, we obtain the following:

$$\left\{ \begin{array}{c} -AE\dfrac{du}{dx}\bigg|_{x=x_i} \\ AE\dfrac{du}{dx}\bigg|_{x=x_j} \end{array} \right\} - \dfrac{A_e E_e}{L_e} \begin{bmatrix} 1 & -1 \\ -1 & 1 \end{bmatrix} \left\{ \begin{array}{c} u_i \\ u_j \end{array} \right\} + \int_{X_e} \left\{ \begin{array}{c} N_1 \\ N_2 \end{array} \right\} T(x)\, dx = \{0\} \quad (9.21)$$

We can now "expand" each element equation and sum the equations over the two-element region.

$$\left\{ \begin{array}{c} -AE\dfrac{du}{dx}\bigg|_{x=0} \\ 0 \\ AE\dfrac{du}{dx}\bigg|_{x=L} \end{array} \right\} - \dfrac{A_1 E_1}{L_1} \begin{bmatrix} 1 & -1 & 0 \\ -1 & 1 & 0 \\ 0 & 0 & 0 \end{bmatrix} \left\{ \begin{array}{c} u_1 \\ u_2 \\ u_3 \end{array} \right\} - \dfrac{A_2 E_2}{L_2} \begin{bmatrix} 0 & 0 & 0 \\ 0 & 1 & -1 \\ 0 & -1 & 1 \end{bmatrix} \left\{ \begin{array}{c} u_1 \\ u_2 \\ u_3 \end{array} \right\}$$

$$+ \int_0^{L_1} \left\{ \begin{array}{c} N_1 T(x) \\ N_2 T(x) \\ 0 \end{array} \right\} dx + \int_{L_1}^{L} \left\{ \begin{array}{c} 0 \\ N_1 T(x) \\ N_2 T(x) \end{array} \right\} dx = \{0\} \quad (9.22)$$

We recognize the boundary terms to be the forces at nodes 1 and 3.

$$F_1 = -AE\dfrac{du}{dx}\bigg|_{x=0}$$

$$F_3 = AE\dfrac{du}{dx}\bigg|_{x=L}$$

We now have an expression of the (familiar) form

$$\{R\}_{NF} - [K]\{r\} + \{R\}_T = \{0\}$$

and we have added the capacity to handle jump conditions of system properties.

9.7 LAPLACE'S EQUATION: TWO DIMENSIONS

The finite element formulation of Laplace's equation,

$$\nabla^2 \phi = 0$$

is an important step, because this equation governs many physical problems, examples being steady-state problems dealing with:

1. Temperature distribution in solids without sources or sinks.
2. Irrotational flow of ideal incompressible fluids without sources or sinks.
3. Flow of electric current in solids.

Our finite element formulation will be for a two-dimensional continuum in the xy plane and of thickness t. We will develop the formulation with the Galerkin method in the usual manner, multiplication of the differential equation by a weighting factor followed by integration over the entire region of the continuum.

$$\int_A W_i \nabla^2 \phi \, t \, dx \, dy = 0 \qquad (9.23)$$

We wish to put this in different form. Noting that

$$\overline{\nabla} \cdot W_i \overline{\nabla} \phi = \overline{\nabla} W_i \cdot \overline{\nabla} \phi + W_i \nabla^2 \phi$$

we solve for $W_i \nabla^2 \phi$:

$$W_i \nabla^2 \phi = \overline{\nabla} \cdot W_i \overline{\nabla} \phi - \overline{\nabla} W_i \cdot \overline{\nabla} \phi$$

Substituting this into (9.23) gives us

$$t \int_A \overline{\nabla} \cdot W_i \overline{\nabla} \phi \, dx \, dy - t \int_A \overline{\nabla} W_i \cdot \overline{\nabla} \phi \, dx \, dy = 0$$

Using Gauss's theorem on the first term yields

$$t \int_S W_i \cdot \overline{\nabla} \phi \cdot \hat{n} \, ds - t \int_A \overline{\nabla} W_i \cdot \overline{\nabla} \phi \, dx \, dy = 0$$

or

$$t \int_S W_i \frac{\partial \phi}{\partial n} \, ds - t \int_A \left[\frac{\partial W_i}{\partial x} \frac{\partial \phi}{\partial x} + \frac{\partial W_i}{\partial y} \frac{\partial \phi}{\partial y} \right] dx \, dy = 0 \qquad (9.24)$$

See Figure 9.8.

Using the interpolation formula to express the dependent variable, ϕ, in terms of the values of $\{\phi_i\}$ at the element nodes,

$$\phi = [N]\{\phi_i\} \qquad \text{and thus} \qquad W_i = N_i$$

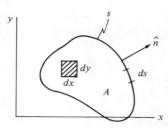

FIGURE 9.8 Region, boundary, s, and unit normal, $\hat{n}$.

we can place (9.24) in terms of the nodal values and the shape functions.

$$\int_S [N]^T \frac{\partial \phi}{\partial n} ds$$

$$- \int_A \left[\frac{\partial [N]^T}{\partial x} \frac{\partial [N]}{\partial x} + \frac{\partial [N]^T}{\partial y} \frac{\partial [N]}{\partial y} \right] dx\, dy\, \{\phi_i\} = \{0\} \quad (9.25)$$

The first term is a column vector representing the natural boundary conditions. The second term is a square matrix. The interpretation of each term will be made as the equation is applied to a particular problem. Note that (9.25) is independent of thickness t.

9.8 POISSON'S EQUATION: TWO DIMENSIONS

The formulation of the Poisson equation

$$\nabla^2 \phi = P(x, y)$$

is a simple extension of the Laplace equation form, since all that needs to be done, additionally, is to multiply the function $P(x, y)$ by the weighting functions, integrate over the region, and place the result on the right side of equation (9.25). This is demonstrated in Chapter 10 with applications.

PROBLEMS

9.1 Given the following differential equation and boundary conditions:

$$\frac{d^2 u}{dx^2} - u = 1$$

$$u(0) = 0$$

$$u(1) = 1$$

use the Galerkin method to obtain an approximate solution function over the region defined by the stated boundaries. Compare with the exact solution.

9.2 Use the finite element method to solve Problem 9.1. Try two, four, and eight elements.

PROBLEMS

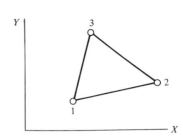

FIGURE P9.3 Triangle element.

9.3 For the element shown in Figure P9.3, assume a variable ϕ that is dependent on coordinates X and Y in a manner governed by Laplace's equation. Assume also a linear variation of this variable across a triangle element:

$$\phi = \alpha_1 + \alpha_2 X + \alpha_3 Y$$

Solve for ϕ in terms of the nodal values ϕ_i (in other words, derive the $[N]$ matrix), substitute into (9.20), and proceed as far as possible in constructing a matrix equation in terms of the unknown nodal values ϕ_i.

9.4 Using the Galerkin method, formulate the finite element equations for the one-dimensional elasticity problem, and check the results for the example given in Section 3.10.

9.5 Using the Galerkin method, formulate the finite element equations for the one-dimensional heat flow problem, and check the solution of the Example 3.25.

9.6 Estimate the solution to the following differential equation in the region 0 to 1. Use the least-squares and/or the Galerkin methods.

$$\frac{dy}{dx} + 2y - x = 0 \qquad y(x=0) = 0$$

The exact solution is given by

$$y = \tfrac{1}{4}(2x - 1 + e^{-2x})$$

CHAPTER 10

Heat Transfer and Fluid Flow in Two Dimensions

10.1 GOVERNING EQUATIONS FOR STEADY-STATE HEAT FLOW

The governing differential equation for steady-state heat flow in the xy plane in a homogeneous isotropic continuum without internal heat sources is the Laplace equation,

$$\nabla^2 T = 0 \tag{10.1}$$

where T is the temperature at a point and

$$\nabla^2 = \frac{\partial^2}{\partial x^2} + \frac{\partial^2}{\partial y^2}$$

The conduction heat flow rate in the x and y directions is dependent on the temperature gradients in those directions and is calculated using the Fourier equations. For heat conduction in the x direction, per unit area normal to the heat flow,

$$q_{cd_x} = -k\frac{\partial T}{\partial x} \tag{10.2a}$$

For conduction per unit area in the y direction,

$$q_{cd_y} = -k\frac{\partial T}{\partial y} \tag{10.2b}$$

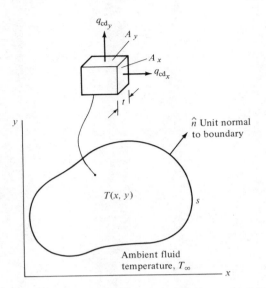

FIGURE 10.1 Continuum with heat flow in xy plane and convection on boundary, s.

where k is the thermal conductivity of the solid with SI units of watts per meter per degree Celsius and English units of Btu per hour per foot per degree Fahrenheit.

At the boundary of the region (Figure 10.1) we will consider only two possible mechanisms of heat flow, conduction and convection; radiation will not be included.

For conduction at the boundary, the heat flow is expressed in terms of the derivative of the temperature in the direction defined by the normal unit vector $\hat{n}$

$$q_n = -k \frac{\partial T}{\partial n} \tag{10.3}$$

For convection at the boundary, the heat flow q_n will be replaced by convection heat flow, q_{cv}, and defined by Newton's law of cooling:

$$q_{cv} = hA(T - T_\infty) \tag{10.4}$$

where h is the convection heat transfer coefficient with English units of Btu per hour per square foot per degree Fahrenheit, A the surface area, T the temperature of the boundary surface, and T_∞ the temperature of the surrounding fluid.

Note: In earlier chapters a distinction was made between local and global coordinates by using upper case XY for global and lower case xy for local coordinates. In all future work we will use only global and

10.2 THE ELEMENT EQUATION

The Galerkin approximation formula for the problem governed by equation (10.1) is given by equation (9.25). For the heat flow problem we simply replace the arbitrary symbol ϕ by the temperature T. Recall that the integration indicated in (9.25) is over the entire region. However, the equation will also be valid for any subregion coming about from a discretization of the region. We will view these subregions as individual elements; instead of creating a single trial function for the entire region, we select trial functions for each element.

We now rewrite equation (9.25) in terms of the scalar temperature, T, and the vector of nodal temperatures, $\{q_t\}$, which was defined initially in Chapter 3. We also place limits on the integration for a single element. This yields the following:

$$\int_{S_e} [N]^T \frac{\partial T}{\partial n} \, ds$$

$$- \int_{A_e} \left[\frac{\partial [N]^T}{\partial x} \frac{\partial [N]}{\partial x} + \frac{\partial [N]^T}{\partial y} \frac{\partial [N]}{\partial y} \right] dx \, dy \, \{q_t\} = \{0\} \quad (10.5)$$

The first term on the left, the boundary integral term, is to be modified to account for convection at a boundary. The heat flow to and from a boundary of the region is shown in Figure 10.2. On the basis of energy balance, the heat, q_{cd_n}, conducted normal to the boundary, S'_e, must

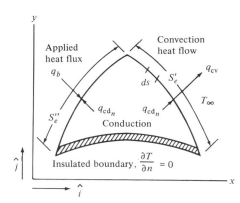

FIGURE 10.2 Heat flow at the boundary.

equal the heat removed by convection, q_{cv}, (radiation effects are not being considered).

$$q_{cd_n} = q_{cv} \quad (10.6)$$

$$\left(q_{cd_x}\hat{i} + q_{cd_y}\hat{j}\right) \cdot \hat{n} = q_{cv}$$

Applying equations (10.2) and (10.4) (continuum thickness is t) yields

$$-tk\frac{\partial T}{\partial n}\,ds = ht(ds)(T - T_\infty)$$

or

$$-\frac{\partial T}{\partial n}\,ds = \frac{h}{k}(T - T_\infty)\,ds \quad (10.7)$$

Substituting the element interpolation formula, $T = [N]\{q_t\}$, we obtain the following:

$$-\frac{\partial T}{\partial n}\,ds = \frac{h}{k}[N]\{q_t\}\,ds - \frac{h}{k}T_\infty\,ds \quad (10.8)$$

Substituting (10.8) into the boundary integral of (10.5), we have the working form of the element equation:

$$-h\int_{S'_e}[N]^T[N]\,ds\{q_t\} + h\int_{S'_e}[N]^T T_\infty\,ds + k\int_{S''_e}[N]^T\frac{\partial T}{\partial n}\,ds$$

$$- k\int_{A_e}\left[\frac{\partial[N]^T}{\partial x}\frac{\partial[N]}{\partial x} + \frac{\partial[N]^T}{\partial y}\frac{\partial[N]}{\partial y}\right]dx\,dy\{q_t\} = \{0\}$$

$$(10.9)$$

where it is assumed that the convection coefficient, h, and the conduction coefficient, k, are constant on the element boundary and throughout the element, respectively. This element equation can be written in the familiar form:

$$[k_T]\{q_t\} = \{Q_{cv}\} + \{Q_b\} \quad (10.10)$$

where the "thermal stiffness" matrix

$$[k_T] = [k_{cd_x}] + [k_{cd_y}] + [k_{cv}].$$

The x-direction conduction matrix is

$$[k_{cd_x}] = k\int_{A_e}\frac{\partial[N]^T}{\partial x}\frac{\partial[N]}{\partial x}\,dx\,dy \quad (10.11a)$$

The y-direction conduction matrix is

$$[k_{cd_y}] = k\int_{A_e}\frac{\partial[N]^T}{\partial y}\frac{\partial[N]}{\partial y}\,dx\,dy \quad (10.11b)$$

The boundary convection matrix is

$$[k_{cv}] = h\int_{S'_e}[N]^T[N]\,ds \quad (10.11c)$$

10.3 SELECTING THE ELEMENT

Note: This matrix is evaluated only for elements with a side on the region convection boundary, S'_e.

The convection boundary, S'_e, "force" vector is

$$\{Q_{cv}\} = h \int_{S'_e} [N]^T T_\infty \, ds \qquad (10.11d)$$

Note: This vector is evaluated only for elements with a side on the region convection boundary, S'_e.

The applied heat boundary, S''_e, "force" vector is

$$\{Q_b\} = k \int_{S''_e} [N]^T \frac{\partial T}{\partial n} \, ds \qquad (10.11e)$$

10.3 SELECTING THE ELEMENT

To illustrate the application of the general matrix equations, we will select the three-node isoparametric triangle element. This element has been discussed in detail in Chapter 6 in its application to the stress analysis formulation. It will be seen that the procedures for deriving the element functions and many of the functions themselves are very similar.

The element trial function for the Galerkin solution is the interpolation formula. For the three-node triangle (Figure 10.3) this can be expressed in terms of the natural coordinates as

$$T = L_1 T_1 + L_2 T_2 + L_3 T_3 = [N]\{q_t\} \qquad (10.12)$$

Similarly, the global coordinates are defined with the same interpolation function:

$$x = L_1 x_1 + L_2 x_2 + L_3 x_3 = [N]\{x_i\} \qquad (10.13a)$$

and

$$y = [N]\{y_i\} \qquad (10.13b)$$

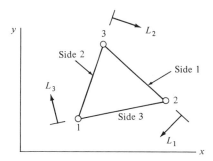

FIGURE 10.3 Triangle natural coordinates and side designation.

10.4 DETERMINATION OF ELEMENT MATRICES

In evaluating the element matrices in equation (10.10), we need the following functions:

$$[N] \quad \frac{\partial [N]}{\partial x} \quad \frac{\partial [N]}{\partial y}$$

we know that

$$[N] = [L_1 \quad L_2 \quad L_3] \tag{10.14}$$

Refer to Chapter 6. Compare equations (6.2a) and (6.3) with equations (10.12) and (10.13). Equation (6.6) defines the spatial derivatives of the displacement, u:

$$\frac{\partial u}{\partial x} = \frac{\partial L_1}{\partial x} u_1 + \frac{\partial L_2}{\partial x} u_2 + \frac{\partial L_3}{\partial x} u_3$$

and we are thus provided with the derivatives of the interpolation functions:

$$\frac{\partial [N]}{\partial x} = \frac{1}{|J|}[y_2 - y_3 \quad y_3 - y_1 \quad y_1 - y_2] = \frac{1}{|J|}[b_1 \quad b_2 \quad b_3] \tag{10.15}$$

$$\frac{\partial [N]}{\partial y} = \frac{1}{|J|}[x_3 - x_2 \quad x_1 - x_3 \quad x_2 - x_1] = \frac{1}{|J|}[c_1 \quad c_2 \quad c_3] \tag{10.16}$$

where

$$|J| = (x_1 - x_3)(y_2 - y_3) - (x_2 - x_3)(y_1 - y_3) = 2 \text{ (area)}$$

Observe that the derivatives are constant for the three-node element. Just as this element provided constant strain throughout the element, in this application the heat flow *per unit area* is constant throughout the triangle, since

$$q_{cd_x} = -k\frac{\partial T}{\partial x} = -k\frac{\partial [N]}{\partial x}\{q_t\} \quad \text{etc.}$$

We may now evaluate the element terms.

10.4 DETERMINATION OF ELEMENT MATRICES

The heat conduction matrices $[k_{cd_x}]$ and $[k_{cd_y}]$:

$$[k_{cd_x}] = k\int_{A_e} \frac{\partial[N]^T}{\partial x}\frac{\partial[N]}{\partial x}\,dx\,dy = k\frac{\partial[N]^T}{\partial x}\frac{\partial[N]}{\partial x}\int_{A_e} dx\,dy$$

$$= \frac{k}{2|J|}\begin{bmatrix} b_1 b_1 & b_1 b_2 & b_1 b_3 \\ b_2 b_1 & b_2 b_2 & b_2 b_3 \\ b_3 b_1 & b_3 b_2 & b_3 b_3 \end{bmatrix} \quad (10.17)$$

$$[k_{cd_y}] = k\int_{A_e} \frac{\partial[N]^T}{\partial y}\frac{\partial[N]}{\partial y}\,dx\,dy$$

$$= \frac{k}{2|J|}\begin{bmatrix} c_1 c_1 & c_1 c_2 & c_1 c_3 \\ c_2 c_1 & c_2 c_2 & c_2 c_3 \\ c_3 c_1 & c_3 c_2 & c_3 c_3 \end{bmatrix} \quad (10.18)$$

The boundary convection matrix $[k_{cv}]$:

$$[k_{cv}] = h\int_{S_e} [N]^T[N]\,ds$$

or

$$[k_{cv}] = h\int_{S_e} \begin{bmatrix} L_1 L_1 & L_1 L_2 & L_1 L_3 \\ L_2 L_1 & L_2 L_2 & L_2 L_3 \\ L_3 L_1 & L_3 L_2 & L_3 L_3 \end{bmatrix} ds \quad (10.19)$$

The natural coordinates are functions of x and y so may not be taken outside the integral sign. Instead, to evaluate the integral we will employ formula (6.13):

$$\int_S L^\alpha L^\beta\,ds = \frac{\alpha!\beta!}{(\alpha + \beta + 1)!} S \quad (6.13)$$

The integration of this element matrix is to be performed only for elements that are on the boundary of the region. Figure 10.4 shows an element with side 1 on the boundary.

Substituting $L_1 = 0$ into (10.19), we have

$$[k_{cv}] = h\int_{S_e} \begin{bmatrix} 0 & 0 & 0 \\ 0 & L_2 L_2 & L_2 L_3 \\ 0 & L_3 L_2 & L_3 L_3 \end{bmatrix} ds$$

Evaluating the 2, 2 term, we obtain

$$\int_{S_e} L_2 L_2\,ds = \frac{2!}{(2+1)!} S_1$$

$$= \frac{1}{3} S_1$$

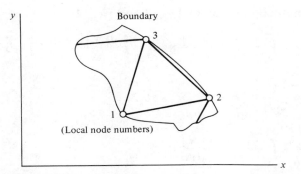

FIGURE 10.4 Side 1 (local node number) on continuum boundary.

Evaluating all terms in the same manner yields:

FOR SIDE 1 ON BOUNDARY

$$[k_{cv}]_1 = \frac{hS_1}{6}\begin{bmatrix} 0 & 0 & 0 \\ 0 & 2 & 1 \\ 0 & 1 & 2 \end{bmatrix} \qquad (10.20)$$

where S_1 is the length of side 1.

FOR SIDE 2 ON BOUNDARY

$$[k_{cv}]_2 = \frac{hS_2}{6}\begin{bmatrix} 2 & 0 & 1 \\ 0 & 0 & 0 \\ 1 & 0 & 2 \end{bmatrix} \qquad (10.21)$$

FOR SIDE 3 ON BOUNDARY

$$[k_{cv}]_3 = \frac{hS_3}{6}\begin{bmatrix} 2 & 1 & 0 \\ 1 & 2 & 0 \\ 0 & 0 & 0 \end{bmatrix} \qquad (10.22)$$

The boundary convection vector $\{Q_{cv}\}$:

$$\{Q_{cv}\} = h\int_{S_e} [N]^T T_\infty \, ds$$

This vector is also to be evaluated only for the boundary elements.

$$\{Q_{cv}\} = h\int_{S_e} \begin{Bmatrix} L_1 \\ L_2 \\ L_3 \end{Bmatrix} T_\infty \, ds$$

Using formula (6.13), evaluate the vector considering, in turn, each side of the triangle being on the boundary.

10.4 DETERMINATION OF ELEMENT MATRICES

FOR SIDE 1 ON BOUNDARY

$$\{Q_{cv}\}_1 = hT_\infty \int_S \begin{Bmatrix} 0 \\ L_2 \\ L_3 \end{Bmatrix} ds$$

$$= \frac{hT_\infty S_1}{2} \begin{Bmatrix} 0 \\ 1 \\ 1 \end{Bmatrix} \quad (10.23)$$

where S_1 is the length of side 1.

FOR SIDE 2 ON BOUNDARY

$$\{Q_{cv}\}_2 = \frac{hT_\infty S_2}{2} \begin{Bmatrix} 1 \\ 0 \\ 1 \end{Bmatrix} \quad (10.24)$$

FOR SIDE 3 ON BOUNDARY

$$\{Q_{cv}\}_3 = \frac{hT_\infty S_3}{2} \begin{Bmatrix} 1 \\ 1 \\ 0 \end{Bmatrix} \quad (10.25)$$

The boundary applied heat vector $\{Q_b\}$:

$$\{Q_b\} = k \int_{S_e''} [N]^T \frac{\partial T}{\partial n} ds$$

This vector represents the heat flow normal to the boundary on all boundary surfaces that are not thermally insulated or do not have convection. The vector defines the heat flow required at each node to yield the equivalent distributed heat flow on the element side. Refer to Figure 10.2.

An energy balance at the region boundary shows that the unit area applied boundary heat flux, q_b, must equal the negative of the region conduction heat flow toward the boundary.

$$q_{b_n} = -q_{cd_n} = -\left(-k\frac{\partial T}{\partial n}\right)$$

Thus we have

$$\{Q_b\} = \int_{S_e''} [N]^T q_{b_n} ds$$

This expression is very similar to the convection force vector of (10.11d) with the hT_∞ product replaced by the boundary heat flux q_b. Assume, for example, that the boundary flux is constant over side 1 of the triangle element of Figure 10.4. The heat flow vector becomes

$$\{Q_b\}_1 = q_{b_n} \frac{S_1}{2} \begin{Bmatrix} 0 \\ 1 \\ 1 \end{Bmatrix}$$

The vector for other sides on the boundary are seen to be equivalent to the convection counterparts given by (10.24) and (10.25).

Note the special case of an insulated boundary. Zero heat flow normal to the boundary yields $\{Q_b\} = \{0\}$.

Example 10.1 The single three-node triangle element shown in Figure 10.5 has side 3 against a heat source and sides 1 and 2 exposed to an ambient fluid temperature T of 50°C.

Assume the following constants:

$$\text{thermal conductivity, } k = 25 \text{ W/cm} \cdot \text{°C}$$

$$\text{convection heat transfer coefficient, } h = 5 \text{ W/cm}^2 \cdot \text{°C}.$$

Develop the matrix equation for this element.

Solution. 1. First, we will calculate the heat conduction matrices defined by (10.17) and (10.18).

$$|J| = (x_1 - x_3)(y_2 - y_3) - (x_2 - x_3)(y_1 - y_3)$$
$$= (-2.5)(-2) - (2.5)(-2)$$
$$= 10$$

$$b_1 = y_2 - y_3 = -2 \qquad c_1 = x_3 - x_2 = -2.5$$
$$b_2 = y_3 - y_1 = 2 \qquad c_2 = x_1 - x_3 = -2.5$$
$$b_3 = y_1 - y_2 = 0 \qquad c_3 = x_2 - x_1 = 5$$

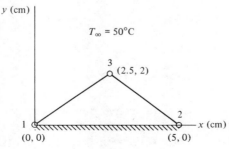

FIGURE 10.5 Single element with convection on two sides.

10.4 DETERMINATION OF ELEMENT MATRICES

Then we substitute into (10.17) and (10.18):

$$[k_{cd_x}] = \frac{25}{(2)(10)} \begin{bmatrix} 4 & -4 & 0 \\ -4 & 4 & 0 \\ 0 & 0 & 0 \end{bmatrix},$$

$$[k_{cd_y}] = \frac{25}{2(10)} \begin{bmatrix} 6.25 & 6.25 & -12.5 \\ 6.25 & 6.25 & -12.5 \\ -12.5 & -12.5 & 25.0 \end{bmatrix}$$

2. Second, calculate the boundary convection matrices defined by (10.20), (10.21), and (10.22). For this triangle, side 3 is not on a boundary exposed to convection. Thus (10.22) is not used. However, side 1 is on the boundary, so we use (10.20) with the length of side 1 equal to 3.2 cm.

$$[k_{cv}]_1 = \frac{5(3.2)}{6} \begin{bmatrix} 0 & 0 & 0 \\ 0 & 2 & 1 \\ 0 & 1 & 2 \end{bmatrix}$$

Similarly, side 2 is on the boundary and for this side 2 we use (10.21) with $S_2 = 3.2$ cm:

$$[k_{cv}]_2 = \frac{5(3.2)}{6} \begin{bmatrix} 2 & 0 & 1 \\ 0 & 0 & 0 \\ 1 & 0 & 2 \end{bmatrix}$$

3. Sum the conduction and convection matrices to obtain the thermal stiffness matrix.

$$[k_T] = \underbrace{[k_{cd_x}] + [k_{cd_y}]}_{} + \underbrace{[k_{cv}]_1 + [k_{cv}]_2}_{}$$

$$= \frac{25}{20} \begin{bmatrix} 10.25 & 2.25 & -12.5 \\ & 10.25 & -12.5 \\ \text{symmetric} & & 25.0 \end{bmatrix} + \frac{16}{6} \begin{bmatrix} 2 & 0 & 1 \\ & 2 & 1 \\ \text{symmetric} & & 4 \end{bmatrix}$$

$$[k_T] = \begin{bmatrix} 18.14 & 2.81 & -12.96 \\ & 18.14 & -12.96 \\ \text{symmetric} & & 41.92 \end{bmatrix}$$

4. Now determine the convection "force" vector. Again, side 3 does not have convection, so (10.25) is not used. For side 1 we use (10.23):

$$\{Q_{cv}\}_1 = \frac{5(50)(3.2)}{2} \begin{Bmatrix} 0 \\ 1 \\ 1 \end{Bmatrix} = \begin{Bmatrix} 0 \\ 400 \\ 400 \end{Bmatrix}$$

For side 2, use (10.24):

$$\{Q_{cv}\}_2 = \frac{5(50)(3.2)}{2} \begin{Bmatrix} 1 \\ 0 \\ 1 \end{Bmatrix} = \begin{Bmatrix} 400 \\ 0 \\ 400 \end{Bmatrix}$$

Sum the two vectors:

$$\{Q_{cv}\} = \{Q_{cv}\}_1 + \{Q_{cv}\}_2$$

$$\{Q_{cv}\} = \begin{Bmatrix} 400 \\ 400 \\ 800 \end{Bmatrix}$$

5. Determine the applied heat vector. Side 3 is the only side that has a nonconvective-type heat flow. The applied heat at this boundary is unknown (not given in the problem statement), but we will assume it to be uniformly distributed so that the vector can be written as:

$$\{Q_b\} = q_{b_n} \frac{S_3}{2} \begin{Bmatrix} 1 \\ 1 \\ 0 \end{Bmatrix} = \begin{Bmatrix} Q_1 \\ Q_2 \\ 0 \end{Bmatrix}$$

From equation (10.10), the resulting matrix equation for the single element is

$$[k_T]\{q_t\} = \{Q_{cv}\} + \{Q_b\}$$

$$\begin{bmatrix} 18.14 & 2.81 & -12.96 \\ & 18.14 & -12.96 \\ \text{symmetric} & & 41.92 \end{bmatrix} \begin{Bmatrix} T_1 \\ T_2 \\ T_3 \end{Bmatrix} = \begin{Bmatrix} 400 \\ 400 \\ 800 \end{Bmatrix} + \begin{Bmatrix} Q_1 \\ Q_2 \\ 0 \end{Bmatrix}$$

Note that the solution of this equation requires the specification of the heat flow quantities Q_1 and Q_2 or, alternatively, specification of the nodal temperatures T_1 and T_2 or any combination of the heat flow, the natural boundary condition, or temperature, the essential boundary condition. For example, if we assume side 3 to be perfectly insulated so that $Q_1 = 0 = Q_2$, the solution of the equation yields

$$T_1 = 50°C = T_2 = T_3 \qquad \square$$

10.5 ASSEMBLING THE ELEMENT EQUATIONS

It may be helpful to review the assembly technique presented in Section 4.5. For the heat flow equations, the matrix assembly procedure is similar to, although simpler than, the procedure presented for the displacement equations. For the displacement problem, each of the two nodal displacements was numbered using a formula based on the node number. For the heat flow problem, there is only one scalar temperature for each node. Thus the local and global node temperature numbers are simply taken to be the same as the local and global node numbers, respectively.

If the global node numbers of the element local nodes 1, 2, and 3 are identified (Figure 10.6), this defines the location of the local element matrix entry in the global assemblage matrix.

10.5 ASSEMBLING THE ELEMENT EQUATIONS

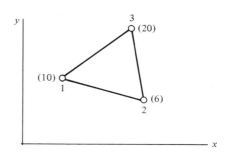

FIGURE 10.6 Global node numbers, ( ), of local nodes 1, 2, and 3.

For example, given the thermal stiffness matrix and force vectors for one element in an assemblage, we obtain

$$
[k_T] = \begin{matrix} \overset{(10)}{1} & \overset{(6)}{2} & \overset{(20)}{3} \\ \begin{bmatrix} - & - & - \\ - & - & - \\ - & - & - \end{bmatrix} & \begin{matrix} 1 & (10) \\ 2 & (6) \\ 3 & (20) \end{matrix} \end{matrix} \qquad \{Q_{cv}\} = \begin{Bmatrix} - \\ - \\ - \end{Bmatrix} \begin{matrix} 1 & (10) \\ 2 & (6) \\ 3 & (20) \end{matrix}
$$

— global numbers
— local numbers

Local entry (1, 3) of the element $[k_T]$ matrix would be placed in the global matrix in location (10, 20); entry 1 of the element $\{Q_{cv}\}$ vector would be placed in row 10 of the global vector; and so on.

Example 10.2 In Figure 10.7, a second triangle has been added to the triangle of Example 10.1, and the vertical wall is insulated. Assume the same constants, and determine the two-element assemblage equation.

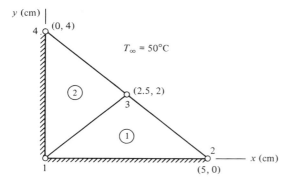

FIGURE 10.7 Node coordinates, global numbers, and element numbers.

350 HEAT TRANSFER AND FLUID FLOW IN TWO DIMENSIONS

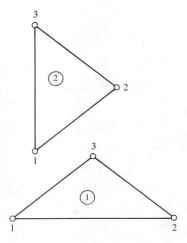

FIGURE 10.8 Element local nodes.

Solution. The local node numbering of the triangle is arbitrarily selected as shown in Figure 10.8. Solving for the element 2 matrices, we have, from (10.17) and (10.18),

$$[k_{cd_x}] = \frac{25}{20}\begin{bmatrix} 4 & -8 & 4 \\ & 16 & -8 \\ \text{symmetric} & & 4 \end{bmatrix}$$

$$[k_{cd_y}] = \frac{25}{20}\begin{bmatrix} 6.25 & 0 & -6.25 \\ & 0 & 0 \\ \text{symmetric} & & 6.25 \end{bmatrix}$$

For the convection matrix and force vector, only side 1 (local) is involved in the integration.

$$[k_{cv}]_1 = \begin{bmatrix} 0 & 0 & 0 \\ & 5.33 & 2.67 \\ \text{symmetric} & & 5.33 \end{bmatrix} \quad \{Q_{cv}\}_1 = \begin{Bmatrix} 0 \\ 400 \\ 400 \end{Bmatrix}$$

Summing the matrices, we have for element 2:

$$[k_T]_{②} = \begin{bmatrix} 12.81 & -10.0 & -2.81 \\ & 25.33 & -7.33 \\ \text{symmetric} & & 18.14 \end{bmatrix} \begin{matrix} ① \\ ③ \\ ④ \end{matrix} \quad \{Q_{cv}\}_{②} = \begin{Bmatrix} 0 \\ 400 \\ 400 \end{Bmatrix} \begin{matrix} ① \\ ③ \\ ④ \end{matrix}$$

— global numbers —

From Example 10.1 we have the matrices for element 1. However, we have to subtract the convection matrix and vector associated with side (local) 2 because it no longer is on the boundary. Having done that,

10.7 SOLVING THE EQUATIONS

we present the following matrix and vector for element 1:

$$[k_T]_① = \begin{bmatrix} 12.81 & 2.81 & -15.63 \\ & 18.14 & -12.96 \\ \text{symmetric} & & 36.59 \end{bmatrix} \begin{matrix} ① \\ ② \\ ③ \end{matrix} \quad \{Q_{cv}\}_① = \begin{Bmatrix} 0 \\ 400 \\ 400 \end{Bmatrix} \begin{matrix} ① \\ ② \\ ③ \end{matrix}$$

where ①, ②, ③ are global numbers.

$$\{Q_b\}_① = \begin{Bmatrix} Q_1 \\ Q_2 \\ 0 \end{Bmatrix}$$

Combining, we have the two-element assemblage equation:

$$\begin{bmatrix} 25.62 & 2.81 & -25.63 & -2.81 \\ & 18.14 & -12.96 & 0 \\ & & 61.92 & -7.33 \\ \text{symmetric} & & & 18.14 \end{bmatrix} \begin{Bmatrix} T_1 \\ T_2 \\ T_3 \\ T_4 \end{Bmatrix} = \begin{Bmatrix} 0 \\ 400 \\ 800 \\ 400 \end{Bmatrix} + \begin{Bmatrix} Q_1 \\ Q_2 \\ 0 \\ 0 \end{Bmatrix}$$

□

10.6 COMMENT ON BOUNDARY CONDITIONS

In the two-element problem of Example 10.2, it was stated that the vertical wall was insulated and thus the natural boundary condition, the heat flow normal to the boundary, was known to be zero. This meant that the boundary integral along the insulated boundary was zero.

Alternatively, it could have been stated that the vertical wall was conducting power of a known magnitude. This would result in known nonzero quantities in the applied heat flux vector.

The final possibility would involve specification of the temperature along the wall. The heat flow across the boundary in this case would be unknown, but the equation-solving procedures, discussed in the next Section, would remove these unknowns from the simultaneous equations.

Either the temperature or the heat flow must be specified over the entire boundary of the region. Heat flow is the boundary condition for insulated and convection boundaries; remaining boundaries must have known temperature or known applied heat flux.

10.7 SOLVING THE EQUATIONS

Looking at the assemblage matrix equation of Example 10.2, we have four unknown node temperatures and two unknown applied heat flux quantities. In order to completely define the problem, we must have the heat flux quantities Q_1 and Q_2 or the temperature of nodes 1 and 2 or some combination of boundary values. Specification of the heat flux values would lead to an obvious equation solving procedure. Specification of the temperature, the essential boundary condition, requires special consideration of the computer program algorithm for solving the equations.

In applying the essential nonzero boundary conditions, we should recall the discussion in Section 3.13 concerning the two techniques of data entry. If maintaining the symmetry of the stiffness matrix K_T is unimportant (for the equation-solving procedure), we would elect to use the first technique. If we must maintain matrix symmetry, the second technique is to be used. The first technique was used in Example 3.3; the second technique will be illustrated in the next example.

Example 10.3 Let the global node 1 temperature of Example 10.2 be given as 100°C and assume that the boundary applied heat flux is concentrated at node 1. This implies that Q_1 is unknown and $Q_2 = 0$. Solve for the temperatures of the other nodes and then solve for the heat flow in each element.

Solution. Applying the technique described above to the matrix equation developed in Example 10.2, we have

$$\begin{bmatrix} 25.62(E16) & 2.81 & -25.63 & -2.81 \\ & 18.14 & -12.96 & 0 \\ & & 61.92 & -7.33 \\ & & & 18.14 \end{bmatrix} \begin{Bmatrix} T_1 \\ T_2 \\ T_3 \\ T_4 \end{Bmatrix} = \begin{Bmatrix} 100(25.62)(E16) \\ 400 \\ 800 \\ 400 \end{Bmatrix}$$

Solving for the temperatures, we obtain

$$\begin{Bmatrix} T_1 \\ T_2 \\ T_3 \\ T_4 \end{Bmatrix} = \begin{Bmatrix} 100.0 \\ 60.0 \\ 74.9 \\ 67.8 \end{Bmatrix} \text{ °C}$$

Having determined the nodal temperatures, we can calculate the heat flow per unit area for each element using equations (10.2a) and (10.2b).

$$q_{cd_x} = -k \frac{\partial T}{\partial x} = -k \frac{\partial [N]}{\partial x} \{q_t\}$$

$$= \frac{k}{|J|} [b_1 \quad b_2 \quad b_3] \{q_t\} \quad (10.26)$$

and

$$q_{cd_y} = \frac{k}{|J|} [c_1 \quad c_2 \quad c_3] \{q_t\} \quad (10.27)$$

10.7 SOLVING THE EQUATIONS

For element 1:

$$q_{cd_x} = \frac{-25}{10}[-2 \quad 2 \quad 0]\begin{Bmatrix} 100 \\ 60 \\ 74.9 \end{Bmatrix} \qquad q_{cd_x} = 200 \text{ W/cm}^2$$

$$q_{cd_y} = \frac{-25}{10}[-2.5 \quad -2.5 \quad 5]\begin{Bmatrix} 100 \\ 60 \\ 74.9 \end{Bmatrix} \qquad q_{cd_y} = 63.8 \text{ W/cm}^2$$

For element 2:

$$q_{cd_x} = \frac{-25}{10}[-2 \quad 4 \quad -2]\begin{Bmatrix} 100 \\ 74.9 \\ 67.8 \end{Bmatrix} \qquad q_{cd_x} = 90 \text{ W/cm}^2$$

$$q_{cd_y} = \frac{-25}{10}[-2.5 \quad 0 \quad 2.5]\begin{Bmatrix} 100 \\ 74.9 \\ 67.8 \end{Bmatrix} \qquad q_{cd_y} = 201 \text{ W/cm}^2$$

The inadequacy of using only two three-node triangle elements becomes apparent when we observe the large differences in heat flow at the common element boundary and the violation of the boundary condition of zero heat flow normal to the insulated walls. A better model will require more elements, and this will necessitate use of the computer. □

Appendix 7 contains a program in Applesoft Basic for heat flow in the xy plane in a region that may have boundary convection but no radiation or internal heat sources. The next example will involve its application.

Example 10.4 In Figure 10.9, the left side of the square plate is maintained at a temperature of 100°C while the other three sides are at a temperature of 0°C.

In keeping with the assumption of our derived finite element equations, heat flow occurs in the XY plane only. The thermal conductivity of the plate is 10 W/cm · °C. Determine the temperature distribution in the plate along the line $Y = 4$ for a variety of element mesh patterns. Compare the results with the exact solution of this problem.

Solution. The computer program listed and described in Appendix 7 was used to determine the temperatures. The four different mesh patterns used for the analysis are shown in Figure 10.10, and the results for each mesh are plotted for comparison in Figure 10.11. The temperature values were obtained by averaging the three node temperatures for each element, and this average temperature was located at the centroid of the particular element.

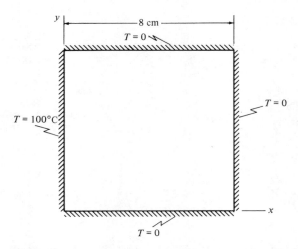

FIGURE 10.9 Plate with applied temperatures on all sides.

With the exception of the very coarse four-element model, there is little difference between the results, and the agreement with the exact solution is excellent.

One obvious dilemma confronts the analyst when applying boundary conditions: That is, what value should be specified at a corner node where two different boundary conditions may be indicated? In this problem, for example, the left side is at one temperature and the top and bottom are at a different temperature. What should be specified at the node located at the intersection of the two sides? (You may be aware that this problem is not encountered in the application of the finite difference technique.) We address this question in the next example. □

Example 10.5 The boundary condition of $0°C$ temperature on three sides of the square plate in Example 10.4 is now replaced with convection conditions with the ambient temperature at $0°C$ and a convection coefficient of 10 W/cm^2 · $°C$ (Figure 10.12). The left side remains at $100°C$.

Determine the temperature distribution along the line $Y = 4$ using each of the four element meshes in Figure 10.4. Examine the convergence rate at three points A, B, and C. Investigate the effect of the particular specification of the boundary value for the corner node at $X = 0$, $Y = 8$, and recommend the best procedure to use when a node is common to two different boundary conditions.

10.7 SOLVING THE EQUATIONS

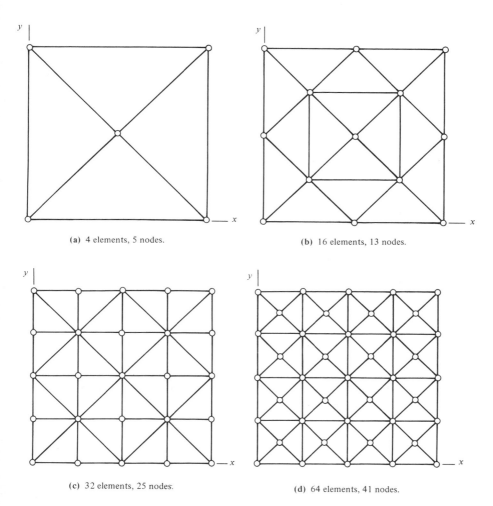

FIGURE 10.10 The four-element mesh patterns used in Examples 10.4 and 10.5.

Solution. The results for the four mesh patterns is shown in Figure 10.13. We note that the convergence of this convection problem is a bit slower than that of the pure conduction problem.

The 16-element solution yields temperatures approximately 20% lower than the 32-element solution, but there appears to be no significant difference between the 32- and 64-element solutions. The convergence rate at the three points A, B, and C is shown in Figure 10.14. It should again be mentioned that orderly (monotonic) convergence is best achieved with mesh refinement that contains all the nodes of any previous mesh.

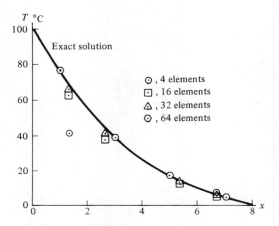

FIGURE 10.11 Element node-average temperatures plotted at each element centroid along the line $Y = 4$ for the conduction problem of Example 10.4.

One more mesh pattern, shown in Figure 10.15, is included to suggest once again the time- and memory-saving bonus that comes with taking advantage of geometric and loading symmetry.

Without any increase in real degrees of freedom, we can effectively double the number of elements in the original model. The boundary condition at the centerline, $Y = 4$, is the natural condition of zero heat flow in the Y direction. The results of this 64-element half-plate mesh

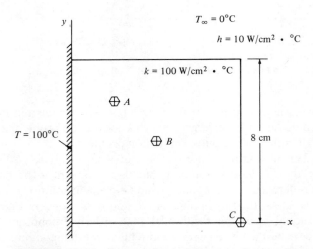

FIGURE 10.12 Square plate with convection on three sides and temperature specified at left side.

10.7 SOLVING THE EQUATIONS 357

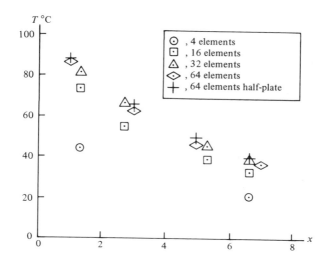

FIGURE 10.13 Element node-average temperatures plotted at element centroids along line $Y = 4$ for conduction/convection Example 10.5.

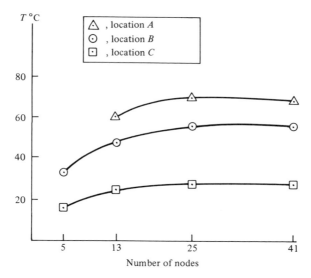

FIGURE 10.14 Convergence behavior at locations A, B, and C for Example 10.5.

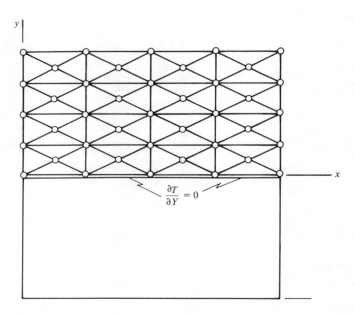

FIGURE 10.15 Element modeling of one half of the region taking advantage of geometric and "loading" symmetry.

pattern is shown in Figure 10.13, and we note very little difference in results from the 64-element full-plate model.

The question of what to do with essential boundary conditions at a corner node is discussed in terms of results from three examples involving the 64-element half-plate model. We will examine the heat flow in the plate in the X direction at two cross sections; $X = 1$ and $X = 3$.

In the first example, the corner node temperature is assigned as the average of the 100°C left boundary and the 0°C ambient. The resulting prediction of the heat flow at the two cross sections is shown in Figure 10.16(a). At the top edge of the plate in the region of the corner node, the heat flow is negative, indicating heat flow in the direction of the region of highest temperature. This does not make good sense physically, but the model predicts this behavior because the temperature specification for the corner node happened to be lower than the calculated temperature of the adjacent node in the X direction. This suggests specification of a higher temperature for the corner node, but what value? Is iteration required?

Results from a second example, which specifies the corner node temperature as the higher temperature at the two boundaries, are shown in Figure 10.16(b). The heat flow is positive for all locations in the cross section, but the pattern at $X = 1$ of maximum heat flow

10.7 SOLVING THE EQUATIONS   359

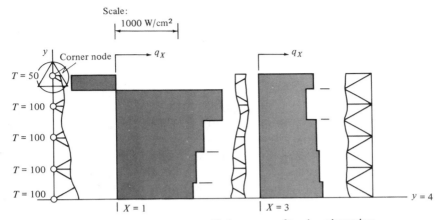

(a) Corner node temperature specified as average of two boundary values.

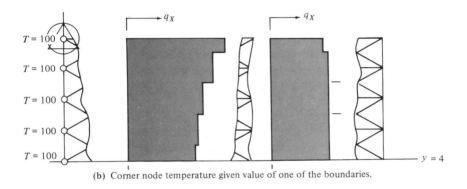

(b) Corner node temperature given value of one of the boundaries.

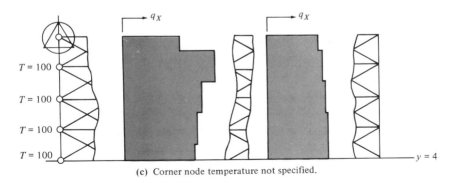

(c) Corner node temperature not specified.

FIGURE 10.16 Effect of the specification of unequal boundary values at a corner node for problem of Example 10.5. Heat flow q_x plotted to scale as indicated at top of figure.

near the boundary and minimum at the centerline is questionable because this implies a higher-temperature gradient in the X direction at the outer boundary. The first example, as well as this example, shows that at the section $X = 3$ the temperature gradient (or heat flow) is less at the boundary than in the interior.

A third example shows the effect of not specifying the corner node temperature but leaving it as one of the unknowns to be solved. The plot of the heat flow in Figure 10.16(c) shows the minimum heat flow near the boundary at both $X = 1$ and $X = 3$. The pattern is a bit irregular at $X = 1$, suggesting the need for mesh refinement for further definition, but the general pattern is the best of the three examples. This suggests that perhaps the best treatment of the corner node at differing boundary values is to leave it as an unknown. ☐

10.8 TWO-DIMENSIONAL STEADY-STATE IRROTATIONAL INCOMPRESSIBLE FLOW

The equations defining the steady-state irrotational flow of an incompressible and inviscid fluid can be expressed in terms of the stream function or the velocity potential function. For simplicity, and to bring out the similarity with the heat transfer problem formulation, we will use only the velocity potential function.

The velocity of the fluid is defined in terms of the velocity potential function ϕ as follows:

$$\text{velocity in the } X \text{ direction, } v_x = -\frac{\partial \phi}{\partial x} \qquad (10.28a)$$

$$\text{velocity in the } Y \text{ direction } v_y = -\frac{\partial \phi}{\partial y} \qquad (10.28b)$$

In the absence of sources and sinks, the governing differential equation is

$$\frac{\partial^2 \phi}{\partial x^2} + \frac{\partial^2 \phi}{\partial y^2} = \nabla^2 \phi = 0 \qquad (10.29)$$

Comparison of these equations with the heat flow equations given in Section 10.1 reveals that the two problems are of identical form. The element matrix equation can be written directly from (10.5) with a substitution of the potential function ϕ for T and the vector of nodal potential functions $\{\phi\}$ for the temperature vector $\{q_t\}$.

$$\int_{S_e} [N]^T \frac{\partial \phi}{\partial n} ds - \int_{A_e} \left[\frac{\partial [N]^T}{\partial x} \frac{\partial [N]}{\partial x} + \frac{\partial [N]^T}{\partial y} \frac{\partial [N]}{\partial y} \right] dx\, dy \{\phi\} = \{0\} \qquad (10.30)$$

The computer program for heat transfer can be used with minor modification if we correlate the variables properly. From a comparison of

10.8 TWO-DIMENSIONAL STEADY-STATE IRROTATIONAL INCOMPRESSIBLE FLOW

the differential equations of the heat and fluid flow problems, we note that the fluid velocity potential function is equivalent to temperature and the fluid velocity is equivalent to the heat flux with a conduction coefficient of unity. Where the modification of the analysis procedure occurs is in the treatment of boundary conditions.

We recall that the heat transfer problem was limited to boundary heat flow by conduction or convection on any given portion of the boundary, and the solution procedure was formulated to accept specification of the boundary node temperature for the essential condition and ambient temperature of the surrounding region for the natural boundary condition. Continuing with the correlation of the variables of the two problem types, we note that the essential boundary condition of the fluid flow problem would be the velocity potential, ϕ. The natural boundary condition, however, is the velocity normal to the boundary, which is defined in terms of the directional derivative as

$$\frac{\partial \phi}{\partial n} = -v_n \qquad (10.31)$$

If we now compare this boundary expression with the natural boundary condition of the heat flow problem, equation (10.8), it is observed that the fluid boundary velocity has an equivalence to the heat problem ambient temperature or applied heat flux. The major difference in the formulation of the problem rests in the fact that the fluid natural boundary condition does not yield matrices comparable to the convection matrices that combine with the conduction matrix. Modification of the heat transfer computer program for fluid flow then simply requires removal of the element convection matrices.

Examining the flow natural boundary condition expression, we note that a positive $\partial \phi / \partial n$ represents a positive change of the velocity potential as we proceed *outward* from the region (Figure 10.17). From the definition of the boundary flow velocity in terms of ϕ, a positive $\partial \phi / \partial n$ defines a velocity normal to the boundary and directed *toward* the region. As sign convention of the boundary velocity can be confusing, we present a simple example that may help to clarify it.

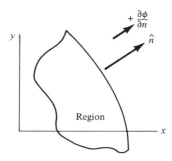

FIGURE 10.17 Normal directional derivative sign convention.

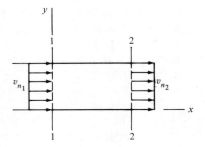

FIGURE 10.18 Uniform velocity distributions on two boundaries.

Consider a simple flow in the x direction as shown in Figure 10.18. By definition, the flow velocity is related to the velocity potential by

$$v_x = -\frac{\partial \phi}{\partial x}$$

At boundary 1, the flow is normal to the boundary and the directional derivative $\partial \phi/\partial n$, being positive outward from the region, is opposite in sign to the derivative of ϕ with respect to x. Thus we have

$$\frac{\partial \phi}{\partial n_1} = -\frac{\partial \phi}{\partial x_1} = v_{x_1} = v_{n_1}$$

At boundary 2, the derivatives ϕ/n and ϕ/x are of the same sign. Thus we can write

$$\frac{\partial \phi}{\partial n_2} = \frac{\partial \phi}{\partial x_2} = -v_{x_2} = -v_{n_2}$$

This can be generalized into the following convention: The boundary flow velocity will be positive if directed toward the region.

In determining the magnitudes of the boundary values, we require that at an impermeable boundary, the flow velocity and thus the directional derivative of the velocity potential function be zero normal to that boundary. At a boundary of uniform velocity any convenient magnitude of velocity potential may be specified. Remember that the velocity is dependent only on the gradient of the potential function; the particular magnitude has no significance. If the velocity at a boundary is not uniform, the required nodal values of the potential function can be determined from equations (10.28) up to the arbitrary constant (which can be assigned any value).

Example 10.6 Assume steady-state irrotational flow of an inviscid and incompressible fluid through the closed channel defined in Figure 10.19. Given a uniform velocity distribution in the x direction at section 1–1, determine the velocity profile at the minimum point of the reduced section and the variation of the potential function value with coordinate x.

10.8 TWO-DIMENSIONAL STEADY-STATE IRROTATIONAL INCOMPRESSIBLE FLOW

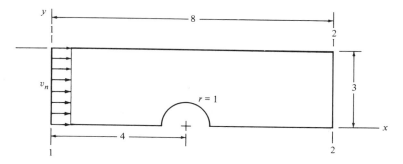

FIGURE 10.19 Flow in a closed channel.

Solution. From equation (10.28a) we obtain the following for the left boundary:

$$\phi = -v_x X + f(Y)$$

From equation (10.28b), $f(Y)$ must be equal to a constant for zero velocity in the y direction at the boundary. Thus, for a constant x and constant v_n at the boundary, the value of ϕ will be a constant for all nodes on this boundary of the finite element mesh. Repeating the same steps at the right boundary (downstream), we will apply a different constant value (a lower value) to all nodes on that boundary of the mesh. The value of the potential function on the left boundary is established at 8, the value at the right at 0. This yields a theoretical uniform average flow velocity of

$$v_{x_{\text{avg}}} = -\frac{\Delta\phi}{\Delta x} = 1 \quad \text{acting in the } +x \text{ direction}$$

Figure 10.20 shows a suggested mesh for the analysis. The uniform mesh is used in the area of uniform velocity and a refined pattern in the region where the gradient of the potential function is expected to vary nonlinearly. For convenience, we will let $\phi = 8$ at the left boundary and 0 at the right boundary.

The results were determined from the computer program in Appendix 7 and velocity profiles and variation of the potential function are plotted in Figure 10.21 and compared with the theoretical result obtained from the superposition of the velocity potential for a uniform flow to the right and a doublet issuing in the $-x$ direction. This theoretical velocity potential expression is

$$\phi = -v_x(x) - v_x\frac{R^2 \cos\theta}{r} + C$$

By placing the origin at the left boundary of the region and specifying $\phi(0) = 8$ and $\phi(8) = 0$, the expression for a unit radius and unit

364   HEAT TRANSFER AND FLUID FLOW IN TWO DIMENSIONS

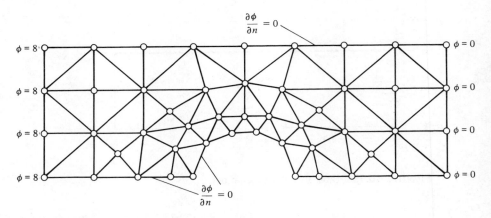

FIGURE 10.20   Finite element mesh pattern for Example 10.6.

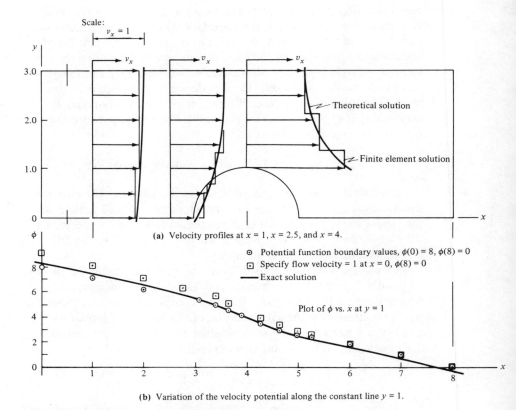

FIGURE 10.21   Results of Example 10.6.

average flow velocity becomes

$$\phi = 4 + (x - 4)\left(1 + \frac{1}{r^2}\right)$$

where

$$r^2 = (x - 4)^2 + y^2.$$

Two solutions are compared. One involves specification of velocity potentials on the boundaries; the other involves specification of the flow velocity of unity on the left boundary and the potential of zero on the right boundary. Comparing the results, we observe remarkable agreement with the theoretical prediction even with the rather coarse mesh involving the lowly constant-velocity triangle element. □

10.9 HEAT AND FLUID FLOW WITH APPLIED SOURCE: POISSON'S EQUATION

The steady-state heat conduction equation for a continuum with an internal applied heat source, $Q(x, y)$ is governed by

$$k \nabla^2 T + Q(x, y) = 0 \qquad (10.32)$$

This is Poisson's equation. Applying Galerkin's method to the equation, we would discover that the finite element formulation is equation (10.11) plus the added term

$$\{f_Q\} = \int_{A_e} [N]^T Q(x, y) \, dx \, dy \qquad (10.33)$$

The element thermal stiffness equation becomes

$$[k_T]\{q_t\} = \{Q_{cv}\} + \{Q_b\} + \{f_Q\} \qquad (10.34)$$

Assume, for example, a linearly distributed heat flux $Q(x, y) = L_1 Q_1 + L_2 Q_2 + L_3 Q_3$. This heat source has English units of Btu per hour per cubic foot and SI units of watts per cubic meter.

Substituting the linear flux into (10.33), we have

$$\{f_Q\} = \int_{A_e} \begin{bmatrix} L_1 L_1 Q_1 + L_1 L_2 Q_2 + L_1 L_3 Q_3 \\ L_2 L_1 Q_1 + L_2 L_2 Q_2 + L_2 L_3 Q_3 \\ L_3 L_1 Q_1 + L_3 L_2 Q_2 + L_3 L_3 Q_3 \end{bmatrix} dx \, dy$$

Using the integration formula (6.14), we obtain the following result in terms of the heat flux at nodes (local) 1, 2, and 3:

$$\{f_Q\} = \frac{A}{12} \begin{bmatrix} 2Q_1 + Q_2 + Q_3 \\ Q_1 + 2Q_2 + Q_3 \\ Q_1 + Q_2 + 2Q_3 \end{bmatrix} \qquad (10.35)$$

A special case of (10.35) occurs when the flux is constant. In this case

$Q_1 = Q_2 = Q_3$, and the flux vector becomes

$$\{f_Q\} = \frac{QA}{3} \begin{Bmatrix} 1 \\ 1 \\ 1 \end{Bmatrix} \quad (10.36)$$

This indicates that the heat flux should be applied equally to the element nodes.

Extension of this discussion to the fluid flow problem is trivial. The source quantity, Q, would represent the volume flow rate per unit volume of the region.

10.10 CONCLUSION

There are other steady-state problems that are governed by the Poisson or Laplace equation. Electromagnetism and torsion, for example, are two additional and perhaps familiar problem types. An interesting project for the reader would be the extension of the material of this chapter to a different type of problem.

Also, we have concentrated exclusively on the application of the three-node triangular element. An excellent exercise would involve the development of a computer program for any problem type using a different element, such as the four-node quadrilateral. This should not present unsurmountable difficulties. The one major difference you will encounter is the resulting nonconstant derivatives of the shape functions, and numerical integration will be used.

PROBLEMS

10.1 Beginning with (10.19), fill in all the details for the derivation of the conduction matrix (10.21)

10.2 Verify (10.23).

10.3 Develop the matrix equation for the triangle of Example 10.1 if convection occurs on side 1 only (Figure P10.3).

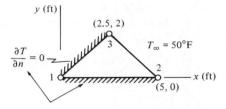

FIGURE P10.3 Triangle of Example 10.1.

10.4 In Figure P10.4, the two elements of Example 10.2 are against the insulated x and y axes. Solve for the temperature of node 1 if there is no convection and the temperatures of nodes 2, 3, and 4 are 100, 200, and 300°F, respectively.

PROBLEMS

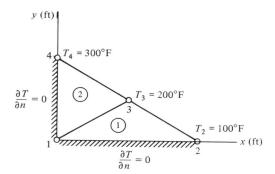

FIGURE P10.4 Elements of Example 10.2.

10.5 Modify the heat flow computer program in Appendix 7 to make it applicable to analysis of steady-state problems of ideal fluid flow.

10.6 Derive the general expression for the conduction matrix of the four-node quadrilateral element and establish the procedure for determining its numerical value.

10.7 Write a computer program for numerically evaluating the conduction matrix of the four-node quadrilateral (two-point quadrature).

10.8 Solve for the steady-state nodal temperatures and heat flow in a square steel plate with insulated boundaries parallel to the x axis and $T(0, y) = 100°F$ and $T(10, y) = 0°F$ (Figure P10.8). Use four elements. Calculate the conduction matrices with the computer program (requested in Problem 10.7), assemble (by hand), and solve.

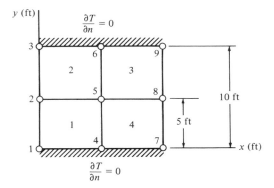

FIGURE P10.8 Square plate with four elements.

10.9 Develop the boundary convection matrices and force vectors for the quadrilateral.

CHAPTER 11

Axisymmetric Stress Analysis

11.1 INTRODUCTION

A class of problems exists that in reality involves three-dimensional continua and forces, but which reduces mathematically to two dimensions. These problems are called *axisymmetric problems*, and they are characterized by solids of revolution and material properties and loads that are unchanged along the circumference of revolution.

Figure 11.1 illustrates a solid of revolution. Shown also is the element that we will use for the discretization of the continuum; it is a toroid with a triangular cross section.

It is vital to realize that for the axisymmetric problem, displacements in the continuum can occur in the radial and axial direction only; displacements cannot occur in the circumferential direction. As a result of this, it becomes natural to use the cylindrical coordinate system in the development of the general element equations. Figure 11.2 illustrates this coordinate system, shows a cross section of the toroidal element, and defines the nodal displacements of the element.

The stress components at a point in the continuum for the axisymmetric case in terms of the cylindrical coordinate reference system are shown in the Figure 11.3.

370 AXISYMMETRIC STRESS ANALYSIS

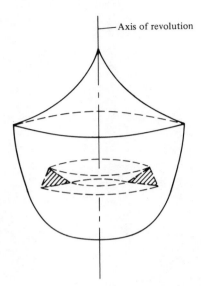

FIGURE 11.1 Body of revolution and toroidal element.

Note: These components are the reduced number from the general three-dimensional cylindrical components. For more detail, consult a textbook on elasticity.

11.2 THE BASIC EQUATIONS FOR THE ELEMENT

The general element equation for the stress analysis of a three-dimensional continuum is identical in form to (8.32).

$$\int_\Omega [B]^T[C][B]\,d\Omega\{q\} = \int_\Omega [B]^T[C]\{\epsilon_T\}\,d\Omega$$
$$+\{Q\}_{\text{NF}}+\{Q\}_T+\{Q\}_{\text{BF}} \quad (8.32)$$

Although the application of this equation for three-dimensional elements is identical in concept to that of the two-dimensional element, the effort is greater because of the additional displacement at each node and the variable third dimension. The area and line integrals of the plane problem elements now become volume and surface integrals.

The terms in equation (8.32), when applied to a three-dimensional continuum, are redefined as follows:

Stiffness matrix

$$[k] = \int_\Omega [B]^T[C][B]\,d\Omega \quad (11.1a)$$

11.2 THE BASIC EQUATIONS FOR THE ELEMENT

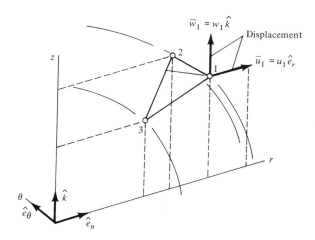

FIGURE 11.2 Coordinate system, cross section of element, node numbering, and displacements.

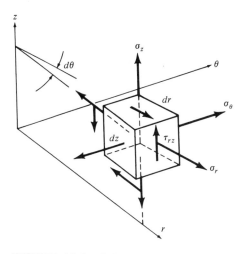

FIGURE 11.3 Cylindrical coordinate stress components for the axisymmetric case.

Temperature nodal load vector

$$\{Q\}_{\text{temp}} = \int_{\Omega} [B]^T [C]\{\epsilon_T\}\, d\Omega \qquad (11.1\text{b})$$

Nodal force vector

$$\{Q\}_{\text{NF}} = \text{forces applied to nodes} \qquad (11.1\text{c})$$

Surface traction vector

$$\{Q\}_T = \int_A [N]^T \{T\}\, dA \qquad (11.1\text{d})$$

Body force vector

$$\{Q\}_{\text{BF}} = \int_{\Omega} [N]^T \{B\}\, d\Omega \qquad (11.1\text{e})$$

11.3 AXISYMMETRIC ELASTICITY EQUATIONS

For axisymmetry, all equations must be independent of θ and all displacements must be in the rz plane. The strain–displacement relations in cylindrical coordinates for this particular problem are as follows:

$$\epsilon_r = \frac{\partial u}{\partial r}$$

$$\epsilon_z = \frac{\partial w}{\partial z}$$

$$\epsilon_\theta = \frac{u}{r}$$

$$\gamma_{rz} = \frac{\partial w}{\partial r} + \frac{\partial u}{\partial z}$$

Note that although the strains are independent of the θ coordinate, there is a strain, ϵ_θ, in the circumferential direction. Putting these relationships in matrix form gives us

$$\{\epsilon\} = \begin{Bmatrix} \epsilon_r \\ \epsilon_z \\ \epsilon_\theta \\ \gamma_{rz} \end{Bmatrix} = \begin{bmatrix} \dfrac{\partial}{\partial r} & 0 \\ 0 & \dfrac{\partial}{\partial z} \\ \dfrac{1}{r} & 0 \\ \dfrac{\partial}{\partial z} & \dfrac{\partial}{\partial r} \end{bmatrix} \begin{Bmatrix} u \\ w \end{Bmatrix} = [\mathcal{L}]\begin{Bmatrix} u \\ w \end{Bmatrix} \qquad (11.2)$$

Also, for this case of symmetry, we have the following stress–strain relationship for isotropic materials [recall (8.31)]:

$$\begin{Bmatrix} \sigma_r \\ \sigma_z \\ \sigma_\theta \\ \tau_{rz} \end{Bmatrix} = \frac{E}{(1+\nu)(1-2\nu)} \begin{bmatrix} 1-\nu & \nu & \nu & 0 \\ \nu & 1-\nu & \nu & 0 \\ \nu & \nu & 1-\nu & 0 \\ 0 & 0 & 0 & \frac{1-2\nu}{2} \end{bmatrix}$$

$$\times \left(\begin{Bmatrix} \epsilon_r \\ \epsilon_z \\ \epsilon_\theta \\ \gamma_{rz} \end{Bmatrix} - \alpha \Delta T \begin{Bmatrix} 1 \\ 1 \\ 1 \\ 0 \end{Bmatrix} \right) \qquad (11.3a)$$

or

$$\{\sigma\} = [C](\{\epsilon\} - \{\epsilon\}_T) \qquad (11.3b)$$

The thermal strain vector to be substituted into (11.1b) is defined as follows:

$$\{\epsilon_T\} = \begin{Bmatrix} \epsilon_r \\ \epsilon_z \\ \epsilon_\theta \\ \gamma_{rz} \end{Bmatrix}_T = \alpha \Delta T \begin{Bmatrix} 1 \\ 1 \\ 1 \\ 0 \end{Bmatrix} \qquad (11.4)$$

11.4 THE ELEMENT DISPLACEMENT FUNCTIONS

The node of our toroidal element is actually the concentric circle passing through the vertex of the triangular cross section. Its coordinates are r and z. The specification of the radial displacement, u, the axial displacement, w, the radial position, r, and the axial position, z, of any point in the toroid will be defined by the linear interpolation formulas in terms of the triangle natural coordinates and the nodal properties.

$$u = L_1 u_1 + L_2 u_2 + L_3 u_3 \qquad (11.5a)$$
$$w = L_1 w_1 + L_2 w_2 + L_3 w_3 \qquad (11.5b)$$
$$r = L_1 r_1 + L_2 r_2 + L_3 r_3 \qquad (11.6a)$$
$$z = L_1 z_1 + L_2 z_2 + L_3 z_3 \qquad (11.6b)$$

where you will recall,

$$L_1 + L_2 + L_3 = 1$$

Referring to the strain–displacement relationships (11.2), we realize that the derivatives of the displacements with respect to the coordinates r

and z are required. Since the displacements are in terms of the natural coordinates, we could follow a differentiation procedure similar to that defined in earlier chapters for the isoparametric elements. For a change of pace, we will try a different approach.

Combining the position equations above with the natural coordinate identity, we can write the following matrix equation:

$$\begin{Bmatrix} 1 \\ r \\ z \end{Bmatrix} = \begin{bmatrix} 1 & 1 & 1 \\ r_1 & r_2 & r_3 \\ z_1 & z_2 & z_3 \end{bmatrix} \begin{Bmatrix} L_1 \\ L_2 \\ L_3 \end{Bmatrix} \qquad (11.7)$$

Inverting the matrix and solving for the natural coordinate vector yields

$$\begin{Bmatrix} L_1 \\ L_2 \\ L_3 \end{Bmatrix} = \frac{1}{\det} \begin{bmatrix} a_1 & b_1 & c_1 \\ a_2 & b_2 & c_2 \\ a_3 & b_3 & c_3 \end{bmatrix} \begin{Bmatrix} 1 \\ r \\ z \end{Bmatrix} \qquad (11.8)$$

where

$$a_1 = r_2 z_3 - r_3 z_2 \qquad a_2 = r_3 z_1 - r_1 z_3 \qquad a_3 = r_1 z_2 - r_2 z_1$$
$$b_1 = z_2 - z_3 \qquad b_2 = z_3 - z_1 \qquad b_3 = z_1 - z_2$$
$$c_1 = r_3 - r_2 \qquad c_2 = r_1 - r_3 \qquad c_3 = r_2 - r_1$$

and

$$\det = (r_1 - r_3)(z_2 - z_3) - (r_2 - r_3)(z_1 - z_3) = 2(\text{triangle area})$$

Comparing the transpose of the square matrix with (4.4) shows them to be the same form, identical if the x is equated to r, and y to z. Thus we have the interpolation coefficients in terms of the physical coordinates, the required differentiation is trivial, and we can proceed to determine the $[B]$ matrix.

From (11.5),

$$\begin{Bmatrix} u \\ w \end{Bmatrix} = \begin{bmatrix} L_1 & 0 & L_2 & 0 & L_3 & 0 \\ 0 & L_1 & 0 & L_2 & 0 & L_3 \end{bmatrix} \begin{Bmatrix} u_1 \\ w_1 \\ u_2 \\ w_2 \\ u_3 \\ w_3 \end{Bmatrix} = [N]\{q\}$$

(11.9)

Substitution of (11.9) into (11.2) defines the $[B]$ matrix:

$$\{\epsilon\} = [\mathscr{L}][N]\{q\}$$
$$= [B]\{q\} \qquad (11.10)$$

The required derivatives of the natural coordinates can be obtained using (11.8). For example,

$$\frac{\partial L_1}{\partial r} = \frac{\partial}{\partial r}\left(\frac{a_1 + rb_1 + zc_1}{\det}\right) = \frac{b_1}{\det}$$

The $[B]$ matrix becomes

$$[B] = \frac{1}{\det} \begin{bmatrix} b_1 & 0 & b_2 & 0 & b_3 & 0 \\ 0 & c_1 & 0 & c_2 & 0 & c_3 \\ \frac{L_1^*}{r} & 0 & \frac{L_2^*}{r} & 0 & \frac{L_3^*}{r} & 0 \\ c_1 & b_1 & c_2 & b_2 & c_3 & b_3 \end{bmatrix} \quad (11.11)$$

where

$$L_1^* = a_1 + rb_1 + zc_1 \qquad L_2^* = a_2 + rb_2 + zc_2 \qquad L_3^* = a_3 + rb_3 + zc_3$$

11.5 EVALUATION OF THE ELEMENT STIFFNESS MATRIX

The stiffness matrix is defined by

$$[k] = \int_\Omega [B]^T [C][B] \, d\Omega \quad (11.1a)$$

The constitutive matrix, $[C]$, is defined by (11.3) and the $[B]$ matrix by (11.11). Unlike the three-node triangle in the plane analysis, the $[B]$ matrix for the three-node triangular cross section toroid is not constant; the entries in row 3 involve the r and z coordinates. One approach to the evaluation of this integral would involve the explicit integration of each term of the matrix product; another procedure would employ quadrature formulas.

A simple approximate method (Zienkiewicz, 1971) that yields excellent results for a fairly fine subdivision is to evaluate the $[B]$ matrix at the centroid of the triangle.

$$[\bar{B}] = [B(\bar{r}, \bar{z})]$$

where

$$\bar{r} = \frac{r_1 + r_2 + r_3}{3} \qquad \bar{z} = \frac{z_1 + z_2 + z_3}{3}$$

Note that

$$\bar{B}_{3,(2i-1)} = \frac{L_i}{\bar{r}} = \frac{1}{r_1 + r_2 + r_3} \qquad i = 1, 2, 3$$

The $[\bar{B}]$ matrix can now be considered constant and the matrix product can be moved outside of the integral. Noting also that the volume of the toroid, for a fine subdivision, can be approximated by

$$V = 2\pi \bar{r} A$$

we can write the element stiffness matrix as

$$[k] = 2\pi \bar{r} A [\bar{B}]^T [C][\bar{B}] \quad (11.12)$$

11.6 THE NODAL LOAD VECTORS

Temperature Nodal Load Vector. Using the centroidal approximation, we have

$$\{Q\}_{\text{temp}} = \int_{\Omega} [\overline{B}]^T [C] \{\epsilon_T\} \, d\Omega = 2\pi \bar{r} A \alpha \Delta T [\overline{B}]^T [C] \begin{Bmatrix} 1 \\ 1 \\ 1 \\ 0 \end{Bmatrix}$$

Substituting the constitutive matrix yields

$$\{Q\}_{\text{temp}} = \frac{2\pi \bar{r} A E \alpha \Delta T}{(1 - 2\nu)} [\overline{B}]^T \begin{Bmatrix} 1 \\ 1 \\ 1 \\ 0 \end{Bmatrix} \qquad (11.13)$$

Nodal Force Vector. The nodal force vector is simply

$$\{Q\}_{\text{NF}} = \begin{bmatrix} F_{1r} & F_{1z} & F_{2r} & F_{2z} & F_{3r} & F_{3z} \end{bmatrix}^T$$

Surface Traction Vector

$$\{Q\}_T = \int_A [N]^T \{T\} \, dA$$

$$= 2\pi \int_S r [N]^T \begin{Bmatrix} T_r \\ T_z \end{Bmatrix} ds$$

where r is the coordinate to the side and ds is the differential length of the side in the rz plane. The $[N]^T$ matrix will remain in terms of the natural coordinates.

$$\{Q\}_T = 2\pi \int_S \begin{bmatrix} rL_1 & 0 \\ 0 & rL_1 \\ rL_2 & 0 \\ 0 & rL_2 \\ rL_3 & 0 \\ 0 & rL_3 \end{bmatrix} \begin{Bmatrix} T_r \\ T_z \end{Bmatrix} ds \qquad (11.14)$$

Writing r in terms of the natural coordinates, we have

$$r = L_1 r_1 + L_2 r_2 + L_3 r_3$$

This can be substituted into the matrix above and integrated using the natural coordinate integration formula (6.13). The integration will be over the particular side exposed to the stress traction, so there are three

11.6 THE NODAL LOAD VECTORS

possibilities. The result of the integration over the sides for *uniform* T_r and T_z is

Side 1:

$$\{Q\}_{T_1} = \frac{2\pi S_1}{6} \begin{Bmatrix} 0 \\ 0 \\ (2r_2 + r_3)T_r \\ (2r_2 + r_3)T_z \\ (r_2 + 2r_3)T_r \\ (r_2 + 2r_3)T_z \end{Bmatrix} \quad (11.15)$$

where side 1 is the side opposite node 1 and S_1 is the length of side 1.

Side 2:

$$\{Q\}_{T_2} = \frac{2\pi S_2}{6} \begin{Bmatrix} (2r_1 + r_3)T_r \\ (2r_1 + r_3)T_z \\ 0 \\ 0 \\ (r_1 + 2r_3)T_r \\ (r_1 + 2r_3)T_z \end{Bmatrix} \quad (11.16)$$

Side 3:

$$\{Q\}_{T_3} = \frac{2\pi S_3}{6} \begin{Bmatrix} (2r_1 + r_2)T_r \\ (2r_1 + r_2)T_z \\ (r_1 + 2r_2)T_r \\ (r_1 + 2r_2)T_z \\ 0 \\ 0 \end{Bmatrix} \quad (11.17)$$

Body Force Vector

$$\{Q\}_{BF} = \int_\Omega [N]^T \begin{Bmatrix} B_r \\ B_z \end{Bmatrix} d\Omega$$

$$= 2\pi \int_A r[N]^T \begin{Bmatrix} B_r \\ B_z \end{Bmatrix} dA$$

This is very similar to the surface traction. Coordinate r is again expressed in terms of the natural coordinates and multiplies each term in the $[N]$ matrix. The area integration is performed using the natural coordinate area integration formula (6.14).

Assuming a *uniform* body force distribution, we obtain the following:

$$\{Q\}_{BF} = \frac{2\pi A}{12} \begin{Bmatrix} (2r_1 + r_2 + r_3)B_r \\ (2r_1 + r_2 + r_3)B_z \\ (r_1 + 2r_2 + r_3)B_r \\ (r_1 + 2r_2 + r_3)B_z \\ (r_1 + r_2 + 2r_3)B_r \\ (r_1 + r_2 + 2r_3)B_z \end{Bmatrix} \quad (11.18)$$

11.7 ASSEMBLY, BOUNDARY CONDITIONS, UNSYMMETRICAL LOADS, AND ANISOTROPY

Having constructed the stiffness matrix and nodal load vector relationships, the solution requires assembly and substitution of the boundary conditions. The assembly of the element equations can follow the same algorithm as that used for the three-node plane triangle. For boundary conditions, be sure to remove rigid-body displacement in the z direction by setting w equal to zero at one node at least; points lying on the z axis have displacement u equal to zero.

We have worked strictly with symmetric loading and isotropic materials. Gallagher (1975) and Cook (1974) discuss nonsymmetrical loading of axisymmetric solids, and Zienkiewicz (1971) discusses modeling anisotropic stratified materials.

PROBLEMS

11.1 When calculating the $[B]$ matrix with coordinate r assigned the constant centroidal value $\bar{r}$, the following relationship was presented for the entries in the third row:

$$\bar{B}_{3,(2i-1)} = \frac{L_i}{r} = \frac{1}{r_1 + r_2 + r_3} \qquad i = 1, 2, 3$$

Verify this result.

11.2 Solve for the nodal load vector $\{Q\}_T$ for the uniform stress traction on side 1 of the element in Figure P11.2.

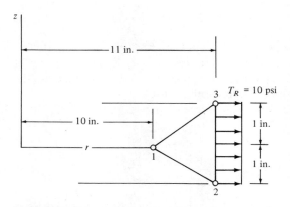

FIGURE P11.2 Uniform stress traction on side 1.

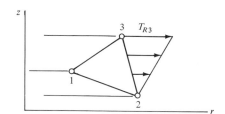

FIGURE P11.4 Linearly distributed stress traction on side 1.

11.3 Solve for the weight body force vector $\{Q\}_{BF}$ for the element in Figure P11.2 if it is of uniform mass density and the gravity is in the negative z direction.

11.4 For the element shown in Figure P11.4, derive the nodal force vector $\{Q\}_T$ for a linearly distributed stress traction acting on side 1.

CHAPTER 12

Three-Dimensional Stress Elements

12.1 INTRODUCTION

This final chapter will include the development of two solid elements for three-dimensional analysis of isotropic materials. This effort is of interest because we become involved with the general three-dimensional elasticity equations and we can observe the similarities and differences encountered when another spatial dimension is added.

The general element equation for the stress analysis of a three-dimensional continuum is identical in form to the equation we have seen many times. Writing once again the equation in symbolic form, and for convenience, including the definition of each term, we have

$$[k]\{q\} = \{Q\}_{\text{temp}} + \{Q\}_{\text{NF}} + \{Q\}_T + \{Q\}_{\text{BF}} \quad (12.1)$$

Stiffness Matrix

$$[k] = \int_\Omega [B]^T [C][B] \, d\Omega \quad (12.2a)$$

381

Temperature Nodal Load Vector

$$\{Q\}_{\text{temp}} = \int_{\Omega} [B]^T [C] \{\epsilon_T\} \, d\Omega \qquad (12.2b)$$

Nodal Force Vector

$$\{Q\}_{\text{NF}} = \text{forces applied to nodes} \qquad (12.2c)$$

Surface Traction Vector

$$\{Q\}_T = \int_A [N]^T \begin{Bmatrix} T_x \\ T_y \\ T_z \end{Bmatrix} dA \qquad (12.2d)$$

Body Force Vector

$$\{Q\}_{\text{BF}} = \int_{\Omega} [N]^T \begin{Bmatrix} B_x \\ B_y \\ B_z \end{Bmatrix} d\Omega \qquad (12.2e)$$

12.2 THE ELASTICITY EQUATIONS

With the points in a continuum having displacement components u, v, and w in the x, y, and z directions, respectively, the linear strain-displacement relationships, in matrix form, are as follows:

$$\begin{Bmatrix} \epsilon_x \\ \epsilon_y \\ \epsilon_z \\ \gamma_{xy} \\ \gamma_{yz} \\ \gamma_{xz} \end{Bmatrix} = \begin{bmatrix} \dfrac{\partial}{\partial x} & 0 & 0 \\ 0 & \dfrac{\partial}{\partial y} & 0 \\ 0 & 0 & \dfrac{\partial}{\partial z} \\ \dfrac{\partial}{\partial y} & \dfrac{\partial}{\partial x} & 0 \\ 0 & \dfrac{\partial}{\partial z} & \dfrac{\partial}{\partial y} \\ \dfrac{\partial}{\partial z} & 0 & \dfrac{\partial}{\partial x} \end{bmatrix} \begin{Bmatrix} u \\ v \\ w \end{Bmatrix} \qquad (12.3)$$

or

$$\{\epsilon\} = [\mathscr{L}] \begin{Bmatrix} u \\ v \\ w \end{Bmatrix}$$

The constitutive matrix for an isotropic material is

$$[C] = \frac{E}{(1+\nu)(1-2\nu)} \times \begin{bmatrix} 1-\nu & \nu & \nu & 0 & 0 & 0 \\ \nu & 1-\nu & \nu & 0 & 0 & 0 \\ \nu & \nu & 1-\nu & 0 & 0 & 0 \\ 0 & 0 & 0 & \frac{1-2\nu}{2} & 0 & 0 \\ 0 & 0 & 0 & 0 & \frac{1-2\nu}{2} & 0 \\ 0 & 0 & 0 & 0 & 0 & \frac{1-2\nu}{2} \end{bmatrix}$$

(12.4)

The stress is defined in terms of the constitutive matrix, $[C]$, the total strain vector, $\{\epsilon\}$, and the thermal strain vector, $\{\epsilon_T\}$, by

$$\{\sigma\} = [C](\{\epsilon\} - \{\epsilon_T\}) \qquad (12.5)$$

12.3 THE FOUR-NODE TETRAHEDRON

The tetrahedron element is the easiest element to construct mathematically but perhaps not the easiest or best to use in practice. This element is similar to the three-node triangle in that the displacement function is linear and the resulting strain and stress are constant throughout the element.

The natural coordinates for this element are ratios of volumes in a manner similar to the ratios of areas for the triangle. Figure 12.1 shows a tetrahedron with the local node numbering sequence; label node 1, then move counter clockwise viewed from node 1 to label nodes 2, 3, and 4. One interior tetrahedron is drawn to point p in the element, and the natural coordinate defined by the ratio of these volumes, or altitudes, is L_1.

$$L_1 = \frac{\text{interior } V}{\text{total volume}}$$

There are four sides to the tetrahedron, so there are four interior tetrahedra to an interior point and this means there are four natural coordinates, only three of which are independent. The sum equals unity.

$$L_1 + L_2 + L_3 + L_4 = 1 \qquad (12.6)$$

Similar to the triangle, the natural coordinates will be the interpolation coefficients for the displacement functions and the global coordinate

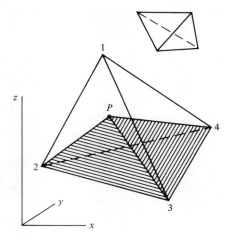

FIGURE 12.1 Tetrahedron with interior tetrahedron and local node numbers.

functions for points in the tetrahedron.

$$u = L_1 u_1 + L_2 u_2 + L_3 u_3 + (1 - L_1 + L_1 + L_3) u_4$$
$$v = L_1 v_1 + L_2 v_2 + L_3 v_3 + (1 - L_1 + L_2 + L_3) v_4 \quad (12.7)$$
$$w = L_1 w_1 + L_2 w_2 + L_3 w_3 + (1 - L_1 + L_2 + L_3) w_4$$

$$x = L_1 x_1 + L_2 x_2 + L_3 x_3 + (1 - L_1 + L_2 + L_3) x_4$$
$$y = L_1 y_1 + L_2 y_2 + L_3 y_3 + (1 - L_1 + L_2 + L_3) y_4 \quad (12.8)$$
$$z = L_1 z_1 + L_2 z_2 + L_3 z_3 + (1 - L_1 + L_2 + L_3) z_4$$

Writing these in matrix form defines the $[N]$ matrix.

$$\begin{Bmatrix} u \\ v \\ w \end{Bmatrix} = \begin{bmatrix} L_1 & 0 & 0 & L_2 & 0 & 0 & L_3 & 0 & 0 & L_4 & 0 & 0 \\ 0 & L_1 & 0 & 0 & L_2 & 0 & 0 & L_3 & 0 & 0 & L_4 & 0 \\ 0 & 0 & L_1 & 0 & 0 & L_2 & 0 & 0 & L_3 & 0 & 0 & L_4 \end{bmatrix} \{q\}$$

or

$$\begin{Bmatrix} u \\ v \\ w \end{Bmatrix} = [N]\{q\} \quad (12.9)$$

where

$$\{q\} = [u_1 \; v_1 \; w_1 \; u_2 \; v_2 \; w_2 \; u_3 \; v_3 \; w_3 \; u_4 \; v_4 \; w_4]^T$$

The next step is the determination of the strain in terms of the nodal displacement vector $\{q\}$. The displacements given by (12.9) are in terms of the natural coordinates, so direct substitution into (12.3) is not possible. We will thus resort to the procedure followed in Section 6.2 in the derivation of the strain for the isoparametric triangle.

12.3 THE FOUR-NODE TETRAHEDRON

The derivatives of the displacement and coordinate functions with respect to the natural coordinates can be obtained explicitly. Writing the equations

$$\frac{\partial u}{\partial L_1} = \frac{\partial u}{\partial x}\frac{\partial x}{\partial L_1} + \frac{\partial u}{\partial y}\frac{\partial y}{\partial L_1} + \frac{\partial u}{\partial z}\frac{\partial z}{\partial L_1}$$

$$\frac{\partial u}{\partial L_2} = \frac{\partial u}{\partial x}\frac{\partial x}{\partial L_2} + \frac{\partial u}{\partial y}\frac{\partial y}{\partial L_2} + \frac{\partial u}{\partial z}\frac{\partial z}{\partial L_2} \quad (12.10)$$

$$\frac{\partial u}{\partial L_3} = \frac{\partial u}{\partial x}\frac{\partial x}{\partial L_3} + \frac{\partial u}{\partial y}\frac{\partial y}{\partial L_3} + \frac{\partial u}{\partial z}\frac{\partial z}{\partial L_3}$$

and solving these equations for the spatial derivatives of the displacements, we obtain the following:

$$\begin{Bmatrix} \dfrac{\partial u}{\partial x} \\ \dfrac{\partial u}{\partial y} \\ \dfrac{\partial u}{\partial z} \end{Bmatrix} = \frac{1}{|J|} \begin{bmatrix} a_1 & a_2 & a_3 & a_4 \\ b_1 & b_2 & b_3 & b_4 \\ c_1 & c_2 & c_3 & c_4 \end{bmatrix} \begin{Bmatrix} u_1 \\ u_2 \\ u_3 \\ u_4 \end{Bmatrix} \quad (12.11)$$

If the same steps are followed for the derivatives of displacements v and w, we can then combine the results to express the strain vector in terms of the nodal displacement vector with the $[B]$ matrix.

$$\{\epsilon\} = [B]\{q\} \quad (12.12a)$$

or

$$\begin{Bmatrix} \epsilon_x \\ \epsilon_y \\ \epsilon_z \\ \tau_{xy} \\ \tau_{yz} \\ \tau_{xz} \end{Bmatrix} = \frac{1}{|J|} \begin{bmatrix} a_1 & 0 & 0 & a_2 & 0 & 0 & a_3 & 0 & 0 & a_4 & 0 & 0 \\ 0 & b_1 & 0 & 0 & b_2 & 0 & 0 & b_3 & 0 & 0 & b_4 & 0 \\ 0 & 0 & c_1 & 0 & 0 & c_2 & 0 & 0 & c_3 & 0 & 0 & c_4 \\ b_1 & a_1 & 0 & b_2 & a_2 & 0 & b_3 & a_3 & 0 & b_4 & a_4 & 0 \\ 0 & c_1 & b_1 & 0 & c_2 & b_2 & 0 & c_3 & b_3 & 0 & c_4 & b_4 \\ c_1 & 0 & a_1 & c_2 & 0 & a_2 & c_3 & 0 & a_3 & c_4 & 0 & a_4 \end{bmatrix} \begin{Bmatrix} u_1 \\ v_1 \\ w_1 \\ u_2 \\ v_2 \\ w_2 \\ u_3 \\ v_3 \\ w_3 \\ u_4 \\ v_4 \\ w_4 \end{Bmatrix}$$

$$(12.12b)$$

where

$$a_1 = (y_2 - y_4)(z_3 - z_4) - (y_3 - y_4)(z_2 - z_4)$$
$$a_2 = (y_3 - y_4)(z_1 - z_4) - (y_1 - y_4)(z_3 - z_4)$$
$$a_3 = (y_1 - y_4)(z_2 - z_4) - (y_2 - y_4)(z_1 - z_4)$$
$$b_1 = (x_3 - x_4)(z_2 - z_4) - (x_2 - x_4)(z_3 - z_4)$$
$$b_2 = (x_1 - x_4)(z_3 - z_4) - (x_3 - x_4)(z_1 - z_4)$$
$$b_3 = (x_2 - x_4)(z_1 - z_4) - (x_1 - x_4)(z_2 - z_4)$$
$$c_1 = (x_2 - x_4)(y_3 - y_4) - (x_3 - x_4)(y_2 - y_4)$$
$$c_2 = (x_3 - x_4)(y_1 - y_4) - (x_1 - x_4)(y_3 - y_4)$$
$$c_3 = (x_1 - x_4)(y_2 - y_4) - (x_2 - x_4)(y_1 - y_4)$$

$$a_4 = -(a_1 + a_2 + a_3)$$
$$b_4 = -(b_1 + b_2 + b_3)$$
$$c_4 = -(c_1 + c_2 + c_3)$$

$$|J| = (x_1 - x_4)(a_1) + (y_1 - y_4)(b_1) + (z_1 - z_4)(c_1)$$

With the $[B]$ matrix and the $[N]$ matrix, the required element matrix equation terms (12.2) can be calculated. Since the $[B]$ matrix is constant, the stiffness matrix is simply

$$[k] = [B]^T[C][B]\int_\Omega d\Omega$$
$$= [B]^T[C][B](\text{vol})$$

where the volume of the element is $\frac{1}{6}|J|$. Thus

$$[k] = [B]^T[C][B]\frac{|J|}{6} \qquad (12.13)$$

Since the $[N]$ matrix is in terms of volume (natural) coordinates, the volume integration required for some of the equivalent nodal load vectors can be simplified by using the following formula:

$$\int_{\text{vol}} L_1^\alpha L_2^\beta L_3^\gamma L_4^\delta \, dx\, dy\, dz = \frac{\alpha!\beta!\gamma!\delta!\, 6(\text{vol})}{(\alpha + \beta + \gamma + \delta + 3)!} \qquad (12.14)$$

Assembly of the element equations can follow the same procedures as those discussed for other elements. The global displacement numbers

12.4 THE ISOPARAMETRIC EIGHT-NODE SOLID

(defining global matrix addresses) for a node would be defined as follows: For a node with global node number (i),

global displacement number for $u = 3(i) - 2$

global displacement number for $v = 3(i) - 1$

global displacement number for $w = 3(i)$

The boundary conditions must be sufficient to prevent rigid-body motion.

Zienkiewicz discusses the visualization difficulty in using the tetrahedron and proposes the construction of an eight-node brick element with five tetrahedra.

12.4 THE ISOPARAMETRIC EIGHT-NODE SOLID

This solid element is the three-dimensional counterpart of the four-node plane element described in Chapter 7. It is a much easier element to use in discretizing a continuum than is the tetrahedron, and the element allows a strain gradient throughout the element. The main reason for including a summary of the development of the matrices for this element is to illustrate the similarity with the previously used procedures and to show how complex the algebra can become even for this relatively simple element.

A sketch of the element is shown in Figure 12.2. The natural (s, t, r) coordinate system is similar to the (s, t) system of the four-node element but with the added dimension shown in Figure 12.3. The local node

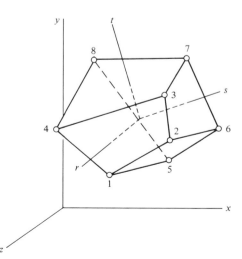

FIGURE 12.2 Eight-node isoparametric solid.

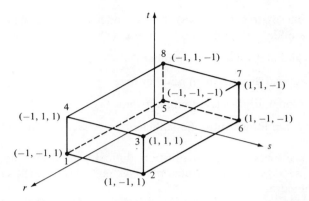

FIGURE 12.3 Solid element in the natural coordinate system (r, s, t).

numbering sequence is also defined, and this numbering order must be adhered to when identifying the global node numbers in an assemblage of elements.

The displacements and coordinates are to be defined with the interpolation polynomials. We could write them in matrix form, such as (12.9), but with eight nodes the matrix with entries becomes too large. Therefore, we stick to the summation form.

$$u = \sum_{j=1}^{8} N_j u_j$$

$$v = \sum_{j=1}^{8} N_j v_j$$

$$w = \sum_{j=1}^{8} N_j w_j$$

$$x = \sum_{j=1}^{8} N_j x_j \quad (12.15)$$

$$y = \sum_{j=1}^{8} N_j y_j$$

$$z = \sum_{j=1}^{8} N_j z_j$$

12.4 THE ISOPARAMETRIC EIGHT-NODE SOLID

The interpolation functions are

$$N_1 = \tfrac{1}{8}(1-s)(1-t)(1+r)$$
$$N_2 = \tfrac{1}{8}(1+s)(1-t)(1+r)$$
$$N_3 = \tfrac{1}{8}(1+s)(1+t)(1+r)$$
$$N_4 = \tfrac{1}{8}(1-s)(1+t)(1+r)$$
$$N_5 = \tfrac{1}{8}(1-s)(1-t)(1-r) \quad (12.16)$$
$$N_6 = \tfrac{1}{8}(1+s)(1-t)(1-r)$$
$$N_7 = \tfrac{1}{8}(1+s)(1+t)(1-r)$$
$$N_8 = \tfrac{1}{8}(1-s)(1+t)(1-r)$$

We must now determine the strain in terms of the nodal displacements. Writing the equations

$$\begin{Bmatrix} \dfrac{\partial u}{\partial s} \\ \dfrac{\partial u}{\partial t} \\ \dfrac{\partial u}{\partial r} \end{Bmatrix} = \begin{bmatrix} \dfrac{\partial x}{\partial s} & \dfrac{\partial y}{\partial s} & \dfrac{\partial z}{\partial s} \\ \dfrac{\partial x}{\partial t} & \dfrac{\partial y}{\partial t} & \dfrac{\partial z}{\partial t} \\ \dfrac{\partial x}{\partial r} & \dfrac{\partial y}{\partial r} & \dfrac{\partial z}{\partial r} \end{bmatrix} \begin{Bmatrix} \dfrac{\partial u}{\partial x} \\ \dfrac{\partial u}{\partial y} \\ \dfrac{\partial u}{\partial z} \end{Bmatrix} \quad (12.17)$$

and solving for the spatial derivatives, we have

$$\begin{Bmatrix} \dfrac{\partial u}{\partial x} \\ \dfrac{\partial u}{\partial y} \\ \dfrac{\partial u}{\partial z} \end{Bmatrix} = \dfrac{1}{|J|} \begin{bmatrix} \dfrac{\partial y}{\partial t}\dfrac{\partial z}{\partial r} - \dfrac{\partial z}{\partial t}\dfrac{\partial y}{\partial r} & \dfrac{\partial z}{\partial t}\dfrac{\partial x}{\partial r} - \dfrac{\partial x}{\partial t}\dfrac{\partial z}{\partial r} & \dfrac{\partial x}{\partial t}\dfrac{\partial y}{\partial r} - \dfrac{\partial y}{\partial t}\dfrac{\partial x}{\partial r} \\ \dfrac{\partial z}{\partial s}\dfrac{\partial y}{\partial r} - \dfrac{\partial y}{\partial s}\dfrac{\partial z}{\partial r} & \dfrac{\partial x}{\partial s}\dfrac{\partial z}{\partial r} - \dfrac{\partial z}{\partial s}\dfrac{\partial x}{\partial r} & \dfrac{\partial y}{\partial s}\dfrac{\partial x}{\partial r} - \dfrac{\partial x}{\partial s}\dfrac{\partial y}{\partial r} \\ \dfrac{\partial y}{\partial s}\dfrac{\partial z}{\partial t} - \dfrac{\partial z}{\partial s}\dfrac{\partial y}{\partial t} & \dfrac{\partial z}{\partial s}\dfrac{\partial x}{\partial t} - \dfrac{\partial x}{\partial s}\dfrac{\partial z}{\partial t} & \dfrac{\partial x}{\partial s}\dfrac{\partial y}{\partial t} - \dfrac{\partial y}{\partial s}\dfrac{\partial x}{\partial t} \end{bmatrix} \begin{Bmatrix} \dfrac{\partial u}{\partial s} \\ \dfrac{\partial u}{\partial t} \\ \dfrac{\partial u}{\partial r} \end{Bmatrix}$$

(12.18)

Similar equations can be written for the v and w displacements, and we will have the same form of the spatial derivative.

$$\begin{Bmatrix} \dfrac{\partial \phi}{\partial x} \\ \dfrac{\partial \phi}{\partial y} \\ \dfrac{\partial \phi}{\partial z} \end{Bmatrix} = \dfrac{1}{|J|} \begin{bmatrix} a_{11} & a_{12} & a_{13} \\ a_{21} & a_{22} & a_{23} \\ a_{31} & a_{32} & a_{33} \end{bmatrix} \begin{Bmatrix} \dfrac{\partial \phi}{\partial s} \\ \dfrac{\partial \phi}{\partial t} \\ \dfrac{\partial \phi}{\partial r} \end{Bmatrix} \quad \phi = u, v, w$$

(12.19)

where the a_{ij} are defined as follows:

$$a_{11} = \sum_{i=1}^{8} \frac{\partial N_i}{\partial t} y_i \sum_{i=1}^{8} \frac{\partial N_i}{\partial r} z_i - \sum_{i=1}^{8} \frac{\partial N_i}{\partial t} z_i \sum_{i=1}^{8} \frac{\partial N_i}{\partial r} y_i$$

$$a_{12} = \sum_{i=1}^{8} \frac{\partial N_i}{\partial t} z_i \sum_{i=1}^{8} \frac{\partial N_i}{\partial r} x_i - \sum_{i=1}^{8} \frac{\partial N_i}{\partial t} x_i \sum_{i=1}^{8} \frac{\partial N_i}{\partial r} z_i$$

$$a_{13} = \sum_{i=1}^{8} \frac{\partial N_i}{\partial t} x_i \sum_{i=1}^{8} \frac{\partial N_i}{\partial r} y_i - \sum_{i=1}^{8} \frac{\partial N_i}{\partial t} y_i \sum_{i=1}^{8} \frac{\partial N_i}{\partial r} x_i$$

$$a_{21} = \sum_{i=1}^{8} \frac{\partial N_i}{\partial s} z_i \sum_{i=1}^{8} \frac{\partial N_i}{\partial r} y_i - \sum_{i=1}^{8} \frac{\partial N_i}{\partial s} y_i \sum_{i=1}^{8} \frac{\partial N_i}{\partial r} z_i$$

$$a_{22} = \sum_{i=1}^{8} \frac{\partial N_i}{\partial s} x_i \sum_{i=1}^{8} \frac{\partial N_i}{\partial r} z_i - \sum_{i=1}^{8} \frac{\partial N_i}{\partial s} z_i \sum_{i=1}^{8} \frac{\partial N_i}{\partial r} x_i \quad (12.20)$$

$$a_{23} = \sum_{i=1}^{8} \frac{\partial N_i}{\partial s} y_i \sum_{i=1}^{8} \frac{\partial N_i}{\partial r} x_i - \sum_{i=1}^{8} \frac{\partial N_i}{\partial s} x_i \sum_{i=1}^{8} \frac{\partial N_i}{\partial r} y_i$$

$$a_{31} = \sum_{i=1}^{8} \frac{\partial N_i}{\partial s} y_i \sum_{i=1}^{8} \frac{\partial N_i}{\partial t} z_i - \sum_{i=1}^{8} \frac{\partial N_i}{\partial s} z_i \sum_{i=1}^{8} \frac{\partial N_i}{\partial t} y_i$$

$$a_{32} = \sum_{i=1}^{8} \frac{\partial N_i}{\partial s} z_i \sum_{i=1}^{8} \frac{\partial N_i}{\partial t} x_i - \sum_{i=1}^{8} \frac{\partial N_i}{\partial s} x_i \sum_{i=1}^{8} \frac{\partial N_i}{\partial t} z_i$$

$$a_{33} = \sum_{i=1}^{8} \frac{\partial N_i}{\partial s} x_i \sum_{i=1}^{8} \frac{\partial N_i}{\partial t} y_i - \sum_{i=1}^{8} \frac{\partial N_i}{\partial s} y_i \sum_{i=1}^{8} \frac{\partial N_i}{\partial t} x_i$$

and the Jacobian determinant by

$$|J| = a_{11} \sum_{i=1}^{8} \frac{\partial N_i}{\partial s} x_i + a_{12} \sum_{i=1}^{8} \frac{\partial N_i}{\partial s} y_i + a_{13} \sum_{i=1}^{8} \frac{\partial N_i}{\partial s} z_i \quad (12.21)$$

Note:

$$y = \sum_{i=1}^{8} N_i y_i, \qquad \frac{\partial y}{\partial s} = \sum_{i=1}^{8} \frac{\partial N_i}{\partial s} y_i, \text{ etc.}$$

12.4 THE ISOPARAMETRIC EIGHT-NODE SOLID

In addition, the derivatives of the interpolation functions are

$$\frac{\partial N_1}{\partial s} = \tfrac{1}{8}(t-1)(1+r) \qquad \frac{\partial N_1}{\partial t} = \tfrac{1}{8}(s-1)(1+r) \qquad \frac{\partial N_1}{\partial r} = \tfrac{1}{8}(1-s)(1-t)$$

$$\frac{\partial N_2}{\partial s} = \tfrac{1}{8}(1-t)(1+r) \qquad \frac{\partial N_2}{\partial t} = \tfrac{1}{8}(-1-s)(1+r) \qquad \frac{\partial N_2}{\partial r} = \tfrac{1}{8}(1+s)(1-t)$$

$$\frac{\partial N_3}{\partial s} = \tfrac{1}{8}(1+t)(1+r) \qquad \frac{\partial N_3}{\partial t} = \tfrac{1}{8}(1+s)(1+r) \qquad \frac{\partial N_3}{\partial r} = \tfrac{1}{8}(1+s)(1+t)$$

$$\frac{\partial N_4}{\partial s} = \tfrac{1}{8}(-1-t)(1+r) \qquad \frac{\partial N_4}{\partial t} = \tfrac{1}{8}(1-s)(1+r) \qquad \frac{\partial N_4}{\partial r} = \tfrac{1}{8}(1-s)(1+t)$$

$$\frac{\partial N_5}{\partial s} = \tfrac{1}{8}(t-1)(1-r) \qquad \frac{\partial N_5}{\partial t} = \tfrac{1}{8}(s-1)(1-r) \qquad \frac{\partial N_5}{\partial r} = \tfrac{1}{8}(s-1)(1-t)$$

$$\frac{\partial N_6}{\partial s} = \tfrac{1}{8}(1-t)(1-r) \qquad \frac{\partial N_6}{\partial t} = \tfrac{1}{8}(1+s)(r-1) \qquad \frac{\partial N_6}{\partial r} = \tfrac{1}{8}(1+s)(t-1)$$

$$\frac{\partial N_7}{\partial s} = \tfrac{1}{8}(1+t)(1-r) \qquad \frac{\partial N_7}{\partial t} = \tfrac{1}{8}(1+s)(1-r) \qquad \frac{\partial N_7}{\partial r} = \tfrac{1}{8}(-1-s)(1+t)$$

$$\frac{\partial N_8}{\partial s} = \tfrac{1}{8}(1+t)(r-1) \qquad \frac{\partial N_8}{\partial t} = \tfrac{1}{8}(1-s)(1-r) \qquad \frac{\partial N_8}{\partial r} = \tfrac{1}{8}(s-1)(1+t)$$

(12.22)

Now we can assemble the $[B]$ matrix for the strain vector in terms of the nodal displacement vector.

$$\begin{Bmatrix} \epsilon_x \\ \epsilon_y \\ \epsilon_z \\ \gamma_{xy} \\ \gamma_{yz} \\ \gamma_{xz} \end{Bmatrix} = \begin{bmatrix} B_{1,1} & 0 & 0 & B_{1,4} & \cdots & B_{1,22} & 0 & 0 \\ 0 & B_{2,2} & 0 & 0 & \cdots & 0 & B_{2,23} & 0 \\ 0 & 0 & B_{3,3} & 0 & \cdots & 0 & 0 & B_{3,24} \\ B_{4,1} & B_{4,2} & 0 & B_{4,4} & \cdots & B_{4,22} & B_{4,23} & 0 \\ 0 & B_{5,2} & B_{5,3} & 0 & \cdots & 0 & B_{5,23} & B_{5,24} \\ B_{6,1} & 0 & B_{6,3} & B_{6,4} & \cdots & B_{6,22} & 0 & B_{6,24} \end{bmatrix} \begin{Bmatrix} u_1 \\ v_1 \\ w_1 \\ u_2 \\ v_2 \\ w_2 \\ u_3 \\ v_3 \\ \vdots \\ u_8 \\ v_8 \\ w_8 \end{Bmatrix}$$

(12.23)

or

$$\{\epsilon\} = [B]\{q\}$$

The nonzero entries in the [B] matrix are

$$B_{1,(3i-2)} = \frac{1}{|J|}\left(a_{11}\frac{N_i}{\partial s} + a_{12}\frac{N_i}{\partial t} + a_{13}\frac{N_i}{\partial r}\right)$$

$$B_{2,(3i-1)} = \frac{1}{|J|}\left(a_{21}\frac{\partial N_i}{\partial s} + a_{22}\frac{\partial N_i}{\partial t} + a_{23}\frac{\partial N_i}{\partial r}\right)$$

$$B_{3,(3i)} = \frac{1}{|J|}\left(a_{31}\frac{\partial N_i}{\partial s} + a_{32}\frac{\partial N_i}{\partial t} + a_{33}\frac{\partial N_i}{\partial r}\right)$$

$$B_{4,(3i-2)} = B_{2,(3i-1)}$$

$$B_{4,(3i-1)} = B_{1,(3i-2)} \qquad i = 1, 2, 3, \ldots, 7, 8 \tag{12.24}$$

$$B_{5,(3i-1)} = B_{3,(3i)}$$

$$B_{5,3i} = B_{2,(3i-1)}$$

$$B_{6,(3i-2)} = B_{3,(3i)}$$

$$B_{6,(3i)} = B_{1,(3i-2)}$$

With the [B] matrix and the [N] matrix defined, all terms in the element equation can be determined, but it will require a bit more effort than for the tetrahedron because now both the [N] and the [B] matrix are functions of the natural coordinates. Consider first the element stiffness matrix.

$$[k] = \int_\Omega [B]^T [C] [B] \, dx \, dy \, dz$$

Since the [B] matrix is a function of coordinates s, t, and r, we perform a change of variable on the differential volume from xyz to str according to the relationship defined in Chapter 5.

$$dx \, dy \, dz = |J| \, ds \, dt \, dr$$

Now instead of integrating each entry, we will approximate the integral with a quadrature formula. The two-point formula is sufficient for this particular element and, with a cube-shaped region in the natural coordinate system, we may approximate the integral with a weighted summation of the integrand evaluated at the eight points defined in Table 12.1.

$$[k] \cong \sum_{i=1}^{8} [B(s_i, t_i, r_i)]^T [C] [B(s_i, t_i, r_i)] |J(s_i, t_i, r_i)| W_i$$

The same general procedure is followed for the equivalent nodal force vectors, in that the two-point quadrature formula is used to approximate the integrals. We note, however, that the area integral for the surface

12.4 THE ISOPARAMETRIC EIGHT-NODE SOLID

TABLE 12.1 Quadrature Coordinates and Weighting Values

Point, i	Natural Coordinates			Weighting Value
	s	t	r	
1	$-\dfrac{1}{\sqrt{3}}$	$-\dfrac{1}{\sqrt{3}}$	$\dfrac{1}{\sqrt{3}}$	1
2	$\dfrac{1}{\sqrt{3}}$	$-\dfrac{1}{\sqrt{3}}$	$\dfrac{1}{\sqrt{3}}$	1
3	$\dfrac{1}{\sqrt{3}}$	$\dfrac{1}{\sqrt{3}}$	$\dfrac{1}{\sqrt{3}}$	1
4	$-\dfrac{1}{\sqrt{3}}$	$\dfrac{1}{\sqrt{3}}$	$\dfrac{1}{\sqrt{3}}$	1
5	$-\dfrac{1}{\sqrt{3}}$	$-\dfrac{1}{\sqrt{3}}$	$-\dfrac{1}{\sqrt{3}}$	1
6	$\dfrac{1}{\sqrt{3}}$	$-\dfrac{1}{\sqrt{3}}$	$-\dfrac{1}{\sqrt{3}}$	1
7	$\dfrac{1}{\sqrt{3}}$	$\dfrac{1}{\sqrt{3}}$	$-\dfrac{1}{\sqrt{3}}$	1
8	$-\dfrac{1}{\sqrt{3}}$	$\dfrac{1}{\sqrt{3}}$	$-\dfrac{1}{\sqrt{3}}$	1

traction vector implies a summation over four quadrature points in the side of the element.

To obtain good results with this element, the mesh will have to be rather fine. Therefore, distributed loadings can in many cases be assumed uniform on and in the element. The result for a uniform stress traction on a side is an equal distribution of the load among the four nodes on that side, a uniform body force also is distributed equally among the eight nodes.

Warning: Do not get caught in thinking that these distributions always conform to "common sense." Higher-order elements have strange nodal load vectors.

All that remains now is the programming. Based on your experience with the simpler elements, you should be able to construct a program to evaluate the element matrices and assemble, reduce, and solve the global equation.

PROBLEMS

12.1 Derive the expression defining a_1 in (12.12b).

12.2 Derive the expression for the determinant, $|J|$, shown with (12.12b).

12.3 Specify the coordinates of a tetrahedron of your choice. Calculate the volume of this tetrahedron, and show that it is equal to $\frac{1}{6}$ of $|J|$ defined in (12.12b).

12.4 Derive the equivalent nodal forces for the case of uniform stress traction on any face of a tetrahedron.

12.5 Derive a_{11} of (12.18) beginning with the matrix equation (12.17).

12.6 Derive the expression for a_{11} of (12.20) from a_{11} as defined in (12.18).

12.7 Beginning with (12.17), derive the expression for the determinant $|J|$ given in (12.21).

12.8 Prove that in (12.23), $B_{6,1}$ must equal $B_{3,3}$.

12.9 Derive the equivalent nodal forces for the case of uniform stress traction on any side of the eight-node solid.

APPENDIX 1

Matrix Definitions and Operations

Definition of a Matrix. A matrix is a rectangular array of elements. Applications in this book deal only with matrices whose elements are real numbers. For example,

$$[A] = \begin{bmatrix} a_{11} & a_{12} & a_{13} \\ a_{21} & a_{22} & a_{23} \end{bmatrix} = \begin{bmatrix} 2 & 3 & 5 \\ 6 & 8 & 1 \end{bmatrix}$$

is a rectangular array of dimension two rows by three columns, thus called a 2×3 matrix. The element a_{ij} is the number in the ith row and jth column.

The notation for a single-column array, commonly called a *vector*, is slightly different. For example,

$$\{V\} = \begin{Bmatrix} v_1 \\ v_2 \\ v_3 \end{Bmatrix}$$

Addition of Matrices. The addition of matrices simply involves the summation of elements having the identical "address" in each matrix. The matrices to be summed must have identical dimensions. The addition of matrices of different dimension is not defined. Matrix addition is associative and commutative. For example,

$$\begin{bmatrix} 2 & 1 \\ 5 & 3 \end{bmatrix} + \begin{bmatrix} -1 & 5 \\ 3 & -2 \end{bmatrix} = \begin{bmatrix} 1 & 6 \\ 8 & 1 \end{bmatrix}$$

Multiplication by a Constant. If a matrix is to be multiplied by a constant, every element in the matrix is multiplied by that constant. Also, if a constant is factored out of a matrix, it is factored out of each element.

Multiplication of Two Matrices. Assume that $[C] = [A][B]$, where $[A]$, $[B]$, and $[C]$ are matrices. Element c_{ij} in matrix $[C]$ is defined as follows:

$$c_{ij} = \sum_{m=1}^{k} a_{im} b_{mj} = a_{i1} b_{1j} + a_{i2} b_{2j} + \cdots + a_{ik} b_{kj}$$

Note that the summing index m defines the column of $[A]$ and the row of $[B]$, and its maximum value, therefore, is the number of columns of $[A]$ and the number of rows of $[B]$. Thus, since the m index defines the same number for $[A]$ and $[B]$, matrix multiplication requires "conformable" matrices. That is, matrix $[A]$ must have the same number of columns as $[B]$ has rows. For example,

$$(3 \times 2)\,(2 \times 1) \quad (3 \times 1)$$

$$\begin{bmatrix} 1 & 2 \\ 3 & 1 \\ 2 & 0 \end{bmatrix} \begin{Bmatrix} 1 \\ -1 \end{Bmatrix} = \begin{Bmatrix} -1 \\ 2 \\ 2 \end{Bmatrix}$$

Matrix multiplication is associative but *not* commutative.

Square Matrix. A square matrix is a matrix having equal numbers of rows and columns.

Column Matrix (Row Matrix). A column (row) matrix is a matrix having one column (row).

Diagonal Matrix. A diagonal matrix is a square matrix with nonzero elements only along the diagonal of the matrix.

Identity Matrix. An identity matrix is a diagonal matrix with each diagonal element equal to unity.

Transpose of a Matrix. The transpose of a matrix is a matrix obtained by interchanging rows and columns. Every matrix has a transpose. The transpose of a column matrix (vector) is a row matrix, the transpose of a row matrix is a column matrix. For example,

$$[A] = \begin{bmatrix} 2 & 4 & 5 \\ 3 & 9 & 10 \end{bmatrix} \quad [A]^T = \begin{bmatrix} 2 & 3 \\ 4 & 9 \\ 5 & 10 \end{bmatrix}$$

Inverse of a Square Matrix. A square matrix *may* have an inverse. The product of a matrix and its inverse matrix yields the identity matrix. The reader is referred to the following techniques for matrix inversion:

1. Gauss–Jordan elimination.
2. Gauss–Seidel iteration.

These techniques are popular and are discussed in most texts on numerical techniques.

For matrices of small dimension, 3×3 or less, the inverse of the matrix can be obtained easily with the following formula:

$$[K]^{-1} = \frac{[K^a]}{|K|} = \frac{[K^c]^T}{|K|}$$

where $|K|$ is the determinant of $[K]$ and $[K^a]$ is called the *adjoint matrix*, which is the transpose of the cofactor matrix $[K^c]$. The cofactor matrix is a matrix of minor determinants. The minor determinant $|M_{ij}|$ is the determinant of the array obtained from $[K]$ by removing row i and column j. The entries K_{ij}^c in the cofactor matrix $[K^c]$ are calculated as follows:

$$K_{ij}^c = (-1)^{(i+j)} |M_{ij}|$$

Example A1.1 Solve for the inverse of

$$[K] = \begin{bmatrix} 1 & 2 & 0 \\ 2 & 3 & 1 \\ 0 & 1 & 2 \end{bmatrix}$$

Solution. The determinant of the matrix is: $|K| = -3$. The 1, 1 entry in the cofactor matrix is

$$k_{11}^c = \begin{vmatrix} 3 & 1 \\ 1 & 2 \end{vmatrix} = 5$$

the 1, 2 entry is

$$k_{12}^c = (-1) \begin{vmatrix} 2 & 1 \\ 0 & 2 \end{vmatrix} = -4$$

The cofactor matrix divided by the determinant is

$$\frac{[K^c]}{|K|} = \frac{1}{3}\begin{bmatrix} 5 & 4 & 2 \\ -4 & 2 & -1 \\ 2 & -1 & -1 \end{bmatrix} = [K]^{-1}$$

(Since the matrix is symmetric, the cofactor matrix equals the adjoint matrix.)

Matrix Product Transpose. The transpose of a product of two matrices is equal to the product of transposed matrices in reverse order.

$$([A][B])^T = [B]^T[A]^T$$

APPENDIX 2

Computer Programs in Applesoft BASIC and TURBO Pascal for Plane Frame Analysis

A2.1 THE ELEMENT

The structural element used in this program is the general axially loaded beam element described in Section 1.6. This element models structural members that are in bending and/or simple tension and compression. The element may be given any orientation in the XY plane (Figure A2.1).

A2.2 MAJOR CHARACTERISTICS OF THE PROGRAM

This is not a sophisticated program. The main purpose of the program is to provide one example of an assemblage of algorithms implementing the structural analysis techniques described in Chapter 1. You should discover, however, that the program can be useful for serious analysis on a scale permitted by the microcomputer's limited memory.

The program is completely interactive; all input data are requested during execution so that the user does not have to remember any input format. As you use the program you will note that care is required in inputting the data. Although there is an editing routine in the program that allows corrections of element node numbers and global coordinates,

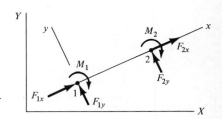

FIGURE A2.1 Element and coordinate systems.

other data cannot be edited if an error is made and you have to begin again. Many more error traps should be added, but these lengthen the program, and refinement of that nature was not a major goal. You are encouraged, however, to make the program more "user-friendly" to suit your own needs.

The program has the following features and limitations:

1. Element node numbers and node coordinates can be stored on the disk and later read from the disk for future reruns.
2. Nodes can be identified as constrained or can be given a nonzero displacement in the X direction and/or the Y direction; node constraint release along a direction inclined to the axes cannot be specified. An inclined displacement freedom can be achieved by adding the rigid two-force member as shown in Figure A2.2. Node rotation about the Z axis can be specified as fixed or given a nonzero value.
3. Structure loading must be applied to nodes only. No provision is made for specifying distributed loading.
4. Beam elements at a common node must have the same linear and angular displacement. For example, beam elements can not be joined together with a "pin joint." Note, however, that two-force members are always "pinned" and always have zero rotational displacement.
5. The user has the option of viewing results on the screen or printing hard copy. Displacement results may be stored on disk for use with a distortion plot program mentioned below.

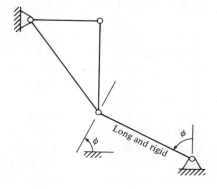

FIGURE A2.2 Modeling a constraint release along an inclined direction.

MAJOR CHARACTERISTICS OF THE PROGRAM 401

Two additional BASIC programs are supplied in this appendix that should aid the user in running the FRAME-2D analysis program. The program FRAME-2D FILE EDITOR reads the input node and coordinate information, allows editing of node numbers and coordinates, and plots the corrected frame model. This step is often helpful in identifying incorrect data. The program FRAME-2D DISTORTION PLOT reads the node X and Y displacement results stored from the analysis program and plots the distorted structure. Both of these programs are completely interactive.

The TURBO Pascal program contains all of the features of the three BASIC programs and is slightly modified but very similar in execution.

A2.3 SAMPLE PROGRAM RUN

Solve for the support reactions and the force and couple at the bend of the continuous 1 inch diameter rod supported and loaded as shown in Figure A2.3.1.

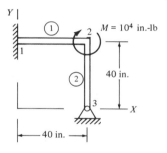

FIGURE A2.3.1 Continuous rod "built in" at wall and pinned at ground.

```
RUN
*****************************************
**                                     **
**    STRUCTURAL ANALYSIS IN XY PLANE  **
**                                     **
**  TWO-FORCE AND BEAM/COLUMN ELEMENT  **
**                                     **
*****************************************
             DATA INPUT OPTIONS
             ---- ----- -------

   1 - INPUT NEW MODEL DATA FROM KEYBOARD
   2 - INPUT OLD MODEL DATA FROM DISC

         (ELEMENT NODE NUMBERS AND
            NODE COORDINATES)

     ENTER CHOICE - (1 OR 2) 1

HOW MANY 'TWO-FORCE' AND 'BEAM' ELEMENTS IN THE STRUCTURE?   2

HOW MANY NODES?   3
```

```
ENTER ELEMENT NUMBER AND GLOBAL NUMBERS OF LOCAL NODES 1 & 2
ENTER IN FOLLOWING FORMAT:

EL #,    #1,    #2    <RETURN>

?1, 2, 3
?2, 3, 2

ENTER GLOBAL X AND Y COORDINATES OF THE NODES
ENTER IN FOLLOWING FORMAT:

NODE#,    X,    Y    <RETURN>

?1,  0, 40
?2, 40, 40
?3, 40,  0


DEFINE ELEMENT TYPE

          TWO-FORCE ELEMENT = 1
          BEAM/COLUMN ELEMENT = 2

ELEMENT 1 = ?2
ELEMENT 2 = ?2

******* CALCULATED HALF-BANDWIDTH = 6 *******

DO YOU WISH TO EDIT NODE NUMBERING?  (Y/N) N

DO YOU WISH TO EDIT NODE COORDINATES? (Y/N) N

DO YOU WANT DATA STORED ON DISK ?  (Y/N) N

HOW MANY NODES HAVE A SPECIFIED DISPLACEMENT (INCLUDING ZERO)
  IN THE X DIRECTION? 2

HOW MANY NODES HAVE A SPECIFIED DISPLACEMENT (INCLUDING ZERO)
  IN THE Y DIRECTION? 2

HOW MANY NODES HAVE A SPECIFIED ROTATION (INCLUDING ZERO)? 1

INPUT NODE NUMBER AND THE KNOWN X DISPLACEMENT IN FOLLOWING
  FORMAT:

NODE #,    U    <RETURN>

?1, 0
?3, 0

INPUT NODE NUMBER AND THE KNOWN Y DISPLACEMENT IN FOLLOWING
  FORMAT:

NODE #,    V    <RETURN>

?1, 0
?3, 0

INPUT NODE NUMBER AND THE KNOWN ROTATIONAL DISPLACEMNT IN
  FOLLOWING FORMAT:

NODE #,    THETA    <RETURN>

?1, 0

HOW MANY NODES CARRY APPLIED (KNOWN MAGNITUDE) FORCES AND /OR
  COUPLES ? 1

ENTER GLOBAL NODE NUMBER, X FORCE, Y FORCE AND COUPLE IN THE
```

SAMPLE PROGRAM RUN

```
    FOLLOWING FORMAT:

 NODE #,   FX,   FY,   C    <RETURN>

 ?2, 0, 0, 10000

 DO ALL ELEMENTS HAVE SAME CROSS SECTION AND MATERIAL
   PROPERTIES? (Y/N) Y

 ENTER ELEMENT MODULUS OF ELASTICITY,
              AREA,
                    AREA MOMENT OF INERTIA

 USE THE FOLLOWING FORMAT:

 E,  A,   I    <RETURN>

 ?30E6,   0.7854,   0.04909
```

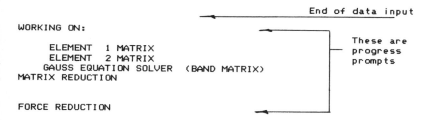

```
 WORKING ON:

      ELEMENT   1 MATRIX
      ELEMENT   2 MATRIX
      GAUSS EQUATION SOLVER   (BAND MATRIX)
 MATRIX REDUCTION


 FORCE REDUCTION


 DO YOU WANT HARDCOPY? (Y/N)

 ?Y

 ARE YOU CONNECTED TO A PRINTER ?    (Y/N)

 ?Y


 PLANE FRAME ANALYSIS --- INPUT DATA

 ***** ELEMENT GLOBAL NODE NUMBERS ****
```

ELEMENT NUMBER	GLOBAL NODE NUMBERS OF LOCAL 1	LOCAL 2	ELEMENT TYPE
1	1	2	2
2	3	2	2

**** NODE GLOBAL COORDINATES AND BOUNDARY DISPLACEMENTS ****

NODE	X	Y	BOUNDARY DISPLACEMENTS U	V	THETA
1	0	40	0	0	0
2	40	40			
3	40	0	0	0	

**** NODAL FORCES AND COUPLES ****

NODE	X FORCE	Y FORCE	COUPLE
2	0	0	10000

**** MATERIAL AND SECTION PROPERTIES ****

ELEMENT	ELASTIC MODULUS	AREA	AREA MOMENT OF INERTIA
1	30000000	.7854	.04909
2	30000000	.7854	.04909

PLANE FRAME ANALYSIS --- RESULTS

```
********************************************
**              DISPLACEMENTS             **
********************************************
```

NODE	XDISPL	YDISPL	THETA, RAD.
1	0	0	0
2	1.81E-04	-3.64E-04	.038811
3	0	0	-.019399

```
********************************************
**           ELEMENT NODE FORCES          **
********************************************
```

--NOTE--NODE NUMBERS ARE GLOBAL, LISTED
IN ASCENDING LOCAL NUMBER ORDER

EL#	NODE	AXIAL LOAD	SHEAR LOAD	COUPLE
1	1	-108	-215	2855
1	2	107	214	5713
2	3	214	-108	0
2	2	-215	107	4286

INTERPRETATION OF RESULTS
----------------- -- -------
    Refer to the assembly of the elements in the original structure
shown in Figure A2.3.1. Identify the local nodes of each element
and read the results from the program. You should identify the
magnitudes and directions of the forces as defined in Figure A2.3.2.

Note: It is always a good idea to check that the results satisfy
equilibrium and that the displacements appear to be reasonable in
direction and magnitude.

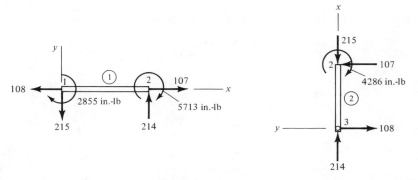

FIGURE A2.3.2 Forces and couples acting on each node of each element in the structure of EXAMPLE A2.3.1.

A2.4 PROGRAM LISTINGS IN APPLESOFT BASIC

FRAME-2D

```
10    TEXT : HOME
15    PRINT "*****************************************"
20    PRINT "**                                      **"
25    PRINT "**    STRUCTURAL ANALYSIS IN XY PLANE   **"
30    PRINT "**                                      **"
35    PRINT "** TWO-FORCE AND BEAM/COLUMN ELEMENT    **"
40    PRINT "**                                      **"
45    PRINT "*****************************************"
100   REM    < NE > NUMBER OF ELEMENTS
102   REM    < NN > NUMBER OF NODES
104   REM    < FX > NUMBER OF X DIRECTION CONSTRAINTS
106   REM    < FY > NUMBER OF Y DIRECTION CONSTRAINTS
108   REM    < FT > NUMBER OF ROTATION CONSTRAINTS
110   REM    < BC(I) > BOUNDARY CONDITION CODE
112   REM    < TC(I) >   ELEMENT TYPE CODE
114   REM    < XG(I),YG(I) > GLOBAL COORDINATES OF NODE I
116   REM    < GN(I,J) > GLOBAL NODE NUMBER OF NODE (J) IN ELEMENT (I)
118   REM    < NG(I,J) > GLOBAL DISPLACEMENT NUMBER OF LOCAL DISPLACEMENT
120   REM                NUMBER (J) OF ELEMENT (I)
122   REM    < K(I,J) > ELEMENT LOCAL COORDINATE STIFFNESS MATRIX
124   REM    < KT(I,J) > ELEMENT GLOBAL COORDINATE STIFFNESS MATRIX
126   REM    < KG(I,J) > ASSEMBLED GLOBAL STIFFNESS MATRIX
128   REM    < FC(I) > GLOBAL NODE FORCE VECTOR
130   REM    < FL(I) > LOCAL ELEMENT NODE FORCE VECTOR
132   REM    < U(I) > GLOBAL REFERENCE DISPLACEMENT VALUE
134   REM              OF GLOBAL DISPLACEMENT NUMBER I
136   REM    < UL(I) > LOCAL REFERENCE COORDINATE DISPLACEMENT VALUE
138   REM              OF LOCAL DISPLACEMENT NUMBER I
140   REM    < UG(I) > GLOBAL REFERENCE COORDINATE DISPLACEMENT VALUE
142   REM              OF LOCAL DISPLACEMENT NUMBER I
300   VTAB 20: PRINT "HIT ANY KEY TO CONTINUE"
302   GET G$: PRINT
304   REM   --------------------------------------------------
306   REM       LINES 310 THRU 348 ARE MINI FRONT END PROGRAM
308   REM   --------------------------------------------------
310   ONERR  GOTO 318
312   POKE 768,104: POKE 769,168: POKE 770,104: POKE 771,166: POKE
      772,223: POKE 773,154: POKE 774,72: POKE 775,152: POKE 776,72: POKE
      777,96
314   D$ =  CHR$ (4): REM     CHR$(4) IS CTRL-D
316   GOTO 356: REM      GO TO NORMAL PROGRAM START
318    CALL 768: PRINT D$;"PR#0":  TEXT : IF  PEEK (222) = 255 THEN  PRINT
      "CNTL C CAUSED A BREAK IN ";  PEEK (218) +  PEEK (219) * 256: END
320    GOSUB 340: END
322   REM            DETERMINE FREE MEMORY
324   TP =  PEEK (175) +  PEEK (176) * 256: REM     TOP OF PROGRAM
326   EA =  PEEK (109) +  PEEK (110) * 256: REM     TOP OF LOMEM POINTER
328   BS =  PEEK (111) +  PEEK (112) * 256: REM     HIMEM POINTER
330    PRINT "FREE DATA SPACE IS ";BS - EA;" BYTES"
332    PRINT :  PRINT "FREE PROGRAM SPACE:":  PRINT
334   TP = 8192 - TP:  IF TP > 0 THEN  PRINT TP;" BYTES TO HGR": RETURN
336   TP = TP + 8192:  IF TP > 0 THEN  PRINT TP;" BYTES TO HGR2": RETURN
338    PRINT "        YOU ARE ";  ABS (TP);" BYTES INTO HGR2!": RETURN
340    PRINT "ERROR # ";  PEEK (222);" AT LINE # ";  PEEK (218) +  PEEK (219)
       * 256
342    PRINT :  PRINT "     (ERROR CODES IN POOLE PG. 331)"
344    PRINT :  PRINT "YOUR OPTIONS NOW ARE:":  PRINT "  (1) HIT ESCAPE TO
      RESUME AT ERROR STATEMENT":  PRINT "  (2) HIT ANY OTHER KEY TO END
      PROGRAM"
346    GET G$:  PRINT :  IF G$ =  CHR$ (27) THEN  RESUME
348    RETURN
350   REM   --------------------------------------------------
352   REM        START PROGRAM HERE
354   REM   --------------------------------------------------
356    HOME
```

```
420  I$ =   CHR$ (9): REM   CHR$(9) IS CTRL-I
430  R$ =   CHR$ (13): REM  CHR$(13) IS CARRIAGE RETURN
440  VTAB 10
450  PRINT "          DATA INPUT OPTIONS"
460  PRINT "          ---- ----- -------"
470  PRINT
480  PRINT "  1 - INPUT NEW MODEL DATA FROM KEYBOARD"
490  PRINT "  2 - INPUT OLD MODEL DATA FROM DISC": PRINT
500  PRINT "          (ELEMENT NODE NUMBERS AND"
510  PRINT "              NODE COORDINATES)"
520  PRINT
530  INPUT "  ENTER CHOICE - (1 OR 2) ";IN$
540  IF IN$ = "1" OR IN$ = "2" GOTO 560
550  PRINT : PRINT "BAD RESPONSE, RE-ENTER": GOTO 450
560  IF IN$ = "1" GOTO 630
570  RT$ = ".GEOM"
580  REM  ---------------------------------------------------
590  GOSUB 10000: REM    OPEN AND READ THE GEOMETRY FILE  ----
600  REM  ---------------------------------------------------
610  REM              DATA INPUT FROM KEYBOARD OR DISC    ----
620  REM  ---------------------------------------------------
630  HOME : INPUT "HOW MANY 'TWO-FORCE' AND 'BEAM' ELEMENTS IN THE
     STRUCTURE? ";NE: PRINT
640  INPUT "HOW MANY NODES? ";NN: PRINT
650  DIM
     XG(NN),YG(NN),A2(6,6),X(2),Y(2),GN(NE,2),K(6,6),R(6,6),KT(6,6),A(NE),
     E(NE),I(NE),NG(NE,6),FC(3 * NN),BC(3 * NN),U(3 * NN),Q(3 *
     NN),UL(6),FL(6),UG(6),FR(3 * NN),TC(NE)
660  PRINT "ENTER ELEMENT NUMBER AND GLOBAL NUMBERS OF LOCAL NODES 1 &
     2": PRINT "ENTER IN FOLLOWING FORMAT:": PRINT
670  PRINT "EL #, #1 , #2  <RETURN>": PRINT
680  FOR J = 1 TO NE
690  INPUT I,GN(I,1),GN(I,2)
700  NEXT J
710  PRINT : PRINT "ENTER GLOBAL X AND Y COORDINATES OF THE NODES": PRINT
     "ENTER IN FOLLOWING FORMAT:": PRINT
720  PRINT "NODE #,   X,    Y    <RETURN>": PRINT
730  FOR J = 1 TO NN
740  INPUT I,XG(I),YG(I)
750  NEXT J
760  IF IN$ = "1" GOTO 800
770  REM  ---------------------------------------------------
780  GOSUB 10110: REM    CLOSE THE GEOMETRY DATA FILE     ----
790  REM  ---------------------------------------------------
800  PRINT : PRINT "DEFINE ELEMENT TYPE"
810  PRINT : PRINT "          TWO-FORCE ELEMENT = 1"
820  PRINT : PRINT "          BEAM/COLUMN ELEMENT = 2": PRINT
830  FOR J = 1 TO NE
840  PRINT "ELEMENT ";J;" = ";: INPUT TC(J)
850  NEXT J
860  REM  CALCULATE THE SEMI-BANDWIDTH OF THE STIFFNESS MATRIX
870  BW = 0
880  FOR I = 1 TO NE
890  DI =   ABS (GN(I,1) - GN(I,2))
900  IF DI > BW THEN BW = DI
910  NEXT I
920  BW = 3 * BW + 3
930  PRINT : PRINT "****** CALCULATED HALF-BANDWIDTH =";BW;" ******":
     PRINT
940  REM  ---------------------------------------------------
950  REM     EDITING OPTIONS TO REVIEW AND/OR CHANGE       ----
960  REM     ELEMENT NODE NUMBERING AND NODE COORDINATES   ----
970  REM  ---------------------------------------------------
980  PRINT : INPUT "DO YOU WISH TO EDIT NODE NUMBERING ?  (Y/N) ";RE$
990  IF RE$ = "Y" OR RE$ = "N" THEN 1020
1000 REM
1010 PRINT : PRINT "RE-ENTER -": GOTO 980
1020 IF RE$ = "N" THEN 1100
1030 PRINT : INPUT "WHICH ELEMENT ?";N
1040 IF N > NE THEN 1030
1050 PRINT "CURRENT VALUES: ";GN(N,1);",   ";GN(N,2)
```

PROGRAM LISTINGS IN APPLESOFT BASIC

```
1060  PRINT : INPUT "ENTER THE NEW VALUES:";GN(N,1),GN(N,2)
1070  PRINT : INPUT "MORE CHANGES ?  (Y/N)";RE$
1080  IF RE$ = "Y" THEN 1030
1090  GOTO 870
1100  DIM KG(3 * NN,BW)
1104  PRINT : PRINT "------------------------------------"
1105  GOSUB 322: REM    TEST FOR FREE MEMORY
1106  PRINT "------------------------------------"
1110  PRINT : PRINT : INPUT "DO YOU WISH TO EDIT NODE COORDINATES ?
      (Y/N) ";RE$
1120  IF RE$ = "Y" OR RE$ = "N" THEN 1140
1130  PRINT : PRINT "RE-ENTER -": GOTO 1110
1140  IF RE$ = "N" THEN 1210
1150  PRINT : INPUT "WHICH NODE ?";N
1160  IF N > NN THEN 1150
1170  PRINT "CURRENT VALUES: ";XG(N);", ";YG(N)
1180  PRINT : INPUT "ENTER THE NEW VALUES- ";XG(N),YG(N)
1190  PRINT : INPUT "MORE CHANGES ?  (Y/N)";RE$
1200  IF RE$ = "Y" THEN 1150
1210  PRINT : INPUT "DO YOU WANT DATA STORED ON DISK ?  (Y/N) ";DK$
1220  IF DK$ < > "Y" THEN 1370
1230  RT$ = ".GEOM"
1240  REM  ------------------------------------------------
1250  GOSUB 10160: REM    OPEN AND WRITE TO GEOMETRY FILE ----
1260  REM  ------------------------------------------------
1270  PRINT NE;R$;NN
1280  FOR I = 1 TO NE
1290  PRINT I;R$;GN(I,1);R$;GN(I,2)
1300  NEXT
1310  FOR I = 1 TO NN
1320  PRINT I;R$;XG(I);R$;YG(I)
1330  NEXT
1340  REM  ------------------------------------------------
1350  GOSUB 10110: REM    CLOSE THE GEOMETRY FILE        ----
1360  REM  ------------------------------------------------
1370  FOR I = 1 TO NN
1380  REM    INITIALIZE BOUNDARY CODE AND DISPLACEMENT VARIABLE
1390  BC(3 * I - 2) = 0:BC(3 * I - 1) = 0:BC(3 * I) = 0
1400  U(3 * I - 2) = 0:U(3 * I - 1) = 0:U(3 * I) = 0
1410  NEXT I
1420  REM  ------------------------------------------------
1430  REM    DETERMINE THE GLOBAL DISPLACEMENT NUMBERS OF  ----
1440  REM    LOCAL DISPLACEMENTS 1 THRU 6 OF EACH ELEMENT  ----
1450  REM  ------------------------------------------------
1460  FOR M = 1 TO NE
1470  FOR J = 1 TO 2
1480  NG(M,3 * J - 2) = 3 * GN(M,J) - 2
1490  NG(M,3 * J - 1) = 3 * GN(M,J) - 1
1500  NG(M,3 * J) = 3 * GN(M,J)
1510  NEXT J
1520  NEXT M
1530  REM  ------------------------------------------------
1540  REM    INPUT THE DISPLACEMENT BOUNDARY CONDITIONS    ----
1550  REM  ------------------------------------------------
1560  REM
1570  PRINT : INPUT "HOW MANY NODES HAVE A SPECIFIED DISPLACEMENT
      (INCLUDING ZERO) IN THE X DIRECTION? ";FX: PRINT
1580  IF FX = 0 THEN  PRINT : PRINT "    WARNING!": PRINT "AT LEAST 1
      NODE MUST BE FIXED": PRINT "    IN THE X DIRECTION.  TRY AGAIN":
      PRINT : GOTO 1570
1590  PRINT : INPUT "HOW MANY NODES HAVE A SPECIFIED DISPLACEMENT
      (INCLUDING ZERO) IN THE Y DIRECTION? ";FY: PRINT
1600  IF FY = 0 THEN  PRINT : PRINT "    WARNING!": PRINT "AT LEAST ONE
      NODE MUST BE FIXED": PRINT "    IN THE Y DIRECTION.  TRY AGAIN":
      PRINT : GOTO 1590
1610  PRINT : INPUT "HOW MANY NODES HAVE A SPECIFIED ROTATION (INCLUDING
      ZERO)? ";FT: PRINT
1620  PRINT "INPUT NODE NUMBER AND THE KNOWN X DISPLACEMENT IN FOLLOWING
      FORMAT:": PRINT : PRINT "NODE #,  U   <RETURN>": PRINT
1630  FOR J = 1 TO FX
```

```
1640  INPUT I,U(3 * I - 2)
1650  BC(3 * I - 2) = 1
1660  NEXT J
1670  PRINT : PRINT "INPUT NODE NUMBER AND THE KNOWN Y DISPLACEMENT IN
      FOLLOWING FORMAT:": PRINT : PRINT "NODE #,   V      <RETURN>": PRINT
1680  FOR J = 1 TO FY
1690  INPUT I,U(3 * I - 1)
1700  BC(3 * I - 1) = 1
1710  NEXT J
1720  IF FT = 0 GOTO 1780
1730  PRINT : PRINT "INPUT NODE NUMBER AND THE KNOWN ROTATIONAL
      DISPLACEMENT IN FOLLOWING FORMAT:": PRINT : PRINT "NODE #,   THETA
      <RETURN>": PRINT
1740  FOR J = 1 TO FT
1750  INPUT I,U(3 * I)
1760  BC(3 * I) = 1
1770  NEXT J
1780  FOR I = 1 TO 3 * NN
1790  FC(I) = 0
1800  NEXT I
1810  REM ------------------------------------------------------
1820  REM           INPUT THE APPLIED FORCES AND COUPLES     ----
1830  REM ------------------------------------------------------

1840  PRINT : INPUT "HOW MANY NODES CARRY APPLIED (KNOWN MAGNITUDE)
      FORCES AND/OR COUPLES ? ";AP: PRINT
1850  PRINT : PRINT "ENTER GLOBAL NODE NUMBER, X FORCE, Y FORCE AND
      COUPLE IN THE FOLLOWING FORMAT:": PRINT : PRINT "NODE#,   FX,   FY,   C
      <RETURN>": PRINT
1860  FOR I = 1 TO AP
1870  INPUT N,FC(3 * N - 2),FC(3 * N - 1),FC(3 * N)
1880  NEXT I
1890  REM ------------------------------------------------------
1900  REM              INPUT THE ELEMENT PROPERTIES          ----
1910  REM ------------------------------------------------------
1920  PRINT : INPUT "DO ALL ELEMENTS HAVE SAME CROSS SECTION AND MATERIAL
      PROPERTIES? (Y/N) ";AM$: PRINT
1930  IF AM$ = "Y" GOTO 1990
1940  PRINT : PRINT "ENTER ELEMENT NUMBER,": PRINT "   ELASTIC MODULUS,":
      PRINT "     AREA,": PRINT "       AREA MOMENT OF INERTIA": PRINT :
      PRINT
1950  PRINT "USE FOLLOWING FORMAT:": PRINT : PRINT "ELEMENT #,  E,  A,  I
      <RETURN>": PRINT
1960  FOR I = 1 TO NE
1970  INPUT M,E(M),A(M),I(M)
1980  NEXT I: GOTO 2040
1990  PRINT : PRINT "ENTER ELEMENT MODULUS OF ELASTICITY,": PRINT "
      AREA,": PRINT "              AREA MOMENT OF INERTIA": PRINT : PRINT
      "USE THE FOLLOWING FORMAT:": PRINT : PRINT "E,  A,  I    <RETURN>":
      PRINT
2000  INPUT E(1),A(1),I(1): IF NE = 1 GOTO 2040
2010  FOR I = 2 TO NE
2020  E(I) = E(1):A(I) = A(1):I(I) = I(1): NEXT I
2030  REM   INITIALIZE GLOBAL STIFFNESS MATRIX
2040  FOR I = 1 TO 3 * NN
2050  FOR J = 1 TO BW
2060  KG(I,J) = 0
2070  NEXT J
2080  NEXT I
2090  REM    SET TWO-FORCE MEMBER MOMENT OF INERTIA = 0
2100  FOR J = 1 TO NE
2110  IF TC(J) = 1 THEN I(J) = 0
2120  NEXT J
2130  REM ------------------------------------------------------
2140  REM                BEGIN CALCULATIONS                  ----
2150  REM            PROGRESS PROMPTED ON TERMINAL           ----
2160  REM ------------------------------------------------------
2170  FLASH : PRINT : PRINT : PRINT "WORKING ON:": PRINT
2180  NORMAL
```

```
2190   FOR EL = 1 TO NE
2200   PRINT "       ELEMENT ";EL;" MATRIX"
2210   GOSUB 4350: REM    ROTATION MATRIX
2220   GOSUB 4600: REM    LOCAL STIFFNESS MATRIX
2230   GOSUB 4900: REM    STIFFNESS MATRIX ROTATION
2240   REM  ----------------------------------------------------
2250   REM         ASSEMBLING GLOBAL STIFFNESS MATRIX IN     ----
2260   REM                    BAND MATRIX FORMAT             ----
2270   REM  ----------------------------------------------------
2280   FOR I = 1 TO 6
2290   FOR J = 1 TO 6
2300   IF NG(EL,J) < NG(EL,I) GOTO 2330
2310 C = NG(EL,J) - NG(EL,I) + 1
2320 KG(NG(EL,I),C) = KT(I,J) + KG(NG(EL,I),C)
2330   NEXT J
2340   NEXT I
2350   NEXT EL
2360   REM  ----------------------------------------------------
2370   REM      TEST FOR TWO-FORCE MEMBER ZERO ROTATIONAL    ----
2380   REM      STIFFNESS ON DIAGONAL OF STIFFNESS MATRIX    ----
2390   REM  ----------------------------------------------------
2400   FOR I = 1 TO 3 * NN
2410   IF KG(I,1) = 0 THEN BC(I) = 1:U(I) = 0
2420   NEXT I
2430   REM  ----------------------------------------------------
2440   REM         MODIFY STIFFNESS MATRIX AND FORCE VECTOR  ----
2450   REM         TO INCLUDE BOUNDARY DISPLACEMENT DATA     ----
2460   REM  ----------------------------------------------------
2470   FOR I = 1 TO 3 * NN
2480 Q(I) = FC(I)
2490   IF BC(I) = 0 GOTO 2520
2500 KG(I,1) = KG(I,1) * 1E10
2510 FC(I) = U(I) * KG(I,1)
2520   NEXT I
2530   REM  ----------------------------------------------------
2540   REM         SOLVE FOR DISPLACEMENTS USING GAUSS       ----
2550   REM         ELIMINATION ROUTINE FOR BANDED MATRIX     ----
2560   REM  ----------------------------------------------------
2570   PRINT "    GAUSS EQUATION SOLVER (BAND MATRIX)"
2580 N = 3 * NN
2590   FOR M = 1 TO N
2600   IF KG(M,1) = 0 GOTO 2730
2610   FOR L = 2 TO BW
2620   IF KG(M,L) = 0 GOTO 2720
2630 I = M + L - 1
2640 QU = KG(M,L) / KG(M,1)
2650 J = 0
2660   FOR K = L TO BW
2670 J = J + 1
2680   IF I > N GOTO 2700
2690 KG(I,J) = KG(I,J) - QU * KG(M,K)
2700   NEXT K
2710 KG(M,L) = QU
2720   NEXT L
2730   NEXT M
2740   PRINT "MATRIX REDUCTION": PRINT
2750   FOR M = 1 TO N
2760   IF KG(M,1) = 0 GOTO 2840
2770   FOR L = 2 TO BW
2780   IF KG(M,L) = 0 GOTO 2820
2790 I = M + L - 1
2800   IF I > N GOTO 2820
2810 FC(I) = FC(I) - KG(M,L) * FC(M)
2820   NEXT L
2830 FC(M) = FC(M) / KG(M,1)
2840   NEXT M
2850   PRINT : PRINT "FORCE REDUCTION": PRINT
2860 U(N) = FC(N)
2870   FOR KK = 2 TO N
2880 M = N + 1 - KK
```

```
2890  FOR L = 2 TO BW
2900  IF KG(M,L) = 0 GOTO 2940
2910  K = M + L - 1
2920  IF K > N GOTO 2960
2930  FC(M) = FC(M) - KG(M,L) * U(K)
2940  NEXT L
2950  U(M) = FC(M)
2960  NEXT KK
2970  REM     IDENTIFY NEGLIGIBLE DISPLACEMENTS
2980  UM = 0
2990  FOR I = 1 TO N
3000  UI =  ABS (U(I))
3010  IF UI > UM THEN UM = UI
3020  NEXT I
3030  FOR I = 1 TO N
3040  IF  ABS (U(I)) / UM < 1E - 4 THEN U(I) = 0
3050  NEXT I
3060  FOR I = 1 TO 3 * NN
3070  FC(I) = Q(I)
3080  NEXT I
3090  REM
3100  REM  ------------------------------------------------
3110  REM          OPTIONAL HARDCOPY OF INPUT DATA     ----
3120  REM  ------------------------------------------------
3130  PRINT : PRINT "DO YOU WANT HARDCOPY? (Y/N)": PRINT
3140  INPUT H$: IF H$ <  > "Y" GOTO 3680
3150  PRINT : PRINT "ARE YOU CONNECTED TO A PRINTER ?     (Y/N)": PRINT
3160  INPUT PC$: IF PC$ <  > "Y" THEN H$ = "N"
3170  IF H$ <  > "Y" GOTO 3680
3180  PRINT D$;"PR#1": REM     ACTIVATE PRINTER
3190  PRINT I$;"ML 10": REM        LEFT MARGIN SETTING (APPLE DMP COMMAND)
3200  PRINT : PRINT "PLANE FRAME ANALYSIS --- INPUT DATA": PRINT : PRINT
      "***** ELEMENT GLOBAL NODE NUMBERS ****"
3210  PRINT : PRINT "ELEMENT";: POKE 36,20: PRINT "GLOBAL NODE NUMBERS
      OF";: POKE 36,50: PRINT "ELEMENT TYPE"
3220  PRINT "NUMBER";: POKE 36,20: PRINT "LOCAL 1";: POKE 36,32: PRINT
      "LOCAL 2": PRINT
3230  FOR J = 1 TO NE
3240  POKE 36,3: PRINT J;: POKE 36,23: PRINT GN(J,1);: POKE 36,35: PRINT
      GN(J,2);: POKE 36,55: PRINT TC(J)
3250  NEXT J
3260  REM
3270  PRINT : PRINT : PRINT "**** NODE GLOBAL COORDINATES AND BOUNDARY
      DISPLACEMENTS ****"
3280  REM
3290  PRINT : PRINT "NODE";: POKE 36,9: PRINT "X";: POKE 36,15: PRINT
      "Y";: POKE 36,25: PRINT "BOUNDARY DISPLACEMENTS": PRINT ;: POKE
      36,25: PRINT "U";: POKE 36,35: PRINT "V";: POKE 36,45: PRINT
      "THETA": PRINT
3300  FOR I = 1 TO NN
3310  IF BC(3 * I - 2) = 1 AND BC(3 * I - 1) = 1 AND BC(3 * I) = 1 GOTO
      3520
3320  IF BC(3 * I - 2) = 1 AND BC(3 * I - 1) = 1 GOTO 3500
3330  IF BC(3 * I - 2) = 1 AND BC(3 * I) = 1 GOTO 3480
3340  IF BC(3 * I - 1) = 1 AND BC(3 * I) = 1 GOTO 3460
3350  IF BC(3 * I - 2) = 1 GOTO 3400
3360  IF BC(3 * I - 1) = 1 GOTO 3420
3370  IF BC(3 * I) = 1 GOTO 3440
3380  PRINT I;: POKE 36,9: PRINT XG(I);: POKE 36,15: PRINT YG(I)
3390  GOTO 3530
3400  PRINT I;: POKE 36,9: PRINT XG(I);: POKE 36,15: PRINT YG(I);: POKE
      36,25: PRINT U(3 * I - 2)
3410  GOTO 3530
3420  PRINT I;: POKE 36,9: PRINT XG(I);: POKE 36,15: PRINT YG(I);: POKE
      36,35: PRINT U(3 * I - 1)
3430  GOTO 3530
3440  PRINT I;: POKE 36,9: PRINT XG(I);: POKE 36,15: PRINT YG(I);: POKE
      36,45: PRINT U(3 * I)
3450  GOTO 3530
```

```
3460  PRINT I;: POKE 36,9: PRINT XG(I);: POKE 36,15: PRINT YG(I);: POKE
      36,35: PRINT U(3 * I - 1);: POKE 36,45: PRINT U(3 * I)
3470  GOTO 3530
3480  PRINT I;: POKE 36,9: PRINT XG(I);: POKE 36,15: PRINT YG(I);: POKE
      36,25: PRINT U(3 * I - 2);: POKE 36,45: PRINT U(3 * I)
3490  GOTO 3530
3500  PRINT I;: POKE 36,9: PRINT XG(I);: POKE 36,15: PRINT YG(I);: POKE
      36,25: PRINT U(3 * I - 2);: POKE 36,35: PRINT U(3 * I - 1)
3510  GOTO 3530
3520  PRINT I;: POKE 36,9: PRINT XG(I);: POKE 36,15: PRINT YG(I);: POKE
      36,25: PRINT U(3 * I - 2);: POKE 36,35: PRINT U(3 * I - 1);: POKE
      36,45: PRINT U(3 * I)
3530  NEXT I
3540  PRINT : PRINT : PRINT "****NODAL FORCES AND COUPLES ****": PRINT
3550  PRINT "NODE";: POKE 36,10: PRINT "X FORCE";: POKE 36,25: PRINT "Y
      FORCE";: POKE 36,40: PRINT "COUPLE": PRINT
3560  FOR N = 1 TO NN
3570  IF (FC(3 * N - 2) = 0 AND FC(3 * N - 1) = 0 AND FC(3 * N) = 0) GOTO
      3590
3580  PRINT N;: POKE 36,10: PRINT FC(3 * N - 2);: POKE 36,25: PRINT FC(3
      * N - 1);: POKE 36,40: PRINT FC(3 * N)
3590  NEXT N
3600  PRINT : PRINT : PRINT "**** MATERIAL AND SECTION PROPERTIES ****":
      PRINT
3610  PRINT "ELEMENT";: POKE 36,10: PRINT "ELASTIC MODULUS";: POKE 36,30:
      PRINT "AREA";: POKE 36,37: PRINT "AREA MOMENT OF INERTIA": PRINT
3620  FOR I = 1 TO NE
3630  POKE 36,3: PRINT I;: POKE 36,10: PRINT E(I);: POKE 36,30: PRINT
      A(I);: POKE 36,40: PRINT I(I)
3640  NEXT I
3650  REM ------------------------------------------------------
3660  REM              OUTPUT OF DISPLACEMENTS        ----
3670  REM ------------------------------------------------------
3680  PRINT : PRINT : PRINT
3690  PRINT "PLANE FRAME ANALYSIS --- RESULTS": PRINT : PRINT : PRINT
3700  PRINT : PRINT
3710  PRINT "******************************************"
3720  PRINT "**              DISPLACEMENTS           **"
3730  PRINT "******************************************": PRINT : PRINT :
      PRINT
3740  PRINT : PRINT "NODE";: POKE 36,7: PRINT "XDISPL";: POKE 36,18:
      PRINT "YDISPL";: POKE 36,30: PRINT "THETA, RAD.": PRINT
3750  FOR I = 1 TO NN
3760  PRINT I;: POKE 36,6: PRINT  INT (1E6 * U(3 * I - 2)) / 1E6;: POKE
      36,17: PRINT  INT (1E6 * U(3 * I - 1)) / 1E6;: POKE 36,29: PRINT
      INT (1E6 * U(3 * I)) / 1E6

3770  NEXT I
3780  PRINT : PRINT
3790  PRINT "******************************************"
3800  PRINT "**            ELEMENT NODE FORCES       **"
3810  PRINT "******************************************": PRINT : PRINT :
      PRINT
3820  PRINT "--NOTE--NODE NUMBERS ARE GLOBAL,LISTED": PRINT "IN ASCENDING
      LOCAL NUMBER ORDER": PRINT : PRINT : PRINT
3830  PRINT "EL#";: POKE 36,5: PRINT "NODE";: POKE 36,11: PRINT "AXIAL
      LOAD";: POKE 36,22: PRINT "SHEAR LOAD";: POKE 36,34: PRINT "COUPLE":
      PRINT
3840  REM
3850  REM ------------------------------------------------------
3860  REM   TRANSFORM GLOBAL DISPLACEMENTS, UG, TO LOCAL ----
3870  REM   COORDINATE COMPONENTS, UL, FOR EACH ELEMENT   ----
3880  REM ------------------------------------------------------
3890  FOR EL = 1 TO NE
3900  FOR I = 1 TO 6
3910  UG(I) = U(NG(EL,I))
3920  UL(I) = 0:FL(I) = 0
3930  NEXT I
3940  GOSUB 4350: REM   ROTATION MATRIX
```

```
3950    FOR I = 1 TO 6
3960     FOR J = 1 TO 6
3970  UL(I) = A2(J,I) * UG(J) + UL(I)
3980     NEXT J
3990     NEXT I
4000     REM  ------------------------------------------------------
4010     REM          MULTIPLY LOCAL STIFFNESS MATRIX AND       ----
4020     REM          LOCAL DISPLACEMENTS TO OBTAIN FORCES      ----
4030     REM  ------------------------------------------------------
4040     GOSUB 4600: REM     ELEMENT LOCAL STIFFNESS MATRIX
4050     FOR I = 1 TO 6
4060     FOR J = 1 TO 6
4070  FL(I) = K(I,J) * UL(J) + FL(I)
4080     NEXT J
4090     NEXT I
4100     REM  ------------------------------------------------------
4110     REM                 OUTPUT OF LOCAL FORCES             ----
4120     REM  ------------------------------------------------------
4130     FOR I = 1 TO 2
4140     PRINT EL;: POKE 36,5: PRINT GN(EL,I);: POKE 36,11: PRINT  INT (FL(3
       * I - 2));: POKE 36,23: PRINT  INT (FL(3 * I - 1));: POKE 36,33:
         PRINT  INT (FL(3 * I))
4150     NEXT I
4160     NEXT EL
4170     IF H$ < > "Y" GOTO 4190
4180     PRINT D$;"PR#0"
4190     PRINT : INPUT "DO YOU WANT DISPLACEMENTS STORED ON DISK?   (Y/N)
         ";UK$
4200     IF UK$ < > "Y" GOTO 4270
4210     RT$ = ".DISPL"
4220     GOSUB 10160: REM      OPEN AND WRITE FILE
4230     FOR I = 1 TO 3 * NN
4240     PRINT U(I)
4250     NEXT I
4260     GOSUB 10110: REM      CLOSE FILE
4270     PRINT : PRINT "DO YOU WISH ANOTHER RUN ?  (Y/N)"
4280     INPUT RU$: IF RU$ = "Y" GOTO 1110
4290     IF RU$ < > "N" GOTO 4270
4300     END
4310     REM  ------------------------------------------------------
4320     REM                     SUBROUTINES                    ----
4330     REM  ------------------------------------------------------
4340     REM
4350     REM  SUBROUTINE FOR ROTATION MATRIX
4360  X(1) = XG(GN(EL,1))
4370  X(2) = XG(GN(EL,2))
4380  Y(1) = YG(GN(EL,1))
4390  Y(2) = YG(GN(EL,2))
4400  L =  SQR ((X(2) - X(1)) ^ 2 + (Y(2) - Y(1)) ^ 2)
4410  ST = (Y(2) - Y(1)) / L
4420  CT = (X(2) - X(1)) / L
4430     FOR I = 1 TO 6
4440     FOR J = 1 TO 6
4450  A2(I,J) = 0
4460     NEXT J
4470     NEXT I
4480  A2(1,1) = CT
4490  A2(1,2) =  - ST
4500  A2(2,1) = ST
4510  A2(2,2) = CT
4520  A2(3,3) = 1
4530  A2(4,4) = CT
4540  A2(4,5) =  - ST
4550  A2(5,4) = ST
4560  A2(5,5) = CT
4570  A2(6,6) = 1
4580     RETURN
4590     REM  ------------------------------------------------------
4600     REM  SUBROUTINE FOR ELEMENT LOCAL  STIFFNESS MATRIX
4610  T1 = A(EL) * E(EL) / L
```

```
 4620  T2 = E(EL) * I(EL) / L ^ 3
 4630   FOR I = 1 TO 6
 4640   FOR J = 1 TO 6
 4650  K(I,J) = 0
 4660   NEXT J
 4670   NEXT I
 4680  K(1,1) =  T1
 4690  K(1,4) =  - T1
 4700  K(2,2) = 12 * T2
 4710  K(2,3) =  - 6 * L * T2
 4720  K(2,5) =  - 12 * T2
 4730  K(2,6) =  - 6 * L * T2
 4740  K(3,2) =  - 6 * L * T2
 4750  K(3,3) = 4 * L ^ 2 * T2
 4760  K(3,5) = 6 * L * T2
 4770  K(3,6) = 2 * L ^ 2 * T2
 4780  K(4,1) =  - T1
 4790  K(4,4) = T1
 4800  K(5,2) =  - 12 * T2
 4810  K(5,3) = 6 * L * T2
 4820  K(5,5) = 12 * T2
 4830  K(5,6) = 6 * L * T2
 4840  K(6,2) =  - 6 * L * T2
 4850  K(6,3) = 2 * L ^ 2 * T2
 4860  K(6,5) = 6 * L * T2
 4870  K(6,6) = 4 * L ^ 2 * T2
 4880   RETURN
 4890   REM  ------------------------------------------------------
 4900   REM   SUBROUTINE FOR STIFFNESS MATRIX ROTATION--LOCAL TO GLOBAL
        REFERENCE
 4910   FOR I = 1 TO 6
 4920   FOR J = 1 TO 6
 4930  R(I,J) = 0
 4940  KT(I,J) = 0
 4950   NEXT J
 4960   NEXT I
 4970   FOR I = 1 TO 6
 4980   FOR J = 1 TO 6
 4990   FOR N = 1 TO 6
 5000  R(I,J) = R(I,J) + K(I,N) * A2(J,N)
 5010   NEXT N
 5020   NEXT J
 5030   NEXT I
 5040   FOR I = 1 TO 6
 5050   FOR J = 1 TO 6
 5060   FOR N = 1 TO 6
 5070  KT(I,J) = KT(I,J) + A2(I,N) * R(N,J)
 5080   NEXT N
 5090   NEXT J
 5100   NEXT I
 5110   RETURN
 5120   REM  ------------------------------------------------------
10000   REM   SUBROUTINE FOR OPENING AND READING A FILE
10010   PRINT : INPUT "WHAT IS THE NAME OF THE FILE ?";NA$
10015  F$ = NA$ + RT$
10020   PRINT : INPUT "WHICH DISK DRIVE ?";DN
10030   IF DN = 1 OR DN = 2 THEN 10050
10040   PRINT : PRINT " TRY AGAIN-": GOTO 10020
10050   PRINT D$;"MONCIO"
10060   PRINT D$;"OPEN";F$;",D";DN
10070   PRINT D$;"UNLOCK";F$
10080   PRINT D$;"OPEN";F$
10090   PRINT D$;"READ";F$
10100   RETURN
10102   REM  ------------------------------------------------------
10110   REM   SUBROUTINE FOR CLOSING A FILE
10120   PRINT D$;"LOCK";F$
10130   PRINT D$;"CLOSE";F$
10140   PRINT D$;"NOMONCIO"
10150   RETURN
```

```
10152  REM    ------------------------------------------------------
10160  REM     SUBROUTINE FOR OPENING AND WRITING TO A FILE
10170  PRINT : INPUT "NAME OF FILE ?";NA$
10175  F$ = NA$ + RT$
10180  PRINT : INPUT "WHICH DISK DRIVE ?";DN
10190  IF DN = 1 OR DN = 2 THEN 10210
10200  PRINT : PRINT " TRY AGAIN-": GOTO 10180
10210  PRINT D$;"MONCIO"
10220  PRINT D$;"OPEN";F$;",D";DN
10230  PRINT D$;"UNLOCK";F$
10240  PRINT D$;"DELETE";F$
10250  PRINT D$;"OPEN";F$
10260  PRINT D$;"WRITE";F$
10270  RETURN
```

FRAME-2D FILE EDITOR

```
100  LOMEM: 16384
110  TEXT : HOME
120  PRINT "*****************************************"
130  PRINT "**                                     **"
140  PRINT "**                                     **"
150  PRINT "**        FRAME-2D FILE EDITOR         **"
160  PRINT "**        ----- -- ---- ------         **"
170  PRINT "**                                     **"
180  PRINT "**   STRUCTURAL ANALYSIS IN XY PLANE   **"
190  PRINT "**                                     **"
200  PRINT "** TWO-FORCE AND BEAM/COLUMN ELEMENT   **"
210  PRINT "**                                     **"
220  PRINT "**                                     **"
230  PRINT "**                                     **"
240  PRINT "*****************************************"
250  PRINT
260  PRINT
300  VTAB 20: PRINT "HIT ANY KEY TO CONTINUE"
302  GET G$: PRINT
304  REM    ------------------------------------------------------
306  REM      LINES 310 THRU 348 ARE MINI FRONT END PROGRAM
308  REM    ------------------------------------------------------
310  ONERR  GOTO 318
312  POKE 768,104: POKE 769,168: POKE 770,104: POKE 771,166: POKE
     772,223: POKE 773,154: POKE 774,72: POKE 775,152: POKE 776,72: POKE
     777,96
314  D$ = CHR$ (4): REM    CHR$(4) IS CTRL-D
316  GOTO 356: REM    GO TO NORMAL PROGRAM START
318  CALL 768: PRINT D$;"PR#0": TEXT : IF  PEEK (222) = 255 THEN  PRINT
     "CNTL C CAUSED A BREAK IN "; PEEK (218) +  PEEK (219) * 256: END
320  GOSUB 340: END
322  PRINT : PRINT : REM          DETERMINE FREE MEMORY
324  TP =  PEEK (175) +  PEEK (176) * 256: REM    TOP OF PROGRAM
326  EA =  PEEK (109) +  PEEK (110) * 256: REM    TOP OF LOMEM POINTER
328  BS =  PEEK (111) +  PEEK (112) * 256: REM    HIMEM POINTER
330  PRINT "FREE DATA SPACE IS ";BS - EA;" BYTES"
332  PRINT : PRINT "FREE PROGRAM SPACE:": PRINT
334  TP = 8192 - TP: IF TP > 0 THEN  PRINT TP;" BYTES TO HGR": RETURN
336  TP = TP + 8192: IF TP > 0 THEN  PRINT TP;" BYTES TO HGR2": RETURN
338  PRINT "       YOU ARE ";  ABS (TP);" BYTES INTO HGR21": RETURN
340  PRINT "ERROR # ";  PEEK (222);" AT LINE # ";  PEEK (218) +  PEEK (219)
     * 256
342  PRINT : PRINT "    (ERROR CODES IN POOLE PG. 331)"
344  PRINT : PRINT "YOUR OPTIONS NOW ARE:": PRINT "  (1) HIT ESCAPE TO
     RESUME AT ERROR STATEMENT": PRINT "  (2) HIT ANY OTHER KEY TO END
     PROGRAM"
346  GET G$: PRINT : IF G$ =  CHR$ (27) THEN  RESUME
348  RETURN
350  REM    ------------------------------------------------------
352  REM       START PROGRAM HERE
354  REM    ------------------------------------------------------
```

```
356  HOME
400  I$ = CHR$ (9): REM   CHR$(9) IS CTRL-I
410  R$ = CHR$ (13): REM  CHR$(13) IS CARRIAGE RETURN
420  DIM XG(100),YG(100),GN(100,2),PX(100),PY(100)
440  REM  --------------------------------------------------
450  PRINT "         1 - CREATE A NEW FILE"
460  PRINT
470  PRINT "         2 - EDIT AN EXISTING FILE"
480  PRINT
490  INPUT "         ENTER CHOICE - (1 OR 2)";IN$
500  REM  --------------------------------------------------
510  IF IN$ = "1" GOTO 540
520  RT$ = ".GEOM"
530  GOSUB 10000: REM    OPEN AND READ GEOMETRY DATA FILE
540  PRINT : INPUT "HOW MANY ELEMENTS? ";NE: PRINT
550  PRINT : INPUT "HOW MANY NODES? ";NN: PRINT
560  PRINT "ENTER ELEMENT NUMBER AND GLOBAL NUMBER OF LOCAL NODES 1 AND 2
     IN FOLLOWING FORMAT:": PRINT : PRINT "EL#,  #1,  #2    <RETURN> ":
     PRINT
570  FOR J = 1 TO NE
580  INPUT I,GN(I,1),GN(I,2)
590  NEXT J
600  PRINT : PRINT "ENTER GLOBAL NODE NUMBER AND ITS X AND Y COORDINATES
     IN FOLLOWING FORMAT:": PRINT : PRINT "NODE#, X,  Y    <RETURN>":
     PRINT
610  FOR J = 1 TO NN
620  INPUT I,XG(I),YG(I)
630  NEXT J
640  IF IN$ = "1" GOTO 710
650  GOSUB 10110: REM   CLOSE THE GEOMETRY FILE
660  REM  *********** EDITING OPTIONS ***********
670  PRINT : PRINT : INPUT "DO YOU WISH TO ALTER NUMBER OF ELEMENTS
     AND/OR NODES ?  (Y/N)";EN$
680  IF EN$ < > "Y" GOTO 720
690  PRINT : PRINT "CURRENT VALUES -  NE =";NE;",   NN =";NN
700  PRINT : INPUT "ENTER NEW VALUES - NE = ";NE: INPUT "
     NN = ";NN
710  PRINT : PRINT
720  PRINT : PRINT : INPUT "DO YOU WISH TO EDIT NODE NUMBERING ?
     (Y/N)";RE$
730  IF RE$ = "Y" OR RE$ = "N" THEN 760
740  REM
750  PRINT : PRINT "RE-ENTER -": GOTO 720
760  IF RE$ = "N" THEN 840
770  PRINT : INPUT "WHICH ELEMENT ?";N
780  IF N > NE THEN 770
790  PRINT "CURRENT VALUES: ";GN(N,1);",   ";GN(N,2)
800  PRINT : INPUT "ENTER THE NEW VALUES:";GN(N,1),GN(N,2)
810  PRINT : INPUT "MORE CHANGES ?  (Y/N)";RE$
820  IF RE$ = "Y" THEN 770
830  PRINT : PRINT
840  PRINT : PRINT : INPUT "DO YOU WISH TO EDIT NODE COORDINATES ?
     (Y/N)";RE$
850  IF RE$ = "Y" OR RE$ = "N" THEN 870
860  PRINT : PRINT "RE-ENTER -": GOTO 840
870  IF RE$ = "N" THEN 940
880  PRINT : INPUT "WHICH NODE ?";N
890  IF N > NN THEN 880
900  PRINT "CURRENT VALUES: ";XG(N);",   ";YG(N)
910  PRINT : INPUT "ENTER THE NEW VALUES- ";XG(N),YG(N)
920  PRINT : INPUT "MORE CHANGES ?  (Y/N)";RE$
930  IF RE$ = "Y" THEN 880
940  GOSUB 1150: REM    PLOT MESH
950  PRINT : PRINT : PRINT "HIT RETURN"
960  GET GO$
970  REM  NO PRINT AFTER GET
980  PRINT "OK TO STORE DATA ON DISC ?  (Y/N) "
990  PRINT "NOTE- AN (N) RESPONSE WILL RESTART EDITING : PRESS 'E' TO
     STOP"
1000 INPUT DK$
```

```
1010   IF DK$ = "E" GOTO 1130
1020   IF DK$ < > "Y" THEN 660
1030 RT$ = ".GEOM"
1040   GOSUB 10160: REM     OPEN AND WRITE TO THE GEOMETRY FILE
1050   PRINT NE;R$;NN
1060   FOR I = 1 TO NE
1070   PRINT I;R$;GN(I,1);R$;GN(I,2)
1080   NEXT
1090   FOR I = 1 TO NN
1100   PRINT I;R$;XG(I);R$;YG(I)
1110   NEXT
1120   GOSUB 10110: REM    CLOSE GEOMETRY FILE
1130   END
1140   REM  ------------------------------------------------------
1150   REM                    PLOTTING SUBROUTINE            ----
1160   REM  ------------------------------------------------------
1170   FLASH : PRINT : PRINT : PRINT "SCALING FOR PLOT": NORMAL
1180   REM   SEARCH FOR MOST NEGATIVE X AND Y COORDINATES
1190   XP = 0:YP = 0:SX = 0:SY = 0
1200   FOR I = 1 TO NN
1210   IF XG(I) < XP THEN XP = XG(I)
1220   IF YG(I) < YP THEN YP = YG(I)
1230   NEXT I
1240   REM  SHIFT COORDINATE AXES
1250   FOR I = 1 TO NN
1260 PX(I) = XG(I) - XP:PY(I) = YG(I) - YP
1270   NEXT
1280   REM  DETERMINE SIZE OF STRUCTURE IN X AND Y
1290   FOR I = 2 TO NN
1300   FOR J = 1 TO NN
1310   BX =   ABS (PX(I) - PX(J - 1)):BY =   ABS (PY(I) - PY(J - 1))
1320   IF BX > SX THEN SX = BX
1330   IF BY > SY THEN SY = BY
1340   NEXT J
1350   NEXT I
1360   REM   CALCULATE THE SCALE FACTOR, SF
1370 SF = 250 / SX
1380   IF SF * SY > 150 THEN SF = 150 / SY
1390   HGR
1400   HCOLOR= 3
1410   FOR I = 1 TO NE
1420 X1 = PX(GN(I,1)) * SF:Y1 = 150 - PY(GN(I,1)) * SF * 0.8
1430 X2 = PX(GN(I,2)) * SF:Y2 = 150 - PY(GN(I,2)) * SF * 0.8
1440    HPLOT 25 + X1,Y1 TO 25 + X2,Y2
1450   NEXT I
1460   RETURN
1470   REM  ------------------------------------------------------
1480   REM                FILE READ/WRITE SUBROUTINES        ----
1490   REM  ------------------------------------------------------
10000   REM  FILE READ
10010   PRINT : INPUT "WHAT IS THE NAME OF THE FILE ?";NA$
10015 F$ = NA$ + RT$
10020   PRINT : INPUT "WHICH DISK DRIVE ?";DN
10030   IF DN = 1 OR DN = 2 THEN 10050
10040   PRINT : PRINT " TRY AGAIN-": GOTO 10020
10050   PRINT D$;"MONCIO"
10060   PRINT D$;"OPEN";F$;",D";DN
10070   PRINT D$;"UNLOCK";F$
10080   PRINT D$;"OPEN";F$
10090   PRINT D$;"READ";F$
10100   RETURN
10110   REM  FILE CLOSE
10120   PRINT D$;"LOCK";F$
10130   PRINT D$;"CLOSE";F$
10140   PRINT D$;"NOMONCIO"
10150   RETURN
10160   REM  FILE WRITE ROUTINE
10170   PRINT : INPUT "NAME OF FILE ?";NA$
10175 F$ = NA$ + RT$
10180   PRINT : INPUT "WHICH DISK DRIVE ?";DN
```

```
10190   IF DN = 1 OR DN = 2 THEN 10210
10200   PRINT : PRINT " TRY AGAIN-": GOTO 10180
10210   PRINT D$;"MONCIO"
10220   PRINT D$;"OPEN";F$;",D";DN
10230   PRINT D$;"UNLOCK";F$
10240   PRINT D$;"DELETE";F$
10250   PRINT D$;"OPEN";F$
10260   PRINT D$;"WRITE";F$
10270   RETURN
```

FRAME-2D DISTORTION PLOT

```
100   LOMEM: 16384
110   TEXT : HOME
120   PRINT "****************************************"
130   PRINT "**                                    **"
140   PRINT "**     FRAME-2D DISTORTION PLOT       **"
150   PRINT "**     ----- -- ---------- ----       **"
160   PRINT "**                                    **"
170   PRINT "**   STRUCTURAL ANALYSIS IN XY PLANE  **"
180   PRINT "**                                    **"
190   PRINT "** TWO-FORCE AND BEAM/COLUMN ELEMENT  **"
200   PRINT "**                                    **"
210   PRINT "**                                    **"
220   PRINT "****************************************"
230   PRINT
240   PRINT
250   PRINT
300   VTAB 20: PRINT "HIT ANY KEY TO CONTINUE"
302   GET G$: PRINT
304   REM -----------------------------------------------
306   REM     LINES 310 THRU 348 ARE MINI FRONT END PROGRAM
308   REM -----------------------------------------------
310   ONERR GOTO 318
312   POKE 768,104: POKE 769,168: POKE 770,104: POKE 771,166: POKE
      772,223: POKE 773,154: POKE 774,72: POKE 775,152: POKE 776,72: POKE
      777,96
314   D$ = CHR$ (4): REM   CHR$(4) IS CTRL-D
316   GOTO 356: REM   GO TO NORMAL PROGRAM START
318   CALL 768: PRINT D$;"PR#0": TEXT : IF  PEEK (222) = 255 THEN  PRINT
      "CNTL C CAUSED A BREAK IN ";  PEEK (218) +  PEEK (219) * 256: END
320   GOSUB 340: END
322   PRINT : PRINT : REM         DETERMINE FREE MEMORY
324   TP =  PEEK (175) +  PEEK (176) * 256: REM   TOP OF PROGRAM
326   EA =  PEEK (109) +  PEEK (110) * 256: REM   TOP OF LOMEM POINTER
328   BS =  PEEK (111) +  PEEK (112) * 256: REM   HIMEM POINTER
330   PRINT "FREE DATA SPACE IS ";BS - EA;" BYTES"
332   PRINT : PRINT "FREE PROGRAM SPACE:": PRINT
334   TP = 8192 - TP: IF TP > 0 THEN  PRINT TP;" BYTES TO HGR": RETURN
336   TP = TP + 8192: IF TP > 0 THEN  PRINT TP;" BYTES TO HGR2": RETURN
338   PRINT "      YOU ARE ";  ABS (TP);" BYTES INTO HGR2!": RETURN
340   PRINT "ERROR # ";  PEEK (222);" AT LINE # ";  PEEK (218) +  PEEK (219)
      * 256
342   PRINT : PRINT "    (ERROR CODES IN POOLE PG. 331)"
344   PRINT : PRINT "YOUR OPTIONS NOW ARE:": PRINT "   (1) HIT ESCAPE TO
      RESUME AT ERROR STATEMENT": PRINT "   (2) HIT ANY OTHER KEY TO END
      PROGRAM"
346   GET G$: PRINT : IF G$ =  CHR$ (27) THEN  RESUME
348   RETURN
350   REM -----------------------------------------------
352   REM        START PROGRAM HERE
354   REM -----------------------------------------------
356   HOME
400   I$ =  CHR$ (9): REM   CHR$(9) IS CTRL-I
410   R$ =  CHR$ (13): REM   CHR$(13) IS CARRIAGE RETURN
420   RU$ = "N"
430   PRINT "INPUT THE GEOMETRY DATA FROM DISK FILE": PRINT
440   REM  OPEN AND READ THE DATA FILE
```

COMPUTER PROGRAMS FOR PLANE FRAME ANALYSIS

```
450 RT$ = ".GEOM"
460 DIM U(200),XG(100),YG(100),GN(100,3)
470 GOSUB 10000: REM   OPEN AND READ FILE
480 INPUT NE,NN
490 FOR J = 1 TO NE
500 INPUT I,GN(I,1),GN(I,2)
510 NEXT J
520 FOR J = 1 TO NN
530 INPUT I,XG(I),YG(I)
540 NEXT J
550 GOSUB 10110: REM   CLOSE FILE
560 PRINT : PRINT "INPUT THE DISPLACEMENT DATA FROM DISK FILE": PRINT
570 RT$ = ".DISPL"
580 GOSUB 10000: REM   OPEN AND READ FILE
590 FOR I = 1 TO 3 * NN
600 INPUT U(I)
610 NEXT I
620 GOSUB 10110: REM   CLOSE FILE
630 REM ------------------------------------------------------
640 REM                 PLOTTING SUBROUTINE              ----
650 REM ------------------------------------------------------
660 DI = 0
670 FOR I = 1 TO 2 * NN
680 IF  ABS (U(I)) > DI THEN DI =  ABS (U(I))
690 NEXT I
700 REM  SEARCH FOR MOST NEGATIVE COORDINATES
710 XP = 0:YP = 0
720 FOR I = 1 TO NN
730 IF XG(I) < XP THEN XP = XG(I)
740 IF YG(I) < YP THEN YP = YG(I)
750 NEXT I
760 REM  SHIFT COORDINATE AXES
770 FOR I = 1 TO NN
780 XG(I) = XG(I) - XP
790 YG(I) = YG(I) - YP
800 NEXT I
810 REM  PLOT OF MESH
820 SX = 0:SY = 0
830  FLASH : PRINT : PRINT : PRINT "SCALING FOR PLOT ": PRINT : PRINT :
     PRINT : PRINT : NORMAL
840 XG(0) = 0:YG(0) = 0
850 REM  DETERMINE SIZE OF STRUCTURE
860 FOR I = 2 TO NN
870 FOR J = 1 TO NN + 1
880 BX =  ABS (XG(I) - XG(J - 1)):BY =  ABS (YG(I) - YG(J - 1))
890 IF BX > SX THEN SX = BX
900 IF BY > SY THEN SY = BY
910 NEXT J
920 NEXT I
930 REM  CALCULATE SCALE FACTOR, SF
940 SF = 200 / SX
950 IF SF * SY > 130 THEN SF = 130 / SY
970 HCOLOR= 3
980 MF = 0: REM  <MF> IS USER CONTROLLED MAGNIFICATION FACTOR
990  HGR : FOR I = 1 TO NE
1000 X1 = (XG(GN(I,1)) + U(3 * GN(I,1) - 2) * MF) * SF
1010 X2 = (XG(GN(I,2)) + U(3 * GN(I,2) - 2) * MF) * SF
1020 Y1 = 130 - (YG(GN(I,1)) + U(3 * GN(I,1) - 1) * MF) * SF * 0.8
1030 Y2 = 130 - (YG(GN(I,2)) + U(3 * GN(I,2) - 1) * MF) * SF * 0.8
1040  HPLOT 40 + X1,Y1 TO 40 + X2,Y2
1050 NEXT I
1060 INPUT "TO MAGNIFY DISTORTION TYPE AN INTEGER BETWEEN 1 AND 10.
     TYPE '0' TO END PROGRAM . ";K
1070 MF = K / DI
1080 IF K < > 0 GOTO 990
1090 END
1100 REM ------------------------------------------------------
1110 REM            FILE READ/WRITE SUBROUTINES           ----
1120 REM ------------------------------------------------------
10000  REM  SUBROUTINE FOR OPENING AND READING A FILE
```

```
10001   REM
10010   PRINT : INPUT "WHAT IS THE NAME OF THE FILE ?";NA$
10015   F$ = NA$ + RT$
10020   PRINT : INPUT "WHICH DISK DRIVE ?";DN
10030   IF DN = 1 OR DN = 2 THEN 10050
10040   PRINT : PRINT " TRY AGAIN-": GOTO 10020
10050   PRINT D$;"MONCIO"
10060   PRINT D$;"OPEN";F$;",D";DN
10070   PRINT D$;"UNLOCK";F$
10080   PRINT D$;"OPEN";F$
10090   PRINT D$;"READ";F$
10100   RETURN
10102   REM
10110   REM  SUBROUTINE FOR CLOSING A FILE
10120   PRINT D$;"LOCK";F$
10130   PRINT D$;"CLOSE";F$
10140   PRINT D$;"NOMONCIO"
10150   RETURN
10152   REM
10160   REM  SUBROUTINE FOR OPENING AND WRITING TO A FILE
10170   PRINT : INPUT "NAME OF FILE ?";NA$
10175   F$ = NA$ + RT$
10180   PRINT : INPUT "WHICH DISK DRIVE ?";DN
10190   IF DN = 1 OR DN = 2 THEN 10210
10200   PRINT : PRINT " TRY AGAIN-": GOTO 10180
10210   PRINT D$;"MONCIO"
10220   PRINT D$;"OPEN";F$;",D";DN
10230   PRINT D$;"UNLOCK";F$
10240   PRINT D$;"DELETE";F$
10250   PRINT D$;"OPEN";F$
10260   PRINT D$;"WRITE";F$
10270   RETURN
10272   REM
---
```

A2.5 PROGRAM LISTINGS IN TURBO PASCAL

```
(*****************************************************)
PROGRAM FRAME2D(INPUT,OUTPUT);
(*****************************************************)
    CONST
        ArrSize1 {No. of elements}       = 40;
        ArrSize2 {Degrees of freedom}    = 120;
        ArrSize3 {Half-Bandwidth}        = 50;
        ArrSize4 {No. of Nodes}          = 40;
    TYPE
        RARY1 = ARRAY[1..2] OF REAL;
        RARY2 = ARRAY[1..6] OF REAL;
        RARY3 = ARRAY[1..ArrSize1] OF REAL;
        RARY4 = ARRAY[1..ArrSize2] OF REAL;
        IARY5 = ARRAY[1..ArrSize1] OF INTEGER;
        RARY6 = ARRAY[1..6,1..6] OF REAL;
        IARY7 = ARRAY[1..ArrSize1,1..2] OF INTEGER;
        IARY8 = ARRAY[1..ArrSize1,1..6] OF INTEGER;
        RARY9 = ARRAY[1..ArrSize2,1..ArrSize3] OF REAL;
        RARY10 = ARRAY[1..ArrSize4] OF REAL;
    VAR
        X,Y                              :RARY1;
        UL,UG,FL                         :RARY2;
        A,E,I                            :RARY3;
        BC,FC,FR,Q,U                     :RARY4;
        A2,K,KT,R                        :RARY6;
        GN                               :IARY7;
        NG                               :IARY8;
        KG                               :RARY9;
        XG,YG                            :RARY10;
```

```
            TC                              :IARY5;
            I1,FX,FY,FT,AP,M,C,BW,NN,NE     :INTEGER;
            SF, MF                          :REAL;
            LT,T1,T2,CT,ST,L,QU,UM,UI       :REAL;
            MINX,MINY                       :REAL;
            RESP,CC,DRV                     :CHAR;
            MEM,LI,KI,IIN,J,EL,N,KK         :INTEGER;
            SHIFTX,SHIFTY                   :INTEGER;
            OUTFIL                          :TEXT;
            ELPROPFILE                      :TEXT;
            FNAME                           :STRING[20];
            DONE,QUIT                       :BOOLEAN;

{ NE = number of elements }
{ NN = number of nodes }
{ FX = number of x direction constraints }
{ FY = number of y directional constraints }
{ FT = number of rotation constraints }
{ BC(I) = boundary condition code }
{ TC(I) = element type code }
{ XG(I),YG(I) = global coordinates of node I }
{ GN(I,J) = global node number of local node number J of }
{            element I }
{ NG(I,J) = global displacement number of local displacement number J }
{            of element I }
{ K(I,J) = element local coordinate stiffness matrix }
{ KT(I,J) = element global coordinate stiffness matrix }
{ KG(I,J) = assembled global stiffness matrix }
{ FC(I) = applied global node force vector }
{ FL(I) = local element node force vector }
{ U(I) = global reference displacement value of global displacement }
{         number I }
{ UL(I) = local reference coordinate displacement value of local }
{         displacement number I }
{ UG(I) = global reference coordinate displacement value of local }
{         displacement number I }
{*****************************************************************}
{                     PROGRAM PROCEDURES                          }
{*****************************************************************}
{Element Coordinate Transformation Matrix [A2]}
PROCEDURE ROT_MATRIX;

    VAR
        I1, J                           :INTEGER;

    BEGIN
        X[1]:=XG[GN[EL,1]];
        X[2]:=XG[GN[EL,2]];
        Y[1]:=YG[GN[EL,1]];
        Y[2]:=YG[GN[EL,2]];
        L:=SQRT(SQR(X[2]-X[1])+SQR(Y[2]-Y[1]));
        ST:=(Y[2]-Y[1])/L;
        CT:=(X[2]-X[1])/L;
        FOR I1:= 1 TO 6 DO
            BEGIN
                FOR J:= 1 TO 6 DO
                    A2[I1,J]:=0
            END;
        A2[1,1]:=CT;
        A2[1,2]:=-ST;
        A2[2,1]:=ST;
        A2[2,2]:=CT;
        A2[3,3]:=1;
        A2[4,4]:=CT;
        A2[4,5]:=-ST;
        A2[5,4]:=ST;
        A2[5,5]:=CT;
        A2[6,6]:=1;
    END;
```

PROGRAM LISTINGS IN TURBO PASCAL

```pascal
(****************************************************************)
(Element Local Stiffness Matrix [k])
PROCEDURE STIFF_MAT;

    VAR
        I1, J                           :INTEGER;

    BEGIN
        T1:=A[EL]*E[EL]/L;
        T2:= E[EL]*I[EL]/((SQR(L))*L);
        FOR I1:=1 TO 6 DO
            BEGIN
                FOR J:=1 TO 6 DO
                    K[I1,J]:=0
            END;
        K[1,1]:=T1;
        K[1,4]:=-T1;
        K[2,2]:=12*T2;
        K[2,3]:=-6*L*T2;
        K[2,5]:=-12*T2;
        K[2,6]:=K[2,3];
        K[3,2]:=K[2,3];
        K[3,3]:=4*L*L*T2;
        K[3,5]:=-K[2,3];
        K[3,6]:=2*L*L*T2;
        K[4,1]:=K[1,4];
        K[4,4]:=K[1,1];
        K[5,2]:=-12*T2;
        K[5,3]:=6*L*T2;
        K[5,5]:=12*T2;
        K[5,6]:=K[5,3];
        K[6,2]:= K[3,2];
        K[6,3]:=2*L*L*T2;
        K[6,5]:=K[5,3];
        K[6,6]:=K[3,3];

    END;


(****************************************************************)

(Element Stiffness Matrix Transformation to Global Reference)
PROCEDURE STIF_MAT_ROT;

    VAR
        I1, J, N                        :INTEGER;

    BEGIN
        FOR I1:= 1 TO 6 DO
            BEGIN
                FOR J:= 1 TO 6 DO
                    BEGIN
                        R[I1,J]:=0;
                        KT[I1,J]:=0
                    END
            END;
        FOR I1:= 1 TO 6 DO
            BEGIN
                FOR J:= 1 TO 6 DO
                    FOR N:=1 TO 6 DO
                        R[I1,J]:=R[I1,J]+K[I1,N]*A2[J,N]
            END;
        FOR I1:=1 TO 6 DO
            BEGIN
                FOR J:=1 TO 6 DO
                    FOR N:=1 TO 6 DO
                        KT[I1,J]:=KT[I1,J]+A2[I1,N]*R[N,J]
            END
    END;
```

```
{*******************************************************************}
{Procedure to Scale and Center Structure for Graphics}
  PROCEDURE CENTER( VAR XG            :RARY10;
                    VAR YG            :RARY10;
                    VAR SF            :REAL;
                    VAR MINX          :REAL;
                    VAR MINY          :REAL;
                    VAR SHIFTX        :INTEGER;
                    VAR SHIFTY        :INTEGER);
  VAR
       MAXX, MAXY, SFXINV, SFYINV, XRANGE, YRANGE :REAL;

  BEGIN
     SFXINV:=0;
     SFYINV:=0;
     SF:=0;
     MINX:=0;
     MINY:=0;
     MAXX:=0;
     MAXY:=0;
     FOR J:=1 TO NN DO
         BEGIN
             IF XG[J] > MAXX  THEN
                MAXX:=XG[J];
             IF XG[J] < MINX  THEN
                MINX:=XG[J];
             IF YG[J] > MAXY  THEN
                MAXY:=YG[J];
             IF YG[J] < MINY  THEN
                MINY:=YG[J];
         END;
     XRANGE := ABS(MAXX-MINX);
     YRANGE := ABS(MAXY-MINY);
     SFXINV := XRANGE / 250;
     SFYINV := YRANGE / 150;
     IF (SFXINV = 0.0)  AND  (SFYINV = 0.0) THEN
         SF := 1.0
     ELSE IF (SFXINV > SFYINV) THEN
         SF := 1.0 / SFXINV

     ELSE
         SF := 1.0 / SFYINV;
     SHIFTX:=TRUNC((300-SF*(MAXX-MINX))/2);
     SHIFTY:=TRUNC((200-SF*(MAXY-MINY))/2);
  END;
{*******************************************************************}
{Display of Structure using the Turbo Graphics; Calls PROC. "center"}
  PROCEDURE GRAPH( VAR XG             :RARY10;
                   VAR YG             :RARY10;
                   VAR SF             :REAL;
                   VAR MINX           :REAL;
                   VAR MINY           :REAL;
                   VAR U              :RARY4;
                   VAR SHIFTX         :INTEGER;
                   VAR SHIFTY         :INTEGER;
                   VAR MF             :REAL);
  VAR
       X1,X2,Y1,Y2                    :INTEGER;
       C                              :CHAR;

  BEGIN
      CENTER(XG,YG,SF,MINX,MINY,SHIFTX,SHIFTY);
      GRAPHMODE;
      PALETTE(1);
      FOR J:=1 TO NE DO
         BEGIN
             X1:=SHIFTX+TRUNC(SF*(XG[GN[J,1]]-MINX)+MF*U[3*GN[J,1]-2]);
             X2:=SHIFTX+TRUNC(SF*(XG[GN[J,2]]-MINX)+MF*U[3*GN[J,2]-2]);
             Y1:=200-SHIFTY-TRUNC(SF*(YG[GN[J,1]]-MINY)+MF*U[3*GN[J,1]-1]);
```

PROGRAM LISTINGS IN TURBO PASCAL   423

```
                Y2:=200-SHIFTY-TRUNC(SF*(YG[GN[J,2]]-MINY)+MF*U[3*GN[J,2]-1]);
                DRAW (X1,Y1, X2,Y2,3);
            END;
        GOTOXY(1,24);
        WRITE ('HIT <CR> TO CONTINUE.');
        READLN;
        TEXTMODE
    END;
{*****************************************************************}
(Optional Edit of Input Node Numbering Data and Element Type Codes)
PROCEDURE EDIT_NN (VAR GN              : IARY7;
                   VAR NE              : INTEGER;
                   VAR TC              : IARY5);

    VAR
        N                              : INTEGER;
        REPLY                          : CHAR;
        DONE                           : BOOLEAN;
    BEGIN
      REPEAT
        DONE := TRUE;
        WRITE('WHICH ELEMENT WOULD YOU LIKE TO EDIT? ');
        READLN(N);
        IF N<=NE THEN
            BEGIN
               WRITELN;
               WRITELN ('    <CR> FOR NO CHANGE');
               WRITELN;
               WRITE   ('CURRENT VALUES:    LOCAL 1 = ',GN[N,1]);
               WRITE   (',    LOCAL 2 = ',GN[N,2]);
               WRITELN (',    TYPE CODE = ',TC[N]);
               WRITE   ('ENTER NEW VALUES:   LOCAL  1 = ');
               READLN (GN[N,1]);
               WRITE   ('                    LOCAL  2 = ');
               READLN (GN[N,2]);
               WRITE   ('                    TYPE CODE = ');
               READLN (TC[N]);
               WRITE('INPUT MORE CHANGES?  [Y/N] ');
               READLN(REPLY);
               IF UPCASE(REPLY) = 'Y' THEN
                    DONE := FALSE;
            END
        ELSE
            BEGIN
              WRITELN('THERE ARE NOT THAT MANY ELEMENTS, TRY AGAIN!');
              DONE := FALSE;
            END;
      UNTIL DONE;
      GRAPH(XG,YG,SF,MINX,MINY,U,SHIFTX,SHIFTY,MF);
    END;
{*****************************************************************}

(Optional Edit of Node Coordinates Data)
PROCEDURE EDIT_NC (VAR NN              : INTEGER;
                   VAR XG              : RARY10;
                   VAR YG              : RARY10);

    VAR
        N                              : INTEGER;
        DONE                           : BOOLEAN;

    BEGIN
      REPEAT
        DONE := TRUE;
        WRITE('WHICH NODE? ');
        READLN(N);
        IF N<=NN  THEN
            BEGIN
               WRITELN;
               WRITELN('    <CR> FOR NO CHANGE');
               WRITELN;
```

```
                        WRITE    ('CURRENT VALUES:      X = ',XG[N]:10:2);
                        WRITELN (',         Y = ',YG[N]:10:2);
                        WRITE    ('ENTER NEW VALUES:    X = ');
                        READLN(XG[N]);
                        WRITE    ('                     Y = ');
                        READLN(YG[N]);
                        WRITE('WOULD YOU LIKE TO MAKE MORE CHANGES?  (Y/N) ');
                        READLN(RESP);
                        IF UPCASE(RESP) = 'Y'  THEN
                            DONE := FALSE;
                    END
                ELSE
                    BEGIN
                        WRITELN('THERE ARE NOT THAT MANY NODES, TRY AGAIN!');
                        DONE := FALSE;
                    END;
        UNTIL DONE;
        GRAPH(XG,YG,SF,MINX,MINY,U,SHIFTX,SHIFTY,MF)
    END;
{*******************************************************************}

{Calculate Half-Bandwidth of Global Stiffness Matrix}
PROCEDURE BANDWIDTH;

    VAR
        I1, DI                              :INTEGER;

    BEGIN
        BW:=0;
        FOR I1:= 1 TO NE DO
            BEGIN
                DI:= ABS(GN[I1,1]-GN[I1,2]);
                IF DI>BW   THEN
                    BW:=DI
            END;
        BW:=3*BW+3;
        WRITELN;
        WRITELN('**********CALCULATED HALF-BANDWIDTH= ',BW,'**********');
        DELAY(1000);
        WRITELN;
    END;
{*******************************************************************}

{All Input Data from Keyboard}
PROCEDURE TERM_STRUCTURE  (VAR NE    : INTEGER;
                           VAR NN    : INTEGER;
                           VAR GN    : IARY7;
                           VAR XG    : RARY10;
                           VAR YG    : RARY10;
                           VAR TC    : IARY5);
    VAR
        I1,J                            :INTEGER;
    BEGIN
        WRITELN('HOW MANY "TWO-FORCE" AND "BEAM" ELEMENTS ');
        WRITE('   IN THE STRUCTURE? ');
        READLN(NE);
        WRITELN;
        WRITE('HOW MANY NODES? ');
        READLN(NN);
        WRITELN;
        WRITELN('ENTER ELEMENT NUMBER,');
        WRITELN('         GLOBAL NUMBERS OF LOCAL NODES 1 & 2, ');
        WRITELN('              AND ELEMENT TYPE CODE:');
        WRITELN('                  TWO-FORCE ELEMENT = "1" ');
        WRITELN('                  BEAM/COLUMN ELEMENT = "2" ');
        WRITELN;
        WRITELN('ENTER IN THE FOLLOWING FORMAT:');
        WRITELN('   EL#    #1   #2  TYPE CODE  <RETURN>');
        FOR J:=1 TO NE DO
            READLN(I1,GN[I1,1],GN[I1,2],TC[I1]);
```

PROGRAM LISTINGS IN TURBO PASCAL 425

```
            WRITELN;
            WRITELN('ENTER GLOBAL X AND Y COORDINATES OF THE NODES');
            WRITELN('ENTER IN THE FOLLOWING FORMAT:');
            WRITELN('   NODE#    X[1]     Y[1]    <RETURN>');
            FOR J:=1 TO NN DO
                READLN(I1,XG[I1],YG[I1]);
            WRITELN;
    END;
{********************************************************************}
{Write Structure Geometry to Disk}
PROCEDURE WRITE_STRUCTURE  (VAR NE      : INTEGER;
                            VAR NN      : INTEGER;
                            VAR GN      : IARY7;
                            VAR XG      : RARY10;
                            VAR YG      : RARY10;
                            VAR TC      : IARY5);

    TYPE
        CHRSTRING=STRING[20];

    VAR
        I1                          :INTEGER;
        OUTFILE                     :TEXT;
        FNAME                       :CHRSTRING;

    BEGIN
        WRITE ('ENTER FILE NAME (DEFAULT = STRUCT.DAT): ');
        READLN (FNAME);
        IF (FNAME = '') THEN FNAME := 'STRUCT.DAT';
        ASSIGN(OUTFILE,FNAME);
        REWRITE(OUTFILE);

        WRITELN(OUTFILE,NE,' ',NN);
        FOR I1:=1 TO NE DO
            BEGIN
                WRITELN(OUTFILE,I1,' ',GN[I1,1],' ',GN[I1,2],' ',TC[I1])
            END;
        FOR I1:=1 TO NN DO
            BEGIN
                WRITELN(OUTFILE,I1,' ',XG[I1],' ',YG[I1])
            END;
        CLOSE(OUTFILE)
    END;
{********************************************************************}
{Read Structure Geometry Data from Disk}
PROCEDURE READ_STRUCTURE   (VAR NE      : INTEGER;
                            VAR NN      : INTEGER;
                            VAR GN      : IARY7;
                            VAR XG      : RARY10;
                            VAR YG      : RARY10;
                            VAR TC      : IARY5);

    TYPE
        CHRSTRING=STRING[20];

    VAR
        I1                  :INTEGER;
        INPFILE             :TEXT;
        FNAME               :CHRSTRING;

    BEGIN
        WRITE('ENTER FILENAME (DEFAULT = STRUCT.DAT): ');
        READLN(FNAME);
        IF(FNAME = '') THEN FNAME:='STRUCT.DAT';
        ASSIGN(INPFILE,FNAME);
        RESET(INPFILE);

        WRITELN('READING NUMBER OF ELEMENTS ');
        READ(INPFILE,NE);
```

```
            WRITELN('READING # OF NODES ');
            READ(INPFILE,NN);
            WRITELN('READING NODE NUMBERS');
            FOR J:=1 TO NE DO
                BEGIN
                    READ(INPFILE,I1);
                    READ(INPFILE,GN[I1,1]);
                    READ(INPFILE,GN[I1,2]);
                    READ(INPFILE,TC[I1]);
                END;
            WRITELN('READING NODE COORDINATES');
            FOR J:=1 TO NN DO
                BEGIN
                    READ(INPFILE,I1);
                    READ(INPFILE,XG[I1]);
                    READ(INPFILE,YG[I1])
                END;
            WRITELN;
            CLOSE(INPFILE)
    END;

{*****************************************************************}

{Output of Input Data and Results}
PROCEDURE RESULTS;

    VAR
        DEV                          : TEXT;
        DONE                         : BOOLEAN;
        CHOICE                       : INTEGER;
        FNAME                        : STRING[20];

  BEGIN
    REPEAT
    BEGIN
      DONE:=TRUE;
      WRITELN;
      WRITELN;
      WRITELN;
      WRITELN ('              DATA OUTPUT OPTIONS');
      WRITELN ('              ---------------------');
      WRITELN;
      WRITELN ('                 1 - TERMINAL');
      WRITELN ('                 2 - PRINTER');
      WRITELN ('                 3 - AUX PORT');
      WRITELN ('                 4 - TEXTFILE');
      WRITELN;
      WRITE   ('             ENTER CHOICE- (1 TO 4) ');
      READLN(CHOICE);
      CASE CHOICE OF
         1: FNAME:='TRM:';
         2: FNAME:='LST:';
         3: FNAME:='AUX:';
         4: BEGIN
               WRITE('ENTER FILENAME (DEFAULT = RESULTS.DAT): ');
               READLN(FNAME);
               IF(FNAME = '') THEN FNAME:='RESULTS.DAT';
            END;
         ELSE
            DONE:=FALSE;
      END; {CASE}
  END;
UNTIL DONE;
ASSIGN (DEV, FNAME);
REWRITE (DEV);

WRITELN(DEV,'PLANE FRAME ANALYSIS--INPUT DATA');
WRITELN(DEV);
WRITELN(DEV,'*********  ELEMENT GLOBAL NODE NUMBERS  *********');
```

PROGRAM LISTINGS IN TURBO PASCAL   427

```pascal
WRITELN(DEV);
WRITELN(DEV,'    ELEMENT    GLOBAL NODE NUMBERS    ELEMENT  ');
WRITELN(DEV,'    NUMBER            OF              TYPE     ');
WRITELN(DEV,'                 LOCAL 1    LOCAL 2            ');
WRITELN(DEV);
  FOR J:=1 TO NE DO
    WRITELN(DEV,J:7,GN[J,1]:11,GN[J,2]:11,TC[J]:10);
WRITELN(DEV);
WRITELN(DEV,'***NODE GLOBAL COORDINATES AND BOUNDARY DISPLACEMENTS***');
WRITELN(DEV);
WRITELN(DEV,'                                      BOUNDARY DISPLACEMENTS');
WRITELN(DEV,'  NODE      X           Y          U           V       THETA  ');
WRITELN(DEV);
  FOR J:=1 TO NN DO
    BEGIN
      WRITE(DEV,J:5,XG[J]:10:3,YG[J]:10:3);
        IF (BC[3*J-2]=0) AND (BC[3*J-1]=0) AND (BC[3*J]=0) THEN
           WRITELN(DEV);
        IF (BC[3*J-2]=1) AND (BC[3*J-1]=1) AND (BC[3*J]=1) THEN
           WRITELN(DEV,U[3*J-2]:10:4,U[3*J-1]:10:4,U[3*J]:10:4);
        IF (BC[3*J-2]=1) AND (BC[3*J-1]=1) AND (BC[3*J]=0) THEN
           WRITELN(DEV,U[3*J-2]:10:4,U[3*J-1]:10:4);
        IF (BC[3*J-2]=1) AND (BC[3*J-1]=0) AND (BC[3*J]=1) THEN
           WRITELN(DEV,U[3*J-2]:10:4,'          ',U[3*J]:10:4);
        IF (BC[3*J-2]=1) AND (BC[3*J-1]=0) AND (BC[3*J]=0) THEN
           WRITELN(DEV,U[3*J-2]:10:4);
        IF (BC[3*J-2]=0) AND (BC[3*J-1]=1) AND (BC[3*J]=1) THEN
           WRITELN(DEV,'          ',U[3*J-1]:10:4,U[3*J]:10:4);
        IF (BC[3*J-2]=0) AND (BC[3*J-1]=1) AND (BC[3*J]=0) THEN
           WRITELN(DEV,'          ',U[3*J-1]:10:4);
        IF (BC[3*J-2]=0) AND (BC[3*J-1]=0) AND (BC[3*J]=1) THEN
           WRITELN(DEV,'                    ',U[3*J]:10:4);
    END;
WRITELN(DEV);
WRITELN(DEV,'********** NODAL FORCES AND COUPLES ****************');
WRITELN(DEV);
WRITELN(DEV,'  NODE      X FORCE        Y FORCE         COUPLE    ');
WRITELN(DEV);
FOR N:=1 TO NN DO
    BEGIN
      IF (FC[3*N-2]<>0) OR (FC[3*N-1]<>0) OR (FC[3*N]<>0) THEN
         WRITELN(DEV,N:5,FC[3*N-2]:14:3,FC[3*N-1]:14:3,FC[3*N]:14:3)
    END;
WRITELN(DEV);
WRITELN(DEV,'******* MATERIAL AND SECTIONAL PROPERTIES *********');
WRITELN(DEV);
WRITELN(DEV,'ELEMENT      ELASTIC         AREA       AREA MOMENT  ');
WRITELN(DEV,'             MODULUS                    OF INERTIA   ');
WRITELN(DEV);
FOR J:=1 TO NE DO
    WRITELN(DEV,J:5,E[J]:16:1,A[J]:10:4,I[J]:14:4);
{------------------------------------------------------------------}
{                    OUTPUT OF DISPLACEMENTS                       }
{------------------------------------------------------------------}
  WRITELN(DEV);
  WRITELN(DEV,'PLANE FRAME ANALYSIS----- RESULTS ');
  WRITELN(DEV);
  WRITELN(DEV,'*************************************************');
  WRITELN(DEV,'                  DISPLACEMENTS              ');
  WRITELN(DEV,'*************************************************');
  WRITELN(DEV);
  WRITELN(DEV,'  NODE       XDISPL        YDISPL       THETA,RAD');
  WRITELN(DEV);
  FOR J:=1 TO NN DO
      WRITELN(DEV,J:5,U[3*J-2]:14:6,U[3*J-1]:14:6,U[3*J]:14:6);
  WRITELN(DEV);
  WRITELN(DEV,'*************************************************');
  WRITELN(DEV,'              ELEMENT NODE FORCES ');
  WRITELN(DEV,'*************************************************');
  WRITELN(DEV);
  WRITELN(DEV,'--NOTE--NODE NUMBERS ARE GLOBAL, LISTED ');
  WRITELN(DEV,'            IN ASCENDING LOCAL NUMBER ORDER ');
```

```
    WRITELN(DEV);
    WRITELN(DEV);
    WRITELN(DEV,'  ELEMENT      NODE      AXIAL LOAD       SHEAR LOAD           COUPLE');
    WRITELN(DEV);
{----------------------------------------------------------}
{   TRANSFORM GLOBAL DISPLACEMENTS, UG, TO LOCAL----------}
{   COORDINATE COMPONENTS, UL, FOR EACH ELEMENT ----------}
{----------------------------------------------------------}
    FOR EL:=1 TO NE DO
        BEGIN
            FOR J:=1 TO 6 DO
                BEGIN
                    UG[J]:=U[NG[EL,J]];
                    UL[J]:=0;
                    FL[J]:=0
                END;
            ROT_MATRIX;
            FOR I1:=1 TO 6 DO
                BEGIN
                    FOR J:=1 TO 6 DO
                        UL[I1]:=A2[J,I1]*UG[J]+UL[I1]
                END;
{----------------------------------------------------------}
{      MULTIPLY LOCAL STIFFNESS MATRIX AND                 }
{      LOCAL DISPLACEMENTS TO OBTAIN FORCES                }
{----------------------------------------------------------}
                STIFF_MAT;
                BEGIN
                    FOR I1:=1 TO 6 DO
                        BEGIN
                            FOR J:=1 TO 6 DO
                                FL[I1]:=K[I1,J]*UL[J]+FL[I1]
                        END
                END;
{----------------------------------------------------------}
{           OUTPUT OF LOCAL FORCES                         }
{----------------------------------------------------------}
                FOR J:=1 TO 2 DO
                    BEGIN
                        WRITE(DEV,EL:5,GN[EL,J]:9,FL[3*J-2]:14:1,FL[3*J-1]:14:1);
                        WRITELN(DEV,FL[3*J]:14:1)
                    END;
        END;
        CLOSE (DEV);
    END;
{*************************************************************************}
{Set All Variables to 0}
PROCEDURE INITIALIZEVARIABLES;
    VAR
        LOOPI, LOOPJ                           : INTEGER;

    BEGIN
      { Here are the integer variables. }

      NN := 0;         NE := 0;        IIN := 0;       J  := 0;
      EL := 0;         N  := 0;        KK  := 0;       I1 := 0;
      FX := 0;         FY := 0;        FT  := 0;       AP := 0;
      M  := 0;         C  := 0;        MEM := 0;       LI := 0;
      KI := 0;         BW := 0;
      { Here are the real variables. }
      LT := 0;         T1 := 0;        T2   := 0;      CT := 0;
      ST := 0;         L  := 0;        QU   := 0;      UM := 0;
      UI := 0;         SF := 0;        MINX := 0;      MINY := 0;

      { Here are the arrays. }
      FOR LOOPI:=1 TO 2 DO
        BEGIN
            X[LOOPI]:=0;  Y[LOOPI]:=0;
            FOR LOOPJ := 1 to ArrSize1  DO
            GN[LOOPJ,LOOPI]:=0
        END;
```

PROGRAM LISTINGS IN TURBO PASCAL

```
        FOR LOOPI:=1 TO 6 DO
          BEGIN
            UL[LOOPI]:=0;  UG[LOOPI]:=0;  FL[LOOPI]:=0;
            FOR LOOPJ:=1 TO ArrSize1 DO
            NG[LOOPJ,LOOPI]:=0
          END;
        FOR LOOPI:=1 TO ArrSize1 DO
          BEGIN
            A[LOOPI]:=0;  E[LOOPI]:=0;  I[LOOPI]:=0;
            TC[LOOPI]:=0;
          END;
        FOR LOOPI:=1 TO ArrSize2 DO
          BEGIN
            BC[LOOPI]:=0;  FC[LOOPI]:=0;  FR[LOOPI]:=0;
            Q[LOOPI]:=0;  U[LOOPI]:=0
          END;
        FOR LOOPI:=1 TO ArrSize4 DO
          BEGIN
            XG[LOOPI]:=0;
            YG[LOOPI]:=0
          END;
        FOR LOOPI:=1 TO 6 DO
          BEGIN
            FOR LOOPJ:=1 TO 6 DO
            A2[LOOPI,LOOPJ]:=0;  K[LOOPI,LOOPJ]:=0;
            KT[LOOPI,LOOPJ]:=0;  R[LOOPI,LOOPJ]:=0
          END;
        FOR LOOPI:=1 TO ArrSize2 DO
          BEGIN
            FOR LOOPJ:=1 TO ArrSize3 DO
            KG[LOOPI,LOOPJ]:=0
          END;
  END;

(****************************************************************)
(******----------------MAIN PROGRAM---------------------********)

BEGIN
    WRITELN('****************************************************');
    WRITELN('****************************************************');
    WRITELN('**                                                **');
    WRITELN('**        STRUCTURAL ANALYSIS IN XY PLANE         **');
    WRITELN('**                                                **');
    WRITELN('**      TWO-FORCE AND BEAM/COLUMN ELEMENT         **');
    WRITELN('**                                                **');
    WRITELN('****************************************************');
    WRITELN('****************************************************');


            ************ START OF MAIN PROGRAM **************

    MEM:=MEMAVAIL;
    WRITELN('    (AVAILABLE MEMORY IS = ',MEM*16,' BYTES)');
    WRITELN;

    INITIALIZEVARIABLES;

    WRITELN ('              DATA INPUT OPTIONS                 ');
    WRITELN ('              -------------------                ');
    WRITELN;
    WRITELN ('     1 - INPUT NEW MODEL DATA FROM KEYBOARD      ');
    WRITELN ('     2 - INPUT OLD MODEL DATA FROM DISK          ');
    WRITELN ('        (ELEMENT NODE NUMBERS AND NODE COORDINATES) ');
    WRITELN;
    WRITE   ('         ENTER CHOICE- (1 OR 2)     ');
    READLN (IIN);
    WRITELN;
    IF IIN = 2  THEN
         READ_STRUCTURE (NE, NN, GN, XG, YG, TC)
    ELSE
         TERM_STRUCTURE (NE, NN, GN, XG, YG, TC);
```

```
    BANDWIDTH;
    WRITE('DO YOU WISH TO SEE PLOT OF STRUCTURE? (Y/N) ');
    READLN(RESP);
    IF UPCASE(RESP) = 'Y' THEN
         GRAPH(XG,YG,SF,MINX,MINY,U,SHIFTX,SHIFTY,MF);
{------------------------------------------------------------------------}
{     EDITING OPTIONS TO REVIEW AND/OR CHANGE ELEMENT NODE NUMBERING     }
{                       AND NODE COORDINATES                             }
{------------------------------------------------------------------------}
    WRITELN;
    WRITE('DO YOU WISH TO EDIT NODE NUMBERING?  (Y/N) ');
    READLN(RESP);
    IF UPCASE(RESP) = 'Y' THEN
       EDIT_NN (GN, NE, TC);
    WRITELN;
    WRITE('DO YOU WISH TO EDIT NODE COORDINATES?  (Y/N) ');
    READLN(RESP);
    IF UPCASE(RESP) = 'Y' THEN
       EDIT_NC (NN, XG, YG);
    WRITELN;
    WRITE('DO YOU WANT DATA STORED ON DISK?  (Y/N) ');
    READLN(RESP);
    WRITELN;
    IF UPCASE(RESP) = 'Y'  THEN
       WRITE_STRUCTURE (NE, NN, GN, XG, YG, TC);
{----------------------------------------}
{       INPUT ELEMENT PROPERTIES         }
{----------------------------------------}
    WRITELN;
    WRITELN ('DO ALL ELEMENTS HAVE THE SAME CROSS SECTION');
    WRITE   (' AND MATERIAL PROPERTIES?  (Y/N) ');
    READLN(RESP);
    WRITELN;
    IF UPCASE(RESP) = 'N'  THEN
      BEGIN
        WRITE('ARE THE ELEMENT PROPERTIES IN A DISK FILE? (Y/N) ');
          READLN (RESP);
          WRITELN;
          IF UPCASE(RESP) = 'Y' THEN
               BEGIN
                  WRITE ('ENTER FILE NAME (DEFAULT = ELPROP.DAT): ');
                  READLN (FNAME);
                  IF (FNAME = '') THEN FNAME := 'ELPROP.DAT';
                  ASSIGN (ELPROPFILE, FNAME);
                  RESET (ELPROPFILE);
                  FOR I1 := 1 TO NE DO
                     READLN (ELPROPFILE, M, E[M], A[M], I[M]);
               END
          ELSE
             BEGIN
                WRITELN('ENTER ELEMENT NUMBER, ELASTIC MODULUS,E,');
                WRITELN('      AREA,A   AND   AREA MOMENT OF INERTIA,I');
                WRITELN('IN THE FOLLOWING FORMAT:');
                WRITELN('EL #    E    A    I <RETURN>');
                FOR I1:=1 TO NE DO
                    READLN(M,E[M],A[M],I[M]);
                WRITELN;
                WRITELN('DO YOU WISH TO STORE THE ELEMENT PROPERTIES',
                        ' ON DISC?  (Y/N) ');
                READLN (RESP);
                WRITELN;
                IF UPCASE(RESP) = 'Y' THEN
                   BEGIN
                      WRITE ('ENTER FILE NAME (DEFAULT = ELPROP.DAT): ');
                      READLN (FNAME);
                      IF (FNAME = '') THEN FNAME := 'ELPROP.DAT';
                      ASSIGN (ELPROPFILE, FNAME);
                      REWRITE (ELPROPFILE);
                      FOR M := 1 TO NE DO
                         WRITELN (ELPROPFILE, M, E[M], A[M], I[M]);
                      CLOSE(ELPROPFILE);
                   END;
```

```pascal
                          END;
                    END
                ELSE
                    BEGIN
                        WRITELN('ENTER MODULUS OF ELASTICITY,E;   AREA,A; AND ');
                        WRITELN(' AREA MOMENT OF INERTIA, I');
                        WRITELN('USE FOLLOWING FORMAT:');
                        WRITELN('  E    A    I  <RETURN>');
                        READLN(E[1],A[1],I[1]);
                        FOR I1:=2 TO NE DO
                            BEGIN
                                E[I1]:=E[1];
                                A[I1]:=A[1];
                                I[I1]:=I[1]
                            END
                    END;
{------------------------------------------------------------}
{    INITIALIZE BOUNDARY CODE AND DISPLACEMENT VARIABLE       }
{------------------------------------------------------------}
      FOR I1:= 1 TO NN DO
          FOR J:=2 TO O DO
              BEGIN
                  BC[3*I1-J]:=O;
                  U[3*I1-J]:=O
              END;
{--------------------------------------------------------------------------}
{   DETERMINE THE GLOBAL DISPLACEMENT NUMBERS LOCAL DISPLACEMENTS          }
{              1 THRU 6 OF EACH ELEMENT                                    }
{--------------------------------------------------------------------------}
      FOR M:= 1 TO NE DO
          FOR J:=1 TO 2 DO
              BEGIN
                  NG[M,3*J-2]:=3*GN[M,J]-2;
                  NG[M,3*J-1]:=3*GN[M,J]-1;
                  NG[M,3*J]  := 3*GN[M,J]
              END;
{------------------------------------------------------------}
{    INPUT BOUNDARY DISPLACEMENT CONDITIONS                  }
{------------------------------------------------------------}
        REPEAT
            DONE := TRUE;
            WRITELN;
            WRITELN('HOW MANY NODES HAVE A SPECIFIED DISPLACEMENT,');
            WRITE(' (INCLUDING ZERO), IN THE X DIRECTION? ');
            READLN(FX);
            WRITELN;
            IF (FX=O)   THEN
                BEGIN
                    TEXTCOLOR(15+BLINK);
                    WRITELN('WARNING!!! AT LEAST 1 NODE MUST BE FIXED');
                    WRITELN('   IN THE X DIRECTION.    TRY AGAIN!');
                    TEXTCOLOR(14);
                    DONE := FALSE;
                END;
        UNTIL DONE;
        REPEAT
          DONE := TRUE;
          WRITELN('HOW MANY NODES HAVE A SPECIFIED DISPLACEMENT,');
          WRITE(' (INCLUDING ZERO), IN THE Y DIRECTION? ');
          READLN(FY);
          WRITELN;
          IF (FY=O)   THEN
              BEGIN
                  TEXTCOLOR(15+BLINK);
                  WRITELN('WARNING!!! AT LEAST 1 NODE MUST BE FIXED');
                  WRITELN('IN THE Y DIRECTION.    TRY AGAIN');
                  TEXTCOLOR(14);
                  DONE := FALSE;
              END;
        UNTIL DONE;
      WRITE('HOW MANY NODES HAVE A SPECIFIED ROTATION? (INCLUDING O) ');
```

```
        READLN(FT);
        WRITELN;
        WRITELN('INPUT NODE NUMBER AND THE KNOWN X DISPLACEMENT IN');
        WRITELN(' THE FOLLOWING FORMAT:');
        WRITELN('NODE#     U <RETURN>');
        FOR I1:= 1 TO FX DO
            BEGIN
                READLN(J,U[3*J-2]);
                BC[3*J-2]:=1
            END;
        WRITELN;
        WRITELN('INPUT NODE NUMBER AND THE KNOWN Y DISPLACEMENT IN');
        WRITELN(' THE FOLLOWING FORMAT:');
        WRITELN('NODE #    V <RETURN>');
        FOR I1:=1 TO FY DO
            BEGIN
                READLN(J,U[3*J-1]);
                BC[3*J-1]:=1
            END;
        WRITELN;
        IF FT<>0  THEN
          BEGIN
            WRITELN('INPUT NODE NUMBER AND THE KNOWN ROTATIONAL DISPLACEMENT');
            WRITELN(' IN THE FOLLOWING FORMAT:');
            WRITELN(' NODE#    THETA <RETURN>');
                FOR J:=1 TO FT DO
                    BEGIN
                        READLN(I1,U[3*I1]);
                        BC[3*I1]:=1
                    END
          END;
        WRITELN;
        FOR J:=1 TO 3*NN DO
            FC[J]:=0;
{-------------------------------------------------------}
{       INPUT APPLIED FORCES AND COUPLES                }
{-------------------------------------------------------}
        WRITELN('HOW MANY NODES CARRY APPLIED (KNOWN MAGNITUDE) FORCES');
        WRITE(' AND/OR COUPLES? ');
        READLN(AP);
        WRITELN;
        IF AP<>0  THEN
          BEGIN
            WRITELN('ENTER GLOBAL NODE NUMBER, XFORCE, Y FORCE AND COUPLE');
            WRITELN(' IN THE FOLLOWING FORMAT:');
            WRITELN(' NODE#    FX    FY    C <RETURN>');
            FOR J:=1 TO AP DO
                READLN(I1,FC[3*I1-2],FC[3*I1-1],FC[3*I1]);
          END;
        WRITELN;
{-------------------------------------------------------}
{       INITIALIZE GLOBAL STIFFNESS MATRIX              }
{-------------------------------------------------------}
    FOR I1:= 1 TO 3*NN DO
        FOR J:= 1 TO BW DO
            KG[I1,J]:=0;
{-------------------------------------------------------}
{  SET TWO FORCE MEMBER MOMENT OF INERTIA =0            }
{-------------------------------------------------------}
    FOR J:= 1 TO NE DO
        IF TC[J]=1  THEN
            I[J]:=0;
{-------------------------------------------------------}
{       BEGIN CALCULATIONS, PROGRESS PROMPTED ON TERMINAL  }
{-------------------------------------------------------}
    CLRSCR;
    TEXTCOLOR(15+BLINK);
    WRITELN('*********** WORKING, WORKING, WORKING **************');
    TEXTCOLOR(14);
    FOR EL:=1 TO NE DO
```

PROGRAM LISTINGS IN TURBO PASCAL

```pascal
            BEGIN
                ROT_MATRIX;
                STIFF_MAT;
                STIF_MAT_ROT;
{---------------------------------------------------------------------}
{       ASSEMBLING GLOBAL STIFFNESS MATRIX IN BAND MATRIX FORMAT      }
{---------------------------------------------------------------------}
            FOR I1:=1 TO 6 DO
                FOR J:=1 TO 6 DO
                    BEGIN
                        IF (NG[EL,J]>=NG[EL,I1])   THEN
                            BEGIN
                                C:=NG[EL,J]-NG[EL,I1]+1;
                                KG[NG[EL,I1],C]:=KT[I1,J]+KG[NG[EL,I1],C]
                            END
                    END;
        END;
{---------------------------------------------------------------------}
{      TEST FOR TWO-FORCE MEMBER ZERO ROTATIONAL STIFFNES ON DIAGONAL }
{                       OF STIFFNESS MATRIX                           }
{---------------------------------------------------------------------}
     FOR I1:=1 TO 3*NN DO
        IF (KG[I1,1]=0) THEN
            BEGIN
                BC[I1]:=1;
                U[I1]:=0
            END;
{---------------------------------------------------------------------}
{    MODIFY STIFFNESS MATRIX AND FORCE VECTOR TO INCLUDE BOUNDARY     }
{                       DISPLACEMENT DATA                             }
{---------------------------------------------------------------------}
     FOR I1:= 1 TO 3*NN DO
        BEGIN
            Q[I1]:=FC[I1];
            IF BC[I1]<>0   THEN
                BEGIN
                    KG[I1,1]:=KG[I1,1]*1E10;
                    FC[I1]:=U[I1]*KG[I1,1]
                END
        END;
{---------------------------------------------------------------------}
{ SOLVE FOR DISPLACEMENTS USING BANDED MATRIX GAUSS ELIMINATION ROUTINE }
{---------------------------------------------------------------------}
     WRITELN('********  GAUSS EQUATION SOLVER [BAND MATRIX] ******** ');
     N:=3*NN;
     FOR M:=1 TO N DO
        BEGIN
            IF KG[M,1]<>0   THEN
                BEGIN
                    FOR LI:= 2 TO BW DO
                        BEGIN
                            IF KG[M,LI]<>0   THEN
                                BEGIN
                                    I1:=M+LI-1;
                                    QU:=KG[M,LI]/KG[M,1];
                                    J:=0;
                                    FOR KK:=LI TO BW DO
                                        BEGIN
                                            J:=J+1;
                                            IF I1<=N   THEN
                                                KG[I1,J]:=KG[I1,J]-QU*KG[M,KK]
                                        END;
                                      KG[M,LI]:=QU
                                END
                        END
                END
        END;

     WRITELN('***************    MATRIX REDUCTION     **************');
     FOR M:=1 TO N DO
```

```
            IF KG[M,1] <> 0  THEN
                 BEGIN
                    FOR LI:=2 TO BW DO
                        IF KG[M,LI] <> 0   THEN
                              BEGIN
                                 I1:=M+LI-1;
                                 IF I1<=N  THEN
                                    FC[I1]:=FC[I1]-KG[M,LI]*FC[M]
                              END;
                     FC[M]:=FC[M]/KG[M,1]
                 END;
   WRITELN('***************    FORCE REDUCTION    ***************');

   U[N]:=FC[N];
   FOR KK:=2 TO N DO
       BEGIN
          M:=N+1-KK;
          FOR LI:= 2 TO BW DO
              IF KG[M,LI] <> 0   THEN
                    BEGIN
                       KI:=M+LI-1;
                       IF KI<=N  THEN
                          FC[M]:=FC[M]-KG[M,LI]*U[KI]
                    END;
              U[M]:=FC[M];
       END;
{---------------------------------------------------------------}
{           IDENTIFY MAXIMUM DISPLACEMENT                       }
{---------------------------------------------------------------}
   UM:=0;
   FOR I1:=1 TO N DO
       BEGIN
          UI:=ABS(U[I1]);
          IF UI > UM THEN
              UM:=UI
       END;
   FOR I1:=1 TO N DO
       IF ABS(U[I1])/UM < 1E-5   THEN
          U[I1]:=0;
{-------------------------------------------------}
{    RESTORE ORIGINAL APPLIED NODAL FORCES        }
{-------------------------------------------------}
   FOR I1:=1 TO 3*NN DO
       FC[I1]:=Q[I1];
{-------------------------------------------------}
{         OUTPUT OF DATA AND RESULTS              }
{-------------------------------------------------}
   REPEAT
       BEGIN
          RESULTS;
          WRITELN;
          WRITELN;
          WRITE('DO YOU WISH TO REPEAT THE OUTPUT? (Y/N) ');
          READLN(RESP);
          IF UPCASE(RESP) = 'Y' THEN
              QUIT:=FALSE
          ELSE
              QUIT:=TRUE
       END;
   UNTIL QUIT;
   WRITELN;
   WRITE('DO YOU WANT A DISTORTION PLOT OF STRUCTURE?   (Y/N) ');
   READLN(RESP);
   IF (UPCASE(RESP)='Y') THEN
       BEGIN
          MF:= 25/UM;
          GRAPH(XG,YG,SF,MINX,MINY,U,SHIFTX,SHIFTY,MF)
       END
   END.
```

APPENDIX 3

Computer Programs in Applesoft BASIC and TURBO Pascal for Plane Stress Analysis

A3.1 THE ELEMENT

The element used in this program is the three-node constant strain triangle (Figure A3.1). Material properties and thickness can be specified for individual elements. Nodes can be specified as fixed in the X and/or Y directions or can be assigned displacement values in the coordinate directions; inclined releases are not permitted in this program. All applied loads must be node forces. Distributed loading, whether surface traction or body forces, must be replaced with equivalent node forces. All results are relative to the global XY coordinate system.

A3.2 APPLICATION SUGGESTIONS

The program is completely interactive; data input format is prompted during entry. The program has an editing and disk storage option for global node numbers and the node coordinates, but all other data must be entered (carefully) via the keyboard. An editing program is included for the BASIC version in Section A3.4 that allows modification of the

436 COMPUTER PROGRAMS FOR PLANE STRESS ANALYSIS

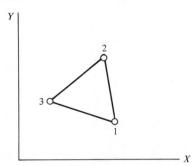

FIGURE A3.1 Triangular element.

triangle node numbering and coordinates and then plots the mesh. The analysis program has an option for disk storage of the node displacements and the BASIC distortion plot program included in Section A3.4 can be used to obtain a plot of the distorted shape of the model. The Pascal program is self-contained.

The program uses the banded matrix storage scheme for the global stiffness matrix. During data entry, the (half) bandwidth is calculated and the user is given the option to renumber nodes if it is desirable and/or possible to reduce the bandwidth.

Again, as in the case of the structures program, you are encouraged to modify the program with error traps, improved convenience of data entry, and so on.

A3.3 SAMPLE PROGRAM RUN

```
    The steel plate  10 inches square and
1.0 inch thick is divided into 4
triangles.  The left side is fixed, and
forces are applied to the nodes as shown
in the sketch.  Run the stress program for
this very coarse mesh.
```

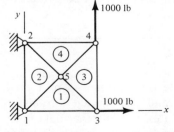

FIGURE A3.3 Simple four-element model for example program run.

SAMPLE PROGRAM RUN

```
RUN
*****************************************
**                                     **
**      PLANE STRESS ANALYSIS          **
**             WITH THE                **
**      CONSTANT STRAIN TRIANGLE       **
**                                     **
*****************************************
HIT ANY KEY TO CONTINUE

           DATA INPUT OPTIONS
           ---- ----- -------

   1 - INPUT NEW MODEL DATA FROM KEYBOARD
   2 - INPUT OLD MODEL DATA FROM DISC

            (ELEMENT NODE NUMBERS AND
              NODE COODINATES)

   ENTER CHOICE - (1 OR 2)   1

HOW MANY 3-NODE ELEMENTS?   4

HOW MANY NODES?   5

ENTER ELEMENT NUMBER AND GLOBAL NUMBERS
OF LOCAL NODES 1, 2, AND 3

      NOTE: USE COUNTER-CLOCKWISE SEQUENCE
        AND FOLLOWING FORMAT:

EL#,   1,   2,   3,    <RETURN>

?1,1,3,5
?2,2,1,5
?3,5,3,4
?4,2,5,4

ENTER GLOBAL NODE NUMBER, N, AND ITS
   X AND Y COORDINATES

  NODE#,  X,   Y    <RETURN>

?1,0,0
?2,0,10
?3,10,0
??4,10,10
?5,5,5

** CALCULATED BANDWIDTH = 10 **


DO YOU WISH TO EDIT
NODE NUMBERING?   (Y/N) N

---------------------------------------

FREE DATA SPACE IS 16172 BYTES

FREE PROGRAM SPACE:

        YOU ARE 3312 BYTES INTO HGR2!

---------------------------------------


DO YOU WISH TO EDIT
NODE COORDINATES    (Y/N) N
```

```
DO YOU WANT DATA STORED
ON DISC? (Y/N)  N

HOW MANY NODES HAVE A SPECIFIED
DISPLACEMENT (INCLUDING  0.0 ) IN
THE  X  DIRECTION? 2

ENTER NODE NUMBERS AND THE KNOWN  X
DISPLACEMENT

 NODE#,   U    <RETURN>

?1,0
?2,0

HOW MANY NODES HAVE A SPECIFIED
DISPLACEMENT  (INCLUDING 0.0) IN
THE   Y  DIRECTION? 2

ENTER NODE NUMBERS AND THE KNOWN
   Y DISPLACEMENT

NODE#,   V    <RETURN>

?1,0
?2,0

INPUT NUMBER OF NØDES SUPPORTING
APPLIED LOADS:  2

ENTER GLOBAL NODE NUMBER AND ITS X
AND  Y  LOAD

NODE #,  FX,  FY     <RETURN>

?3, 1000, 0
?4, 0, 1000


IS MATERIAL HOMOGENEOUS AND
THE THICKNESS UNIFORM?  (Y/N)   Y

ENTER ELASTIC MODULUS, E,
     POISSON'S RATIO, V, AND
          THICKNESS,  T

E,  V,  T    <RETURN>

?30E6, 0.3, 1.0

WORKING ON:                                    ← Progress prompts

     ELEMENT 1 STIFFNESS MATRIX
     ELEMENT 2 STIFFNESS MATRIX
     ELEMENT 3 STIFFNESS MATRIX
     ELEMENT 4 STIFFNESS MATRIX
     GAUSS EQUATION SOLVER (BAND MATRIX)
MATRIX REDUCTION

FORCE REDUCTION

*****************************************
**                                     **
**            RESULTS                  **
**                                     **
*****************************************

HIT RETURN
```

SAMPLE PROGRAM RUN **439**

```
NODE#     XDISPL          YDISPL

1         0               0
2         0               0                   ┐
3         1.428E-04       1.888E-04           ├── output to terminal
4         -9.12E-05       2.185E-04           │
5         1.65E-05        7.36E-05            │

EL #      SIGMAX          SIGMAY      TAUXY   │
                                              │
1         430             4.5         91.1    │
2         108.8           32.6        170     │
3         91.1            116.5       30      │
4         -230.1          144.5       108.8  ─┘
```

DO YOU WANT DISPLACEMENTS STORED
ON DISC?   (Y/N)   N

DO YOU WANT HARDCOPY?   (Y/N)   Y

ARE YOU CONNECTED TO A PRINTER? (Y/N)   Y

DO YOU WANT HARDCOPY OUTPUT OF DISPLACEMENTS ?   (Y/N)   Y

DO YOU WANT HARDCOPY OUTPUT OF STRAIN ?   (Y/N)   Y

PLANE STRESS ANALYSIS---INPUT DATA

**** ELEMENT GLOBAL NODE NUMBERS ****

```
ELEMENT           GLOBAL NODE NUMBERS OF
NUMBER            LOCAL 1      LOCAL 2      LOCAL3

  1                  1            3            5
  2                  2            1            5
  3                  5            3            4
  4                  2            5            4
```

**** NODE GLOBAL COORDINATES AND BOUNDARY DISPLACEMENTS ****

```
NODE    X      Y         BOUNDARY DISPLACEMENTS
                           U            V

 1      0      0           0            0
 2      0      10          0            0
 3      10     0
 4      10     10
 5      5      5
```

**** APPLIED NODAL FORCES ****

```
NODE    X FORCE      Y FORCE

 3      1000         0
 4      0            1000
```

**** MATERIAL PROPERITES AND THICKNESS ****
```
ELEMENT   ELASTIC MODULUS      POISSON'S RATIO      THICKNESS

   1         30000000                .3                 1
```
ALL ELEMENTS HAVE SAME PROPERTIES

*** RESULTS OF PLANE STRESS ANALYSIS ***
    ------- -- ----- ------ --------

```
*******************************************
**                                       **
**            DISPLACEMENTS              **
**                                       **
*******************************************

NODE           XDISPL              YDISPL

1         0                   0
2         0                   0
3         1.42874168E-04      1.88802783E-04
4         -9.11258319E-05     2.18530551E-04
5         1.65154265E-05      7.36666668E-05


*******************************************
**                                       **
**               STRAIN                  **
**                                       **
*******************************************

ELEMENT        EXX                 EYY                 EXY

1         1.42E-05            -4.2E-06            7.8E-06
2         3.3E-06             0                   1.47E-05
3         1.8E-06             2.9E-06             2.6E-06
4         -9.2E-06            7.1E-06             9.4E-06


*******************************************
**                                       **
**               STRESS                  **
**                                       **
*******************************************

ELEMENT        SXX                 SYY                 SXY

1         430                 4                   91
2         108                 32                  170
3         91                  116                 30
4         -231                144                 108
```

A3.4 PROGRAM LISTINGS IN APPLESOFT BASIC

STRESS-3NODE

```
100  TEXT : HOME
102  PRINT "*****************************************"
104  PRINT "**                                     **"
106  PRINT "**        PLANE STRESS ANALYSIS        **"
108  PRINT "**               WITH THE              **"
110  PRINT "**        CONSTANT STRAIN TRIANGLE     **"
112  PRINT "**                                     **"
114  PRINT "*****************************************"
116  REM    < NE > NUMBER OF ELEMENTS
118  REM    < FX > NUMBER OF X DIRECTION CONSTRAINTS
120  REM    < FY > NUMBER OF Y DIRECTION CONSTRAINTS
122  REM    < BC(I) > BOUNDARY CONDITION CODE
124  REM    < DR > GLOBAL DEGREES OF FREEDOM
126  REM    < XG(I),YG(I) > GLOBAL COORDINATES OF NODE (I)
128  REM    < GN(I,J) > GLOBAL NODE NUMBER OF NODE (J) IN ELEMENT (I)
130  REM    < NG(I,J) > GLOBAL DISPLACEMENT NUMBER OF LOCAL DISPLACEMENT
132  REM              NUMBER (J) OF ELEMENT (I)
134  REM    < K(I,J) > ELEMENT GLOBAL COORDINATE STIFFNESS MATRIX
136  REM    < KG(I,J) > ASSEMBLED GLOBAL STIFFNESS MATRIX
138  REM    < C(I,J) >  CONSTITUTIVE MATRIX
140  REM    < FC(I) >  APPLIED NODAL FORCES
142  REM    < U(L) >  NODAL DISPLACEMENTS
144  REM    < EP(I,J) >  STRAIN VECTOR OF ELEMENT (I)
```

PROGRAM LISTINGS IN APPLESOFT BASIC

```
146  REM    < SS(I,J) > STRESS VECTOR OF ELEMENT (I)
300  VTAB 20: PRINT "HIT ANY KEY TO CONTINUE"
302  GET G$: PRINT
304  REM -------------------------------------------------
306  REM       LINES 310 THRU 348 ARE MINI FRONT END PROGRAM
308  REM -------------------------------------------------
310  ONERR  GOTO 318
312  POKE 768,104: POKE 769,168: POKE 770,104: POKE 771,166: POKE
     772,223: POKE 773,154: POKE 774,72: POKE 775,152: POKE 776,72: POKE
     777,96
314  D$ =  CHR$ (4): REM    CHR$(4) IS CTRL-D
316  GOTO 356: REM    GO TO NORMAL PROGRAM START
318  CALL 768: PRINT D$;"PR#0": TEXT : IF  PEEK (222) = 255 THEN  PRINT
     "CNTL C CAUSED A BREAK IN "; PEEK (218) +  PEEK (219) * 256: END
320  GOSUB 340: END
322  PRINT : PRINT : REM           DETERMINE FREE MEMORY
324  TP =  PEEK (175) +  PEEK (176) * 256: REM    TOP OF PROGRAM
326  EA =  PEEK (109) +  PEEK (110) * 256: REM    TOP OF LOMEM POINTER
328  BS =  PEEK (111) +  PEEK (112) * 256: REM    HIMEM POINTER
330  PRINT "FREE DATA SPACE IS ";BS - EA;" BYTES"
332  PRINT : PRINT "FREE PROGRAM SPACE:": PRINT
334  TP = 8192 - TP: IF TP > 0 THEN  PRINT TP;" BYTES TO HGR": RETURN
336  TP = TP + 8192: IF TP > 0 THEN  PRINT TP;" BYTES TO HGR2": RETURN
338  PRINT "         YOU ARE ";  ABS (TP);" BYTES INTO HGR2!": RETURN
340  PRINT "ERROR # ";  PEEK (222);" AT LINE # ";  PEEK (218) +  PEEK (219)
     * 256
342  PRINT : PRINT "     (ERROR CODES IN POOLE PG. 331)"
344  PRINT : PRINT "YOUR OPTIONS NOW ARE:": PRINT "  (1) HIT ESCAPE TO
     RESUME AT ERROR STATEMENT": PRINT "  (2) HIT ANY OTHER KEY TO END
     PROGRAM"
346  GET G$: PRINT : IF G$ =  CHR$ (27) THEN  RESUME
348  RETURN
350  REM -------------------------------------------------
352  REM       START PROGRAM HERE
354  REM -------------------------------------------------
356  HOME
370  I$ =  CHR$ (9): REM    CHR$(9) IS CTRL-I
375  R$ =  CHR$ (13): REM    CHR$(13) IS CARRIAGE RETURN
380  RU$ = "N"
385  BW = 0
390  VTAB 10
400  PRINT "          DATA INPUT OPTIONS"
410  PRINT "          ---- ----- -------"
420  PRINT
430  PRINT "  1 - INPUT NEW MODEL DATA FROM KEYBOARD"
440  PRINT "  2 - INPUT OLD MODEL DATA FROM DISC": PRINT
450  PRINT "          (ELEMENT NODE NUMBERS AND"
460  PRINT "             NODE COORDINATES)"
470  PRINT
480  INPUT "  ENTER CHOICE - (1 OR 2)   ";IN$
490  IF IN$ = "1" OR IN$ = "2" GOTO 510
500  PRINT : PRINT "BAD RESPONSE, RE-ENTER": GOTO 400
510  IF IN$ = "1" GOTO 600
520  REM   OPEN AND READ THE DATA FILE
530  RT$ = ".GEOM"
540  REM -------------------------------------------------
550  GOSUB 10000: REM    OPEN AND READ GEOMETRY FILE       ----
560  REM -------------------------------------------------
570  REM              DATA INPUT FROM KEYBOARD OR DISC     ----
580  REM -------------------------------------------------
590  REM
600  HOME : PRINT : INPUT "HOW MANY 3-NODE ELEMENTS? ";NE: PRINT
610  INPUT "HOW MANY NODES? ";NN: PRINT
620  DIM
     XG(NN),YG(NN),BE(3,6),C(3,3),K(6,6),R(3,6),GN(NE,3),B(3,6),X(3),Y(3),
     NG(NE,6)
630  DIM FC(2 * NN),BC(2 * NN),U(2 *
     NN),E(NE),V(NE),H(NE),EP(NE,3),ST(3),FR(2 * NN),SS(NE,3)
640  PRINT "ENTER ELEMENT NUMBER AND GLOBAL NUMBERS": PRINT "OF LOCAL
     NODES 1, 2, AND 3": PRINT
```

```
650  PRINT "    NOTE:";: INVERSE : PRINT "USE COUNTER-CLOCKWISE
     SEQUENCE": NORMAL
660  PRINT "     AND FOLLOWING FORMAT:": PRINT
670  PRINT "EL#,   1,   2,   3,    <RETURN>": PRINT
680  FOR J = 1 TO NE
690  INPUT I,GN(I,1),GN(I,2),GN(I,3)
700  NEXT J
710  PRINT : PRINT "ENTER GLOBAL NODE NUMBER, N, AND ITS"
720  PRINT "  X AND Y COORDINATES"
730  PRINT : PRINT " NODE#,  X,  Y    <RETURN>": PRINT
740  FOR J = 1 TO NN
750  INPUT I,XG(I),YG(I)
760  NEXT J
770  IF IN$ = "1" GOTO 820
780  REM ---------------------------------------------------------
790  GOSUB 10110: REM    CLOSE THE GEOMETRY FILE          ----
800  REM ---------------------------------------------------------
810  REM    CALCULATE THE BANDWIDTH
820  FOR I = 1 TO NE
830  FOR J = 1 TO 2
835  FOR JJ = 1 TO 3
840  DI = ABS (GN(I,J) - GN(I,JJ))
850  IF DI > BW THEN BW = DI
860  NEXT JJ: NEXT J: NEXT I
870  BW = 2 * BW + 2
880  PRINT : PRINT "** CALCULATED BANDWIDTH = ";BW;" **": PRINT
890  REM ---------------------------------------------------------
900  REM    EDITING OPTIONS TO REVIEW AND/OR CHANGE        ----
910  REM    ELEMENT NODE NUMBERING AND COORDINATES         ----
920  REM ---------------------------------------------------------
930  PRINT : PRINT "DO YOU WISH TO EDIT": INPUT "NODE NUMBERING?  (Y/N)
     ";RE$
940  IF RE$ = "Y" OR RE$ = "N" THEN 970
950  REM
960  PRINT : PRINT "RE-ENTER -": GOTO 930
970  IF RE$ = "N" THEN 1050
980  PRINT : INPUT "WHICH ELEMENT ? ";N
990  IF N > NE THEN 980
1000 PRINT "CURRENT VALUES: ";GN(N,1);",   ";GN(N,2);",   ";GN(N,3)
1010 PRINT : INPUT "ENTER THE NEW VALUES:";GN(N,1),GN(N,2),GN(N,3)
1020 PRINT : INPUT "MORE CHANGES ?  (Y/N) ";RE$
1030 IF RE$ = "Y" THEN 980
1040 BW = 0: GOTO 800
1050 DIM KG(2 * NN,BW)
1054 HOME : PRINT "---------------------------------------"
1055 GOSUB 322: REM    TEST FOR FREE MEMORY
1056 PRINT "---------------------------------------": PRINT
1060 PRINT : PRINT "DO YOU WISH TO EDIT": INPUT "NODE COORDINATES?
     (Y/N) ";RE$
1070 IF RE$ = "Y" OR RE$ = "N" THEN 1090
1080 PRINT : PRINT "RE-ENTER -": GOTO 1060
1090 IF RE$ = "N" THEN 1160
1100 PRINT : INPUT "WHICH NODE ? ";N
1110 IF N > NN THEN 1100
1120 PRINT "CURRENT VALUES: ";XG(N);",   ";YG(N)
1130 PRINT : INPUT "ENTER THE NEW VALUES- ";XG(N),YG(N)
1140 PRINT : INPUT "MORE CHANGES ?  (Y/N) ";RE$
1150 IF RE$ = "Y" THEN 1100
1160 PRINT : PRINT "DO YOU WANT DATA STORED": INPUT "ON DISC?  (Y/N)
     ";DK$
1170 IF DK$ < > "Y" THEN 1320
1180 RT$ = ".GEOM"
1190 REM ---------------------------------------------------------
1200 GOSUB 10160: REM    OPEN AND WRITE TO GEOMETRY FILE   ----
1210 REM ---------------------------------------------------------
1220 PRINT NE;R$;NN
1230 FOR I = 1 TO NE
1240 PRINT I;R$;GN(I,1);R$;GN(I,2);R$;GN(I,3)
1250 NEXT
1260 FOR I = 1 TO NN
```

```
1270    PRINT I;R$;XG(I);R$;YG(I)
1280    NEXT
1290    REM  --------------------------------------------------------
1300    GOSUB 10110: REM      CLOSE THE GEOMETRY FILE            ----
1310    REM  --------------------------------------------------------
1320    REM  INITIALIZE NODE DISPLACEMENT VECTOR
1330    FOR I = 1 TO NN
1340 U(2 * I - 1) = 0:U(2 * I) = 0
1350    NEXT I
1360    REM  --------------------------------------------------------
1370    REM       DETERMINE THE GLOBAL DISPLACEMENT NUMBERS OF   ----
1380    REM       LOCAL DISPLACEMENTS 1 THRU 6 OF EACH ELEMENT   ----
1390    REM  --------------------------------------------------------
1400    FOR M = 1 TO NE
1410    FOR J = 1 TO 3
1420    NG(M,2 * J - 1) = 2 * GN(M,J) - 1
1430    NG(M,2 * J) = 2 * GN(M,J)
1440    NEXT J
1450    NEXT M
1460    IF RU$ < > "Y" GOTO 1530: REM  RU$ IS RERUN CODE
1470    PRINT : INPUT "DO YOU WISH TO CHANGE THE DISPLACEMENT BOUNDARY
        CONDITIONS? (Y/N) ";NC$
1480    IF NC$ = "N" GOTO 1750
1490    REM  --------------------------------------------------------
1500    REM       INPUT DISPLACEMENT BOUNDARY CONDITIONS          ----
1510    REM  --------------------------------------------------------
1520    REM  INITIALIZE BOUNDARY DISPLACEMENT CODE VECTOR, BC(I)
1530    FOR I = 1 TO NN
1540 BC(2 * I - 1) = 0:BC(2 * I) = 0
1550    NEXT I
1560    PRINT : PRINT "HOW MANY NODES HAVE A SPECIFIED": PRINT
        "DISPLACEMENT (INCLUDING 0.0 ) IN": INPUT "THE X   DIRECTION?
        ";FX: PRINT
1570    IF FX = 0 THEN  PRINT "WARNING!": PRINT : PRINT "REMOVE RIGID BODY
        DISPLACEMENT": PRINT : PRINT "TRY AGAIN": PRINT : GOTO 1560
1580    PRINT : PRINT "ENTER NODE NUMBERS AND THE KNOWN   X": PRINT
        "DISPLACEMENT": PRINT
1590    PRINT " NODE#,     U       <RETURN>": PRINT
1600    FOR J = 1 TO FX
1610    INPUT I,U(2 * I - 1)
1620 BC(2 * I - 1) = 1
1630    NEXT J
1640    PRINT : PRINT "HOW MANY NODES HAVE A SPECIFIED": PRINT
        "DISPLACEMENT (INCLUDING 0.0) IN": INPUT "THE  Y   DIRECTION? ";FY:
        PRINT
1650    IF FY = 0 THEN  PRINT "WARNING!": PRINT "  REMOVE RIGID BODY
        DISPLACEMENT": PRINT "TRY AGAIN": PRINT : GOTO 1640
1660    PRINT : PRINT "ENTER NODE NUMBERS AND THE KNOWN": PRINT "  Y
        DISPLACEMENT": PRINT
1670    PRINT "NODE#,      V       <RETURN>": PRINT
1680    FOR J = 1 TO FY
1690    INPUT I,U(2 * I)
1700 BC(2 * I) = 1
1710    NEXT J
1720    REM  --------------------------------------------------------
1730    REM           INPUT THE APPLIED NODAL FORCES             ----
1740    REM  --------------------------------------------------------
1750    PRINT : PRINT "INPUT NUMBER OF NODES SUPPORTING": INPUT "APPLIED
        LOADS: ";AP: PRINT
1760    REM  INITIALIZE LOAD VECTOR
1770    FOR I = 1 TO 2 * NN
1780 FC(I) = 0
1790    NEXT I
1800    IF AP = 0 GOTO 1890
1810    PRINT "ENTER GLOBAL NODE NUMBER AND ITS X": PRINT "AND  Y  LOAD":
        PRINT : PRINT "NODE #,    FX,   FY    <RETURN>": PRINT

1820    PRINT " "
1830    FOR I = 1 TO AP
```

```
1840    INPUT N,FC(2 * N - 1),FC(2 * N)
1850    NEXT I
1860    REM  ------------------------------------------------------
1870    REM                  INPUT MATERIAL PROPERTIES           ----
1880    REM  ------------------------------------------------------
1890    PRINT : PRINT "IS MATERIAL HOMOGENEOUS AND": INPUT "THE THICKNESS
        UNIFORM? (Y/N) ";MC$: PRINT
1910    IF MC$ = "Y" THEN  GOTO 1990
1920    PRINT
1930    PRINT "ENTER ELEMENT NUMBER, ELASTIC MODULUS,E,     POISSON'S
        RATIO,V,  AND THICKNESS,T": PRINT : PRINT "EL#   E,   V,   T
        <RETURN>:?
1940    PRINT
1950    FOR I = 1 TO NE
1960    INPUT M,E(M),V(M),H(M)
1970    NEXT I
1980    GOTO 2060
1990    PRINT : PRINT "ENTER ELASTIC MODULUS, E,": PRINT "     POISSON'S
        RATIO, V, AND": PRINT "              THICKNESS, T": PRINT : PRINT "E,   V,
        T    <RETURN>": PRINT
2000    INPUT EE,VV,HH
2010    FOR I = 1 TO NE
2020    E(I) = EE
2030    V(I) = VV
2040    H(I) = HH
2050    NEXT I
2060    FOR I = 1 TO 2 * NN
2070    FOR J = 1 TO BW
2080    KG(I,J) = 0
2090    NEXT J: NEXT I
2100    REM  ------------------------------------------------------
2110    REM                    BEGIN CALCULATIONS                 ----
2120    REM              PROGRESS PROMPTED ON TERMINAL            ----
2130    REM  ------------------------------------------------------
2140    REM  MONITOR PROGRESS WITH PROMPTS
2150    REM
2160    FLASH : PRINT : PRINT : PRINT "WORKING ON:": PRINT : NORMAL
2170    REM  CALCULATE ELEMENT STIFFNESS MATRIX
2180    FOR M = 1 TO NE
2190    PRINT "     ELEMENT ";M;" STIFFNESS MATRIX"
2200    FOR S = 1 TO 3
2210    FOR T = 1 TO 6
2220    R(S,T) = 0.
2230    NEXT T
2240    NEXT S
2250    FOR S = 1 TO 6
2260    FOR T = 1 TO 6
2270    K(S,T) = 0.
2280    NEXT T
2290    NEXT S
2300    GOSUB 4760
2310    GOSUB 5090
2320    FOR I = 1 TO 3
2330    FOR J = 1 TO 6
2340    FOR N = 1 TO 3
2350    R(I,J) = R(I,J) + C(I,N) * BE(N,J)
2360    NEXT N
2370    NEXT J
2380    NEXT I
2390    FOR I = 1 TO 6
2400    FOR J = 1 TO 6
2410    FOR N = 1 TO 3
2420    K(I,J) = K(I,J) + BE(N,I) * R(N,J)
2430    NEXT N
2440    NEXT J
2450    NEXT I
2460    FOR I = 1 TO 6
2470    FOR J = 1 TO 6
2480    K(I,J) = K(I,J) * H(M) * DET / 2
2490    NEXT J
```

```
2500   NEXT I
2510   REM    ------------------------------------------------
2520   REM          ASSEMBLING GLOBAL STIFFNESS MATRIX IN   ----
2530   REM                  BAND MATRIX FORMAT              ----
2540   REM    ------------------------------------------------
2550   FOR I = 1 TO 6
2560   FOR J = 1 TO 6
2570   IF NG(M,J) < NG(M,I) GOTO 2600
2580 C = NG(M,J) - NG(M,I) + 1
2590 KG(NG(M,I),C) = K(I,J) + KG(NG(M,I),C)
2600   NEXT J
2610   NEXT I
2620   NEXT M
2630   REM    ------------------------------------------------
2640   REM         MODIFY STIFFNESS MATRIX AND FORCE VECTOR  ----
2650   REM           TO INCLUDE BOUNDARY DISPLACEMENT DATA   ----
2660   REM    ------------------------------------------------
2670   FOR I = 1 TO 2 * NN
2680 FR(I) = FC(I)
2690   IF BC(I) = 0 GOTO 2720
2700 KG(I,1) = KG(I,1) * 1E10
2710 FC(I) = U(I) * KG(I,1)
2720   NEXT I
2730   REM    ------------------------------------------------
2740   REM          SOLVE FOR DISPLACEMENTS USING GAUSS      ----
2750   REM          ELIMINATION ROUTINE FOR BANDED MATRIX    ----
2760   REM    ------------------------------------------------
2770   PRINT "    GAUSS EQUATION SOLVER (BAND MATRIX)"
2780 N = 2 * NN
2790   FOR M = 1 TO N
2800    IF KG(M,1) = 0 GOTO 2930
2810    FOR L = 2 TO BW
2820     IF KG(M,L) = 0 GOTO 2920
2830 I = M + L - 1
2840 QU = KG(M,L) / KG(M,1)
2850 J = 0
2860    FOR K = L TO BW
2870 J = J + 1
2880     IF I > N GOTO 2900
2890 KG(I,J) = KG(I,J) - QU * KG(M,K)
2900    NEXT K
2910 KG(M,L) = QU
2920    NEXT L
2930   NEXT M
2940   PRINT "MATRIX REDUCTION": PRINT
2950   FOR M = 1 TO N
2960    IF KG(M,1) = 0 GOTO 3040
2970    FOR L = 2 TO BW
2980     IF KG(M,L) = 0 GOTO 3020
2990 I = M + L - 1
3000     IF I > N GOTO 3020
3010 FC(I) = FC(I) - KG(M,L) * FC(M)
3020    NEXT L
3030 FC(M) = FC(M) / KG(M,1)
3040   NEXT M
3050   PRINT : PRINT "FORCE REDUCTION": PRINT
3060 U(N) = FC(N)
3070   FOR KK = 2 TO N
3080 M = N + 1 - KK
3090    FOR L = 2 TO BW
3100     IF KG(M,L) = 0 GOTO 3140
3110 K = M + L - 1
3120     IF K > N GOTO 3160
3130 FC(M) = FC(M) - KG(M,L) * U(K)
3140    NEXT L
3150 U(M) = FC(M)
3160   NEXT KK
3170   REM    IDENTIFY NEGLIGIBLE DISPLACEMENTS
3180 UM = 0
3190   FOR I = 1 TO N
```

```
3200 UI =  ABS (U(I))
3210  IF UI > UM THEN UM = UI
3220  NEXT I
3230  FOR I = 1 TO N
3240  IF  ABS (U(I)) / UM < 1E - 4 THEN U(I) = 0
3250  NEXT I
3260  FOR I = 1 TO 2 * NN
3270 FC(I) = FR(I)
3280  NEXT I
3290  REM   PROMPT WHEN CALCULATION COMPLETED
3300  REM  --------------------------------------------------------
3310  REM               CALCULATE ELEMENT STRAIN           ----
3320  REM  --------------------------------------------------------
3330  FOR M = 1 TO NE
3340  FOR I = 1 TO 3
3350 EP(M,I) = 0.
3360  NEXT I
3370  GOSUB 4760
3380  FOR I = 1 TO 3
3390  FOR J = 1 TO 6
3400 EP(M,I) = EP(M,I) + BE(I,J) * U(NG(M,J))
3410  NEXT J
3420  NEXT I
3430  NEXT M
3440  HOME
3450  PRINT "****************************************"
3460  PRINT "**                                    **"
3470  PRINT "**              RESULTS               **"
3480  PRINT "**                                    **"
3490  PRINT "****************************************"
3500  REM    PROMPT WHEN CALCULATION COMPLETED
3510  PRINT : PRINT : PRINT "HIT RETURN"
3520  GET GO$: PRINT
3530  PRINT : PRINT
3540  PRINT "NODE#"; TAB( 10);"XDISPL"; TAB( 25);"YDISPL"
3550  PRINT
3560  FOR I = 1 TO NN
3570  PRINT I; TAB( 10);( INT (1E7 * U(2 * I - 1))) / 1E7; TAB( 25);( INT
     (1E7 * U(2 * I))) / 1E7
3580  NEXT I
3590  PRINT : PRINT
3600  PRINT "EL #"; TAB( 8);"SIGMAX"; TAB( 22);"SIGMAY"; TAB(
     32);"TAUXY": PRINT : PRINT
3610  REM  --------------------------------------------------------
3620  REM               CALCULATE ELEMENT STRESS           ----
3630  REM  --------------------------------------------------------
3640  FOR M = 1 TO NE
3650  FOR I = 1 TO 3
3660 ST(I) = 0
3670  NEXT I
3680  GOSUB 5090
3690  FOR I = 1 TO 3
3700  FOR J = 1 TO 3
3710 ST(I) = ST(I) + C(I,J) * EP(M,J)
3720  NEXT J
3730 SS(M,I) = ST(I)
3740  NEXT I
3750  PRINT M; TAB( 8);( INT (ST(1) * 10)) / 10; TAB( 22);( INT (ST(2) *
     10)) / 10; TAB( 32);( INT (ST(3) * 10)) / 10
3760  NEXT M
3770  PRINT : PRINT "DO YOU WANT DISPLACEMENTS STORED": INPUT "ON DISK?
     (Y/N)  ";UK$
3780  IF UK$ < > "Y" GOTO 3880
3790 RT$ = ".DISPL"
3800  GOSUB 10160: REM   OPEN AND WRITE FILE
3810  FOR I = 1 TO 2 * NN
3820  PRINT U(I)
3830  NEXT I
```

```
3840   GOSUB 10110: REM   CLOSE FILE
3850   REM   ------------------------------------------------------
3860   REM      OPTIONAL HARDCOPY OF INPUT DATA AND RESULTS    ----
3870   REM   ------------------------------------------------------
3880   PRINT : INPUT "DO YOU WANT HARDCOPY? (Y/N) ";H$
3890   IF H$ < > "Y" GOTO 4690
3900   PRINT : PRINT " ARE YOU CONNECTED TO A PRINTER?  (Y/N)": PRINT
3910   INPUT PC$: IF PC$ < > "Y" THEN H$ = "N"
3920   IF H$ < > "Y" GOTO 4690
3930   PRINT : INPUT "DO YOU WANT HARDCOPY OUTPUT OF DISPLACEMENTS ?
       (Y/N) ";UI$
3940   PRINT : INPUT "DO YOU WANT HARDCOPY OUTPUT OF STRAIN ?  (Y/N) ";ST$
3950   REM
3960   PRINT D$;"PR#1": REM    ACTIVATE PRINTER
3970   PRINT I$;"ML 10": REM    LEFT MARGIN SETTING (APPLE DMP COMMAND)
3980   REM
3990   PRINT : PRINT "PLANE STRESS ANALYSIS---INPUT DATA": PRINT : PRINT
       "**** ELEMENT GLOBAL NODE NUMBERS ****"
4000   PRINT : PRINT "ELEMENT";: POKE 36,20: PRINT "GLOBAL NODE NUMBERS
       OF": PRINT "NUMBER";: POKE 36,20: PRINT "LOCAL 1";: POKE 36,32:
       PRINT "LOCAL 2";: POKE 36,44: PRINT "LOCAL 3": PRINT
4010   FOR J = 1 TO NE
4020   POKE 36,3: PRINT J;: POKE 36,23: PRINT GN(J,1);: POKE 36,35: PRINT
       GN(J,2);: POKE 36,47: PRINT GN(J,3)
4030   NEXT J
4040   PRINT : PRINT : PRINT "**** NODE GLOBAL COORDINATES AND BOUNDARY
       DISPLACEMENTS ****"
4050   PRINT : PRINT "NODE";: POKE 36,9: PRINT "X";: POKE 36,15: PRINT
       "Y";: POKE 36,25: PRINT "BOUNDARY DISPLACEMENTS": PRINT ;: POKE
       36,30: PRINT "U";: POKE 36,40: PRINT "V": PRINT
4060   FOR I = 1 TO NN
4070   IF BC(2 * I) = 1 AND BC(2 * I - 1) = 1 GOTO 4160
4080   IF BC(2 * I) = 1 GOTO 4120
4090   IF BC(2 * I - 1) = 1 GOTO 4140
4100   PRINT I;: POKE 36,9: PRINT XG(I);: POKE 36,15: PRINT YG(I)
4110   GOTO 4170
4120   PRINT I;: POKE 36,9: PRINT XG(I);: POKE 36,15: PRINT YG(I);: POKE
       36,40: PRINT U(2 * I)
4130   GOTO 4170
4140   PRINT I;: POKE 36,9: PRINT XG(I);: POKE 36,15: PRINT YG(I);: POKE
       36,30: PRINT U(2 * I - 1)
4150   GOTO 4170
4160   PRINT I;: POKE 36,9: PRINT XG(I);: POKE 36,15: PRINT YG(I);: POKE
       36,30: PRINT U(2 * I - 1);: POKE 36,40: PRINT U(2 * I)
4170   NEXT I
4180   PRINT : PRINT : PRINT "**** APPLIED NODAL FORCES ****": PRINT
4190   PRINT "NODE";: POKE 36,10: PRINT "X FORCE";: POKE 36,25: PRINT "Y
       FORCE": PRINT
4200   FOR N = 1 TO NN
4210   IF (FC(2 * N) = 0 AND FC(2 * N - 1) = 0) GOTO 4230
4220   PRINT N;: POKE 36,10: PRINT FC(2 * N - 1);: POKE 36,25: PRINT FC(2
       * N)
4230   NEXT N
4240   PRINT : PRINT : PRINT "**** MATERIAL PROPERTIES AND THICKNESS ****"
4250   PRINT "ELEMENT";: POKE 36,10: PRINT "ELASTIC MODULUS";: POKE 36,30:
       PRINT "POISSON'S RATIO";: POKE 36,50: PRINT "THICKNESS": PRINT
4260   FOR I = 1 TO NE
4270   POKE 36,3: PRINT I;: POKE 36,10: PRINT E(I);: POKE 36,30: PRINT
       V(I);: POKE 36,50: PRINT H(I)
4280   IF MC$ < > "Y" GOTO 4310
4290   PRINT "ALL ELEMENTS HAVE SAME PROPERTIES"
4300   I = NE
4310   NEXT I
4320   PRINT
4330   PRINT "***   RESULTS OF PLANE STRESS ANALYSIS    ***"
4340   PRINT "       ------- -- ----- ------ --------    "
```

```
4350 REM
4360 IF UI$ = "N" GOTO 4470
4370 PRINT
4380 PRINT "****************************************"
4390 PRINT "**                                    **"
4400 PRINT "**          DISPLACEMENTS             **"
4410 PRINT "**                                    **"
4420 PRINT "****************************************"
4430 PRINT : PRINT "NODE";: POKE 36,14: PRINT "XDISPL";: POKE 36,29:
     PRINT "YDISPL": PRINT
4440 FOR I = 1 TO NN
4450 PRINT I;: POKE 36,9: PRINT U(2 * I - 1);: POKE 36,27: PRINT U(2 *
     I)
4460 NEXT
4470 IF ST$ = "N" GOTO 4580
4480 PRINT
4490 PRINT "****************************************"
4500 PRINT "**                                    **"
4510 PRINT "**             STRAIN                 **"
4520 PRINT "**                                    **"
4530 PRINT "****************************************"
4540 PRINT : PRINT "ELEMENT";: POKE 36,14: PRINT "EXX";: POKE 36,29:
     PRINT "EYY";: POKE 36,46: PRINT "EXY": PRINT
4550 FOR M = 1 TO NE
4560 PRINT M;: POKE 36,12: PRINT  INT (1E7 * EP(M,1)) / 1E7;: POKE
     36,27: PRINT  INT (1E7 * EP(M,2)) / 1E7;: POKE 36,44: PRINT  INT
     (1E7 * EP(M,3)) / 1E7
4570 NEXT M
4580 PRINT
4590 PRINT "****************************************"
4600 PRINT "**                                    **"
4610 PRINT "**             STRESS                 **"
4620 PRINT "**                                    **"
4630 PRINT "****************************************"
4640 PRINT : PRINT "ELEMENT";: POKE 36,14: PRINT "SXX";: POKE 36,29:
     PRINT "SYY";: POKE 36,46: PRINT "SXY": PRINT
4650 FOR M = 1 TO NE
4660 PRINT M;: POKE 36,12: PRINT  INT (SS(M,1));: POKE 36,27: PRINT  INT
     (SS(M,2));: POKE 36,44: PRINT  INT (SS(M,3))
4670 NEXT M
4680 PRINT D$;"PR#0"
4690 PRINT : INPUT "DO YOU WISH ANOTHER RUN? (Y/N) ";RU$
4700 IF RU$ = "Y" GOTO 1060
4710 IF RU$ < > "N" GOTO 4690
4720 END
4730 REM ------------------------------------------------
4740 REM                    SUBROUTINES                ----
4750 REM ------------------------------------------------
4760 REM  SUBROUTINE FOR ELEMENT B MATRIX
4770 X(1) = XG(GN(M,1))
4780 X(2) = XG(GN(M,2))
4790 X(3) = XG(GN(M,3))
4800 Y(1) = YG(GN(M,1))
4810 Y(2) = YG(GN(M,2))
4820 Y(3) = YG(GN(M,3))
4830 DET = (X(1) - X(3)) * (Y(2) - Y(3)) - (X(2) - X(3)) * (Y(1) - Y(3))
4840 B(1,1) = Y(2) - Y(3)
4850 B(2,1) = 0.
4860 B(3,1) = X(3) - X(2)
4870 B(1,2) = 0.
4880 B(2,2) = B(3,1)
4890 B(3,2) = B(1,1)
4900 B(1,3) = Y(3) - Y(1)
4910 B(2,3) = 0.
4920 B(3,3) = X(1) - X(3)
4930 B(1,4) = 0.
4940 B(2,4) = B(3,3)
4950 B(3,4) = B(1,3)
4960 B(1,5) = Y(1) - Y(2)
4970 B(2,5) = 0.
```

PROGRAM LISTINGS IN APPLESOFT BASIC

```
4980 B(3,5) = X(2) - X(1)
4990 B(1,6) = 0.
5000 B(2,6) = B(3,5)
5010 B(3,6) = B(1,5)
5020 FOR I = 1 TO 3
5030 FOR J = 1 TO 6
5040 BE(I,J) = B(I,J) / DET
5050 NEXT J
5060 NEXT I
5070 RETURN
5080 REM ------------------------------------------
5090 REM   SUBROUTINE FOR ELEMENT C MATRIX
5100 C1 = E(M) / (1 - V(M) * V(M))
5110 C(1,1) = C1
5120 C(1,2) = C1 * V(M)
5130 C(1,3) = 0.
5140 C(2,1) = C(1,2)
5150 C(2,2) = C(1,1)
5160 C(2,3) = 0.
5170 C(3,1) = 0.
5180 C(3,2) = 0.
5190 C(3,3) = C1 * (1 - V(M)) / 2
5200 RETURN
5210 REM ------------------------------------------
10000 REM   SUBROUTINE FOR OPENING AND READING A FILE
10001 REM
10010 PRINT : INPUT "WHAT IS THE NAME OF THE FILE ?";NA$
10015 F$ = NA$ + RT$
10020 PRINT : INPUT "WHICH DISK DRIVE ?";DN
10030 IF DN = 1 OR DN = 2 THEN 10050
10040 PRINT : PRINT " TRY AGAIN-": GOTO 10020
10050 PRINT D$;"MONCIO"
10060 PRINT D$;"OPEN";F$;",D";DN
10070 PRINT D$;"UNLOCK";F$
10080 PRINT D$;"OPEN";F$
10090 PRINT D$;"READ";F$
10100 RETURN
10102 REM ------------------------------------------
10110 REM   SUBROUTINE FOR CLOSING A FILE
10120 PRINT D$;"LOCK";F$
10130 PRINT D$;"CLOSE";F$
10140 PRINT D$;"NOMONCIO"
10150 RETURN
10152 REM ------------------------------------------
10160 REM   SUBROUTINE FOR OPENING AND WRITING TO A FILE
10170 PRINT : INPUT "NAME OF FILE ?";NA$
10175 F$ = NA$ + RT$
10180 PRINT : INPUT "WHICH DISK DRIVE ?";DN
10190 IF DN = 1 OR DN = 2 THEN 10210
10200 PRINT : PRINT " TRY AGAIN-": GOTO 10180
10210 PRINT D$;"MONCIO"
10220 PRINT D$;"OPEN";F$;",D";DN
10230 PRINT D$;"UNLOCK";F$
10240 PRINT D$;"DELETE";F$
10250 PRINT D$;"OPEN";F$
10260 PRINT D$;"WRITE";F$
10270 RETURN
10272 REM ------------------------------------------
----------
```

TRIANGLE FILE EDITOR

```
100 LOMEM: 16384
110 TEXT : HOME
120 PRINT "*********************************"
130 PRINT "**                              **"
140 PRINT "**   3-NODE TRIANGLE FILE EDITOR   **"
150 PRINT "**   - ---- -------- ---- ------   **"
```

```
160   PRINT "**                                        **"
170   PRINT "**       STRESS AND HEAT TRANSFER        **"
180   PRINT "**             ANALYSIS                  **"
190   PRINT "**                                        **"
200   PRINT "*******************************************"
210   PRINT
220   PRINT
300   VTAB 20: PRINT "HIT ANY KEY TO CONTINUE"
302   GET G$: PRINT
304   REM  ----------------------------------------------
306   REM       LINES 310 THRU 348 ARE MINI FRONT END PROGRAM
308   REM  ----------------------------------------------
310   ONERR  GOTO 318
312   POKE 768,104: POKE 769,168: POKE 770,104: POKE 771,166: POKE
      772,223: POKE 773,154: POKE 774,72: POKE 775,152: POKE 776,72: POKE
      777,96
314   D$ =  CHR$ (4): REM    CHR$(4) IS CTRL-D
316   GOTO 356: REM     GO TO NORMAL PROGRAM START
318   CALL 768: PRINT D$;"PR#0": TEXT : IF  PEEK (222) = 255 THEN  PRINT
      "CNTL C CAUSED A BREAK IN ";  PEEK (218) +  PEEK (219) * 256: END
320   GOSUB 340: END
322   PRINT : PRINT : REM         DETERMINE FREE MEMORY
324   TP =  PEEK (175) +  PEEK (176) * 256: REM    TOP OF PROGRAM
326   EA =  PEEK (109) +  PEEK (110) * 256: REM    TOP OF LOMEM POINTER
328   BS =  PEEK (111) +  PEEK (112) * 256: REM    HIMEM POINTER
330   PRINT "FREE DATA SPACE IS ";BS - EA;" BYTES"
332   PRINT : PRINT "FREE PROGRAM SPACE:": PRINT
334   TP = 8192 - TP: IF TP > 0 THEN  PRINT TP;" BYTES TO HGR": RETURN
336   TP = TP + 8192: IF TP > 0 THEN  PRINT TP;" BYTES TO HGR2": RETURN
338   PRINT "       YOU ARE ";  ABS (TP);" BYTES INTO HGR2!": RETURN
340   PRINT "ERROR # ";  PEEK (222);" AT LINE # ";  PEEK (218) +  PEEK (219)
       * 256
342   PRINT : PRINT "      (ERROR CODES IN POOLE PG. 331)"
344   PRINT : PRINT "YOUR OPTIONS NOW ARE:": PRINT "   (1) HIT ESCAPE TO
      RESUME AT ERROR STATEMENT": PRINT "   (2) HIT ANY OTHER KEY TO END
      PROGRAM"
346   GET G$: PRINT : IF G$ =  CHR$ (27) THEN  RESUME
348   RETURN
350   REM  ----------------------------------------------
352   REM       START PROGRAM HERE
354   REM  ----------------------------------------------
356   HOME
400   PRINT "              OPTIONS"
410   PRINT "              -------"
420   PRINT
430   I$ =  CHR$ (9): REM    CHR$(9) IS CTRL-I
440   R$ =  CHR$ (13): REM   CHR$(13) IS CARRIAGE RETURN
450   DIM XG(100),YG(100),GN(100,3),PX(100),PY(100)
460   XP = 0:YP = 0
470   REM  ----------------------------------------------
480   PRINT "       1 - CREATE A NEW FILE"
490   PRINT
500   PRINT "       2 - EDIT AN EXISTING FILE"
510   PRINT
520   INPUT "       ENTER CHOICE - (1 OR 2) ";IN$
530   REM  ----------------------------------------------
540   IF IN$ = "1" GOTO 600
550   RT$ = ".GEOM"
560   GOSUB 10000: REM    OPEN AND READ THE GEOMETRY DATA FILE
570   REM  ----------------------------------------------
580   REM                    DATA INPUT                  ----
590   REM  ----------------------------------------------
600   HOME : INPUT "HOW MANY ELEMENTS? ";NE: PRINT
610   PRINT : INPUT "HOW MANY NODES? ";NN: PRINT
620   IF MS$ = "Y" GOTO 790
630   PRINT "ENTER ELEMENT NUMBER AND GLOBAL NUMBER OF LOCAL NODES 1, 2
      AND 3 IN THE FOLLOWING FORMAT:": PRINT : PRINT "EL#,   #1,   #2,   #3
      <RETURN>": PRINT
```

```
640    FOR J = 1 TO NE
650    INPUT I,GN(I,1),GN(I,2),GN(I,3)
660    NEXT J
670    PRINT : PRINT "ENTER GLOBAL NODE NUMBER AND ITS X AND Y COORDINATES
       IN THE FOLLOWING FORMAT:": PRINT : PRINT "NODE#,   X,   Y
       <RETURN>": PRINT
680    FOR J = 1 TO NN
690    INPUT I,XG(I),YG(I)
700    NEXT J
710    IF IN$ = "1" GOTO 750
720    GOSUB 10110: REM   CLOSE THE GEOMETRY FILE
730    REM  *************** EDITING OPTIONS ***************
740    REM
750    PRINT : PRINT : INPUT "DO YOU WISH TO CHANGE THE NUMBER OF ELEMENTS
       AND/OR NODES? ";MS$: PRINT
760    IF MS$ < > "Y" GOTO 790
770    PRINT : PRINT "CURRENT VALUES:  NE = ";NE;"   NN = ";NN
780    INPUT "ENTER NEW VALUES - NE = ";NE: INPUT "                    NN =
       ";NN
790    PRINT : PRINT : INPUT "DO YOU WISH TO EDIT NODE NUMBERING ?  (Y/N)
       ";RE$
800    IF RE$ = "Y" OR RE$ = "N" THEN 830
810    REM
820    PRINT : PRINT "RE-ENTER -": GOTO 790
830    IF RE$ = "N" THEN 910
840    PRINT : INPUT "WHICH ELEMENT ? ";N
850    IF N > NE THEN 840
860    PRINT "CURRENT VALUES: ";GN(N,1);",   ";GN(N,2);",   ";GN(N,3)
870    PRINT : INPUT "ENTER THE NEW VALUES:";GN(N,1),GN(N,2),GN(N,3)
880    PRINT : INPUT "MORE CHANGES ?  (Y/N)";RE$
890    IF RE$ = "Y" THEN 840
900    PRINT
910    PRINT : PRINT : INPUT "DO YOU WISH TO EDIT NODE COORDINATES ?  (Y/N)
       ";RE$
920    IF RE$ = "Y" OR RE$ = "N" THEN 940
930    PRINT : PRINT "RE-ENTER -": GOTO 910
940    IF RE$ = "N" THEN 1010
950    PRINT : INPUT "WHICH NODE ? ";N
960    IF N > NN THEN 950
970    PRINT "CURRENT VALUES: ";XG(N);",   ";YG(N)
980    PRINT : INPUT "ENTER THE NEW VALUES- ";XG(N),YG(N)
990    PRINT : INPUT "MORE CHANGES ?  (Y/N) ";RE$
1000   IF RE$ = "Y" THEN 950
1010   GOSUB 1220: REM     PLOT MESH
1020   PRINT : PRINT : PRINT "HIT RETURN"
1030   GET GO$
1040   REM  NO PRINTING AFTER GET STATEMENT
1050   PRINT : PRINT "OK TO STORE DATA ? (Y/N)   ": PRINT "AN (N) RESPONSE
       WILL RESTART EDITING"
1052   PRINT "AN 'E' WILL END PROGRAM"
1060   INPUT DK$
1065   IF DK$ = "E" THEN  GOTO 1180
1070   IF DK$ < > "Y" THEN 750
1080   RT$ = ".GEOM"
1090   GOSUB 10160: REM    OPEN FILE
1100   PRINT NE;R$;NN
1110   FOR I = 1 TO NE
1120   PRINT I;R$;GN(I,1);R$;GN(I,2);R$;GN(I,3)
1130   NEXT
1140   FOR I = 1 TO NN
1150   PRINT I;R$;XG(I);R$;YG(I)
1160   NEXT
1170   GOSUB 10110: REM     CLOSE FILE
1180   END
1190   REM  ---------------------------------------------------
1200   REM               PLOTTING SUBROUTINE            ----
1210   REM  ---------------------------------------------------
1220   FLASH : PRINT : PRINT : PRINT "SCALING FOR PLOT": NORMAL
```

```
1230    REM    SEARCH FOR MOST NEGATIVE X AND Y COORDINATE
1240    XP = 0:YP = 0
1250    FOR I = 1 TO NN
1260    IF XG(I) < XP THEN XP = XG(I)
1270    IF YG(I) < YP THEN YP = YG(I)
1280    NEXT I
1290    REM    SHIFT COORDINATE AXES
1300    FOR I = 1 TO NN
1310  PX(I) = XG(I) - XP:PY(I) = YG(I) - YP
1320    NEXT I
1330    REM    DETERMINE SIZE OF SYSTEM
1340    SX = 0:SY = 0
1350    FOR I = 2 TO NN
1360    FOR J = 1 TO NN
1370    BX =  ABS (PX(I) - PX(J)):BY =  ABS (PY(I) - PY(J))
1380    IF BX > SX THEN SX = BX
1390    IF BY > SY THEN SY = BY
1400    NEXT J
1410    NEXT I
1420    REM    CALCULATE SCALE FACTOR, SF
1430    SF = 250 / SX
1440    IF SF * SY > 150 THEN SF = 150 / SY
1450    HGR
1460    HCOLOR= 3
1470    FOR I = 1 TO NE
1480    X1 = PX(GN(I,1)) * SF:Y1 = 150 - PY(GN(I,1)) * SF * 0.8
1490    X2 = PX(GN(I,2)) * SF:Y2 = 150 - PY(GN(I,2)) * SF * 0.8
1500    X3 = PX(GN(I,3)) * SF:Y3 = 150 - PY(GN(I,3)) * SF * 0.8
1510    HPLOT 25 + X1,Y1 TO 25 + X2,Y2 TO 25 + X3,Y3 TO 25 + X1,Y1
1520    NEXT I
1530    RETURN
1540    REM    ------------------------------------------------
1550    REM              FILE READ/WRITE SUBROUTINES           ----
1560    REM    ------------------------------------------------
10000   REM    SUBROUTINE FOR OPENING AND READING A FILE
10001   REM
10010   PRINT : INPUT "WHAT IS THE NAME OF THE FILE ?";NA$
10015  F$ = NA$ + RT$
10020   PRINT : INPUT "WHICH DISK DRIVE ?";DN
10030   IF DN = 1 OR DN = 2 THEN 10050
10040   PRINT : PRINT " TRY AGAIN-": GOTO 10020
10050   PRINT D$;"MONCIO"
10060   PRINT D$;"OPEN";F$;",D";DN
10070   PRINT D$;"UNLOCK";F$
10080   PRINT D$;"OPEN";F$
10090   PRINT D$;"READ";F$
10100   RETURN
10102   REM    ------------------------------------------------
10110   REM    SUBROUTINE FOR CLOSING A FILE
10120   PRINT D$;"LOCK";F$
10130   PRINT D$;"CLOSE";F$
10140   PRINT D$;"NOMONCIO"
10150   RETURN
10152   REM    ------------------------------------------------
10160   REM    SUBROUTINE FOR OPENING AND WRITING TO A FILE
10170   PRINT : INPUT "NAME OF FILE ?";NA$
10175  F$ = NA$ + RT$
10180   PRINT : INPUT "WHICH DISK DRIVE ?";DN
10190   IF DN = 1 OR DN = 2 THEN 10210
10200   PRINT : PRINT " TRY AGAIN-": GOTO 10180
10210   PRINT D$;"MONCIO"
10220   PRINT D$;"OPEN";F$;",D";DN
10230   PRINT D$;"UNLOCK";F$
10240   PRINT D$;"DELETE";F$
10250   PRINT D$;"OPEN";F$
10260   PRINT D$;"WRITE";F$
10270   RETURN
10272   REM    -------
```

TRIANGLE DISTORTION PLOT

```
100  LOMEM: 16384
110  TEXT : HOME
120  PRINT "*******************************"
130  PRINT "**                            **"
140  PRINT "**      DISTORTION PLOT       **"
150  PRINT "**      ---------- ----       **"
160  PRINT "**                            **"
170  PRINT "**     PLANE STRESS ANALYSIS  **"
180  PRINT "**          WITH THE          **"
190  PRINT "**   CONSTANT STRAIN TRIANGLE **"
200  PRINT "**                            **"
210  PRINT "*******************************"
220  PRINT
230  PRINT
240  PRINT
250  D$ = CHR$ (4): REM  CHR$(4) IS CTRL-D
260  I$ = CHR$ (9): REM  CHR$(9) IS CTRL-I
270  REM  ------------------------------------------------
280  REM           DATA INPUT FROM DISC DRIVE         ----
290  REM  ------------------------------------------------
300  PRINT "INPUT THE GEOMETRY DATA FROM DISK FILE": PRINT
310  REM  OPEN AND READ THE DATA FILE
320  RT$ = ".GEOM"
330  DIM U(200),XG(100),YG(100),GN(100,3)
340  GOSUB 10000: REM  OPEN AND READ FILE
350  INPUT NE,NN
360  FOR J = 1 TO NE
370  INPUT I,GN(I,1),GN(I,2),GN(I,3)
380  NEXT J
390  FOR J = 1 TO NN
400  INPUT I,XG(I),YG(I)
410  NEXT J
420  GOSUB 10110: REM   CLOSE FILE
430  PRINT : PRINT "INPUT THE DISPLACEMENT DATA FROM DISK FILE": PRINT
440  RT$ = ".DISPL"
450  GOSUB 10000: REM  OPEN AND READ FILE
460  FOR I = 1 TO 2 * NN
470  INPUT U(I)
480  NEXT I
490  GOSUB 10110: REM   CLOSE FILE
500  REM  ------------------------------------------------
510  REM            PLOTTING SUBROUTINE              ----
520  REM  ------------------------------------------------
530  FLASH : PRINT : PRINT : PRINT "SCALING FOR PLOT ": PRINT : NORMAL
540  DI = 0
550  FOR I = 1 TO 2 * NN
560  IF  ABS (U(I)) > DI THEN DI =  ABS (U(I))
570  NEXT I
580  REM  SEARCH FOR MOST NEGATIVE X AND Y COORDINATES
590  XP = 0:YP = 0
600  FOR I = 1 TO NN
610  IF XG(I) < XP THEN XP = XG(I)
620  IF YG(I) < YP THEN YP = YG(I)
630  NEXT I
640  REM  SHIFT COORDINATE AXES
650  FOR I = 1 TO NN
660  XG(I) = XG(I) - XP:YG(I) = YG(I) - YP
670  NEXT I
680  REM  DETERMINE SIZE OF SYSTEM
690  SX = 0:SY = 0
700  XG(0) = 0:YG(0) = 0
710  FOR I = 2 TO NN
720  FOR J = 1 TO NN + 1
730  BX =  ABS (XG(I) - XG(J - 1)):BY =  ABS (YG(I) - YG(J - 1))
740  IF BX > SX THEN SX = BX
750  IF BY > SY THEN SY = BY
```

```
760   NEXT J
770   NEXT I
780   REM   CALCULATE SCALE FACTOR
790   SF = 230 / SX
800   IF SF * SY > 130 THEN SF = 130 / SY
805   MF = 0
810   HGR
820   HCOLOR= 3
840   FOR I = 1 TO NE
850   X1 = (XG(GN(I,1)) + U(2 * GN(I,1) - 1) * MF) * SF
860   X2 = (XG(GN(I,2)) + U(2 * GN(I,2) - 1) * MF) * SF
870   X3 = (XG(GN(I,3)) + U(2 * GN(I,3) - 1) * MF) * SF
880   Y1 = 130 - (YG(GN(I,1)) + U(2 * GN(I,1)) * MF) * SF * 0.8
890   Y2 = 130 - (YG(GN(I,2)) + U(2 * GN(I,2)) * MF) * SF * 0.8
900   Y3 = 130 - (YG(GN(I,3)) + U(2 * GN(I,3)) * MF) * SF * 0.8
910   HPLOT 30 + X1,Y1 TO 30 + X2,Y2 TO 30 + X3,Y3 TO 30 + X1,Y1
920   NEXT I
930   INPUT "TO MAGNIFY DISTORTION TYPE AN INTEGER BETWEEN 1 AND 10.  TYPE
      '0' TO END PROGRAM . ";K
940   MF = K / DI
950   IF K < > 0 GOTO 810
960   END
970   REM   --------------------------------------------------------
980   REM                  FILE READ/WRITE SUBROUTINES          ----
990   REM   --------------------------------------------------------
10000 REM   SUBROUTINE FOR OPENING AND READING A FILE
10001 REM
10010 PRINT : INPUT "WHAT IS THE NAME OF THE FILE ?";NA$
10015 F$ = NA$ + RT$
10020 PRINT : INPUT "WHICH DISK DRIVE ?";DN
10030 IF DN = 1 OR DN = 2 THEN 10050
10040 PRINT : PRINT " TRY AGAIN-": GOTO 10020
10050 PRINT D$;"MONCIO"
10060 PRINT D$;"OPEN";F$;",D";DN
10070 PRINT D$;"UNLOCK";F$
10080 PRINT D$;"OPEN";F$
10090 PRINT D$;"READ";F$
10100 RETURN
10102 REM   --------------------------------------------------
10110 REM   SUBROUTINE FOR CLOSING A FILE
10120 PRINT D$;"LOCK";F$
10130 PRINT D$;"CLOSE";F$
10140 PRINT D$;"NOMONCIO"
10150 RETURN
10152 REM   --------------------------------------------------
10160 REM   SUBROUTINE FOR OPENING AND WRITING TO A FILE
10170 PRINT : INPUT "NAME OF FILE ?";NA$
10175 F$ = NA$ + RT$
10180 PRINT : INPUT "WHICH DISK DRIVE ?";DN
10190 IF DN = 1 OR DN = 2 THEN 10210
10200 PRINT : PRINT " TRY AGAIN-": GOTO 10180
10210 PRINT D$;"MONCIO"
10220 PRINT D$;"OPEN";F$;",D";DN
10230 PRINT D$;"UNLOCK";F$
10240 PRINT D$;"DELETE";F$
10250 PRINT D$;"OPEN";F$
10260 PRINT D$;"WRITE";F$
10270 RETURN
10272 REM   --------------------------------------------------
```

A3.5 PROGRAM LISTINGS IN TURBO PASCAL

```
{*************************************************************}

PROGRAM STRESS3NODE (INPUT, OUTPUT);

{*************************************************************}
```

PROGRAM LISTINGS IN TURBO PASCAL   455

```pascal
CONST
    ARRSIZE1 (No. of elements)              =  80;
    ARRSIZE2 (degrees of freedom )          = 160;
            ( twice No. of nodes)
    ARRSIZE3 (No. of nodes)                 =  80;
    ARRSIZE4 (Half-Bandwidth)               =  40;
TYPE
    RARY1  = ARRAY [1..3] OF REAL;
    RARY2  = ARRAY [1..3,1..3] OF REAL;
    RARY3  = ARRAY [1..3,1..6] OF REAL;
    RARY4  = ARRAY [1..6,1..6] OF REAL;
    RARY5  = ARRAY [1..ARRSIZE1] OF REAL;
    RARY6  = ARRAY [1..ARRSIZE2] OF REAL;
    RARY7  = ARRAY [1..ARRSIZE3] OF REAL;
    RARY8  = ARRAY [1..ARRSIZE1, 1..3] OF REAL;
    RARY9  = ARRAY [1..ARRSIZE2, 1..ARRSIZE4] OF REAL;
    IARY20 = ARRAY [1..ARRSIZE2] OF INTEGER;
    IARY21 = ARRAY [1..ARRSIZE1, 1..3] OF INTEGER;
    IARY22 = ARRAY [1..ARRSIZE1, 1..6] OF INTEGER;

VAR
    MC, RE, DK, RESP                        :CHAR;
    DET, EE, VV, HH, QU, UM, UI             :REAL;
    MINX, MINY, SF, MF                      :REAL;
    CHOICE, C1, I, J, S, T, L, M, JJ, AP    :INTEGER;
    DI, BW, NE, NN, N, KK, K1, FX, FY       :INTEGER;
    SHIFTX,SHIFTY                           :INTEGER;
    ST                                      :RARY1;
    C                                       :RARY2;
    B, BE, R                                :RARY3;
    K                                       :RARY4;
    E, V, H                                 :RARY5;
    U, FC, FR                               :RARY6;
    XG, YG                                  :RARY7;
    EP, SS                                  :RARY8;
    KG                                      :RARY9;
    BC                                      :IARY20;
    GN                                      :IARY21;
    NG                                      :IARY22;
    QUIT                                    :BOOLEAN;

( NE = number of elements )
( NN = number of nodes )
( FX = number of x direction constraints )
( FY = number of y direction constraints )
( BC[I] = boundary condition code)
( XG[I], YG[I] = global coordinates of node I)
( GN[I,J] = global node number of node J in element I)
( NG[I,J] = global displacement number of local displacement)
          ( number J of element I )
( K[I,J] = element global coordinates stiffness matrix )
( KG[I,J] = assembled global stiffness matrix )
( C[I,J] = constitutive matrix )
( FC[I] = applied nodal forces )
( U[I] = nodal displacements )
( EP[I,J] = strain vector of element I )
( SS[I,J] = strain vector of element I )

(***********************************************************)
(                  PROGRAM PROCEDURES                       )
(***********************************************************)

(Set all variables to 0)
PROCEDURE INITIALIZE_VARIABLES;

    VAR
        I, J, KK                            :INTEGER;
```

```
BEGIN
    { Here are the single value real variables }

    DET:= 0;            EE := 0;            VV := 0;            HH := 0;
    QU := 0;            UM := 0;            UI := 0;

    { Here are the single value integer variables }

    CHOICE := 0;        C1 := 0;            I  := 0;            J  := 0;
     S := 0;            T  := 0;            L  := 0;            M  := 0;
    JJ := 0;            AP := 0;            FX := 0;            FY := 0;
    DI := 0;            BW := 0;            NE := 0;            NN := 0;
    N  := 0;            KK := 0;            K1 := 0;

    { Here are the characters }

    MC := 'n';          RE := 'n';          DK := 'n';

    { Here are the arrays }

    FOR I := 1 TO 3 DO
        BEGIN
            ST[I] := 0;
            FOR J := 1 TO 3 DO
                C[I,J] := 0;
            FOR KK := 1 TO 6 DO
                BEGIN
                    B[I,KK] := 0;
                    BE[I,KK] := 0;
                    R[I,KK] := 0;
                END
        END;
    FOR I := 1 TO 6 DO
        FOR J := 1 TO 6 DO
            K[I,J] := 0;
    FOR I := 1 TO ARRSIZE1 DO
        BEGIN
            E[I] := 0;
            V[I] := 0;
            H[I] := 0;
            FOR J := 1 TO 3 DO
                BEGIN
                    EP[I,J] := 0;
                    SS[I,J] := 0;
                    GN[I,J] := 0
                END;
            FOR KK := 1 TO 6 DO
                NG[I,KK] := 0
        END;
    FOR I := 1 TO ARRSIZE2 DO
        BEGIN
            BC[I] := 0;
            U[I] := 0;
            FC[I] := 0;
            FR[I] := 0;
            FOR J := 1 TO ARRSIZE4 DO
                KG[I,J] := 0
        END;
    FOR I := 1 TO ARRSIZE3 DO
        BEGIN
            XG[I] := 0;
            YG[I] := 0
        END
END;

{*************************************************************}

{Subroutine for element B matrix}
PROCEDURE BMATRIX;
```

```
    VAR
       I, J                         :INTEGER;
       X, Y                         :RARY1;
BEGIN
   X[1] := XG[GN[M,1]];
   X[2] := XG[GN[M,2]];
   X[3] := XG[GN[M,3]];
   Y[1] := YG[GN[M,1]];
   Y[2] := YG[GN[M,2]];
   Y[3] := YG[GN[M,3]];
   DET := (X[1]-X[3])*(Y[2]-Y[3])-(X[2]-X[3])*(Y[1]-Y[3]);
   B[1,1] := Y[2]-Y[3];
   B[2,1] := 0;
   B[3,1] := X[3]-X[2];
   B[1,2] := 0;
   B[2,2] := B[3,1];
   B[3,2] := B[1,1];
   B[1,3] := Y[3]-Y[1];
   B[2,3] := 0;
   B[3,3] := X[1]-X[3];
   B[1,4] := 0;
   B[2,4] := B[3,3];
   B[3,4] := B[1,3];
   B[1,5] := Y[1]-Y[2];
   B[2,5] := 0;
   B[3,5] := X[2]-X[1];
   B[1,6] := 0;
   B[2,6] := B[3,5];
   B[3,6] := B[1,5];
   FOR I:= 1 TO 3 DO
       FOR J := 1 TO 6 DO
           BE[I,J] := B[I,J]/DET
END;

(*************************************************************)

{Subroutine for element C matrix}
PROCEDURE CMATRIX;

   VAR
      C1                           :REAL;

BEGIN
   C1 := E[M]/(1-V[M]*V[M]);
   C[1,1] := C1;
   C[1,2] := C1*V[M];
   C[1,3] := 0;
   C[2,1] := C[1,2];
   C[2,2] := C[1,1];
   C[2,3] := 0;
   C[3,1] := 0;
   C[3,2] := 0;
   C[3,3] := C1*(1-V[M])/2
END;

(*************************************************************)

{All Input Data from Keyboard}
PROCEDURE TERM_INP;

BEGIN
   WRITE('How many 3-node elements? ');
   READLN(NE);
   WRITELN;
   WRITE('How many nodes? ');
   READLN(NN);
   WRITELN;
   WRITELN('Enter element number and global node numbers');
   WRITELN('    of local nodes 1, 2 and 3');
```

```
    WRITELN;
    WRITELN('Note: Use counter-clockwise sequence and ');
    WRITELN('      the following format:');
    WRITELN;
    WRITELN('EL #      1      2      3   <RETURN>');
    WRITELN;
    FOR J := 1 TO NE DO
        READLN(I, GN[I,1],GN[I,2],GN[I,3]);
    WRITELN;
    WRITELN('Enter node number and its global');
    WRITELN('    X and Y coordinates');
    WRITELN;
    WRITELN('Enter in following format:');
    WRITELN;
    WRITELN('  NODE#      X      Y      <RETURN>');
    WRITELN;
    FOR J := 1 TO NN DO
        READLN(I, XG[I],YG[I])
END;

{*************************************************************}

{Read Structure Geometry Data from Disk}
PROCEDURE DISC_INP(VAR NE                       :INTEGER;
                   VAR NN                       :INTEGER;
                   VAR GN                       :IARY21;
                   VAR XG                       :RARY7;
                   VAR YG                       :RARY7);

    TYPE
        CHRSTRING=STRING[20];

    VAR
        I1, J                                   :INTEGER;
        INPFILE                                 :TEXT;
        FNAME                                   :CHRSTRING;

BEGIN
    WRITELN('Input file name');
    READLN(FNAME);
    ASSIGN(INPFILE, FNAME);
    RESET(INPFILE);

    WRITELN('Reading number of elements ');
    READ(INPFILE, NE);
    WRITELN('Reading # of nodes');
    READ(INPFILE, NN);
    WRITELN('Reading node numbers');
    FOR J := 1 TO NE DO
        BEGIN
            READ(INPFILE, I1);
            READ(INPFILE, GN[I1,1]);
            READ(INPFILE, GN[I1,2]);
            READ(INPFILE, GN[I1,3])
        END;
    WRITELN('Reading node coordinates');
    FOR J := 1 TO NN DO
        BEGIN
            READ(INPFILE, I1);
            READ(INPFILE, XG[I1]);
            READ(INPFILE, YG[I1])
        END;
    WRITELN;
    CLOSE(INPFILE)
END;

{*************************************************************}
{Procedure to Scale and Center Structure for Graphics}
    PROCEDURE CENTER( VAR XG            :RARY7;
                      VAR YG            :RARY7;
```

PROGRAM LISTINGS IN TURBO PASCAL

```
                        VAR SF              :REAL;
                        VAR MINX            :REAL;
                        VAR MINY            :REAL;
                        VAR SHIFTX          :INTEGER;
                        VAR SHIFTY          :INTEGER);
    VAR
        MAXX,MAXY,SFX,SFY               :REAL;

    BEGIN
      SFX:=0;
      SFY:=0;
      SF:=0;
      MINX:=0;
      MINY:=0;
      MAXX:=0;
      MAXY:=0;
      FOR J:=1 TO NN DO
          BEGIN
               IF XG[J] > MAXX  THEN
                  MAXX:=XG[J];
               IF XG[J] < MINX  THEN
                  MINX:=XG[J];
               IF YG[J] > MAXY  THEN
                  MAXY:=YG[J];
               IF YG[J] < MINY  THEN
                  MINY:=YG[J];
          END;
      SFX:=250/ABS(MAXX-MINX);
      SFY:=150/ABS(MAXY-MINY);
      IF SFX < SFY   THEN
         SF:=SFX
      ELSE SF:=SFY;
      SHIFTX:=TRUNC((300-SF*(MAXX-MINX))/2);
      SHIFTY:=TRUNC((200-SF*(MAXY-MINY))/2);
    END;

(*****************************************************************)
 (Display of Structure using the Turbo Graphics; Calls PROC. "center")
 PROCEDURE GRAPH(  VAR   XG              :RARY7;
                   VAR   YG              :RARY7;
                   VAR   SF              :REAL;
                   VAR   MINX            :REAL;
                   VAR   MINY            :REAL;
                   VAR   U               :RARY6;
                   VAR   SHIFTX          :INTEGER;
                   VAR   SHIFTY          :INTEGER;
                   VAR   MF              :REAL);
     VAR
        X1,X2,X3,Y1,Y2,Y3               :INTEGER;
        C                               :CHAR;

BEGIN
    CENTER(XG,YG,SF,MINX,MINY,SHIFTX,SHIFTY);
    GRAPHMODE;
    PALETTE(1);
    FOR J:=1 TO NE DO
        BEGIN
            X1:=SHIFTX+TRUNC(SF*(XG[GN[J,1]]-MINX)+MF*U[2*GN[J,1]-1]);
            X2:=SHIFTX+TRUNC(SF*(XG[GN[J,2]]-MINX)+MF*U[2*GN[J,2]-1]);
            X3:=SHIFTX+TRUNC(SF*(XG[GN[J,3]]-MINX)+MF*U[2*GN[J,3]-1]);
            Y1:=200-SHIFTY-TRUNC(SF*(YG[GN[J,1]]-MINY)+MF*U[2*GN[J,1]]);
            Y2:=200-SHIFTY-TRUNC(SF*(YG[GN[J,2]]-MINY)+MF*U[2*GN[J,2]]);
            Y3:=200-SHIFTY-TRUNC(SF*(YG[GN[J,3]]-MINY)+MF*U[2*GN[J,3]]);
            DRAW (X1,Y1, X2,Y2,3);
            DRAW (X2,Y2, X3,Y3,3);
            DRAW (X3,Y3, X1,Y1,3);
        END;
    GOTOXY(1,24);
    WRITE ('HIT <CR> TO CONTINUE.');
    READLN;
```

```
      TEXTMODE;
   END;

{************************************************************}

{Write Structure Geometry to Disk}
PROCEDURE WRITE_DISK (VAR NE                    :INTEGER;
                      VAR NN                    :INTEGER;
                      VAR GN                    :IARY21;
                      VAR XG                    :RARY7;
                      VAR YG                    :RARY7);

   TYPE
      CHRSTRING=STRING[20];

   VAR
      I1, J                                     :INTEGER;
      OUTFILE                                   :TEXT;
      FNAME                                     :CHRSTRING;

BEGIN
   WRITELN('Input a name for the new file to be created? ');
   READLN(FNAME);
   WRITELN;
   ASSIGN(OUTFILE, FNAME);
   REWRITE(OUTFILE);

   WRITELN(OUTFILE, NE, ' ', NN);
   FOR I1 :=1 TO NE DO
      WRITELN(OUTFILE,I1,' ',GN[I1,1],' ',GN[I1,2],' ',GN[I1,3]);
   FOR I1 := 1 TO NN DO
      WRITELN(OUTFILE,I1,' ',XG[I1],' ',YG[I1]);
   CLOSE(OUTFILE)
END;

{************************************************************}

{Calculate Half-Bandwidth of Global Stiffness Matrix}
PROCEDURE BANDWIDTH;

   VAR
      I, J                                      :INTEGER;

BEGIN
   FOR I := 1 TO NE DO
      FOR J := 1 TO 2 DO
         FOR JJ := 1 TO 3 DO
            BEGIN
               DI := ABS(GN[I,J]-GN[I,JJ]);
               IF DI > BW THEN
                  BW := DI
            END;
   BW := 2*BW+2;
   WRITELN;
   WRITELN('**  Calculated Bandwidth = ', BW:4, '  **');
   WRITELN
END;

{************************************************************}

{Optional Edit of Input Node Numbering Data}
PROCEDURE EDIT_NN  (VAR GN         :IARY21;
                    VAR NE         :INTEGER);

   VAR
      N                                         :INTEGER;

BEGIN
   WHILE UPCASE(RESP) = 'Y' DO
      BEGIN
```

PROGRAM LISTINGS IN TURBO PASCAL

```
                N := NE + 1;
                WHILE N> NE DO
                    BEGIN
                        WRITE('Which element? ');
                        READLN(N)
                    END;
                WRITELN;
                WRITELN('Current values: ',GN[N,1]:3,' ',GN[N,2]:3,' ',GN[N,3]:3);
                WRITE('Enter the new values: ');
                READLN(GN[N,1], GN[N,2], GN[N,3]);
                GRAPH(XG,YG,SF,MINX,MINY,U,SHIFTX,SHIFTY,MF);
                WRITELN;
                WRITE('More changes? (Y/N) ');
                READLN(RESP);
                WRITELN
            END;
END;

{************************************************************}

{Optional Edit of Input Node Coordinates Data}
PROCEDURE EDIT_NC   (VAR NN               :INTEGER;
                     VAR XG               :RARY7;
                     VAR YG               :RARY7);

    VAR
        N                                                  :INTEGER;

BEGIN
    WHILE UPCASE(RESP) = 'Y' DO
        BEGIN
            N := NN+1;
            WHILE N>NN DO
                BEGIN
                    WRITE('Which node? ');
                    READLN(N);
                    WRITELN
                END;
            WRITELN('Current value: ', XG[N]:10:8,' ', YG[N]:10:5);
            WRITE('Enter the new values: ');
            READLN(XG[N], YG[N]);
            GRAPH(XG,YG,SF,MINX,MINY,U,SHIFTX,SHIFTY,MF);
            WRITELN;
            WRITE('More changes? (Y/N) ');
            READLN(RESP);
            WRITELN
        END
END;

{************************************************************}

{Output of Input Data and Results}
PROCEDURE RESULTS;

    VAR
        DEV                                                :TEXT;
        DONE                                               :BOOLEAN;
        CHOICE                                             :INTEGER;
        FNAME                                              :STRING[20];

BEGIN
    REPEAT
        BEGIN
            DONE := TRUE;
            WRITELN;
            WRITELN;
            WRITELN('                    DATA OUTPUT OPTIONS');
            WRITELN('                    -------------------');
            WRITELN;
            WRITELN('                        Terminal = "1" ');
```

```
            WRITELN('                        Printer   = "2" ');
            WRITELN('                        Aux       = "3" ');
            WRITELN('                        Textfile  = "4" ');
            WRITELN;
            WRITE('                ENTER CHOICE - ( 1 TO 4) ');
            READLN(CHOICE);
            CASE CHOICE OF
               1: FNAME := 'TRM:';
               2: FNAME := 'LST:';
               3: FNAME := 'AUX:';
               4: BEGIN
                     WRITE('Enter filename (default = RESULTS.DAT): ');
                     READLN(FNAME);
                     IF FNAME = '' THEN
                        FNAME := 'RESULTS.DAT';
                  END
            ELSE
               DONE := FALSE;
            END (CASE)
         END;
   UNTIL DONE;
   ASSIGN (DEV, FNAME);
   REWRITE (DEV);
   WRITELN(DEV);
   WRITELN(DEV);
   WRITELN(DEV, 'PLANE STRESS ANALYSIS -- INPUT DATA');
   WRITELN(DEV, '----- ------ --------    ----- ----');
   WRITELN(DEV);
   WRITELN(DEV, '**** ELEMENT GLOBAL NODE NUMBERS ****');
   WRITELN(DEV);
   WRITELN(DEV, '                    GLOBAL NODE NUMBERS OF');
   WRITELN(DEV,' ELEMENT      LOCAL 1         LOCAL 2        LOCAL 3');
   WRITELN(DEV);
   FOR J := 1 TO NE DO
      WRITELN(DEV, J:4, GN[J,1]:14, GN[J,2]:12, GN[J,3]:12);
   WRITELN(DEV);
   WRITELN(DEV,'** NODE GLOBAL COORDINATES AND BOUNDARY DISPLACEMENTS **');
   WRITELN(DEV);
   WRITELN(DEV, '                                BOUNDARY DISPLACEMENTS');
   WRITELN(DEV,'   NODE          X              Y             U              V');
   WRITELN(DEV);
   FOR I := 1 TO NN DO
      BEGIN
         WRITE(DEV,I:3,XG[I]:14:3,YG[I]:14:3);
         IF (BC[2*I-1]<>1) AND (BC[2*I]<>1) THEN
            WRITELN(DEV);
         IF (BC[2*I-1]=1) AND (BC[2*I]=1) THEN
            WRITELN(DEV,U[2*I-1]:14:5,U[2*I]:14:5);
         IF (BC[2*I-1]=1) AND (BC[2*I]<>1) THEN
            WRITELN(DEV,U[2*I-1]:14:5);
         IF (BC[2*I-1]<>1) AND (BC[2*I]=1) THEN
            WRITELN(DEV,'              ',U[2*I]:14:5);
      END;
   WRITELN(DEV);
   WRITELN(DEV, '**** APPLIED NODAL FORCES ****');
   WRITELN(DEV);
   WRITELN(DEV, '  NODE     X FORCE        Y FORCE');
   WRITELN(DEV);
   FOR N := 1 TO NN DO
      IF NOT ((FC[2*N] = 0) AND (FC[2*N-1] = 0)) THEN
         WRITELN(DEV, N:3, FC[2*N-1]:16:3, FC[2*N]:15:3);
   WRITELN(DEV);
   WRITELN(DEV, '**** MATERIAL PROPERTIES AND THICKNESS ****');
   WRITELN(DEV);
   WRITELN(DEV, 'ELEMENT      ELASTIC       POISSONS     THICKNESS');
   WRITELN(DEV, '             MODULUS        RATIO');
   WRITELN(DEV);
   I := 1;
```

PROGRAM LISTINGS IN TURBO PASCAL

```pascal
    REPEAT
      BEGIN
        WRITELN(DEV, I:3, E[I]:16:1, V[I]:14:3, H[I]:12:3);
        IF UPCASE(MC) = 'Y' THEN
          BEGIN
            WRITELN(DEV, 'ALL ELEMENTS HAVE SAME PROPERTIES');
            I := NE
          END;
        I := I+1
      END
    UNTIL I > NE;
    WRITELN(DEV);
    WRITELN(DEV);
    WRITELN(DEV, '              *** RESULTS OF PLANE STRESS ANALYSIS ***');
    WRITELN(DEV, '                  ------- -- ----- ------ --------   ');
    WRITELN(DEV);
    WRITELN(DEV, '          ******************************************');
    WRITELN(DEV, '          **                                      **');
    WRITELN(DEV, '          **            DISPLACEMENTS             **');
    WRITELN(DEV, '          **                                      **');
    WRITELN(DEV, '          ******************************************');
    WRITELN(DEV);
    WRITELN(DEV,'    NODE        XDISPL          YDISPL');
    WRITELN(DEV);
    FOR I := 1 TO NN DO
      WRITELN(DEV, I:3, U[2*I-1]:16:6, U[2*I]:16:6);
    WRITELN(DEV);
    WRITELN(DEV, '          ******************************************');
    WRITELN(DEV, '          **                                      **');
    WRITELN(DEV, '          **                STRAIN                **');
    WRITELN(DEV, '          **                                      **');
    WRITELN(DEV, '          ******************************************');
    WRITELN(DEV);
    WRITELN(DEV, 'ELEMENT     EXX           EYY            EXY');
    WRITELN(DEV);
    FOR M := 1 TO NE DO
      WRITELN(DEV, M:3, EP[M,1]:14:6, EP[M,2]:14:6, EP[M,3]:14:6);
    WRITELN(DEV);
    WRITELN(DEV, '          ******************************************');
    WRITELN(DEV, '          **                                      **');
    WRITELN(DEV, '          **                STRESS                **');
    WRITELN(DEV, '          **                                      **');
    WRITELN(DEV, '          ******************************************');
    WRITELN(DEV);
    WRITELN(DEV, 'ELEMENT     SXX           SYY            SXY');
    WRITELN(DEV);
    FOR M := 1 TO NE DO
      WRITELN(DEV, M:3, SS[M,1]:14:3, SS[M,2]:14:3, SS[M,3]:14:3);
    WRITELN(DEV)
END;

{*************************************************************}


{******---------------MAIN PROGRAM---------------**********}

BEGIN
    WRITELN('          *************************************************');
    WRITELN('          *************************************************');
    WRITELN('          **            Plane Stress Analysis          **');
    WRITELN('          **                 with the                 **');
    WRITELN('          **          Constant Strain Triangle         **');
    WRITELN('          *************************************************');
    WRITELN('          *************************************************');

{           ************ Start of Main Program ************       }

    INITIALIZE_VARIABLES;
```

```
WRITELN;
WRITELN('                          Data Input Options            ');
WRITELN('                          ------------------            ');
WRITELN('              1 - input new model data from keyboard');
WRITELN('              2 - input old model data from disc');
WRITELN;
WRITE('             Enter choice - (1 or 2) ');
READLN(CHOICE);
WRITELN;
IF CHOICE = 2 THEN
    DISC_INP(NE, NN, GN, XG, YG)
ELSE
    TERM_INP;
BANDWIDTH;
WRITELN;
WRITE('Do you wish to see plot of structure? (Y/N) ');
READLN(RESP);
IF UPCASE(RESP) = 'Y'  THEN
    GRAPH(XG,YG,SF,MINX,MINY,U,SHIFTX,SHIFTY,MF);
WRITELN;
WRITE('Do you wish to edit node numbering?  (Y/N) ');
READLN(RESP);
IF UPCASE(RESP) = 'Y'  THEN
    EDIT_NN(GN,NE);
WRITELN;
WRITE('Do you wish to edit node coordinates (Y/N) ');
READLN(RESP);
IF UPCASE(RESP) = 'Y'  THEN
    EDIT_NC(NN,XG,YG);
WRITELN;
WRITE('Do you want data stored on disk?   (Y/N) ');
READLN (DK);
WRITELN;
IF UPCASE(DK) = 'Y'   THEN
    WRITE_DISK(NE, NN, GN, XG, YG);
FOR I := 1 TO 2*NN DO
    U[I] := 0;
FOR M := 1 TO NE DO
    FOR J := 1 TO 3 DO
        BEGIN
            NG[M,2*J-1] := 2*GN[M,J]-1;
            NG[M,2*J] := 2*GN[M,J]
        END;
FOR I := 1 TO 2*NN DO
    BC[I] := 0;
    FX := 0;
WHILE FX = 0 DO
    BEGIN
        WRITE('How many nodes have specified X displacement (incl. 0)? ');
        READLN(FX);
        WRITELN;
        IF FX = 0 THEN
            WRITELN('WARNING!  Remove rigid body displacement.  Try again.');
    END;
WRITELN('Enter node numbers and the known X displacement');
WRITELN('Node #        U       <return>');
FOR J := 1 TO FX DO
    BEGIN
        READLN(I, U[2*I-1]);
        BC[2*I-1] := 1
    END;
WRITELN;
FY := 0;
WHILE FY = 0 DO
    BEGIN
        WRITE('How many nodes have specified Y displacement (incl. 0)? ');
        READLN(FY);
        WRITELN;
```

```
            IF FY = O THEN
                WRITELN('WARNING!    Remove rigid body displacement.    Try again.')
        END;
    WRITELN('Enter node numbers and the known Y displacement');
    WRITELN('Node #        V      <return>');
    FOR J := 1 TO FY DO
        BEGIN
            READLN(I, U[2*I]);
            BC[2*I] := 1
        END;
    WRITELN;
    FOR I := 1 TO 2*NN DO
        FC[I] := 0;
    WRITE('How many nodes are supporting applied loads? ');
    READLN(AP);
    WRITELN;
    IF AP <> O THEN
        BEGIN
            WRITELN('Enter global node number and its X and Y load');
            WRITELN('Node #         FX        FY      <return>');
            FOR I := 1 TO AP DO
                READLN(N, FC[2*N-1], FC[2*N]);
            WRITELN
        END;
    WRITE('Is material homogeneous and thickness uniform? (Y/N) ');
    READLN(MC);
    WRITELN;
    IF UPCASE(MC) <> 'Y' THEN
        BEGIN
            WRITELN('Enter element number, elastic modulus, E, Poissons ratio, v,
 and thickness, T.');
            WRITELN('EL #         E        V        T      <return>');
            FOR I := 1 TO NE DO
                READLN(M, E[M], V[M], H[M])
        END
    ELSE
        BEGIN
            WRITELN('Enter elastic modulus,E, Poisson ratio,v, and thickness,T');
            WRITELN('E         V        T     <return>');
            READLN(EE, VV, HH);
            FOR I := 1 TO NE DO
                BEGIN
                    E[I] := EE;
                    V[I] := VV;
                    H[I] := HH
                END
        END;
    WRITELN;
    {-----------------------------------------------------------------}
    {              Calculate Element Stiffness Matrix                 }
    {-----------------------------------------------------------------}
    FOR I := 1 TO 2*NN DO
        FOR J := 1 TO BW DO
            KG[I,J] := O;
    WRITELN('Working on ');
    FOR M := 1 TO NE DO
        BEGIN
            WRITELN('Element ', M, ' stiffness matrix');
            FOR S := 1 TO 3 DO
                FOR T := 1 TO 6 DO
                    R[S,T] := O;
            FOR S := 1 TO 6 DO
                FOR T := 1 TO 6 DO
                    K[S,T] := O;
            BMATRIX;
            CMATRIX;
            FOR I := 1 TO 3 DO
                FOR J := 1 TO 6 DO
```

```
                       FOR N := 1 TO 3 DO
                           R[I,J] := R[I,J]+C[I,N]*BE[N,J];
           FOR I := 1 TO 6 DO
               FOR J := 1 TO 6 DO
                   FOR N := 1 TO 3 DO
                       K[I,J] := K[I,J]+BE[N,I]*R[N,J];
           FOR I := 1 TO 6 DO
               FOR J := 1 TO 6 DO
                   K[I,J] := K[I,J]*H[M]*DET/2;
{------------------------------------------------------------------------}
{          Assemble Global Stiffness Matrix in Banded Format             }
{------------------------------------------------------------------------}
           FOR I := 1 TO 6 DO
               FOR J := 1 TO 6 DO
                   IF NG[M,J] >= NG[M,I] THEN
                       BEGIN
                           C1 := NG[M,J]-NG[M,I]+1;
                           KG[NG[M,I],C1] := K[I,J]+KG[NG[M,I],C1]
                       END
       END;
   FOR I := 1 TO 2*NN DO
       BEGIN
{------------------------------------------------------------------------}
{     Save Boundary Forces in Temporary Array FR[I]    and               }
{     Input Boundary Displacements in FC[I], Modify KG[I,J]              }
{------------------------------------------------------------------------}
           FR[I] := FC[I];
           IF BC[I] <> 0 THEN
               BEGIN
                   KG[I,1] := KG[I,1]*1E10;
                   FC[I] := U[I]*KG[I,1]
               END
       END;
   WRITELN('GAUSS EQUATION SOLVER (BAND MATRIX)');
   N := 2*NN;
   FOR M := 1 TO N DO
       IF KG[M,1] <> 0 THEN
           FOR L := 2 TO BW DO
               IF KG[M,L] <> 0 THEN
                   BEGIN
                       I := M+L-1;
                       QU := KG[M,L]/KG[M,1];
                       J := 0;
                       FOR KK := L TO BW DO
                           BEGIN
                               J := J+1;
                               IF I <= N THEN
                                   KG[I,J] := KG[I,J]-QU*KG[M,KK]
                           END;
                       KG[M,L] := QU
                   END;
   FOR M := 1 TO N DO
       BEGIN
           IF KG[M,1] <> 0 THEN
               BEGIN
                   FOR L := 2 TO BW DO
                       IF KG[M,L] <> 0 THEN
                           BEGIN
                               I := M+L-1;
                               IF I <= N THEN
                                   FC[I] := FC[I]-KG[M,L]*FC[M]
                           END;
                   FC[M] := FC[M]/KG[M,1];
               END
       END;
   WRITELN('Back substitution');
   U[N] := FC[N];
   FOR KK := 2 TO N DO
```

PROGRAM LISTINGS IN TURBO PASCAL

```pascal
        BEGIN
          M := N+1-KK;
          FOR L:=2 TO BW DO
              IF KG[M,L] <> 0   THEN
                 BEGIN
                    K1:=M+L-1;
                    IF K1<=N THEN
                       FC[M]:=FC[M]-KG[M,L]*U[K1]
                 END;
              U[M] := FC[M]
        END;
{---------------------------------------------------------------------}
{         Determine Maximum Displacment Value for Graph scaling       }
{---------------------------------------------------------------------}
  UM := 0;
  FOR I := 1 TO N DO
      BEGIN
         UI := ABS(U[I]);
         IF UI > UM THEN
            UM := UI;
      END;
{---------------------------------------------------------------------}
{                 Set Negligible Displacements = 0                    }
{---------------------------------------------------------------------}
  FOR I:= 1 TO N DO
      IF ABS(U[I])/UM < 1E-6   THEN
         U[I]:= 0;
{---------------------------------------------------------------------}
{             Restore Boundary Loads to Array FC[I]                   }
{---------------------------------------------------------------------}
  FOR I := 1 TO 2*NN DO
      FC[I] := FR[I];
{---------------------------------------------------------------------}
{                       Calculate Element Strain                      }
{---------------------------------------------------------------------}
  FOR M := 1 TO NE DO
      BEGIN
         FOR I:= 1 TO 3 DO
            EP[M,I]:= 0;
         BMATRIX;
         FOR I:= 1 TO 3 DO
            FOR J := 1 TO 6 DO
               EP[M,I] := EP[M,I]+BE[I,J]*U[NG[M,J]]
      END;
{---------------------------------------------------------------------}
{                       Calculate Element Stress                      }
{---------------------------------------------------------------------}
    FOR M := 1 TO NE DO
        BEGIN
           FOR I:= 1 TO 3 DO
              SS[M,I]:= 0;
           CMATRIX;
           FOR I := 1 TO 3 DO
               BEGIN
                  FOR J := 1 TO 3 DO
                     SS[M,I] := SS[M,I]+C[I,J]*EP[M,J];
               END;
        END;
  REPEAT
    BEGIN
       RESULTS;
       WRITELN;
       WRITELN;
       WRITE('Do you wish to repeat the output?   (Y/N)');
       READLN(RESP);
       IF UPCASE(RESP) = 'Y'   THEN
          QUIT:=FALSE
       ELSE
          QUIT:=TRUE
    END;
```

```
        UNTIL QUIT;
        WRITELN;
        WRITE('Do you want a distortion plot of model?   (Y/N)');
        READLN(RESP);
        IF UPCASE(RESP) = 'Y'   THEN
            BEGIN
                MF:=25/UM;
                GRAPH(XG,YG,SF,MINX,MINY,U,SHIFTX,SHIFTY,MF)
            END
END.
```

APPENDIX 4

Input File and Results for ICES STRUDL-II Program

```
****************************************
*                                       *
*           ICES STRUDL-II              *
*    THE STRUCTURAL DESIGN LANGUAGE     *
*                                       *
*   CIVIL ENGINEERING SYSTEMS LABORATORY*
*   MASSACHUSETTS INSTITUTE OF TECHNOLOGY*
*       CAMBRIDGE, MASSACHUSETTS        *
*         V2 M2      JUNE, 1972         *
*                                       *
*         UNIVAC VS/9-ICES-1.0          *
*      WORCESTER POLYTECHNIC INSTITUTE  *
*                                       *
*         13:47:26      01/04/85        *
*                                       *
****************************************
```

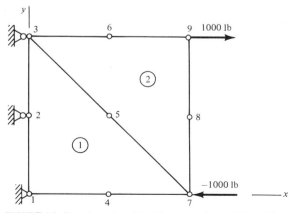

FIGURE A4 Two-element model of the square-plate problem of Example 8.2.

469

```
THE FOLLOWING AREAS HAVE BEEN TESTED.
    FRAME ANALYSIS, FINITE ELEMENTS, PLOTTING, REINFORCED CONCRETE
THE FOLLOWING AREAS HAVE SURVIVED A SAMPLE PROGRAM.
    DYNAMICS, MEMBER SELECT, OPTIMIZE, NONLINEAR ANALYSIS
THIS TEXT MAY BE SUPPRESSED BY: MESSAGE 'OFF'
    ON THE STRUDL CARD        18 SEPT 1973
UNITS LB INCHES
JOINT COORDINATES
1 0. 0. S
2 0. 4. S
3 0. 8. S
4 4. 0.
5 4. 4.
6 4. 8.
7 8. 0.
8 8. 4.
9 8. 8.
TYPE PLANE STRESS
ELEMENT INCIDENCES
1 1 7 3 4 5 2
2 3 7 9 5 8 6
JOINT RELEASES
2 FORCE Y
3 FORCE Y
ELEMENT PROPERTIES
1 TYPE 'LST' THICKNESS 0.125
2 TYPE 'LST' THICKNESS 0.125
CONSTANTS
E 10.E+06 ALL
POISSON 0.3 ALL
LOADING 1
JOINT LOADS
9 FORCE X 1000.
7 FORCE X -1000.
PRINT DATA


****************************************
*   PROBLEM DATA FROM INTERNAL STORAGE  *
****************************************

JOB ID -  BODY        JOB TITLE - 031249362
ACTIVE UNITS -  LENGTH        WEIGHT        ANGLE    TEMPERATURE    TIME
                INCH          LB            RAD      DEGF           SEC

            ********** STRUCTURAL DATA **********


ACTIVE STRUCTURE TYPE - PLANE     STRESS
ACTIVE COORDINATE AXES   X Y
```

INPUT FILE AND RESULTS FOR ICES STRUDL-II PROGRAM

```
JOINT COORDINATES------------------------------------/ STATUS---/
JOINT        X              Y              Z     CONDITION

 1         0.000          0.000          0.000   SUPPORT   ACTIVE   GLOBAL
 2         0.000          4.000          0.000   SUPPORT   ACTIVE   GLOBAL
 3         0.000          8.000          0.000   SUPPORT   ACTIVE   GLOBAL
 4         4.000          0.000          0.000             ACTIVE   GLOBAL
 5         4.000          4.000          0.000             ACTIVE   GLOBAL
 6         4.000          8.000          0.000             ACTIVE   GLOBAL
 7         8.000          0.000          0.000             ACTIVE   GLOBAL
 8         8.000          4.000          0.000             ACTIVE   GLOBAL
 9         8.000          8.000          0.000             ACTIVE   GLOBAL


JOINT RELEASES---------------------------------/ELASTIC SUPPORT RELEASES------------------------------------/
JOINT  FORCE  MOMENT  THETA 1  THETA 2  THETA 3   KFX     KFY     KFZ     KMX     KMY     KMZ
 2      Y              0.000    0.000    0.000   0.000   0.000   0.000   0.000   0.000   0.000
 3      Y              0.000    0.000    0.000   0.000   0.000   0.000   0.000   0.000   0.000


ELEMENT INCIDENCES---------------------------------------------------------------/
ELEMENT       NODES
 1        1    7    3    4    5    2                             ACTIVE
 2        3    7    9    5    8    6                             ACTIVE


ELEMENT PROPERTIES---------------------------------------------------------------/
ELEMENT  TYPE  THICKNESS  /--------CURVATURES--------/ /--------------THERMAL EXPANSION COEFFICIENTS---------------/
                             K1      K2     K12     CAX     CAY     CAZ     CSXY    CSXZ    CSYZ

 1       LST    0.125
 2       LST    0.125


MEMBER CONSTANTS--------------------------------------------------------------------------------------/
CONSTANT   STANDARD VALUE   DOMAIN,        VALUE      MEMBER LIST


   E        0.100000E 08    ALL
   G        0.000000E 00    ALL
 DENSITY    0.100000E 01    ALL
   CTE      0.100000E 01    ALL
  BETA      0.000000E 00    ALL
 POISSON    0.300000E 00    ALL


        ********** LOADING DATA **********


LOADING - 1                                                                STATUS - ACTIVE
MEMBER AND ELEMENT LOADS---------------------------------------------------------------------/
MEMBER/ELEMENT
JOINT LOADS-----------------------------------/ /-----------------------------/
JOINT   STEP   FORCE X       Y        Z     MOMENT X     Y        Z
 7            -1000.000    0.000    0.000     0.000    0.000    0.000
 9             1000.000    0.000    0.000     0.000    0.000    0.000
JOINT DISPLACEMENTS-------------------------/ /-------------------------/
JOINT   STEP   DISP. X       Y        Z     ROT.  X      Y        Z


****************************************
*  END OF DATA FROM INTERNAL STORAGE  *
****************************************


STIFFNESS ANALYSIS

LIST DISPLACEMENTS STRESSES STRAINS FORCES ALL


    *************************
    *RESULTS OF LATEST ANALYSES*
    *************************


   PROBLEM - BODY        TITLE - 031249362

   ACTIVE UNITS   INCH   LB   RAD   DEGF   SEC

   ACTIVE STRUCTURE TYPE   PLANE   STRESS

   ACTIVE COORDINATE AXES   X  Y
```

LOADING - 1

ELEMENT STRESSES

ELEMENT								
1	NODE	1	SXX	-0.600004E 04	SYY	-0.268555E-02	SXY	0.447751E-03
1	NODE	7	SXX	-0.600000E 04	SYY	-0.732422E-03	SXY	0.107460E-01
1	NODE	3	SXX	0.600000E 04	SYY	-0.122070E-02	SXY	-0.313426E-02
2	NODE	3	SXX	0.600001E 04	SYY	-0.219727E-02	SXY	-0.447751E-03
2	NODE	7	SXX	-0.600000E 04	SYY	-0.708008E-02	SXY	-0.358201E-02
2	NODE	9	SXX	0.599999E 04	SYY	0.119629E-01	SXY	0.000000E 00

ELEMENT STRAINS

ELEMENT								
1	NODE	1	EXX	-0.600004E-03	EYY	0.180001E-03	EXY	0.116415E-09
1	NODE	7	EXX	-0.600000E-03	EYY	0.180000E-03	EXY	0.279397E-08
1	NODE	3	EXX	0.600001E-03	EYY	-0.180000E-03	EXY	-0.814907E-09
2	NODE	3	EXX	0.600001E-03	EYY	-0.180001E-03	EXY	-0.116415E-09
2	NODE	7	EXX	-0.600001E-03	EYY	0.180000E-03	EXY	-0.931323E-09
2	NODE	9	EXX	0.599999E-03	EYY	-0.179999E-03	EXY	0.000000E 00

LOADING - 1

MEMBER FORCES

MEMBER	JOINT	/--------- FORCE ---------/			/--------- MOMENT ---------/		
		AXIAL	SHEAR Y	SHEAR Z	TORSIONAL	BENDING Y	BENDING Z

RESULTANT JOINT DISPLACEMENTS - SUPPORTS

JOINT		/---------DISPLACEMENT---------/		
		X DISP.	Y DISP.	Z DISP.
1	GLOBAL	0.0000000	0.0000000	
2	GLOBAL	0.0000000	0.0003600	
3	GLOBAL	0.0000000	0.0000000	

RESULTANT JOINT DISPLACEMENTS - FREE JOINTS

JOINT		/---------DISPLACEMENT---------/		
		X DISP.	Y DISP.	Z DISP.
4	GLOBAL	-0.0024000	-0.0012000	
5	GLOBAL	-0.0000000	-0.0008400	
6	GLOBAL	0.0024000	-0.0012000	
7	GLOBAL	-0.0048000	-0.0048000	
8	GLOBAL	-0.0000000	-0.0044400	
9	GLOBAL	0.0048000	-0.0048000	

APPENDIX 5

Derivatives of the Shape Functions

A5.1 THE SIX-NODE ISOPARAMETRIC TRIANGLE

The $[B]$ matrix entries defined in equation (8.10) require the evaluation of b_{ij}, which is a function of the derivatives of the shape functions N_i:

$$b_{ij} = \frac{\partial N_i}{\partial L_2}\frac{\partial N_j}{\partial L_1} - \frac{\partial N_i}{\partial L_1}\frac{\partial N_j}{\partial L_2} \qquad (8.6)$$

The shape functions, N_i, are defined in (8.3). The required derivatives are as follows:

i	N_i	$\dfrac{\partial N_i}{\partial L_1}$	$\dfrac{\partial N_i}{\partial L_2}$
1	$L_1(2L_1 - 1)$	$4L_1 - 1$	0
2	$L_2(2L_2 - 1)$	0	$4L_2 - 1$
3	$L_1(2L_1 - 3)$ $+ L_2(2L_2 - 3)$ $+ 4L_1L_2 + 1$	$4L_1 + 4L_2 - 3$	$4L_1 + 4L_2 - 3$

DERIVATIVES OF THE SHAPE FUNCTIONS

4	$4L_1L_2$	$4L_2$	$4L_1$
5	$4L_2(1-L_2) - 4L_2L_1$	$-4L_2$	$4 - 4L_1 - 8L_2$
6	$4L_1(1-L_1) - 4L_2L_1$	$4 - 8L_1 - 4L_2$	$-4L_1$

In application, the derivative expression can be expanded in (8.6), or the expressions can be programmed and the b_{ij} entries generated during computation.

A5.2 THE EIGHT-NODE ISOPARAMETRIC ELEMENT

The $[B]$ matrix entries defined in equation (8.21) require the evaluation of g_{ij}, which are functions of the derivatives of the shape functions N:

$$g_{ij} = \frac{\partial N_i}{\partial t}\frac{\partial N_j}{\partial s} - \frac{\partial N_i}{\partial s}\frac{\partial N_j}{\partial t} \qquad (8.20)$$

The shape functions, N_i, are defined in (8.14). The required derivatives are as follows:

i	N_i	$\dfrac{\partial N_i}{\partial s}$	$\dfrac{\partial N_i}{\partial t}$
1	$\dfrac{-(1-s)(1-t)(1+s+t)}{4}$	$\dfrac{(1-t)(2s+t)}{4}$	$\dfrac{(1-s)(2t+s)}{4}$
2	$\dfrac{-(1+s)(1-t)(1-s+t)}{4}$	$\dfrac{(1-t)(2s-t)}{4}$	$\dfrac{(1+s)(2t-s)}{4}$
3	$\dfrac{-(1+s)(1+t)(1-s-t)}{4}$	$\dfrac{(1+t)(2s+t)}{4}$	$\dfrac{(1+s)(2t+s)}{4}$
4	$\dfrac{-(1-s)(1+t)(1+s-t)}{4}$	$\dfrac{(1+t)(2s-t)}{4}$	$\dfrac{(1-s)(2t-s)}{4}$
5	$\dfrac{(1-s^2)(1-t)}{2}$	$-s(1-t)$	$\dfrac{-(1-s^2)}{2}$
6	$\dfrac{(1+s)(1-t^2)}{2}$	$\dfrac{1-t^2}{2}$	$-t(1+s)$
7	$\dfrac{(1-s^2)(1+t)}{2}$	$-s(1+t)$	$\dfrac{1-s^2}{2}$
8	$\dfrac{(1-s)(1-t^2)}{2}$	$\dfrac{-(1-t^2)}{2}$	$-t(1-s)$

APPENDIX 6

Nodal Body Forces for the Four-Node Quadrilateral

```
10   PRINT "*****************************************"
12   PRINT "**                                     **"
14   PRINT "**   4-NODE QUADRILATERAL BODY FORCE   **"
16   PRINT "**                                     **"
18   PRINT "*****************************************"
20   PRINT : PRINT : PRINT "       MULTIPLY RESULTS BY ELEMENT": PRINT
     "          THICKNESS AND DENSITY": PRINT : PRINT
30   DIM X(4),Y(4),A(4,4),B(4),C(4)
40   PRINT "INPUT THE X COORDINATES OF THE 4 NODES": PRINT
50   FOR I = 1 TO 4
60   INPUT X(I)
70   NEXT
80   PRINT : PRINT "INPUT THE Y COORDINATES OF THE 4 NODES": PRINT
90   FOR I = 1 TO 4
100  INPUT Y(I)
110  NEXT
120  FOR ND = 1 TO 4: REM    CALCULATION FOR NODE 'ND'
130  W = 0
140  GQ = 0.57735: REM    GAUSS POINTS COORDINATES
150  FOR N = 1 TO 4: REM    GAUSS POINT 'N'
160   IF N = 1 THEN S =  - GQ:T =  - GQ
170   IF N = 2 THEN S = GQ:T =  - GQ
180   IF N = 3 THEN S = GQ:T = GQ
190   IF N = 4 THEN S =  - GQ:T = GQ
200   IF ND = 1 THEN SH = (1 - S) * (1 - T) * 0.25
210   IF ND = 2 THEN SH = (1 + S) * (1 - T) * 0.25
220   IF ND = 3 THEN SH = (1 + S) * (1 + T) * 0.25
230   IF ND = 4 THEN SH = (1 - S) * (1 + T) * 0.25
240  REM    MATRIX LOWER CASE A(I,J)
250  A(1,1) = 0:A(2,2) = 0:A(3,3) = 0:A(4,4) = 0
260  A(1,2) = 1 - T:A(2,1) =  - A(1,2)
270  A(1,3) = T - S:A(3,1) =  - A(1,3)
```

475

```
280 A(1,4) = S - 1:A(4,1) =   - A(1,4)
290 A(2,3) = S + 1:A(3,2) =   - A(2,3)
300 A(2,4) =   - (T + S):A(4,2) =   - A(2,4)
310 A(3,4) = T + 1:A(4,3) =   - A(3,4)
320   FOR I = 1 TO 4
330 B(I) = 0: NEXT I
340   FOR I = 1 TO 4
350   FOR J = 1 TO 4
360 B(I) = (A(I,J) / 8) * Y(J) + B(I)
370  NEXT J: NEXT I
380 JA = 0: REM   JACOBIAN 'JA'
390   FOR I = 1 TO 4
400 JA = JA + X(I) * B(I)
410   NEXT
420 W = W + SH * JA
430   NEXT N
440   PRINT : PRINT "NODE ";ND
450   PRINT "BODY FORCE VECTOR = ";W
460   NEXT ND
470   END
```

APPENDIX 7

Computer Programs in Applesoft BASIC and TURBO Pascal for Two-Dimensional Heat Transfer

A7.1 THE ELEMENT

The element used in the program is the three-node triangle (Figure A7.1). The thermal conductivity and heat transfer coefficient can be specified for each element. The continuum must be of uniform thickness, but the thickness is not required in the program. The heat flow rate value from the program is per unit of cross-sectional area of the continuum, so thickness must be used to calculate total heat flow per unit time.

A7.2 PROGRAM CAPABILITY

The program models the steady-state heat flow rate in the XY plane only. It determines temperatures at element nodes and the heat flow rate per unit area in each element with temperature or convection specified on any portion on the boundary of the continuum. Radiation effects, internal heat sources and applied boundary flux are not included. Do not forget that the heat flow rate is a constant in the element, and in a manner similar to the constant-strain element, high gradients require many elements.

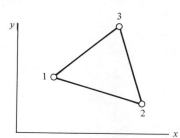

FIGURE A7.1 Programmed element.

A7.3 SAMPLE PROGRAM RUN

The square plate in Figure A7.3.1 has known temperature on the left side and convection on the remaining sides. Run the program using a (overly simplified) four element model and obtain an estimate of the temperature at the middle of the plate and the heat flow.

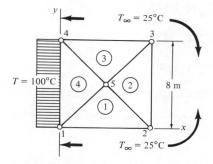

FIGURE A7.3 Four-element model of a square region.

```
RUN
*****************************************
**                                     **
**          HEAT TRANSFER              **
**             IN THE                  **
**           X-Y PLANE                 **
**                                     **
*****************************************
HIT ANY KEY TO CONTINUE

            DATA INPUT OPTIONS
            ---- ----- -------

   1 - INPUT NEW MODEL DATA FROM KEYBOARD
   2 - INPUT OLD MODEL DATA FROM DISC

          (ELEMENT NODE NUMBERS AND
            NODE COORDINATES)

ENTER CHOICE - (1 OR 2)    1
```

SAMPLE PROGRAM RUN

HOW MANY ELEMENTS?

?4

HOW MANY NODES?

?5

ENTER ELEMENT NUMBER, N, AND GLOBAL
NUMBER OF LOCAL NODES #1, #2, & #3

       CAUTION

USE COUNTER-CLOCKWISE SEQUENCE AND
THE FOLLOWING FORMAT:

N, #1, #2, #3   <RETURN>

?1,1,2,5
?2,2,3,5
?3,3,4,5
?4,4,1,5

ENTER GLOBAL NODE NUMBER, N, AND ITS
  X AND Y COORDINATES

USE FOLLOWING FORMAT:

N, X, Y   <RETURN>

?1,0,0
?2,8,0
?3,8,8
?4,0,8
?5,4,4

FREE DATA SPACE IS 19267 BYTES

FREE PROGRAM SPACE:

      YOU ARE 1985 BYTES INTO HGR2

** CALCULATED HALF- BANDWIDTH = 5 **

DO YOU WISH TO EDIT NODE NUMBERING ?
  (Y/N) N

DO YOU WISH TO EDIT NODE COORDINATES ?
  (Y/N) N

DO YOU WANT DATA STORED ON DISK ?
  (Y/N) N

HOW MANY BOUNDARY ELEMENT SIDES ARE
   EXPOSED TO CONVECTION?...3

HOW MANY ELEMENT NODES HAVE KNOWN
   TEMPERATURE?...2

IDENTIFY THE ELEMENT NUMBER
  AND ITS BOUNDARY NODES FOR EACH OF
    THE 3 BOUNDARY ELEMENTS

BOUNDARY ELEMENT # = 1
ONE BOUNDARY NODE # = 1

```
OTHER BOUNDARY NODE # = 2
AMBIENT TEMPERATURE, T = 25
CONVECTION COEFFICIENT, H = 10

BOUNDARY ELEMENT # = 2
ONE BOUNDARY NODE # = 2
OTHER BOUNDARY NODE # = 3
AMBIENT TEMPERATURE, T = 25
CONVECTION COEFFICIENT, H = 10

BOUNDARY ELEMENT # = 3
ONE BOUNDARY NODE # = 4
OTHER BOUNDARY NODE # = 3
AMBIENT TEMPERATURE, T = 25
CONVECTION COEFFICIENT, H = 10

INPUT THE TEMPERATURE AT EACH OF
    THE 2 NODES

NODE # = 1
TEMPERATURE = 100
NODE # = 4
TEMPERATURE = 100

ENTER MATERIAL CONDUCTION COEFFICIENT

IS MATERIAL HOMOGENEOUS ?   (Y/N)    Y

CONDUCTION COEFFICIENT = 2


WORKING ON:                             ─┐
    ELEMENT 1 CONDUCTION MATRIX          │   Progress
    ELEMENT 2 CONDUCTION MATRIX          │   prompts
    ELEMENT 3 CONDUCTION MATRIX          │
    ELEMENT 4 CONDUCTION MATRIX          │
        SIDE 1 CONVECTION MATRIX         │
        SIDE 2 CONVECTION MATRIX         │
        SIDE 3 CONVECTION MATRIX         │
MATRIX REDUCED                           │

FORCE VECTOR REDUCED ────────────────────┘

DO YOU WANT HARDCOPY?   (Y/N)  Y

ARE YOU CONNECTED TO A PRINTER? (Y/N)   Y


*****************************************
**                                     **
**     STEADY STATE HEAT TRANSFER      **
**            IN THE X-Y PLANE         **
**                                     **
*****************************************

 -------------- INPUT DATA --------------

=========================================
ELEMENT    GLOBAL NODE NUMBERS
NUMBER

    1         1     2    5
    2         2     3    5
    3         3     4    5
    4         4     1    5

=========================================
```

SAMPLE PROGRAM RUN

```
NODE         COORDINATES
NUMBER        X      Y

  1           0      0
  2           8      0
  3           8      8
  4           0      8
  5           4      4
```

==
CONDUCTION COEFFICIENTS

```
ELEMENT      CONDUCTION COEFFICIENT

   1               2
   ALL ELEMENTS HAVE SAME COEFFICIENT
```

==
BOUNDARY CONVECTION COEFFICIENTS

```
ELEMENT      CONVECTION     AMBIENT
             COEFFICIENT    TEMPERATURE

   1             10             25
   2             10             25
   3             10             25
```

==
```
NODE         TEMPERATURE

   1             100
   4             100
```
==

---------------- RESULTS ----------------

```
NODE NUMBER   TEMPERATURE

     1           100
     2           11.33
     3           11.23
     4           100
     5           55.66


ELEMENT NUMBER   ELEMENT TEMPERATURE
                  (NODE AVERAGE)

     1               55.66
     2               26.1
     3               55.66
     4               85.22


ELEMENT    HEAT FLOW IN      HEAT FLOW IN
           X DIRECTION       Y DIRECTION
           HEAT/T/AREA       HEAT/T/AREA

   1          22.16               0
   2          22.16               0
   3          22.16               0
   4          22.16               0
```

A7.4 PROGRAM LISTING OF HEAT-3NODE IN APPLESOFT BASIC

```
100  TEXT : HOME
110  PRINT "******************************************"
120  PRINT "**                                      **"
130  PRINT "**           HEAT TRANSFER              **"
140  PRINT "**              IN THE                  **"
150  PRINT "**            X-Y PLANE                 **"
160  PRINT "**                                      **"
170  PRINT "******************************************"
180  REM
190  REM    --NE-- NUMBER OF ELEMENTS
192  REM    --NN-- NUMBER OF NODES
194  REM    --NB-- NUMBER OF ELEMENT SIDES EXPOSED TO CONVECTION
196  REM -- GN(I,J)-- GLOBAL NODE NUMBER OF LOCAL NODE J OF ELEMENT I
198  REM    --XG(I), YG(I) -- COORDINATES OF GLOBAL NODE I
200  REM    --KR(I,J)-- GLOBAL CONDUCTION/CONVECTION "STIFFNESS" MATRIX
202  REM    --PR(I,J)-- ELEMENT CONDUCTION "STIFFNESS" MATRIX
204  REM    --CV(I,J)-- ELEMENT CONVECTION "STIFFNESS" MATRIX
206  REM      --FR(I)-- ELEMENT CONVECTION "FORCE" VECTOR
208  REM      --QX, QY-- HEAT FLOW IN X AND Y DIRECTION
210  REM      --Q(I)-- NODE (I) TEMPERATURE
212  REM      --QA(I)-- ELEMENT (I) TEMPERATURE (NODE AVERAGE)
214  REM      --TN(I)-- GLOBAL "FORCE" VECTOR
300  VTAB 20: PRINT "HIT ANY KEY TO CONTINUE"
302  GET G$: PRINT
304  REM  ------------------------------------------------------
306  REM      LINES 310 THRU 348 ARE MINI FRONT END PROGRAM
308  REM  ------------------------------------------------------
310  ONERR  GOTO 318
312  POKE 768,104: POKE 769,168: POKE 770,104: POKE 771,166: POKE
     772,223: POKE 773,154: POKE 774,72: POKE 775,152: POKE 776,72: POKE
     777,96
314 D$ = CHR$ (4): REM    CHR$(4) IS CTRL-D
316  GOTO 356: REM    GO TO NORMAL PROGRAM START
318  CALL 768: PRINT D$;"PR#0": TEXT : IF  PEEK (222) = 255 THEN  PRINT
     "CNTL C CAUSED A BREAK IN "; PEEK (218) +  PEEK (219) * 256: END
320  GOSUB 340: END
322  PRINT : PRINT : REM        DETERMINE FREE MEMORY
324 TP =  PEEK (175) +  PEEK (176) * 256: REM    TOP OF PROGRAM
326 EA =  PEEK (109) +  PEEK (110) * 256: REM    TOP OF LOMEM POINTER
328 BS =  PEEK (111) +  PEEK (112) * 256: REM    HIMEM POINTER
330  PRINT "FREE DATA SPACE IS ";BS - EA;" BYTES"
332  PRINT : PRINT "FREE PROGRAM SPACE:": PRINT
334 TP = 8192 - TP: IF TP > 0 THEN  PRINT TP;" BYTES TO HGR": RETURN
336 TP = TP + 8192: IF TP > 0 THEN  PRINT TP;" BYTES TO HGR2": RETURN
338  PRINT "      YOU ARE ";  ABS (TP);" BYTES INTO HGR2!": RETURN
340  PRINT "ERROR # ";  PEEK (222);" AT LINE # ";  PEEK (218) +  PEEK (219)
     * 256
342  PRINT : PRINT "     (ERROR CODES IN POOLE PG. 331)"
344  PRINT : PRINT "YOUR OPTIONS NOW ARE:": PRINT "    (1) HIT ESCAPE TO
     RESUME AT ERROR STATEMENT": PRINT "    (2) HIT ANY OTHER KEY TO END
     PROGRAM"
346  GET G$: PRINT : IF G$ =  CHR$ (27) THEN  RESUME
348  RETURN
350  REM  ------------------------------------------------------
352  REM       START PROGRAM HERE
354  REM  ------------------------------------------------------
356  HOME
380 I$ =  CHR$ (9): REM    CHR$(9) IS CTRL-I
385 R$ =  CHR$ (13): REM    CHR$(13) IS CARRIAGE RETURN
390 RU$ = "N"
395  VTAB 10
400  PRINT "        DATA INPUT OPTIONS"
405  PRINT "        ---- ----- -------"
410  PRINT
```

PROGRAM LISTING OF HEAT-3NODE IN APPLESOFT BASIC

```
415  PRINT "  1 - INPUT NEW MODEL DATA FROM KEYBOARD"
420  PRINT "  2 - INPUT OLD MODEL DATA FROM DISC": PRINT
425  PRINT "           (ELEMENT NODE NUMBERS AND"
430  PRINT "             NODE COORDINATES)"
435  PRINT
440  INPUT "ENTER CHOICE - (1 OR 2)   ";IN$
460  IF IN$ = "1" OR IN$ = "2" GOTO 480
470  PRINT : PRINT "BAD RESPONSE, RE-ENTER": GOTO 400
480  IF IN$ = "1" GOTO 570
490  REM   OPEN AND READ THE DATA FILE
500  RT$ = ".GEOM"
510  REM  ------------------------------------------------------
520  GOSUB 10000: REM    OPEN AND READ GEOMETRY FILE      ----
530  REM  ------------------------------------------------------
540  REM                DATA INPUT FROM KEYBOARD OR DISC      ----
550  REM  ------------------------------------------------------
560  REM
570  HOME : PRINT : PRINT "HOW MANY ELEMENTS?": PRINT
580  INPUT NE: PRINT
590  PRINT "HOW MANY NODES?": PRINT
600  INPUT NN: PRINT
610  DIM
     GN(NE,3),XG(NN),YG(NN),X(3),Y(3),B(2,3),PR(3,3),Q(NN),TN(NN),CV(3,3),
     FR(3),IT(NN),QA(NE)
620  PRINT "ENTER ELEMENT NUMBER, N, AND GLOBAL"
630  PRINT "NUMBER OF LOCAL NODES #1, #2, & #3"
632  PRINT : INVERSE : PRINT "          CAUTION         ": NORMAL :
     PRINT
634  PRINT "USE COUNTER-CLOCKWISE SEQUENCE AND"
635  PRINT "THE FOLLOWING FORMAT:"
640  PRINT : PRINT " N,  #1,  #2,  #3   <RETURN>": PRINT
650  FOR J = 1 TO NE
660  INPUT I,GN(I,1),GN(I,2),GN(I,3)
670  NEXT J
680  PRINT : PRINT "ENTER GLOBAL NODE NUMBER, N, AND ITS"
690  PRINT "  X AND Y COORDINATES"
695  PRINT : PRINT "USE FOLLOWING FORMAT:": PRINT
700  PRINT " N,   X,   Y    <RETURN>": PRINT
710  FOR J = 1 TO NN
720  INPUT I,XG(I),YG(I)
730  NEXT J
740  IF IN$ = "1" GOTO 790
750  REM  ------------------------------------------------------
760  GOSUB 10110: REM    CLOSE THE GEOMETRY FILE         ----
770  REM  ------------------------------------------------------
780  REM    CALCULATE THE BANDWIDTH, BW
790  FOR I = 1 TO NE
800  FOR J = 1 TO 2
810  FOR JJ = 1 TO 3
820  DI = ABS (GN(I,J) - GN(I,JJ))
830  IF DI > BW THEN BW = DI
840  NEXT JJ: NEXT J: NEXT I
850  BW = BW + 1
870  REM  ------------------------------------------------------
880  REM       EDITING OPTIONS TO REVIEW AND/OR CHANGE    ----
890  REM       ELEMENT NODE NUMBERING AND COORDINATES     ----
900  REM  ------------------------------------------------------
904  HOME : PRINT "---------------------------------"
905  GOSUB 322: REM    TEST FOR FREE MEMORY
906  PRINT "---------------------------------": PRINT
908  PRINT : PRINT "** CALCULATED HALF- BANDWIDTH = ";BW;" **": PRINT
910  PRINT : PRINT "DO YOU WISH TO EDIT NODE NUMBERING ?": INPUT "   (Y/N)
     ";RE$
920  IF RE$ = "Y" OR RE$ = "N" THEN 950
930  REM
940  PRINT : PRINT "RE-ENTER -": GOTO 910
950  IF RE$ = "N" THEN 1030
960  PRINT : INPUT "WHICH ELEMENT ? ";N
970  IF N > NE THEN 960
980  PRINT "CURRENT VALUES: ";GN(N,1);",   ";GN(N,2);",   ";GN(N,3)
```

```
 990   PRINT : INPUT "ENTER THE NEW VALUES:";GN(N,1),GN(N,2),GN(N,3)
1000   PRINT : INPUT "MORE CHANGES ?  (Y/N)";RE$
1010   IF RE$ = "Y" THEN 960
1020   BW = 0: GOTO 780
1030   DIM KR(NN,BW)
1040   PRINT : PRINT "DO YOU WISH TO EDIT NODE COORDINATES ?": INPUT "
       (Y/N) ";RE$
1050   IF RE$ = "Y" OR RE$ = "N" THEN 1070
1060   PRINT : PRINT "RE-ENTER -": GOTO 1040
1070   IF RE$ = "N" THEN 1140
1080   PRINT : INPUT "WHICH NODE ? ";N
1090   IF N > NN THEN 1080
1100   PRINT "CURRENT VALUES: ";XG(N);",    ";YG(N)
1110   PRINT : INPUT "ENTER THE NEW VALUES- ";XG(N),YG(N)
1120   PRINT : INPUT "MORE CHANGES ?  (Y/N) ";RE$
1130   IF RE$ = "Y" THEN 1080
1140   PRINT : PRINT "DO YOU WANT DATA STORED ON DISK ?": INPUT "  (Y/N)
       ";DK$
1150   IF DK$ < > "Y" THEN 1290
1160   RT$ = ".GEOM"
1170   REM ---------------------------------------------------
1180   GOSUB 10160: REM     OPEN AND WRITE TO GEOMETRY FILE  ----
1190   REM ---------------------------------------------------
1200   PRINT NE;R$;NN
1210   FOR I = 1 TO NE
1220   PRINT I;R$;GN(I,1);R$;GN(I,2);R$;GN(I,3)
1230   NEXT
1240   FOR I = 1 TO NN
1250   PRINT I;R$;XG(I);R$;YG(I)
1260   NEXT
1270   REM ---------------------------------------------------
1280   GOSUB 10110: REM    CLOSE THE GEOMETRY FILE           ----
1290   REM ---------------------------------------------------
1300   IF RU$ = "Y" GOTO 1340: REM    RU$ IS RERUN CONTROL VARIABLE
1310   PRINT : PRINT "HOW MANY BOUNDARY ELEMENT SIDES ARE": INPUT "
       EXPOSED TO CONVECTION?...";NB
1320   PRINT : PRINT "HOW MANY ELEMENT NODES HAVE KNOWN": INPUT "
       TEMPERATURE?...";KT
1330   DIM BN(NB),N1(NB),N2(NB),KE(NE),TA(NB),HE(NB),BC(NN)
1340   FOR I = 1 TO NN
1350   BC(I) = 0
1360   NEXT I
1370   REM ---------------------------------------------------
1380   REM     IDENTIFY ELEMENT SIDES ON BOUNDARY, AMBIENT  ----
1390   REM     TEMPERATURE AND CONVECTION COEFFICIENTS       ----
1400   REM ---------------------------------------------------
1410   IF NB = 0 THEN  GOTO 1550
1420   PRINT : PRINT "IDENTIFY THE ELEMENT NUMBER": PRINT "  AND ITS
       BOUNDARY NODES FOR EACH OF": PRINT "    THE ";NB;" BOUNDARY
       ELEMENTS": PRINT
1430   FOR I = 1 TO NB
1440   INPUT "BOUNDARY ELEMENT # = ";BN(I)
1450   INPUT "ONE BOUNDARY NODE # = ";N1(I)
1460   INPUT "OTHER BOUNDARY NODE # = ";N2(I)
1470   INPUT "AMBIENT TEMPERATURE, T = ";TA(I)
1480   INPUT "CONVECTION COEFFICIENT, H = ";HE(I)
1490   IF I = NB GOTO 1510
1500   PRINT : PRINT "NEXT ELEMENT SIDE:": PRINT
1510   NEXT I
1520   REM ---------------------------------------------------
1530   REM    INPUT KNOWN NODE TEMPERATURES                  ----
1540   REM ---------------------------------------------------
1550   IF KT = 0 GOTO 1650
1560   PRINT : PRINT "INPUT THE TEMPERATURE AT EACH OF": PRINT "     THE
       ";KT;" NODES": PRINT
1570   FOR I = 1 TO KT
1580   INPUT "NODE # = ";J
1590   INPUT "TEMPERATURE = ";Q(J)
1600   BC(J) = 1
1610   NEXT I
```

PROGRAM LISTING OF HEAT-3NODE IN APPLESOFT BASIC

```
1620 REM --------------------------------------------------------
1630 REM     INPUT THE MATERIAL CONDUCTION COEFFICIENTS      ----
1640 REM --------------------------------------------------------
1650 PRINT : PRINT "ENTER MATERIAL CONDUCTION COEFFICIENT"
1660 PRINT : PRINT "IS MATERIAL HOMOGENEOUS ?   (Y/N) ";: INPUT MC$:
     PRINT
1670 IF MC$ = "Y" GOTO 1720
1680 FOR I = 1 TO NE
1690 PRINT "ELEMENT ";I;: PRINT ", K = ";: INPUT KE(I)
1700 NEXT I
1710 GOTO 1760
1720 PRINT : INPUT "CONDUCTION COEFFICIENT = ";KE
1730 FOR I = 1 TO NE
1740 KE(I) = KE
1750 NEXT I
1760 FOR I = 1 TO NN
1770 FOR J = 1 TO BW
1780 KR(I,J) = 0
1790 NEXT J
1800 NEXT I
1810 REM --------------------------------------------------------
1820 REM                 BEGIN CALCULATIONS                  ----
1830 REM --------------------------------------------------------
1840 PRINT : PRINT : FLASH : PRINT "WORKING ON:": NORMAL
1850 FOR M = 1 TO NE
1860 PRINT "   ELEMENT ";M;" CONDUCTION MATRIX": REM  PROGRESS PROMPT

1870 GOSUB 5000: REM   CALCULATE CONDUCTION MATRIX
1880 REM --------------------------------------------------------
1890 REM       ASSEMBLE GLOBAL CONDUCTION MATRIX             ----
1900 REM             IN BAND MATRIX FORMAT                   ----
1910 REM --------------------------------------------------------
1920 FOR I = 1 TO 3
1930 FOR J = 1 TO 3
1940 IF GN(M,J) < GN(M,I) GOTO 1970
1950 C = GN(M,J) - GN(M,I) + 1
1960 KR(GN(M,I),C) = PR(I,J) + KR(GN(M,I),C)
1970 NEXT J
1980 NEXT I
1990 NEXT M
2000 REM    CALCULATE THE ELEMENT CONVECTION MATRICES
2010 IF NB = 0 THEN  GOTO 2570
2020 FOR I = 1 TO 3
2030 FOR J = 1 TO 3
2040 CV(I,J) = 0
2050 NEXT J
2060 FR(I) = 0
2070 NEXT I
2080 FOR I = 1 TO NN
2090 TN(I) = 0:IT(I) = 0
2100 NEXT I
2110 FOR W = 1 TO NB: REM    NB IS NUMBER OF EXPOSED ELEMENT SIDES
2120 PRINT "     SIDE ";W;" CONVECTION MATRIX": REM  PROGRESS PROMPT
2130 I = W
2140 IF W = 1 THEN  GOTO 2180
2150 REM   BN(I) IS ELEMENT NUMBER OF SIDE I
2160 REM   TEST IF THE NEXT SIDE BELONGS TO THE SAME ELEMENT
2170 IF BN(W) = BN(W - 1) THEN I = W - 1
2180 S =  SQR ((XG(N1(W)) - XG(N2(W))) ^ 2 + (YG(N1(W)) - YG(N2(W))) ^ 2)
2190 CC = (HE(W)) / 6 * S:CF = HE(W) * TA(W) * S
2200 REM    TEST FOR ELEMENT LOCAL SIDE NUMBER
2210 IF GN(BN(I),1) < > N1(W) AND GN(BN(I),1) < > N2(W) THEN  GOTO
     2280
2220 IF GN(BN(I),2) < > N1(W) AND GN(BN(I),2) < > N2(W) THEN  GOTO
     2330
2230 CV(1,1) = 2 * CC:FR(1) = CF / 2
2240 CV(1,2) = CC:FR(2) = CF / 2
2250 CV(2,1) = CC
2260 CV(2,2) = 2 * CC
```

```
2270   GOTO 2420
2280 CV(2,2) = 2 * CC:FR(2) = CF / 2
2290 CV(2,3) = CC:FR(3) = CF / 2
2300 CV(3,2) = CC
2310 CV(3,3) = 2 * CC
2320   GOTO 2420
2330 CV(1,1) = 2 * CC:FR(1) = CF / 2
2340 CV(1,3) = CC:FR(3) = CF / 2
2350 CV(3,1) = CC
2360 CV(3,3) = 2 * CC
2370   REM -----------------------------------------------------
2380   REM        ADD THE CONVECTION MATRIX TO THE GLOBAL   ----
2390   REM                 CONDUCTION MATRIX               ----
2400   REM        AND CONSTRUCT CONVECTION FORCE VECTOR    ----
2410   REM -----------------------------------------------------
2420   FOR L = 1 TO 3
2430   FOR M = 1 TO 3
2440   IF GN(BN(I),M) < GN(BN(I),L) GOTO 2470
2450 C = GN(BN(I),M) - GN(BN(I),L) + 1
2460 KR(GN(BN(I),L),C) = CV(L,M) + KR(GN(BN(I),L),C)
2470   NEXT M
2480 TN(GN(BN(I),L)) = FR(L) + TN(GN(BN(I),L))
2490   NEXT L
2500   FOR II = 1 TO 3
2510   FOR JJ = 1 TO 3
2520 CV(II,JJ) = 0
2530   NEXT JJ
2540 FR(II) = 0
2550   NEXT II
2560   NEXT W
2570   IF KT = 0 GOTO 2720
2580   REM -----------------------------------------------------
2590   REM      APPLY KNOWN NODE TEMPERATURES AND MODIFY    ----
2600   REM       THE STIFFNESS MATRIX 'DIAGONAL' ENTRIES    ----
2610   REM -----------------------------------------------------
2620   FOR J = 1 TO NN
2630   IF BC(J) = 0 GOTO 2670
2640 KR(J,1) = KR(J,1) * 1E10
2650 TN(J) = Q(J) * KR(J,1)
2660 IT(J) = TN(J)
2670   NEXT J
2680   REM -----------------------------------------------------
2690   REM        SOLVE FOR NODE TEMPERATURES USING GAUSS   ----
2700   REM        ELIMINATION ROUTINE FOR BANDED MATRIX     ----
2710   REM -----------------------------------------------------
2720 N = NN
2730   FOR M = 1 TO N
2740   IF KR(M,1) = 0 GOTO 2870
2750   FOR L = 2 TO BW
2760   IF KR(M,L) = 0 GOTO 2860
2770 I = M + L - 1
2780 QU = KR(M,L) / KR(M,1)
2790 J = 0
2800   FOR K = L TO BW
2810 J = J + 1
2820   IF I > N GOTO 2840
2830 KR(I,J) = KR(I,J) - QU * KR(M,K)
2840   NEXT K
2850 KR(M,L) = QU
2860   NEXT L
2870   NEXT M
2880   PRINT "MATRIX REDUCED": PRINT
2890   FOR M = 1 TO N
2900   IF KR(M,1) = 0 GOTO 2980
2910   FOR L = 2 TO BW
2920   IF KR(M,L) = 0 GOTO 2960
2930 I = M + L - 1
2940   IF I > N GOTO 2960
```

PROGRAM LISTING OF HEAT-3NODE IN APPLESOFT BASIC

```
2950  TN(I) = TN(I) - KR(M,L) * TN(M)
2960    NEXT L
2970  TN(M) = TN(M) / KR(M,1)
2980    NEXT M
2990    PRINT : PRINT "FORCE VECTOR REDUCED": PRINT
3000  Q(N) = TN(N)
3010    FOR KK = 2 TO N
3020  M = N + 1 - KK
3030    FOR L = 2 TO BW
3040    IF KR(M,L) = 0 GOTO 3080
3050  K = M + L - 1
3060    IF K > N GOTO 3100
3070  TN(M) = TN(M) - KR(M,L) * Q(K)
3080    NEXT L
3090  Q(M) = TN(M)
3100    NEXT KK
3110    REM ------------------------------------------------------
3120    REM     CALCULATE ELEMENT TEMPERATURE AS AVERAGE OF    ----
3130    REM         NODE TEMPS.  USE QA(I) FOR ELEMENT TEMP    ----
3140    REM ------------------------------------------------------
3150    FOR I = 1 TO NE
3160  QA(I) = 0
3170    FOR J = 1 TO 3
3180  QA(I) = Q(GN(I,J)) / 3 + QA(I)
3190    NEXT J
3200    NEXT I
3210    REM ------------------------------------------------------
3220    REM        OPTIONAL HARDCOPY OF DATA AND RESULTS       ----
3230    REM ------------------------------------------------------
3240    PRINT : PRINT : INPUT "DO YOU WANT HARDCOPY? (Y/N) ";H$
3250    IF H$ < > "Y" GOTO 3740
3260    PRINT : PRINT "ARE YOU CONNECTED TO A PRINTER? (Y/N) ": PRINT
3270    INPUT PC$: IF PC$ < > "Y" THEN H$ = "N"
3280    IF H$ < > "Y" GOTO 3740
3290    PRINT D$"PR#1": REM    ACTIVATE PRINTER
3300    PRINT I$"ML 10": REM    LEFT MARGIN SETTING
3310    PRINT "****************************************"
3320    PRINT "**                                    **"
3330    PRINT "**      STEADY STATE HEAT TRANSFER    **"
3340    PRINT "**            IN THE X-Y PLANE        **"
3350    PRINT "**                                    **"
3360    PRINT "****************************************"
3370    PRINT
3380    PRINT "-------------- INPUT DATA -------------"
3390    PRINT
3400    PRINT "======================================="
3410    PRINT "ELEMENT    GLOBAL NODE NUMBERS": PRINT "NUMBER": PRINT
3420    FOR I = 1 TO NE
3430    POKE 36,4: PRINT I;: POKE 36,12: PRINT GN(I,1);: POKE 36,17: PRINT
        GN(I,2);: POKE 36,22: PRINT GN(I,3)
3440    NEXT I
3450    PRINT : PRINT "======================================="
3460    PRINT "NODE";: POKE 36,12: PRINT "COORDINATES": PRINT "NUMBER";:
        POKE 36,14: PRINT "X";: POKE 36,20: PRINT "Y": PRINT
3470    FOR I = 1 TO NN
3480    POKE 36,3: PRINT I;: POKE 36,13: PRINT XG(I);: POKE 36,19: PRINT
        YG(I)
3490    NEXT I
3500    PRINT : PRINT "======================================="
3510    PRINT "CONDUCTION COEFFICIENTS": PRINT
3520    PRINT "ELEMENT";: POKE 36,12: PRINT "CONDUCTION COEFFICIENT": PRINT
3530    FOR I = 1 TO NE
3540    POKE 36,4: PRINT I;: POKE 36,17: PRINT KE(I)
3550    IF MC$ < > "Y" GOTO 3580
3560    PRINT "    ALL ELEMENTS HAVE SAME COEFFICIENT": PRINT
3570  I = NE
3580    NEXT I
```

```
3590   PRINT : PRINT "======================================="
3600   PRINT "BOUNDARY CONVECTION COEFFICIENTS": PRINT
3610   PRINT "ELEMENT";: POKE 36,12: PRINT "CONVECTION";: POKE 36,25:
       PRINT "AMBIENT": POKE 36,12: PRINT "COEFFICIENT";: POKE 36,25: PRINT
       "TEMPERATURE": PRINT
3620   IF NB = 0 GOTO 3660
3630   FOR I = 1 TO NB
3640   POKE 36,4: PRINT BN(I);: POKE 36,14: PRINT HE(I);: POKE 36,28:
       PRINT TA(I)
3650   NEXT I
3660   PRINT : PRINT "======================================="
3670   PRINT "NODE            TEMPERATURE": PRINT
3680   FOR I = 1 TO NN
3690   IF BC(I) = 0 THEN  GOTO 3720
3700   IT(I) = IT(I) / KR(I,1)
3710   POKE 36,4: PRINT I;: POKE 36,15: PRINT IT(I)
3720   NEXT I
3730   PRINT "======================================="
3740   PRINT : PRINT : PRINT : PRINT
3750   PRINT
3760   PRINT "--------------- RESULTS --------------"
3770   PRINT
3780   PRINT "NODE NUMBER";"  TEMPERATURE": PRINT
3790   FOR I = 1 TO NN
3800   POKE 36,6: PRINT I;: POKE 36,18: PRINT  INT (100 * Q(I)) / 100
3810   NEXT I
3820   PRINT : PRINT : PRINT "ELEMENT NUMBER";"   ELEMENT TEMPERATURE"
3821   PRINT "              (NODE AVERAGE)": PRINT
3830   FOR I = 1 TO NE
3840   POKE 36,6: PRINT I;: POKE 36,25: PRINT  INT (100 * QA(I)) / 100
3850   NEXT I
3860   PRINT : PRINT : PRINT : PRINT "ELEMENT";: POKE 36,10: PRINT "HEAT
       FLOW IN";: POKE 36,25: PRINT "HEAT FLOW IN"
3870   POKE 36,10: PRINT "X DIRECTION";: POKE 36,25: PRINT "Y DIRECTION"
3880   POKE 36,11: PRINT "HEAT/T/AREA";: POKE 36,26: PRINT "HEAT/T/AREA":
       PRINT
3890   PRINT : PRINT
3900   FOR M = 1 TO NE
3910   REM  ----------------------------------------------------
3920   REM         CALCULATE HEAT FLOW/TIME/AREA              ----
3930   REM  ----------------------------------------------------
3940   GOSUB 6000
3950   POKE 36,4: PRINT M;: POKE 36,12: PRINT  INT (100 * QX) / 100;: POKE
       36,28: PRINT  INT (100 * QY) / 100
3960   NEXT M
3970   IF H$ = "Y" THEN  PRINT D$"PR#0"
3980   REM  ----------------------------------------------------
3990   REM      OPTIONAL DISC STORAGE OF ELEMENT TEMPERATURE  ----
4000   REM  ----------------------------------------------------
4010   PRINT : PRINT "DO YOU WANT ELEMENT TEMPERATURES STORED": INPUT "
       ON DISK?  (Y/N)  ";QK$
4020   IF QK$ < > "Y" GOTO 4090


4030   RT$ = ".TEMP"
4040   GOSUB 10160: REM    OPEN AND WRITE FILE
4050   FOR I = 1 TO NE
4060   PRINT QA(I)
4070   NEXT I
4080   GOSUB 10110: REM   CLOSE FILE
4090   PRINT : PRINT : INPUT "DO YOU WISH ANOTHER RUN?   (Y/N) ";RU$
4100   IF RU$ = "Y" GOTO 1040
4110   IF RU$ < > "N" GOTO 4090
4120   END
4130   REM  ----------------------------------------------------
4140   REM                     SUBROUTINES                    ----
4150   REM  ----------------------------------------------------
5000   REM   CALCULATE ELEMENT CONDUCTION STIFFNESS MATRIX
5010   X(1) = XG(GN(M,1))
5020   X(2) = XG(GN(M,2))
```

PROGRAM LISTING OF HEAT-3NODE IN APPLESOFT BASIC

```
5030 X(3) = XG(GN(M,3))
5040 Y(1) = YG(GN(M,1))
5050 Y(2) = YG(GN(M,2))
5060 Y(3) = YG(GN(M,3))
5070 DET = (X(1) - X(3)) * (Y(2) - Y(3)) - (X(2) - X(3)) * (Y(1) - Y(3))
5080 B(1,1) = Y(2) - Y(3)
5090 B(1,2) = Y(3) - Y(1)
5100 B(1,3) = Y(1) - Y(2)
5110 B(2,1) = X(3) - X(2)
5120 B(2,2) = X(1) - X(3)
5130 B(2,3) = X(2) - X(1)
5140  FOR I = 1 TO 3
5150   FOR J = 1 TO 3
5160 PR(I,J) = 0
5170   NEXT J
5180  NEXT I
5190  FOR I = 1 TO 3
5200   FOR J = 1 TO 3
5210    FOR N = 1 TO 2
5220 PR(I,J) = PR(I,J) + B(N,I) * B(N,J)
5230    NEXT N
5240   NEXT J
5250  NEXT I
5260  FOR I = 1 TO 3
5270   FOR J = 1 TO 3
5280 PR(I,J) = KE(M) * PR(I,J) / (2 * DET)
5290   NEXT J
5300  NEXT I
5310  RETURN
5999  REM ---------------------------------------------------
6000  REM   CALCULATE HEAT FLOW PER UNIT AREA
6010 X(1) = XG(GN(M,1))
6020 X(2) = XG(GN(M,2))
6030 X(3) = XG(GN(M,3))
6040 Y(1) = YG(GN(M,1))
6050 Y(2) = YG(GN(M,2))
6060 Y(3) = YG(GN(M,3))
6070 BB(1) = Y(2) - Y(3)
6080 BB(2) = Y(3) - Y(1)
6090 BB(3) = Y(1) - Y(2)
6100 CC(1) = X(3) - X(2)
6110 CC(2) = X(1) - X(3)
6120 CC(3) = X(2) - X(1)
6130 DET = (X(1) - X(3)) * (Y(2) - Y(3)) - (X(2) - X(3)) * (Y(1) - Y(3))
6140 QX = 0
6150 QY = 0
6160  FOR I = 1 TO 3
6170 QX =  - KE(M) / DET * BB(I) * Q(GN(M,I)) + QX
6180 QY =  - KE(M) / DET * CC(I) * Q(GN(M,I)) + QY
6190  NEXT I
6200  RETURN
6300  REM ---------------------------------------------------
10000  REM   FILE READ
10010  PRINT : INPUT "WHAT IS THE NAME OF THE FILE ?";NA$
10015 F$ = NA$ + RT$
10020  PRINT : INPUT "WHICH DISK DRIVE ?";DN
10030  IF DN = 1 OR DN = 2 THEN 10050
10040  PRINT : PRINT " TRY AGAIN-": GOTO 10020
10050  PRINT D$;"MONCIO"
10060  PRINT D$;"OPEN";F$;",D";DN
10070  PRINT D$;"UNLOCK";F$
10080  PRINT D$;"OPEN";F$
10090  PRINT D$;"READ";F$
10100  RETURN
10102  REM ---------------------------------------------------
10110  REM   FILE CLOSE
10120  PRINT D$;"LOCK";F$
10130  PRINT D$;"CLOSE";F$
10140  PRINT D$;"NOMONCIO"
10150  RETURN
```

```
10152 REM ------------------------------------------------
10160 REM   FILE WRITE ROUTINE
10170 PRINT : INPUT "NAME OF FILE ?";NA$
10175 F$ = NA$ + RT$
10180 PRINT : INPUT "WHICH DISK DRIVE ?";DN
10190 IF DN = 1 OR DN = 2 THEN 10210
10200 PRINT : PRINT " TRY AGAIN-": GOTO 10180
10210 PRINT D$;"MONCIO"
10220 PRINT D$;"OPEN";F$;",D";DN
10230 PRINT D$;"UNLOCK";F$
10240 PRINT D$;"DELETE";F$
10250 PRINT D$;"OPEN";F$
10260 PRINT D$;"WRITE";F$
10270 RETURN
10272 REM ------------------------------------------------
```

A7.5 PROGRAM LISTING IN TURBO PASCAL

```
{*****************************************************************}
  PROGRAM HEAT3NODE(INPUT,OUTPUT);
{*****************************************************************}

  CONST
    ArrSize1(Number of Elements)         =100;
    Arrsize2(Number of Nodes)            =100;
    ArrSize3(Half-bandwidth )            = 50;
    ArrSize4(Number of Boundary Sides)   = 40;
  TYPE

    RARY1  = ARRAY[1..3] OF REAL;
    RARY2  = ARRAY[1..ArrSize1] OF REAL;
    RARY3  = ARRAY[1..ArrSize2] OF REAL;
    RARY4  = ARRAY[1..3,1..3] OF REAL;
    RARY5  = ARRAY[1..ArrSize2,1..ArrSize3] OF REAL;
    IARY6  = ARRAY[1..ArrSize1,1..3] OF INTEGER;
    RARY7  = ARRAY[1..2,1..3] OF REAL;
    RARAY8 = ARRAY[1..ArrSize1] OF INTEGER;
    IARY9  = ARRAY[1..ArrSize2] OF INTEGER;
    IARY10 = ARRAY[1..ArrSize4] OF INTEGER;
    RARY11 = ARRAY[1..ArrSize4] OF REAL;

VAR
  FR,X,Y                        :RARY1;
  QA,KE                         :RARY2;
  Q,XG,YG,IT,TN                 :RARY3;
  BC                            :IARY9;
  BN,N1,N2                      :IARY10;
  TA,HE                         :RARY11;
  PR,CV                         :RARY4;
  KR                            :RARY5;
  GN                            :IARY6;
  B                             :RARY7;
  NE,NN,NB,MEM,I,J,K,W          :INTEGER;
  QX,QY,S,CC,KE1,CF,QU          :REAL;
  MINX,MINY,SF                  :REAL;
  INP,BW,KT,M,C                 :INTEGER;
  RESP,DISC,HOMO                :CHAR;
  L,II,JJ,SHIFTX,SHIFTY         :INTEGER;
  OUTT                          :BOOLEAN;
  HOLDFIL                       :TEXT;
  FNAME,NNAME                   :STRING[20];

(WHERE: )
{NE     -- NUMBER OF ELEMENTS}
{NN     -- NUMBER OF NODES}
{NB     -- NUMBER OF ELEMENT SIDES EXPOSED TO CONVECTION}
{GN[I,J]-- GLOBAL NODE NUMBERS OF LOCAL NODE J OF ELEMENT I}
```

PROGRAM LISTING IN TURBO PASCAL

```
{XG,YG    -- COORDINATES OF GLOBAL NODES}
{KR[I,J]-- GLOBAL CONDUCTION/CONVECTION 'STIFFNESS' MATRIX}
{PR[I,J]-- ELEMENT CONDUCTION 'STIFFNESS' MATRIX}
{CV[I,J]-- ELEMENT CONVECTION 'STIFFNESS' MATRIX}
{FR[I]   -- ELEMENT CONVECTION 'FORCE' VECTOR }
{QX,QY   -- HEAT FLOW IN X AND Y DIRECTION}
{Q[I]    -- NODE [I] TEMPERATURE}
{QA[I]   -- ELEMENT I TEMPERATURE (NODE AVERAGE)}
{TN[I]   -- GLOBAL 'FORCE' VECTOR}
{***************************************************************}
{Set All Variables to 0}
 PROCEDURE INITIALIZE_VARIABLES;

     VAR
      LOOPI,LOOPJ            :INTEGER;

     BEGIN

   { Here are the integer variables }

          NN:=0;     NE:=0;     NB:=0;     MEM:=0;     J:=0;      I:=0;
          K:=0;      INP:=0;    BW:=0;     KT:=0;      KE1:=0;    M:=0;
          C:=0;      W:=0;      L:=0;      II:=0;      JJ:=0;

   { Here are the real variables   }

          QX:=0;     QY:=0;     S:=0;      CC:=0;      CF:=0;     QU:=0;

  {Here are the arrays}

          FOR I:=1 TO 3 DO
             BEGIN
                FR[I]:=0;
                X[I]:=0;
                Y[I]:=0
             END;
          FOR I:=1 TO ARRSIZE1 DO
             BEGIN
                QA[I]:=0;
                KE[I]:=0;
             END;
          FOR I:=1 TO ARRSIZE2 DO
             BEGIN
                TN[I]:=0;   IT[I]:=0;    Q[I]:=0;    BC[I]:=0;
                XG[I]:=0;   YG[I]:=0;
             END;
          FOR I:=1 TO ARRSIZE4 DO
             BEGIN
                BN[I]:=0;   N1[I]:=0;    N2[I]:=0;
                TA[I]:=0;   HE[I]:=0;
             END;
          FOR I:=1 TO 3 DO
             FOR J:=1 TO 3 DO
                BEGIN
                   PR[I,J]:=0;
                   CV[I,J]:=0
                END;
          FOR I:=1 TO ARRSIZE2 DO
             FOR J:=1 TO ARRSIZE3 DO
                KR[I,J]:=0;
          FOR I:=1 TO ARRSIZE1 DO
             FOR J:=1 TO 3 DO
                GN[I,J]:=0;
          FOR I:=1 TO 2 DO
             FOR J:=1 TO 3 DO
                B[I,J]:=0
     END;
{***************************************************************}
{Subroutine to input geometry data from keyboard}
 PROCEDURE TERMINAL_INPUT;
```

```
   BEGIN
      WRITELN;
      WRITE('HOW MANY ELEMENTS? ');
      READLN(NE);
      WRITELN;
      WRITE('HOW MANY NODES? ');
      READLN(NN);
      WRITELN;
      WRITELN('ENTER ELEMENT NUMBER AND GLOBAL NUMBERS OF');
      WRITELN('     LOCAL NODES #1 #2 & #3');
      WRITELN;
      WRITELN('CAUTION: USE COUNTER-CLOCKWISE SEQUENCE AND ');
      WRITELN('            THE FOLLOWING FORMAT: ');
      WRITELN('     EL#    #1    #2    #3    <RETURN>');
      WRITELN;
      FOR J:=1 TO NE DO
         READLN(I,GN[I,1],GN[I,2],GN[I,3]);
      WRITELN;
      WRITELN('ENTER GLOBAL NODE NUMBER, N, AND ITS X AND Y COORDINATES');
      WRITELN;
      WRITELN('USE THE FOLLOWING FORMAT: ');
      WRITELN('     N     X     Y      <RETURN>');
      WRITELN;
      FOR J:=1 TO NN DO
         READLN(I,XG[I],YG[I])
   END;
{***************************************************************}
{Read Structure Geometry From Disk }
PROCEDURE DISC_INPUT;

   TYPE
      CHRSTRING = STRING[20];

   VAR
      FNAME                           :CHRSTRING;
      INPFILE                         :TEXT;

   BEGIN
      WRITE('INPUT THE NAME OF YOUR GEOMETRY FILE ');
      READLN(FNAME);

      ASSIGN(INPFILE,FNAME);
      RESET(INPFILE);

      WRITELN('READING NUMBER OF ELEMENTS');
      READLN(INPFILE,NE,NN);
      WRITELN('READING NUMBER OF NODES');
      WRITELN('READING NODE NUMBERS');
      FOR J:=1 TO NE DO
         READLN(INPFILE,I,GN[I,1],GN[I,2],GN[I,3]);
      WRITELN('READING NODE COORDINATES');
      FOR J:=1 TO NN DO
          READLN(INPFILE,I,XG[I],YG[I]);
      WRITELN('FINISHED INPUTTING DATA FROM DISK');
      WRITELN;
      CLOSE(INPFILE)
   END;

{***************************************************************}
{Procedure to Scale and Center Structure for Graphics}
PROCEDURE CENTER( VAR XG             :RARY3;
                  VAR YG             :RARY3;
                  VAR SF             :REAL;
                  VAR MINX           :REAL;
                  VAR MINY           :REAL;
                  VAR SHIFTX         :INTEGER;
                  VAR SHIFTY         :INTEGER);
   VAR
      MAXX,MAXY,SFX,SFY              :REAL;
```

PROGRAM LISTING IN TURBO PASCAL

```
    BEGIN
      SFX:=0;
      SFY:=0;
      SF:=0;
      MINX:=0;
      MINY:=0;
      MAXX:=0;
      MAXY:=0;
      FOR J:=1 TO NN DO
          BEGIN
                IF XG[J] > MAXX   THEN
                    MAXX:=XG[J];
                IF XG[J] < MINX   THEN
                    MINX:=XG[J];
                IF YG[J] > MAXY   THEN
                    MAXY:=YG[J];
                IF YG[J] < MINY   THEN
                    MINY:=YG[J];
          END;
      SFX:=250/ABS(MAXX-MINX);
      SFY:=150/ABS(MAXY-MINY);
      IF SFX < SFY   THEN
          SF:=SFX
      ELSE SF:=SFY;
      SHIFTX:=TRUNC((300-SF*(MAXX-MINX))/2);
      SHIFTY:=TRUNC((200-SF*(MAXY-MINY))/2);
    END;

{*****************************************************************}
  {Display of Structure using the Turbo Graphics; Calls PROC. "center"}
  PROCEDURE GRAPH( VAR   XG            :RARY3;
                   VAR   YG            :RARY3;
                   VAR   SF            :REAL;
                   VAR   MINX          :REAL;
                   VAR   MINY          :REAL;
                   VAR   SHIFTX        :INTEGER;
                   VAR   SHIFTY        : INTEGER);
    VAR
        X1,X2,X3,Y1,Y2,Y3              :INTEGER;
        C                              :CHAR;

    BEGIN
        CENTER(XG,YG,SF,MINX,MINY,SHIFTX,SHIFTY);
        GRAPHMODE;
        PALETTE(1);
        FOR J:=1 TO NE DO
            BEGIN
                X1:=SHIFTX+TRUNC(SF*(XG[GN[J,1]]-MINX));
                X2:=SHIFTX+TRUNC(SF*(XG[GN[J,2]]-MINX));
                X3:=SHIFTX+TRUNC(SF*(XG[GN[J,3]]-MINX));
                Y1:=200-SHIFTY-TRUNC(SF*(YG[GN[J,1]]-MINY));
                Y2:=200-SHIFTY-TRUNC(SF*(YG[GN[J,2]]-MINY));
                Y3:=200-SHIFTY-TRUNC(SF*(YG[GN[J,3]]-MINY));
                DRAW (X1,Y1, X2,Y2,3);
                DRAW (X2,Y2, X3,Y3,3);
                DRAW (X3,Y3, X1,Y1,3);
            END;
        GOTOXY(1,24);
        WRITE ('HIT <CR> TO CONTINUE.');
        READLN;
        TEXTMODE
    END;
{*****************************************************************}
{ Write Structure Geometry to disk}
PROCEDURE WRITE_DISC;
   TYPE
     CHRSTRING = STRING[20];

   VAR OUTFILE       :TEXT;
       FNAME         :CHRSTRING;
```

```
    BEGIN
        WRITE('INPUT THE NAME FOR THE NEW FILE TO BE CREATED ');
        READLN(FNAME);

        ASSIGN(OUTFILE,FNAME);
        REWRITE(OUTFILE);
        WRITELN(OUTFILE,NE,' ',NN);
        FOR I:=1 TO NE DO
            WRITELN(OUTFILE,I,' ',GN[I,1],' ',GN[I,2], ' ',GN[I,3]);
        FOR I:= 1 TO NN DO
            WRITELN(OUTFILE,I,' ',XG[I],' ',YG[I],' ');

        CLOSE(OUTFILE)
    END;
{****************************************************************}
{Procedure Which Calculates Half-Bandwidth of Global Stiffness Matrix}
PROCEDURE BANDWIDTH;

    VAR DI,I1,J1,K1  :INTEGER;

    BEGIN
        FOR I1:=1 TO NE DO
            FOR J1:=1 TO 2 DO
                FOR K1:=1 TO 3 DO
                    BEGIN
                        DI:=ABS(GN[I1,J1]-GN[I1,K1]);
                        IF(DI>BW)THEN
                            BW:=DI
                    END;
        BW:=BW+1
    END;
{****************************************************************}
{Optional Edit of Input Node Numbering Data}
PROCEDURE EDIT_NN;

    VAR
        N         :INTEGER;
        REPLY     :CHAR;
        DONE      :BOOLEAN;

    BEGIN
        REPEAT
            DONE:=TRUE;
            WRITE('WHICH ELEMENT WOULD YOU LIKE TO EDIT?  ');
            READLN(N);
            IF N <= NE  THEN
                BEGIN
                    WRITELN('CURRENT VALUES:   LOCAL 1 = ',GN[N,1]);
                    WRITELN('                  LOCAL 2 = ',GN[N,2]);
                    WRITELN('                  LOCAL 3 = ',GN[N,3]);
                    WRITELN;
                    WRITE('ENTER NEW VALUES: LOCAL 1 = ');
                    READLN(GN[N,1]);
                    WRITE('                  LOCAL 2 = ');
                    READLN(GN[N,2]);
                    WRITE('                  LOCAL 3 = ');
                    READLN(GN[N,3]);
                    GRAPH(XG,YG,SF,MINX,MINY,SHIFTX,SHIFTY);
                    WRITELN;
                    WRITE('INPUT MORE CHANGES?   (Y/N) ');
                    READLN(REPLY);
                    IF UPCASE(REPLY) = 'Y'  THEN
                        DONE:=FALSE;
                END
            ELSE
                BEGIN
                    WRITELN(' THERE ARE NOT THAT MANY ELEMENTS, TRY AGAIN!!');
                    DONE:=FALSE;
```

PROGRAM LISTING IN TURBO PASCAL

```pascal
          END;
      UNTIL DONE;
    END;
{*****************************************************************}
{Optional Edit of Node Coordinates Data}
PROCEDURE EDIT_NC;

    VAR
        N                              :INTEGER;
        DONE                           :BOOLEAN;

    BEGIN
      REPEAT
          DONE:=TRUE;
          WRITE('WHICH NODE # WOULD YOU LIKE TO CHANGE?  ');
          READLN(N);
          IF N <= NN  THEN
              BEGIN
                  WRITELN('CURRENT VALUES: X = ',XG[N]:10:2 ,' Y = ',YG[N]:10:2);
                  WRITELN;
                  WRITE('ENTER NEW VALUES: X = ');
                  READLN(XG[N]);
                  WRITE('                  Y = ');
                  READLN(YG[N]);
                  GRAPH(XG,YG,SF,MINX,MINY,SHIFTX,SHIFTY);
                  WRITELN;
                  WRITE('WOULD YOU LIKE TO MAKE MORE CHANGES?  (Y/N) ');
                  READLN(RESP);
                  IF UPCASE(RESP)='Y'  THEN
                      DONE:=FALSE;
              END
          ELSE
              BEGIN
                  WRITELN('THERE ARE NOT THAT MANY NODES, TRY AGAIN!!');
                  DONE:=FALSE;
              END;
      UNTIL DONE;
    END;
{*****************************************************************}
{Calculate Conduction Matrix}
PROCEDURE CONDUCTION_MATRIX(VAR M:INTEGER);

    VAR
        DET                            :REAL;
        N,J1,I1                        :INTEGER;

    BEGIN
        FOR J1:=1 TO 3 DO
            BEGIN
                X[J1]:=XG[GN[M,J1]];
                Y[J1]:=YG[GN[M,J1]]
            END;
        DET:=(X[1]-X[3])*(Y[2]-Y[3])-(X[2]-X[3])*(Y[1]-Y[3]);
        B[1,1]:=Y[2]-Y[3];
        B[1,2]:=Y[3]-Y[1];
        B[1,3]:=Y[1]-Y[2];
        B[2,1]:=X[3]-X[2];
        B[2,2]:=X[1]-X[3];
        B[2,3]:=X[2]-X[1];
        FOR I1:=1 TO 3 DO
            FOR J1:=1 TO 3 DO
                PR[I1,J1]:=0;
        FOR I1:=1 TO 3 DO
            FOR J1:=1 TO 3 DO
                FOR N:=1 TO 2 DO
                    PR[I1,J1]:=PR[I1,J1]+B[N,I1]*B[N,J1];
        FOR I1:=1 TO 3 DO
```

```
            FOR J1:=1 TO 3 DO
                PR[I1,J1]:=KE[M]*PR[I1,J1]/(2*DET)
END;

{*****************************************************************}
{Calculate Heat Flow per Unit Area}
PROCEDURE HEAT_FLOW(VAR M:INTEGER);

    VAR
        BB,CC                               :RARY1;
        DET                                 :REAL;
        I1                                  :INTEGER;

    BEGIN
        FOR I1:=1 TO 3  DO
            BEGIN
                X[I1]:=XG[GN[M,I1]];
                Y[I1]:=YG[GN[M,I1]]
            END;
        BB[1]:=Y[2]-Y[3];
        BB[2]:=Y[3]-Y[1];
        BB[3]:=Y[1]-Y[2];
        CC[1]:=X[3]-X[2];
        CC[2]:=X[1]-X[3];
        CC[3]:=X[2]-X[1];
        DET:=(X[1]-X[3])*(Y[2]-Y[3])-(X[2]-X[3])*(Y[1]-Y[3]);
        QX:=0;
        QY:=0;
        FOR I1:=1 TO 3 DO
            BEGIN
                QX:= -KE[M]/DET*BB[I1]*Q[GN[M,I1]]+QX;
                QY:= -KE[M]/DET*CC[I1]*Q[GN[M,I1]]+QY
            END
    END;

{*********************************************************************}
{Output of Input Data and Results }
PROCEDURE RESULTS;

    VAR
        DEV                                 :TEXT;
        DONE                                :BOOLEAN;
        CHOICE                              :INTEGER;
        FNAME                               :STRING[20];

    BEGIN
        REPEAT
            BEGIN
                DONE:=TRUE;
                WRITELN;
                WRITELN;
                WRITELN;
                WRITELN('               OUTPUT OPTIONS?');
                WRITELN('               ------ --------');
                WRITELN;
                WRITELN('               TERMINAL = "1"');
                WRITELN('               PRINTER  = "2"');
                WRITELN('               AUX      = "3"');
                WRITELN('               TEXTFILE = "4"');
                WRITELN;
                WRITE  ('          ENTER CHOICE- (1 TO 4) ');
                READLN(CHOICE);
                CASE CHOICE OF
                    1: FNAME:='TRM:';
                    2: FNAME:='LST:';
                    3: FNAME:='AUX:';
                    4:BEGIN
                        WRITE('ENTER FILENAME (DEFAULT = RESULTS.DAT): ');
                        READLN(FNAME);
```

PROGRAM LISTING IN TURBO PASCAL

```pascal
                        IF FNAME='' THEN FNAME:='RESULTS.DAT'
                END;
            ELSE
                DONE:=FALSE;
            END {CASE}
        END;
    UNTIL DONE;

ASSIGN(DEV,FNAME);
REWRITE(DEV);

WRITELN(DEV);
WRITELN(DEV);
WRITELN(DEV,'****************************************');
WRITELN(DEV,'**                                    **');
WRITELN(DEV,'**    STEADY STATE HEAT TRANSFER      **');
WRITELN(DEV,'**          IN THE X-Y PLANE          **');
WRITELN(DEV,'**                                    **');
WRITELN(DEV,'****************************************');
WRITELN(DEV);
WRITELN(DEV);
WRITELN(DEV,'------------ INPUT DATA ---------------');
WRITELN(DEV);
WRITELN(DEV);
WRITELN(DEV,'    ELEMENT         GLOBAL NODE NUMBERS  ');
WRITELN(DEV,'    NUMBER            OF LOCAL NODES     ');
WRITELN(DEV,'                    1       2       3    ');
WRITELN(DEV);
FOR I:=1 TO NE DO
    WRITELN(DEV,I:7,GN[I,1]:15,GN[I,2]:8,GN[I,3]:8 );
WRITELN(DEV);
WRITELN(DEV);
WRITELN(DEV,'     NODE         GLOBAL COORDINATES   ');
WRITELN(DEV,'    NUMBER            X           Y    ');
WRITELN(DEV);
FOR I:=1 TO NN DO
    WRITELN(DEV,I:7,XG[I]:20:3,YG[I]:12:3);
WRITELN(DEV);
WRITELN(DEV);
WRITELN(DEV,'          CONDUCTION COEFFICIENTS         ');
WRITELN(DEV);
WRITELN(DEV,'    ELEMENT        CONDUCTION COEFFICIENT, k');
WRITELN(DEV);
IF UPCASE(HOMO) = 'N'  THEN
    FOR I:=1 TO NE DO
        WRITELN(DEV,I:7, KE[I]:22:2 )
ELSE
        WRITELN(DEV,'    ALL', KE[1]:22:2);
WRITELN(DEV);
WRITELN(DEV);
WRITELN(DEV,'    BOUNDARY CONVECTION COEFFICIENTS          ');
WRITELN(DEV);
WRITELN(DEV,'    ELEMENT     CONVECTION      AMBIENT    ');
WRITELN(DEV,'                COEFFICIENT    TEMPERATURE ');
WRITELN(DEV);
IF NB <> 0  THEN
    FOR I:=1 TO NB DO
        WRITELN(DEV,BN[I]:7,HE[I]:17:4,TA[I]:17:2);
WRITELN(DEV);
WRITELN(DEV);
WRITELN(DEV,'    NODE    APPLIED TEMPERATURE       ');
WRITELN(DEV);
FOR I:=1 TO NN DO
    IF BC[I] <> 0  THEN
        WRITELN(DEV,I:5,IT[I]:18:2);
WRITELN(DEV);
WRITELN(DEV);
WRITELN(DEV);
WRITELN(DEV);
```

```
      WRITELN(DEV,'--------------    RESULTS    -------------------');
      WRITELN(DEV);
      WRITELN(DEV);
      WRITELN(DEV,'    NODE NUMBER            NODE TEMPERATURE      ');
      WRITELN(DEV);
      FOR I:= 1 TO NN DO
          WRITELN(DEV,I:11,Q[I]:25:2);
      WRITELN(DEV);
      WRITELN(DEV);
      WRITELN(DEV,'    ELEMENT NUMBER         ELEMENT TEMPERATURE   ');
      WRITELN(DEV,'                              (NODE AVERAGE)     ');
      WRITELN(DEV);
      FOR I:=1 TO NE DO
          WRITELN(DEV,I:11,QA[I]:25:2);
      WRITELN(DEV);
      WRITELN(DEV);
      WRITELN(DEV,'    ELEMENT          HEAT FLOW IN         HEAT FLOW IN   ');
      WRITELN(DEV,'                      X DIRECTION          Y DIRECTION   ');
      WRITELN(DEV,'                      HEAT/T/AREA          HEAT/T/AREA   ');
      WRITELN(DEV);
{---------------------------------------------------------------------}
{              CALCULATE HEAT FLOW/TIME/AREA                          }
{---------------------------------------------------------------------}
      FOR I:=1 TO NE DO
          BEGIN
              HEAT_FLOW(I);
              WRITELN(DEV,I:7,QX:20:2,QY:20:2)
          END;
      WRITELN;
      WRITELN('      USER OPTIONS:');
      WRITELN;
      WRITELN('            1: REVIEW RESULTS');
      WRITELN('            2: EXIT PROGRAM ');
      READLN(J);
      IF J = 1   THEN
         RESULTS
   END;
{*****************************************************************}

{**********------------- MAIN PROGRAM ----------------------------}
   BEGIN
      WRITELN('*****************************************************');
      WRITELN('*****************************************************');
      WRITELN('***                                               ***');
      WRITELN('***           HEAT TRANSFER                       ***');
      WRITELN('***              IN THE                           ***');
      WRITELN('***             X-Y PLANE                         ***');
      WRITELN('***                                               ***');
      WRITELN('*****************************************************');
      WRITELN('*****************************************************');
{            *** START OF MAIN PROGRAM ***                         }
      WRITELN;
      MEM:=MEMAVAIL;
      WRITELN('       (AVAILABLE MEMORY IS = ',MEM*16,' BYTES)');
      WRITELN;
      WRITELN;
      INITIALIZE_VARIABLES;
      WRITELN('              DATA INPUT OPTIONS           ');
      WRITELN('              -------------------          ');
      WRITELN;
      WRITELN('    1 - INPUT NEW MODEL DATA FROM KEYBOARD');
      WRITELN('    2 - INPUT OLD MODEL DATA FROM DISC   ');
      WRITELN('           (ELEMENT NODE NUMBERS AND ');
      WRITELN('              NODE COORDINATES)');
      WRITELN;
      WRITE('    ENTER CHOICE - (1 OR 2) ');
      READLN(INP);
      IF INP = 1  THEN
         TERMINAL_INPUT
```

PROGRAM LISTING IN TURBO PASCAL 499

```
      ELSE
         DISC_INPUT;
      BANDWIDTH;
      WRITELN;
      WRITELN('****** CALCULATED HALF BANDWIDTH = ',BW,' ******');
      WRITELN;
      WRITELN;
      WRITE('DO YOU WISH TO SEE PLOT OF MESH?  (Y/N) ');
      READLN(RESP);
      IF UPCASE(RESP) = 'Y'  THEN
         GRAPH(XG,YG,SF,MINX,MINY,SHIFTX,SHIFTY);
      WRITELN;
{---------------------------------------------------------}
{-----     EDITING OPTIONS TO REVIEW AND/OR CHANGE   -----}
{-----     ELEMENT NODE NUMBERING AND COORDINATES    -----}
{---------------------------------------------------------}
      WRITE('DO YOU WISH TO EDIT NODE NUMBERING? ');
      READLN(RESP);
      IF UPCASE(RESP)='Y'  THEN
         EDIT_NN;
      WRITELN;
      WRITE('DO YOU WISH TO EDIT NODE COORDINATES? ');
      READLN(RESP);
      IF UPCASE(RESP)='Y'   THEN
         EDIT_NC;
      WRITELN;
      WRITE('DO YOU WANT THE NEW GEOMETRY DATA STORED ON THE DISK? ');
      READLN(RESP);
      IF UPCASE(RESP)='Y'   THEN
         WRITE_DISC;
      WRITELN;
      WRITE('HOW MANY ELEMENT SIDES ARE EXPOSED TO CONVECTION? ');
      READLN(NB);
      WRITELN;
      WRITE('HOW MANY ELEMENT NODES HAVE KNOWN TEMPERATURE? ');
      READLN(KT);
      WRITELN;
      FOR I:=1 TO NN DO
          BEGIN
             BC[I]:=0;
             IT[I]:=0
          END;
{-------------------------------------------------------------------------}
{------          IDENTIFY ELEMENT SIDES ON BOUNDARY, AMBIENT       -------}
{------          TEMPERATURE AND CONVECTION COEFFICIENTS           -------}
{-------------------------------------------------------------------------}

      IF NB <> 0  THEN
         BEGIN
            WRITELN('IDENTIFY THE ELEMENT NUMBER AND ITS BOUNDARY NODES');
            WRITELN('   FOR EACH OF THE ',NB,' BOUNDARY ELEMENTS: ');
            WRITELN;
            FOR I:=1 TO NB DO
                BEGIN
                   WRITE('BOUNDARY ELEMENT # ');
                   READLN(BN[I]);
                   WRITE('ONE BOUNDARY NODE # ');
                   READLN(N1[I]);
                   WRITE('OTHER BOUNDARY NODE # ');
                   READLN(N2[I]);
                   WRITE('AMBIENT TEMPERATURE, T = ');
                   READLN(TA[I]);
                   WRITE('CONVECTION COEFFICIENT, H = ');
                   READLN(HE[I]);
                   WRITELN;
                   IF I <> NB  THEN
                       WRITELN('NEXT ELEMENT SIDE: ')
                END
         END;
```

```
{---------------------------------------------------------}
{-------      INPUT KNOWN TEMPERATURES      ------------}
{---------------------------------------------------------}
    WRITELN;
    IF KT <> 0  THEN
        BEGIN
            WRITELN('INPUT THE TEMPERATURE AT EACH OF THE   ',KT,' NODES ');
            WRITELN;
            FOR I:=1 TO KT DO
                BEGIN
                    WRITE('NODE # = ');
                    READLN(J);
                    WRITE('TEMPERATURE = ');
                    READLN(Q[J]);
                    BC[J]:=1;
                    IT[J]:=Q[J]
                END
        END;
{----------------------------------------------------------------}
{------   INPUT THE MATERIAL CONDUCTION COEFFICIENTS   ---------}
{----------------------------------------------------------------}
    WRITELN;
    WRITELN('ENTER MATERIAL CONDUCTION COEFFICIENT');
    WRITE('IS MATERIAL HOMOGENEOUS? (Y/N) ');
    READLN(HOMO);
    IF UPCASE(HOMO) = 'N'  THEN
        FOR I:=1 TO NE DO
            BEGIN
                WRITE('ELEMENT ',I,' , K = ');
                READLN(KE[I]);
            END
    ELSE
        BEGIN
            WRITE('INPUT COEFFICIENT = ');
            READLN(KE1);
            FOR I:=1 TO NE DO
                KE[I]:=KE1
        END;
    FOR I:=1 TO NN DO
        FOR J:=1 TO BW DO
            KR[I,J]:=0;
{----------------------------------------------------------------------}
{----------           BEGIN CALCULATIONS           -------------------}
{----------------------------------------------------------------------}
    TEXTCOLOR(15+BLINK);
    WRITELN;
    WRITELN('********   WORKING WORKING WORKING ********');
    WRITELN;
    TEXTCOLOR(14);
    FOR M:=1 TO NE DO
        BEGIN
            WRITELN('********* ELEMENT ', M, ' CONDUCTION MATRIX *********');
            CONDUCTION_MATRIX(M);
{-------------------------------------------------------------------}
{----------      ASSEMBLE GLOBAL CONDUCTION MATRIX   -----------}
{----------             IN BAND MATRIX FORMAT        -----------}
{-------------------------------------------------------------------}
            FOR I:=1 TO 3 DO
                BEGIN
                    FOR J:=1 TO 3 DO
                        IF GN[M,J] >= GN[M,I]  THEN
                            BEGIN
                                C:=GN[M,J]-GN[M,I]+1;
                                KR[GN[M,I],C]:=PR[I,J]+KR[GN[M,I],C]
                            END
                END
        END;
```

PROGRAM LISTING IN TURBO PASCAL

```pascal
{----    CALCULATE THE ELEMENT CONVECTION MATRICES -----}
      IF NB <> 0 THEN
        BEGIN
          FOR I:=1 TO 3 DO
              BEGIN
                FR[I]:=0;
                FOR J:=1 TO 3 DO
                    CV[I,J]:=0
              END;
          FOR I:=1 TO NN DO
              BEGIN
                TN[I]:=0;
              END;
{---- NB IS NUMBER OF EXPOSED ELEMENT SIDES    ----}
          FOR W:=1 TO NB DO
              BEGIN
                I:=W;
                IF W <> 1  THEN
                    BEGIN
{---- BN[I] IS THE ELEMENT NUMBER OF SIDE I ----}
{---- TEST IF NEXT SIDE BELONGS TO SAME ELEMENT----}
                      IF BN[W]=BN[W-1]  THEN
                          I:=W-1
                    END;
                S:=SQRT(SQR(XG[N1[W]]-XG[N2[W]])+ SQR(YG[N1[W]]-YG[N2[W]]));
                CC:=HE[W]/6*S;
                CF:=HE[W]*TA[W]*S;
{---- TEST FOR ELEMENT LOCAL SIDE NUMBER ----}
                writeln('side ',bn[i],' convection matrix');
                IF (GN[BN[I],1] <> N1[W]) AND (GN[BN[I],1] <> N2[W])  THEN
                    BEGIN
                      CV[2,2]:=2*CC;
                      FR[2]:=CF/2;
                      CV[2,3]:=CC;
                      FR[3]:=CF/2;
                      CV[3,2]:=CC;
                      CV[3,3]:=2*CC
                    END
                ELSE IF (GN[BN[I],2]<>N1[W]) AND (GN[BN[I],2]<>N2[W])  THEN
                    BEGIN
                      CV[1,1]:=2*CC;
                      FR[1]:=CF/2;
                      CV[1,3]:=CC;
                      FR[3]:=CF/2;
                      CV[3,1]:=CC;
                      CV[3,3]:=2*CC
                    END
                ELSE
                    BEGIN
                      CV[1,1]:=2*CC;
                      FR[1]:=CF/2.;
                      CV[1,2]:=CC;
                      FR[2]:=CF/2;
                      CV[2,1]:=CC;
                      CV[2,2]:=2*CC
                    END;
{---------------------------------------------------------------------}
{--------     ADD THE CONVECTION MATRIX TO THE GLOBAL      -----------}
{--------                CONDUCTION MATRIX                 -----------}
{--------     AND CONSTRUCT CONVECTION FORCE VECTOR        -----------}
{---------------------------------------------------------------------}
                FOR L:=1 TO 3 DO
                    BEGIN
                      FOR M:=1 TO 3 DO
                          IF (GN[BN[I],M]>=GN[BN[I],L])  THEN
                              BEGIN
                                C:=GN[BN[I],M]-GN[BN[I],L]+1;
                                KR[GN[BN[I],L],C]:=CV[L,M]+KR[GN[BN[I],L],C]
                              END;
```

```
                            TN[GN[BN[I],L]]:=FR[L]+TN[GN[BN[I],L]]
                        END;
                    FOR II:=1 TO 3 DO
                        BEGIN
                            FR[II]:=0;
                            FOR JJ:=1 TO 3 DO
                                CV[II,JJ]:=0;
                        END
                END
            END;
{------------------------------------------------------------------------}
{----------    APPLY KNOWN NODE TEMPERATURES AND MODIFY      ------------}
{----------       THE STIFFNESS MATRIX 'DIAGONAL' ENTRIES    ------------}
{------------------------------------------------------------------------}
        IF KT <> 0    THEN
            BEGIN
                FOR J:=1 TO NN DO
                    IF BC[J] <> 0    THEN
                        BEGIN
                            KR[J,1]:=KR[J,1]*1E10;
                            TN[J]:=Q[J]*KR[J,1];
                        END
            END;
{------------------------------------------------------------------------}
{--------    SOLVE FOR NODE TEMPERATURES USING GAUSS       --------------}
{--------      ELIMINATION ROUTINE FOR BANDED MATRIX       --------------}
{------------------------------------------------------------------------}
        II:=NN;
        FOR M:=1 TO II DO
            IF (KR[M,1]<>0)    THEN
                BEGIN
                    FOR L:=2 TO BW DO
                        IF KR[M,L]<>0 THEN
                            BEGIN
                                I:=M+L-1;
                                QU:=KR[M,L]/KR[M,1];
                                J:=0;
                                FOR K:=L TO BW DO
                                    BEGIN
                                        J:=J+1;
                                        IF I<=II THEN
                                            KR[I,J]:=KR[I,J]-QU*KR[M,K]
                                    END;
                                KR[M,L]:=QU
                            END
                END;
WRITELN('********* MATRIX REDUCED *********');

        FOR M:=1 TO II DO
            IF KR[M,1] <> 0    THEN
                BEGIN
                    FOR L:=2 TO BW DO
                        IF KR[M,L] <> 0    THEN
                            BEGIN
                                I:=M+L-1;
                                IF I <= II    THEN
                                    TN[I]:=TN[I]-KR[M,L]*TN[M]
                            END;
                    TN[M]:=TN[M]/KR[M,1]
                END;
WRITELN('********* FORCE VECTOR REDUCED *********');
        Q[II]:=TN[II];
        FOR JJ:=2 TO II DO
            BEGIN
                M:=II+1-JJ;
                FOR L:=2 TO BW DO
                    IF (KR[M,L] <> 0)    THEN
                        BEGIN
                            K:=M+L-1;
```

PROGRAM LISTING IN TURBO PASCAL   **503**

```
                        IF K <= II   THEN
                           TN[M]:=TN[M]-KR[M,L]*Q[K]
                  END;
               Q[M]:=TN[M];
           END;
{------------------------------------------------------------------}
{------    CALCULATE ELEMENT TEMPERATURE AS AVERAGE OF      ------}
{------         NODE TEMPS.  USE QA[I] FOR ELEMENT TEMP     ------}
{------------------------------------------------------------------}
       FOR I:=1 TO NE DO
           BEGIN
              QA[I]:=0;
              FOR J:=1 TO 3 DO
                  QA[I]:=Q[GN[I,J]]/3 + QA[I]
           END;
{------------------------------------------------------------------}
{---------    OPTIONAL HARDCOPY OF DATA AND RESULTS      --------}
{------------------------------------------------------------------}
       RESULTS;
       WRITELN;
   END.
```

APPENDIX 8

Partial List of Commercial Finite Element Programs

The following list provides a reference to a few popular large-scale, general-purpose finite element programs. These programs require a mainframe or minicomputer; most are available with the option of outright purchase or lease. The code purchase costs can range typically from $5000 to $60,000, rental fees from $500 to $2000 per month. Code for the microcomputer is rapidly becoming available, and as the capacity of the micro increases it will become a very satisfactory alternative for many analysis problems. Acquiring the computer and code for numerical analysis is a major investment. The advantages of one system over the other must be carefully weighed against your particular needs.

Program	Source of Information
ABAQUS	Dr. E. P. Sorensen Hibbitt, Karlsson & Sorensen, Inc. 35 Angell Street Providence, RI 02906

PARTIAL LIST OF COMMERCIAL FINITE ELEMENT PROGRAMS

ADINA	ADINA Engineering, Inc. 71 Elton Avenue Watertown, MA 02172
ANSYS	Marketing: Sue Batt Swanson Analysis Systems, Inc. P.O. Box 65 Houston, PA 15342-0065
GIFTS	Dr. H. A. Kamel, Professor Aerospace and Mechanical Engineering Dept. University of Arizona Aero Building, Suite 16 Tucson, AZ 85721
GTSTRUDL	Dr. Leroy Z. Emkin GTICES Systems Laboratory Georgia Institute of Technology School of Civil Engineering Atlanta, GA 30332
MARC	Dr. Fritz Hatt Marketing Manager MARC Software International, Inc. 260 Sheridan Avenue, Suite 200 Palo Alto, CA 94306
MSC/NASTRAN	Perry L. Grant The MacNeal-Schwendler Corp. 815 Colorado Boulevard Los Angeles, CA 90041
PAFEC	Jeff Thompson PAFEC Engineering Consultants, Inc. 601 Concord Street Knoxville, TN 37919
SAP7	Structural Mechanics Computer Laboratory SAP Users Group University of Southern California Denney Research Center University Park Los Angeles, CA 90007

Selected Bibliography

Chapter 1

BEAUFAIT, F. W., W. H. ROWAN, JR., P. G. HOADLEY, and R. M. HACKETT. *Computer Methods of Structural Analysis*. Englewood Cliffs, N.J.: Prentice-Hall, Inc., 1970.

KARDESTUNCER, H. *Elementary Matrix Analysis of Structures*. New York: McGraw-Hill Book Company, 1974.

MARTIN, H. C. *Introduction to Matrix Methods of Structural Analysis*. New York: McGraw-Hill Book Company, 1966.

WANG, C. K. *Matrix Methods of Structural Analysis*. Scranton, Pa.: International Textbook Company, 1970.

Chapter 2

ARPACI, V. S. *Conduction Heat Transfer*. Reading Mass.: Addison-Wesley Publishing Co., Inc., 1966.

DYN, C. L., and I. H. SHAMES. *Solid Mechanics: A Variational Approach*. New York: McGraw-Hill Book Company, 1973.

GALLAGHER, R. H. *Finite Element Analysis Fundamentals*. Englewood Cliffs, N.J.: Prentice-Hall, Inc., 1975.

LANGHAAR, H. L. *Energy Methods in Applied Mechanics*. New York: John Wiley & Sons, Inc., 1962.

LEON, S. J. *Linear Algebra with Applications*. New York: Macmillan Publishing Company, 1980.

LITTLE, R. W. *Elasticity*. Englewood Cliffs, N.J.: Prentice-Hall, Inc., 1973.
TIMOSHENKO, S., and J. N. GOODIER. *Theory of Elasticity*. New York: McGraw-Hill Book Company, 1951.
WANG, C. T. *Applied Elasticity*. New York: McGraw-Hill Book Company, 1953.
WASHIZU, K. *Variational Methods in Elasticity and Plasticity*. Oxford, England: Pergamon Press, Ltd., 1968.
WEINSTOCK, R. *Calculus of Variations*. New York: McGraw-Hill Book Company, 1952.

Chapter 3
ZIENKIEWICZ, O. C. *The Finite Element Method in Engineering Science*. London, England: McGraw-Hill Publishing Company Limited, 1971.

Chapter 5
JAMES, M. L., G. M. SMITH, and J. C. WOLFORD, *Applied Numerical Methods for Digital Computation*. New York: Harper & Row, Publishers, 1977.
MCCORMICK, J. M., and M. G. SALVADORI. *Numerical Methods in Fortran*. Englewood Cliffs, N.J.: Prentice Hall, Inc., 1964.

Chapters 6 through 12
There are many textbooks and articles on the finite element method. This book has taken the approach of filling in the details for relatively limited but basic material. Upon completion of topics of interest in this book, the reader should be well prepared to profit from reading other books that present more general coverage, such as the following textbooks.

BAKER, A. J. *Finite Element Computational Fluid Mechanics*. New York: McGraw-Hill Book Company, 1983.
BATHE, K., and E. L. WILSON. *Numerical Methods in Finite Element Analysis*. Englewood Cliffs, N. J.: Prentice-Hall, Inc., 1976.
CHEUNG, Y. K., and M. F. YEO. *A Practical Introduction to Finite Element Analysis*. Marshfield, Mass.: Pitman Publishing, Inc., 1979.
COOK, R. D. *Concepts and Applications of Finite Element Analysis*. New York: John Wiley & Sons, Inc., 1974.
DAVIES, A. J. *The Finite Element Method: A First Approach*. Oxford, England: Clarendon Press, 1970.
DESAI C. S. *Elementary Finite Element Method*. Englewood Cliffs, N.J.: Prentice-Hall, Inc., 1979.
GALLAGHER, R. H. *Finite Element Analysis Fundamentals*. Englewood Cliffs, N.J.: Prentice-Hall, Inc., 1975.
HUEBNER, K. H. *The Finite Element Method for Engineers*. New York: John Wiley & Sons, Inc., 1975.
MARTIN, H. C., and G. F. CAREY. *Introduction to Finite Element Analysis*. New York: McGraw-Hill Book Company, 1973.
NORRIE, D. H., and G. DEVRIES. *An Introduction to Finite Element Analysis*. New York: Academic Press, Inc., 1978.
RAO, S. S. *The Finite Element Method in Engineering*. Elmsford, N.Y.: Pergamon Press, Inc., 1982.

SEGERLIND, L. J. *Applied Finite Element Analysis.* New York: John Wiley & Sons, Inc., 1976.

ZIENKIEWICZ, O. C. *The Finite Element Method in Engineering Science.* London, England: McGraw-Hill Publishing Company Limited, 1971.

ZIENKIEWICZ, O. C. *The Finite Element Method.* London, England: McGraw-Hill Publishing Company Limited, 1977.

Answers to Selected Exercises

Chapter 1

1.1 $\begin{Bmatrix} 17 \\ 39 \end{Bmatrix}$ **1.2** $\frac{1}{8}\begin{bmatrix} 3 & -2 \\ -2 & 4 \end{bmatrix}$

1.3 (a) $\begin{bmatrix} 1 & 0 & 2 \\ 0 & 6 & 5 \\ 2 & 5 & 7 \end{bmatrix}$ **(b)** $\begin{bmatrix} 3 & 0 & 2 \\ 0 & 6 & 5 \\ 2 & 5 & 5 \end{bmatrix}$

1.4 $\begin{bmatrix} 0.866 & 0.50 \\ -0.50 & 0.866 \end{bmatrix}$ **1.5** $\begin{Bmatrix} F_x \\ F_y \end{Bmatrix} = \begin{Bmatrix} 6.16 \\ 9.33 \end{Bmatrix}$

1.6 (a) $[k_1] = 5(10^6)\begin{bmatrix} 1 & -1 \\ -1 & 1 \end{bmatrix} = [k_2]$

(b) $[K] = 5(10^6)\begin{bmatrix} 1 & -1 & 0 \\ -1 & 2 & -1 \\ 0 & -1 & 1 \end{bmatrix}$

(c) $\{r\}^T = [u_1 \quad u_2 \quad u_3], \{R\}^T = [R_1 \quad R_2 \quad R_3]$

(d) $5(10^6)\begin{bmatrix} 1 & -1 & 0 \\ -1 & 2 & -1 \\ 0 & -1 & 1 \end{bmatrix} \begin{Bmatrix} u_1 \\ u_2 \\ u_3 \end{Bmatrix} = \begin{Bmatrix} R_1 \\ R_2 \\ R_3 \end{Bmatrix}$

(e) $5(10^6)\begin{bmatrix} 1 & -1 & 0 \\ -1 & 2 & -1 \\ 0 & -1 & 1 \end{bmatrix} \begin{Bmatrix} u_1 \\ u_2 \\ 0 \end{Bmatrix} = \begin{Bmatrix} 0 \\ -1000 \\ R_3 \end{Bmatrix}$

(f) $\begin{Bmatrix} u_1 \\ u_2 \end{Bmatrix} = -2(10^{-4}) \begin{Bmatrix} 1 \\ 1 \end{Bmatrix}$

1.7 $[K] = (10^6) \begin{bmatrix} 3.33 & -3.33 & 0 & 0 \\ & 18.33 & -15 & 0 \\ & \text{symmetric} & 20 & -5 \\ & & & 5 \end{bmatrix}$,

$\begin{Bmatrix} u_2 \\ u_3 \end{Bmatrix} = 10^{-5} \begin{Bmatrix} 7 \\ 15.2 \end{Bmatrix}$ in.

1.8 (a) $[k]_{xy} = 7.83(10^5) \begin{bmatrix} 1 & 0 & -1 & 0 \\ & 0 & 0 & 0 \\ & \text{symmetric} & 1 & 0 \\ & & & 0 \end{bmatrix}$

(b) $[A_2] = \begin{bmatrix} 0.832 & 0.555 & 0 & 0 \\ -0.555 & 0.832 & 0 & 0 \\ 0 & 0 & 0.832 & 0.555 \\ 0 & 0 & -0.555 & 0.832 \end{bmatrix}$

(c) $[k]_{XY} = 10^5 \begin{bmatrix} 5.42 & -3.62 & -5.42 & 3.62 \\ & 2.41 & 3.62 & -2.41 \\ & \text{symmetric} & 5.42 & -3.62 \\ & & & 2.41 \end{bmatrix}$

1.9 (a)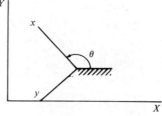

(b) $[A_2] = \begin{bmatrix} -0.832 & -0.555 & 0 & 0 \\ 0.555 & -0.832 & 0 & 0 \\ 0 & 0 & -0.832 & -0.555 \\ 0 & 0 & 0.555 & -0.832 \end{bmatrix}$

(c) Same as Exercise 1.8(c).

1.10 (a) $[k]_{XY} = 10^4 \begin{bmatrix} 77.95 & 58.46 & -77.95 & -58.46 \\ & 43.85 & -58.46 & -43.85 \\ & \text{symmetric} & 77.95 & 58.46 \\ & & & 43.85 \end{bmatrix}$

ANSWERS TO SELECTED EXERCISES

(b) $u_2 = -1.28(10^{-3})$ in.

(c) $\begin{Bmatrix} F_{2x} \\ F_{2y} \end{Bmatrix} = \begin{Bmatrix} -1250 \\ 0 \end{Bmatrix}$ lb

1.11 (a) $[k_1]_{xy} = 7.95(10) \begin{bmatrix} 1 & 0 & -1 & 0 \\ 0 & 0 & 0 & 0 \\ -1 & 0 & 1 & 0 \\ 0 & 0 & 0 & 0 \end{bmatrix}$.

$[k_2]_{xy} = 12(10^5) \begin{bmatrix} 1 & 0 & -1 & 0 \\ 0 & 0 & 0 & 0 \\ -1 & 0 & 1 & 0 \\ 0 & 0 & 0 & 0 \end{bmatrix}$

(b) $[k_1]_{XY} = 3.98(10^5) \begin{bmatrix} 1 & -1 & -1 & 1 \\ & 1 & 1 & -1 \\ & & 1 & -1 \\ & \text{symmetric} & & 1 \end{bmatrix}$

$[k_2]_{XY} = 10^5 \begin{bmatrix} 4.32 & 5.76 & -4.32 & -5.76 \\ & 7.68 & -5.76 & -7.68 \\ & & 4.32 & 5.76 \\ & \text{symmetric} & & 7.68 \end{bmatrix}$

(c) $[K]_{XY} = 10^5 \begin{bmatrix} 3.98 & -3.98 & 0 & 0 & -3.98 & 3.98 \\ & 3.98 & 0 & 0 & 3.98 & -3.98 \\ & & 4.32 & 5.76 & -4.32 & -5.76 \\ & & & 7.68 & -5.76 & -7.68 \\ & \text{symmetric} & & & 8.30 & 1.78 \\ & & & & & 11.66 \end{bmatrix}$

(d) $10^5 \begin{bmatrix} 8.30 & 1.78 \\ 1.78 & 11.66 \end{bmatrix} \begin{Bmatrix} u_3 \\ v_3 \end{Bmatrix} = \begin{Bmatrix} 5000 \\ 0 \end{Bmatrix}$

(e) $\begin{Bmatrix} u_3 \\ v_3 \end{Bmatrix}_{XY} = \begin{Bmatrix} 6.23(10^{-3}) \\ -9.5(10^{-4}) \end{Bmatrix}$ in.

(f) Element 1: $F_{\text{axial}} = 4041$ lb, tension
Element 2: $F_{\text{axial}} = 3571$ lb, compression

1.12 Assembled stiffness matrix:

$[K] = 10^5 \begin{bmatrix} 7.50 & 0 & -7.50 & 0 & 0 & 0 & 0 & 0 \\ & 0 & 0 & 0 & 0 & 0 & 0 & 0 \\ & & 9.21 & -2.56 & 0 & 0 & -1.71 & 2.56 \\ & & & 13.84 & 0 & -10 & 2.56 & -3.84 \\ & & & & 0 & 0 & 0 & 0 \\ & \text{symmetric} & & & & 10 & 0 & 0 \\ & & & & & & 1.71 & -2.56 \\ & & & & & & & 3.84 \end{bmatrix}$

514 ANSWERS TO SELECTED EXERCISES

Element	Axial Force
1	5220 compression
2	2330 tension
3	3220 tension

1.13 (a)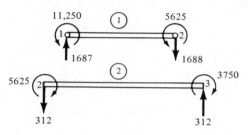

(b)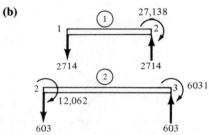

(c)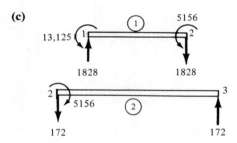

(d)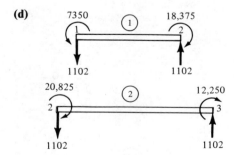

ANSWERS TO SELECTED EXERCISES

1.14
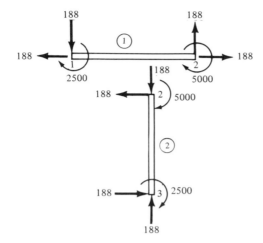

Chapter 2

2.1 $\epsilon_X = C_1, \epsilon_Y = C_5, \gamma_{XY} = C_2 + C_4$

2.2 $\begin{Bmatrix} \sigma_X \\ \sigma_Y \\ \tau_{XY} \end{Bmatrix} = \dfrac{E}{1-\nu^2} \begin{Bmatrix} C_1 + \nu C_5 \\ \nu C_1 + C_5 \\ \dfrac{1-\nu}{2}(C_2 + C_4) \end{Bmatrix}$

2.3 Global coordinates:

$$X_P = 5$$
$$Y_P = 5$$

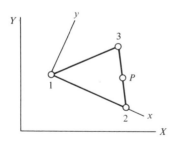

Local coordinates:

$$x_P = 2.24$$
$$y_P = 2.24$$

Natural coordinates:

$$L_1 = 0, \qquad L_2 = \tfrac{1}{2}, \qquad L_3 = \tfrac{1}{2}$$

ANSWERS TO SELECTED EXERCISES

2.4 $u = \begin{bmatrix} \dfrac{X_2 - X}{X_2 - X_1} & \dfrac{X - X_1}{X_2 - X_1} \end{bmatrix} \begin{Bmatrix} u_1 \\ u_2 \end{Bmatrix}$

2.5 $V = -T_0(X_2 - X_1)[\tfrac{1}{2} \quad \tfrac{1}{2}] \begin{Bmatrix} u_1 \\ u_2 \end{Bmatrix}$

2.6 $V = \dfrac{-T_0(X_2 - X_1)}{2} [\tfrac{1}{3} \quad \tfrac{2}{3}] \begin{Bmatrix} u_1 \\ u_2 \end{Bmatrix}$

2.7 $V = -0.28$ in.-lb

2.8 (a) $v_E = \dfrac{Pl^3}{4EI}$ (b) $v_E = \dfrac{32 Pl^3}{\pi^4 EI}$

2.9 (a) $u = C_0 + C_1 X + C_2 X^2$ (b) $U = \dfrac{2EAu_E^2}{3l}$

(c) Weight energy potential:
$$V = -\dfrac{2\rho A l u_E}{3}$$

(d) $\pi = U + V$, $\quad d\pi/du_E = 0$, $\quad u_E = \rho l^2 / 2E$

2.10 $u = 10^{-4} X$, $\quad C_i = 0 = C_3 = C_4 = C_5$, $\quad \epsilon_x = 10^{-4}$,

$\epsilon_Y = 0 = \gamma_{XY}$, $\quad \{\sigma\} = \begin{Bmatrix} 100 \\ 0 \\ 0 \end{Bmatrix}$ psi

2.11 $u = 10^{-8}(189X + 22.8XY)$

$\epsilon_X = 10^{-8}(189 + 22.8Y)$

$\gamma_{XY} = 10^{-8}(22.8x)$

$\{\epsilon\}_{(0,0)} = \begin{Bmatrix} 189 \\ 0 \\ 0 \end{Bmatrix} 10^{-8}$ $\quad \{\sigma\}_{(0,0)} = \begin{Bmatrix} 62.4 \\ 18.8 \\ 0 \end{Bmatrix}$ psi

$\{\epsilon\}_{(10,10)} = \begin{Bmatrix} 4.18 \\ 0 \\ 2.28 \end{Bmatrix} 10^{-6}$ $\quad \{\sigma\}_{(10,10)} = \begin{Bmatrix} 138 \\ 41.2 \\ 26.4 \end{Bmatrix}$ psi

Chapter 3

3.1 $U = 3(10^6)[u_1 \quad u_2] \begin{bmatrix} 1 & -1 \\ -1 & 1 \end{bmatrix} \begin{Bmatrix} u_1 \\ u_2 \end{Bmatrix}$

3.2 $\begin{Bmatrix} F_1 \\ F_2 \end{Bmatrix} = \begin{Bmatrix} 143.3 \\ 226.7 \end{Bmatrix}$

$F_1 \longleftarrow \underset{1}{\circ} \qquad \underset{2}{\circ} \longrightarrow F_2$

ANSWERS TO SELECTED EXERCISES

3.3 $\begin{Bmatrix} F_1 \\ F_2 \end{Bmatrix} = \begin{Bmatrix} 833 \\ 2500 \end{Bmatrix}$

$F_1 \longrightarrow \circ \underset{1}{} \underset{2}{\circ} \longrightarrow F_2$

3.4 $U = \dfrac{1.17AE}{2L}[u_1 \ \ u_2]\begin{bmatrix} 1 & -1 \\ -1 & 1 \end{bmatrix}\begin{Bmatrix} u_1 \\ u_2 \end{Bmatrix}$

3.5 $\begin{Bmatrix} F_1 \\ F_2 \end{Bmatrix} = \begin{Bmatrix} 667 \\ 333 \end{Bmatrix}$

$F_1 \longrightarrow \circ \underset{1}{} \underset{2}{\circ} \longrightarrow F_2$

3.6 Two elements:

$\{Q\}_{T①} = \begin{Bmatrix} 1500 \\ 3000 \end{Bmatrix}$

$\{Q\}_{T②} = \begin{Bmatrix} 4500 \\ 4500 \end{Bmatrix}$ equivalent nodal forces

$u_2 = 8.75(10^{-4})$ in.

$\sigma_{X①} = 875$ psi $\qquad \sigma_{X②} = -875$ psi

3.7 Two elements:

$\{Q\}_{T①} = \begin{Bmatrix} 15 \\ 30 \end{Bmatrix}$ Btu/hr

$\{Q\}_{T②} = \begin{Bmatrix} 45 \\ 45 \end{Bmatrix}$ Btu/hr

$\begin{Bmatrix} T_1 \\ T_2 \\ T_3 \end{Bmatrix} = \begin{Bmatrix} 100 \\ 700 \\ 925 \end{Bmatrix}$ °F

3.9 $\dfrac{2}{\pi}wL \qquad\qquad \dfrac{\pi-2}{\pi}wL$

3.12
$$\begin{bmatrix} 1.0 & -0.2 & 0 & 0 \\ & 0.4 & -0.2 & 0 \\ & & 0.4 & -0.2 \\ & \text{symmetric} & & 4.2 \end{bmatrix} \begin{Bmatrix} T_1 \\ T_2 \\ T_3 \\ T_4 \end{Bmatrix} = \begin{Bmatrix} 32 \\ 50 \\ 0 \\ 80 \end{Bmatrix}$$

$$\begin{Bmatrix} T_1 \\ T_2 \\ T_3 \\ T_4 \end{Bmatrix} = \begin{Bmatrix} 77 \\ 226 \\ 126 \\ 25 \end{Bmatrix} \;°C$$

Chapter 4

4.3 Node 1, $X = 2, Y = 2,$ $[N] = [N_1, \; N_2, \; N_3] = [1, \; 0, \; 0]$
Node 2, $X = 4, Y = 2,$ $[N] = [N_1, \; N_2, \; N_3] = [0, \; 1, \; 0]$
Node 3, $X = 3, Y = 5,$ $[N] = [N_1, \; N_2, \; N_3] = [0, \; 0, \; 1]$
Centroid, $X = 3, Y = 3,$ $[N] = [N_1, \; N_2, \; N_3] = [\tfrac{1}{3}, \; \tfrac{1}{3}, \; \tfrac{1}{3}]$

4.4 $\begin{Bmatrix} u \\ v \end{Bmatrix} = 10^{-3} \begin{Bmatrix} 0.33 \\ 1.00 \end{Bmatrix}$ **4.5** $\begin{Bmatrix} \epsilon_X \\ \epsilon_Y \\ \gamma_{XY} \end{Bmatrix} = 10^{-3} \begin{Bmatrix} 0.5 \\ 0 \\ -0.5 \end{Bmatrix}$

4.6 $V_T = -h\{q\}^T \begin{Bmatrix} 0 \\ 0 \\ 1.5C_1 + 4.5C_2 \\ 0 \\ 1.5C_1 + 6C_2 \\ 0 \end{Bmatrix}$

4.7 $V_{BF} = \dfrac{-h\{q\}^T}{3} \rho_0 g \begin{Bmatrix} 0 \\ -16 \\ 0 \\ -24 \\ 0 \\ -24 \end{Bmatrix} = +\{q\}^T \begin{Bmatrix} 0 \\ \tfrac{1}{4} \\ 0 \\ \tfrac{3}{8} \\ 0 \\ \tfrac{3}{8} \end{Bmatrix}$ (weight)

4.8

Location	N_1	N_2	N_3
Node 1	1	0	0
Node 2	0	1	0
Node 3	0	0	1
Centroid	$\tfrac{1}{3}$	$\tfrac{1}{3}$	$\tfrac{1}{3}$

$$[N] = \begin{bmatrix} N_1 & 0 & N_2 & 0 & N_3 & 0 \\ 0 & N_1 & 0 & N_2 & 0 & N_3 \end{bmatrix}$$

ANSWERS TO SELECTED EXERCISES

4.9 $[k] = h[B]^T[C][B]A$

$$[B]_{XY \text{ system}} = \frac{1}{6}\begin{bmatrix} -3 & 0 & 3 & 0 & 0 & 0 \\ 0 & -1 & 0 & -1 & 0 & 2 \\ -1 & -3 & -1 & 3 & 2 & 0 \end{bmatrix} = [B]_{xy \text{ system}}$$

Local and global matrices are the same.

(a)

$$[k]_{XY} = h(10^6)\begin{bmatrix} 25.7 & 5.36 & -23.8 & -0.41 & -1.92 & -4.95 \\ & 11.40 & 0.41 & -5.91 & -5.77 & -5.49 \\ & & 25.7 & -5.36 & -1.92 & 4.95 \\ & & & 11.40 & 5.77 & -5.49 \\ & \text{symmetric} & & & 3.85 & 0 \\ & & & & & 11.0 \end{bmatrix}$$

(b) $[k]_{xy} = [k]_{XY}$

4.10 (a)

$$[k]_{XY} = (10^6)\begin{bmatrix} 1.54 & 0 & 0 & 1.15 & 1.54 & 1.15 \\ & 4.40 & -0.99 & 0 & 0.99 & -4.40 \\ & & 2.47 & 0 & -2.47 & -0.99 \\ & & & 0.87 & 1.15 & -0.87 \\ & \text{symmetric} & & & 4.01 & -2.14 \\ & & & & & 5.26 \end{bmatrix}$$

(b)

$$[k]_{xy} = (10^6)\begin{bmatrix} 2.57 & 1.37 & -1.03 & -0.38 & -1.54 & -0.99 \\ & 3.37 & -0.22 & 1.03 & -1.15 & -4.40 \\ & & 1.89 & -0.77 & -0.86 & 0.99 \\ & & & 1.44 & 1.15 & -2.47 \\ & \text{symmetric} & & & 2.40 & 0 \\ & & & & & 6.87 \end{bmatrix}$$

(c) $[k]_{XY} = [A_3][k]_{xy}[A_3]^T$
where

$$[A_3] = \frac{1}{5}\begin{bmatrix} 4 & -3 & 0 & 0 & 0 & 0 \\ 3 & 4 & 0 & 0 & 0 & 0 \\ 0 & 0 & 4 & -3 & 0 & 0 \\ 0 & 0 & 3 & 4 & 0 & 0 \\ 0 & 0 & 0 & 0 & 4 & -3 \\ 0 & 0 & 0 & 0 & 3 & 4 \end{bmatrix}$$

4.11

$$K_{17} = 0$$
$$K_{56} = k_{56_1} + k_{34_2} + k_{12_3} + k_{56_4}$$
$$K_{66} = k_{66_1} + k_{44_2} + k_{22_3} + k_{66_4}$$
$$K_{6,10} = k_{24_3} + k_{64_4}$$

4.13 Triangle weight $= 1.2\rho g$. Apply one-third of the weight to each node.

$$\begin{Bmatrix} u_2 \\ v_2 \end{Bmatrix} = 10^{-6} \begin{Bmatrix} 0 \\ -0.46 \end{Bmatrix} \rho g \qquad \begin{Bmatrix} \epsilon_X \\ \epsilon_Y \\ \gamma_{XY} \end{Bmatrix} = \begin{Bmatrix} 0 \\ 0 \\ -0.115 \end{Bmatrix} 10^{-6} \rho g$$

$$\begin{Bmatrix} \sigma_X \\ \sigma_Y \\ \tau_{XY} \end{Bmatrix} = \begin{Bmatrix} 0 \\ 0 \\ 1.33 \end{Bmatrix} \rho g$$

4.14 Displacements:

NODE	XDISPL	YDISPL
1	0	0
2	0	0
3	6.7E−05	1.29E−04
4	−4.6E−05	1.83E−04

Strain:

ELEMENT	EXX	EYY	EXY
1	−4.6E−06	5.4E−06	7E−06
2	6.7E−06	0	1.29E−05

Stress:

ELEMENT	SXX	SYY	SXY
1	−71	129	70
2	270	135	129

4.15 $\{Q\}_T = \begin{Bmatrix} 0 \\ 0 \\ 6.7 \\ 0 \\ 20 \\ 0 \end{Bmatrix}$

Chapter 5

5.1 $y = -\frac{1}{6}(x-2)(x-3)(x-4) + \frac{3}{2}(x-1)(x-3)(x-4)$
$\qquad -(x-1)(x-2)(x-4) + \frac{1}{2}(x-1)(x-2)(x-3)$
$y(1.5) = 2.69 \qquad y(2.5) = 2.50 \qquad y(3.5) = 1.94$

5.2 $\phi = \dfrac{(x-1)(y-1)}{4}\phi_1 - \dfrac{(x+1)(y-1)}{4}\phi_2$
$\qquad + \dfrac{(x+1)(y+1)}{4}\phi_3 - \dfrac{(x-1)(y+1)}{4}\phi_4$

5.3 Two-point formula yields 8.0, which is the exact value of the integral.

5.4 Two-point and three-point formulas yield value of 0, which is the exact value of the integral.

ANSWERS TO SELECTED EXERCISES 521

5.5 Two-point formula, $I_a = -0.1212$
Three-point formula, $I_a = -0.1307$
Exact value of integral $= -0.1315$

5.6 Method: Derive interpolation formula.
Evaluate the function for the Gauss point ± 0.57735.
Two-point formula yields integral value of 4.33.

5.7 Two-point formula, $I_a = 3.59$
Three-point formula, $I_a = 3.60$

5.8 Two-point formula, $I_a = 5.60$
Three-point formula, $I_a = 5.744$
Exact value, $I = 5.763$

5.9 One-point formula, $I_a = 1.11$
Two-point formula, $I_a = 0.996$
Three-point formula, $I_a = 0.999$
Exact value $= 1.00$

Chapter 6

6.1 $\begin{Bmatrix} u \\ v \end{Bmatrix} = \begin{Bmatrix} 4 \\ -5 \end{Bmatrix} 10^{-4}$

6.2 $\begin{Bmatrix} X \\ Y \end{Bmatrix} = \begin{Bmatrix} 3.5 \\ 4.5 \end{Bmatrix}$

6.3 $\begin{Bmatrix} u \\ v \end{Bmatrix} = \begin{Bmatrix} 1.17 \\ 0 \end{Bmatrix} 10^{-3}$

6.4 $\begin{Bmatrix} \epsilon_X \\ \epsilon_Y \\ \gamma_{XY} \end{Bmatrix} = \begin{Bmatrix} 1.36 \\ -2.39 \\ 0.20 \end{Bmatrix} 10^{-4}$

6.5 $\begin{Bmatrix} \sigma_X \\ \sigma_Y \\ \tau_{XY} \end{Bmatrix} = \begin{Bmatrix} 2120 \\ -6534 \\ 257 \end{Bmatrix}$ psi

6.7 $\{Q\}_T = \begin{Bmatrix} 0 \\ 0 \\ 760.5 \\ 0 \\ 1189.5 \\ 0 \end{Bmatrix}$

6.8 $\{Q\}_T = \{Q\}_{T_{\text{side 1}}} + \{Q\}_{T_{\text{side 2}}} = \begin{Bmatrix} 0 \\ 0 \\ 30 \\ 0 \\ 30 \\ 0 \end{Bmatrix} + \begin{Bmatrix} -15 \\ 20 \\ 0 \\ 0 \\ -30 \\ 40 \end{Bmatrix} = \begin{Bmatrix} -15 \\ 20 \\ 30 \\ 0 \\ 0 \\ 40 \end{Bmatrix}$

6.9 $\{Q\}_{BF} = -Ahg \begin{Bmatrix} 0 \\ \frac{\rho_0}{3} + \frac{\rho_1}{6} \\ 0 \\ \frac{\rho_0}{3} + \frac{\rho_1}{12} \\ 0 \\ \frac{\rho_0}{3} + \frac{\rho_1}{12} \end{Bmatrix}$

Chapter 7

7.1 For $s = 0 = t$, $|J| = 7.5$
For $s = 1 = t$, $|J| = 8.75$

7.3 For node 1, $s = -1 = t$

$|J| = 6.25$

$$[B] = \begin{bmatrix} -0.2 & 0 & 0.2 & 0 & 0 & 0 & 0 & 0 \\ 0 & -0.2 & 0 & 0 & 0 & 0 & 0 & 0.2 \\ -0.2 & -0.2 & 0 & 0.2 & 0 & 0 & 0.2 & 0 \end{bmatrix}$$

For node 3, $s = 1 = t$

$|J| = 8.75$

$$[B] = \begin{bmatrix} 0 & 0 & 0.0286 & 0 & 0.1429 & 0 & -0.1714 & 0 \\ 0 & 0 & 0 & -0.1714 & 0 & 0.1429 & 0 & 0.0286 \\ 0 & 0 & -0.1714 & 0.0286 & 0.1429 & 0.1429 & 0.0286 & -0.1714 \end{bmatrix}$$

7.7 $\{Q\}_T = \begin{Bmatrix} F_{1X} \\ F_{1Y} \\ F_{2X} \\ F_{2Y} \\ \vdots \\ F_{4Y} \end{Bmatrix} = \begin{Bmatrix} 0 \\ 0 \\ 333.4 \\ 0 \\ 666.6 \\ 0 \\ 0 \\ 0 \end{Bmatrix} \dfrac{hL}{2}$

where $\dfrac{hL}{2} = \dfrac{(0-1)(6.08)}{2} = 0.30$

7.8 $F_{1Y} = (h\rho/12)(4WH)$
= weight/4.5 versus weight/4 for rectangle

Chapter 8

8.1 $N_1 = -\frac{1}{9} = N_2 = N_3$
$N_4 = \frac{4}{9} = N_5 = N_6$

8.2 For element weight W:

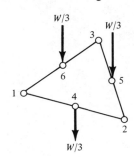

ANSWERS TO SELECTED EXERCISES

8.3 18.4 lb

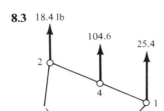

8.4

$$[b] = \begin{bmatrix} 0 & 3 & -3 & -12 & 0 & 12 \\ -3 & 0 & -1 & 0 & 0 & 4 \\ 3 & 1 & 0 & -4 & 0 & 0 \\ 12 & 0 & 4 & 0 & 0 & -16 \\ 0 & 0 & 0 & 0 & 0 & 0 \\ -12 & -4 & 0 & 16 & 0 & 0 \end{bmatrix} \text{ at node 1}$$

$|J| = 60$ at node 1
$B_{11} = 0.4$ at node 1

8.7

$[N] =$
$$-\frac{1}{4}\begin{bmatrix} 1 & 0 & 1 & 0 & 1 & 0 & 1 & 0 & -2 & 0 & -2 & 0 & -2 & 0 & -2 & 0 \\ 0 & 1 & 0 & 1 & 0 & 1 & 0 & 1 & 0 & -2 & 0 & -2 & 0 & -2 & 0 & -2 \end{bmatrix}$$

8.13 Linear elements are not compatible with the quadratic elements. For example, elements 5 and 6 are linear with sides remaining straight during deformation. These two elements are compatible with one another, but they are not compatible with element 1, which has sides that can deform quadratically. Elements 3, 4, 5, 6, 7 are linear. Elements 3 and 4 should not be joined to element 2, and so on.

Chapter 9

9.1 $u = a_0 + a_1 x + a_2 x^2$, $a_1 = 1 - a_2$, $a_2 = 15/22$
Exact solution: $u = 0.6944 e^x + 0.3056 e^{-x} - 1$

9.2 For two elements of equal length, $u(0.5) = u_2 = 0.327$
Exact solution, $u = 0.3302$

9.6 Assume that $y = C_1 x + C_2 x^2$. By least squares, $C_1 = 0.081$ and $C_2 = 0.204$.

Chapter 10

10.3 $[K_T] = \begin{bmatrix} 12.81 & 2.81 & -15.58 \\ & 18.14 & -12.96 \\ \text{symmetric} & & 36.59 \end{bmatrix}$

10.4 $T_1 = 222°\text{F}$

Chapter 11

11.2 $\{Q\}_{T1} = \begin{Bmatrix} 0 \\ 0 \\ 691 \\ 0 \\ 691 \\ 0 \end{Bmatrix}$ lb

11.3 $\{Q\}_{BF} = \rho g \begin{Bmatrix} 0 \\ 22.0 \\ 0 \\ 22.5 \\ 0 \\ 22.5 \end{Bmatrix}$ lb

11.4 $\{Q\}_T = \dfrac{T_{R3} S_1}{2} \begin{Bmatrix} 0 \\ 0 \\ \frac{1}{3} \\ 0 \\ \frac{2}{3} \\ 0 \end{Bmatrix}$

Index

A

Approximate functions, 84, 108, 316
Area coordinates
   3-node triangle in, 188
   6-node triangle in, 281
   definition, 58
   integration formulas in, 209
Assembly of element matrices
   2-node element, 13, 17, 27, 119
   3-node triangle (heat transfer), 348
   3-node triangle (stress), 167
   4-node quadrilateral, 264
Assumed solution functions, 84
   2-node element, 108, 128
   3-node axisymmetric ring, 373
   3-node isoparametric triangle
     displacement, 202
     fluid flow, 360
     temperature, 341
   3-node triangle in global coordinates, 146
   3-node triangle in local coordinates, 160
   4-node quadrilateral, 241
   4-node tetrahedron, 384
   6-node triangle, 281
   8-node quadrilateral, 291
   8-node solid, 388
Axisymmetric, 369

B

B matrix; *see* Strain–displacement matrix
Banded matrix, 45
Basis functions; *see* Shape functions
Beam/column element, 34
Beam element, 30
Bibliography, 507
Body force vector, 155, 372
   3-node triangle, 206, 210
   4-node quadrilateral, 262
   8-node quadrilateral, 295
Boundary conditions, 10
   beam, 100
   Dirichlet (primary, essential), 93, 139

526

displacement, 15, 93
force, 95
heat flow, 97, 351
Neumann (secondary, natural), 93, 324
programming techniques, 139, 351
temperature, 97, 351

C

Calculus of variations, 77
Collocation, 317
Comparison examples
  irrotational fluid flow, 362
  one-dimensional elasticity, 122
  one-dimensional heat flow, 127
    with convection, 133
  two-dimensional elasticity, 215, 265
    cantilever beam, 222, 273, 289, 297, 300
    plate with hole, 233
  two-dimensional heat flow, 354
Compatibility, 300, 322
Computer programs
  frame analysis, 399
  heat transfer, 477
  quadrilateral body force, 475
  stress analysis, 435
Condensation, 276
Conduction matrix, 129, 343
Conductivity, 96, 337
Conservative force, 64
Constant strain triangle, 145
Constitutive relations, 61, 373, 383
Constraints; *see* Boundary conditions
Continuity, 126, 322
Convection, 133, 338
Convection matrix, 133, 343
Convection vector, 133, 344
Convergence, 126, 274, 300
Coordinate systems, definitions of
  global, 56
  local, 56
  natural, 57
Coordinate transformation
  of stiffness matrix, 23, 165, 304
  of vector, 21, 165, 303

D

Data reduction, 230
Degree of freedom, 12
Determinant, 195
Dirac delta function, 118
Dirichlet boundary conditions, 93, 139

Displacement function; *see* Assumed solution functions
Displacement numbering
  3-node triangle, plane elasticity, 169
  beam/column element, 36
  solid element, 387
  two-force element, 27
Displacement vector, 148
  assembled, 15, 168
  element, 11, 158

E

Element force/displacement equation, 11, 20, 34, 117, 158, 214, 305, 370, 381
Element heat flow equation, 135, 340
Element stiffness matrix; *see* Stiffness matrix
Euler equation, 79

F

Fin modeling, 135
Finite element modeling, 117
Fluid flow problem, 98, 360
Functional, 78
  beam bending, 99
  flow through permeable media, 98
  Laplace equation, 101
  uniaxial heat flow example, 95
  uniaxial stress example, 91

G

Galerkin method, 318
Gauss elimination, 43
Gauss quadrature, 189
  triangle region, 284
  unit square region, 192, 194
Generalized coordinates, 147
Global assembled displacement vector, 15
Global assembled load vector, 15
Global coordinates, 19, 146
Global displacement numbering; *see* Displacement numbering
Global node map, 12

H

Heat conduction matrix, 129, 343
Heat convection matrix, 134, 343
Heat flow problem, 95, 337
Heat flow vector, 129, 344

INDEX

H

Higher-order elements
  6-node triangle, 281
  8-node quadrilateral, 291

I

Integration, 189
Integration formulas, triangle natural coordinates, 209
Interior node removal by condensation, 276
Interpolation formulas, 181
  one independent variable, 182
  two independent variables, 184
Interpolation functions, 202, 241, 281; *see also* Shape functions
Isoparametric element
  definition, 202
  quadrilateral, 239, 291
  solid (brick), 387
  tetrahedron, 383
  triangle, 201, 281

J

Jacobian determinant, 195

L

Lagrange interpolation functions, 183
Lagrangian interpolation, 182
Laplace equation, 101, 333, 337
Least-squares approximation, 317
Load vector, equivalent node force
  for body force, 206, 262
  for distributed stress traction, 213, 259, 286, 295
  for heat flux, 130
Local coordinates, 56
  3-node triangle, 160

M

Matrix operations, 395
Mesh refinement, 126
Minimum potential energy principle, 74, 116, 157
Mixed-element-type assembly, 300

N

Natural boundary condition; *see* Boundary condition
Natural coordinate system
  for 2-node element, 57
  for quadrilateral, 240
  for triangle, 58, 202
Neumann boundary conditions, 93

Newton's law of cooling, 97, 338
Nodal force vectors, 20, 35, 117, 129, 158, 205, 262, 285, 295
Node, 10
Node load vector formulas, 213, 260, 286, 295
Node numbering, 10, 12, 146, 231
Numbering
  of displacements, 27, 36, 169
  of elements, 11, 146, 231
  of nodes, 10, 12, 146, 231
Numerical integration, 189

P

Plane strain, 62
Plane stress, 62
Poisson's equation, 334, 365
Potential energy of conservative forces, 68, 83, 112
  body force, 155
  distributed stress traction, 152
  node forces, 152

Q

Quadrature; *see* Gauss quadrature
Quadrilateral element
  4-node, 239
  8-node, 291

R

Rayleigh–Ritz, 84, 108, 121
Reaction loads, 30
Reduction of matrix equation, 15, 30
Residual, 317
Rigid body displacement, 139
Rotation matrix; *see* Coordinate transformation

S

Saint-Venant, 228
Shape function derivatives, 473
Shape functions; *see also* Lagrange interpolation functions
  2-node element, 110
  3-node axisymmetric ring, 373
  3-node triangle, 202
  4-node quadrilateral, 241
  6-node triangle, 281
  8-node quadrilateral, 292
  8-node solid, 389
  tetrahedron, 384
Sign conventions, 10, 35, 93, 97, 99
Skew boundary conditions, 302

Solid elements
  4-node tetrahedron, 383
  8-node solid (brick), 387
Stiffness matrix
  beam/column element, 35
  beam element, 34
  3-node axisymmetric ring element, 375
  3-node triangle, 158, 205
  4-node quadrilateral, 250
  4-node tetrahedron, 386
  6-node triangle, 285
  8-node quadrilateral, 294
  8-node solid, 392
  two-force element, 11, 23
Strain, 60, 372, 382
Strain energy, 70, 83, 111, 150
  in simple beam, 73
Strain–displacement matrix
  2-node element, 111
  3-node axisymmetric ring, 375
  3-node triangle, 149, 162, 204
  4-node quadrilateral, 246
  6-node triangle, 283
  8-node quadrilateral, 293
  solid, 391
  tetrahedron, 385
Strain–displacement relations; *see* Strain
Strain vector, 71
Stress–strain relations, 61, 372
Stress vector, 62, 71

Subdomain method, 317
Surface tractions, 152
  equivalent node load vector, 158
Symmetry advantages, 232

T

Tetrahedron, 383
Thermal nodal load vector, 305, 372
Thermal strain, 304, 373
Torsion problem, 366
Transformation; *see* Coordinate transformation
Triangle element
  3-node, 146, 160, 201, 341
  6-node, 281
Triangle integration formulas, 209
Two-node element (bar element, uniaxial element), 10, 19, 89, 107, 128

V

Variational formulation, 77, 83, 96, 99
Virtual displacement, 81
Virtual work, 83
Volume coordinates, 383

W

Weighted residual methods, 316
Work (force potential energy), 63